VOLUME EIGHTY FOUR

CURRENT TOPICS IN DEVELOPMENTAL BIOLOGY

Mouse Models of Developmental Genetic Disease

VOLUME EIGHTY FOUR

CURRENT TOPICS IN DEVELOPMENTAL BIOLOGY

Mouse Models of Developmental Genetic Disease

Edited by

ROBERT S. KRAUSS
Department of Developmental and Regenerative Biology, Mount Sinai School of Medicine, New York

ELSEVIER

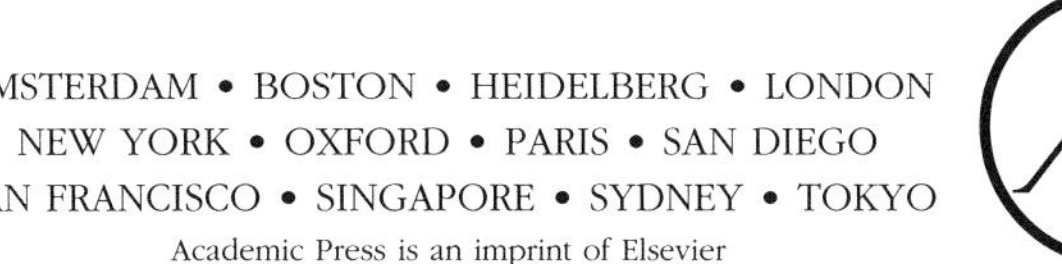

AMSTERDAM • BOSTON • HEIDELBERG • LONDON
NEW YORK • OXFORD • PARIS • SAN DIEGO
SAN FRANCISCO • SINGAPORE • SYDNEY • TOKYO
Academic Press is an imprint of Elsevier

Academic Press is an imprint of Elsevier
525 B Street, Suite 1900, San Diego, CA 92101-4495, USA
30 Corporate Drive, Suite 400, Burlington, MA 01803, USA
32 Jamestown Road, London, NW1 7BY, UK
Linacre House, Jordan Hill, Oxford OX2 8DP, UK

First edition 2008

ISBN: 978-0-12-374454-8
ISSN: 0070-2153

For information on all Academic Press publications
visit our website at elsevierdirect.com

Printed and bound in USA
08 09 10 11 9 8 7 6 5 4 3 2 1

Contents

Contributors

Glen B. Banks
Department of Neurology, University of Washington, Seattle, Washington

Philip L. Beales
Molecular Medicine Unit, Institute of Child Health, University College London, London, WC1N1EH, United Kingdom

Jeffrey S. Chamberlain
Department of Neurology, University of Washington, Seattle, Washington

Tatiana V. Cohen
Center for Genetic Medicine, Children's National Medical Center, N.W. Washington, DC 20010

Amel Gritli-Linde
Department of Oral Biochemistry, Sahlgrenska Academy at the University of Gothenburg, Medicinaregatan 12F, Göteborg, Sweden

Robert S. Krauss
Department of Developmental and Regenerative Biology, Mount Sinai School of Medicine, New York 10029

Michel Leibovici
Institut Pasteur, Unité de Génétique et Physiologie de l'Audition, 25 rue du Dr. Roux, F75015 Paris, France; Inserm UMRS587, Collège de France, UPMC University, Paris, France

Anne Moon
School of Medicine, University of Utah, 15 North 2030 East Room 4160B, EIHG University of Utah, Salt Lake City, UT 84112

Christine Petit
Institut Pasteur, Unité de Génétique et Physiologie de l'Audition, 25 rue du Dr. Roux, F75015 Paris, France; Inserm UMRS587, Collège de France, UPMC University, Paris, France

Robyn J. Quinlan
Molecular Medicine Unit, Institute of Child Health, University College London, London, WC1N1EH, United Kingdom

Saaid Safieddine
Institut Pasteur, Unité de Génétique et Physiologie de l'Audition, 25 rue du Dr. Roux, F75015 Paris, France; Inserm UMRS587, Collège de France, UPMC University, Paris, France

Anjali A. Sarkar
Center for Neuroscience Research, Children's Research Institute, Children's National Medical Center, Washington, District of Columbia

Karen A. Schachter
Department of Developmental and Regenerative Biology, Mount Sinai School of Medicine, New York 10029

Colin L. Stewart
Institute of Medical Biology, 8A Biomedical Grove, Immunos, Singapore 138668

Jonathan L. Tobin
Molecular Medicine Unit, Institute of Child Health, University College London, London, WC1N1EH, United Kingdom

Patricia D. Wilson
Mount Sinai School of Medicine, New York

Irene E. Zohn
Center for Neuroscience Research, Children's Research Institute, Children's National Medical Center, Washington, District of Columbia

Preface

Mouse Models of Developmental Genetic Disease

Approximately 3% of newborn human beings have congenital anomalies with significant cosmetic and/or functional consequences (Maitra and Kumar, 2005). The causes of such anomalies may be genetic, environmental, or multifactorial in nature; however, the etiology is unknown in at least 50% of cases. Much of our ability to understand what has gone awry in these human birth defects rests with the development of animal models for them. Furthermore, such models may lead to identification of genetic and environmental factors for anomalies where etiology is currently unclear. The mouse has emerged as the model organism of choice for these studies (Nguyen and Xu, 2008; Rosenthal and Brown, 2007). Many features of the laboratory mouse converge to make this so, including the relatively short gestation period, facile breeding, existence of numerous inbred strains, ease of genetic manipulation, and genetic kinship with people. Targeted, gene-trap, chemically-induced, and spontaneous mouse mutants exist that mimic many human developmental genetic disorders. These lines of mice have provided penetrating mechanistic insights into basic morphogenetic processes, including pattern formation, specification and differentiation of specific cell lineages, and organogenesis. Furthermore, potential therapeutic interventions will always need appropriate animal models for preclinical analyses.

Hundreds of human developmental genetic diseases are known (*Online Mendelian Inheritance in Man*). Therefore, selection of topics for a single-volume review is inherently idiosyncratic. This volume of *Current Topics in Developmental Biology* employs a limited number of examples to illustrate larger concepts that are important to the field as a whole. These examples reveal both strengths and limitations of the mouse as a model organism. Some of these concepts are described here.

Many specific classes of birth defects can occur as a part of syndromes. While syndromes are complicated, they often reveal shared usage of a few signaling pathways in disparate and diverse developmental events. For example, congenital heart defects and cleft lip/palate, each of which is sometimes associated with broader syndromes, implicate the Sonic hedgehog (Shh) and Wnt pathways as regulators of morphogenetic processes that, when flawed in execution, result in these common anomalies. The ability to construct mouse lines with spatial and temporal conditional mutations has permitted analysis of these reiteratively used pathways in the specific structures and stages of development that are relevant to these birth defects.

While syndromes are informative in this way, congenital anomalies usually occur in an isolated, nonsyndromic form, generally sporadically but also, more rarely, heritably. These cases often reveal extremely variable penetrance and expressivity, leading to the conclusion that many defects have a multifactorial etiology that derives from complex interactions of either multiple genes or a combination of genetic and environmental insults. Mouse models of such anomalies offer strong support for such conjecture; examples include holoprosencephaly, neural tube defects, and cleft lip/palate.

While some of the birth defects mentioned above are relatively common, occurring as often as 1 in 500–1 in 1000 live births, many others occur only rarely. However, rare diseases often illuminate processes of wide general interest. The ciliopathies, as a group, and a ciliopathy in particular, polycystic kidney disease (PKD), are illustrative. Analysis of ciliopathies, for example Bardet–Biedl syndrome (BBS), has played a major role in the surge of interest in cilia as essential structures in developmental signaling. The various forms of BBS or PKD also reveal that phenotypically similar diseases caused by mutations in different genes trigger discovery of protein complexes or signaling pathways in which the proteins encoded by the various disease genes interact to perform the same biological function. Mouse models of BBS and PKD phenocopy many but not all aspects of the human diseases, pointing out potential differences in development of mice and people. It is also important to note that detailed discussion of individual diseases that fall into larger categories broadens understanding. For example, PKD is a ciliopathy, but the PKD gene products are involved with both ciliary and nonciliary functions that likely contribute to the distinctiveness of phenotypes that arise in isolated form versus as part of syndromes.

In contrast to the concept that mutations in many genes give rise to similar phenotypes, as observed with neural tube defects and cleft lip/palate, the laminopathies are a set of disorders in which different mutations in a single gene result in an extremely wide range of distinctive phenotypes. Approximately 200 different mutations have been identified in the human *LMNA* gene (encoding lamin A), and the consequent diseases range from muscular dystrophies to lipodystrophies and skeletal dysplasias to progeroid (premature aging) syndromes. Construction of mouse models with specific disease-associated mutations is able to address the question of how different mutations in the same gene can result in such diverse outcomes.

Many developmental genetic diseases are incompatible with survival or result in early mortality; insight into the underlying causes may lead to preventive strategies, such as the use of folic acid supplementation in prevention of neural tube defects. However, some anomalies are either progressive over many years, or do not alter life span but present lifelong challenges to affected individuals; Duchenne muscular dystrophy (DMD) is an example of the former and congenital deafness of the latter. Such disorders may be amenable to therapies that are guided by studies of mouse models. For example, models of congenital deafness have indicated which types of patients may benefit from cochlear implants. Furthermore, DMD models are at the forefront of assessing novel preclinical therapeutics for genetic diseases, including small molecule, gene, and cell-based therapies.

The Volume

The volume begins with a description of neural tube defects and how they arise from failures in the process of neurulation, in a chapter by Zohn and Sarkar. Analyses of mouse models of these disorders illuminate a multifactorial threshold etiology proposed for human neural tube defects and help explain the variable penetrance associated with predisposing mutations. Furthermore, mouse models have led to identification of candidate genes for both syndromic and nonsyndromic neural tube defects and to approaches for prevention.

Gritli-Linde then focuses in Chapter 2 on the spectrum of anomalies categorized as cleft lip and/or cleft palate and their genetic and environmental causes. The process of palatogenesis is complex and delicate, and a large number of genes that participate in several signaling pathways have been implicated in development of cleft lip, with or without cleft palate, and cleft palate only. Despite the obvious differences in facial structure between humans and mice, palatogenesis is remarkably similar in these two organisms, and mice have provided deep insight into the process and how it can go awry. Nevertheless, Gritli-Linde highlights important differences as well, including the observation that mutations in people that result in cleft lip and palate generally produce cleft palate only in mice.

In Chapter 3, Schachter and Krauss discuss holoprosencephaly, the failure to delineate the midline of the forebrain and midface. Mouse models of this defect are beginning to shed light on the conundrum of the extremely variable clinical expression of holoprosencephaly in humans, even in familial forms. Studies with various mutant mouse lines have linked most of the holoprosencephaly genes together as regulators of either Shh expression or signaling. However, it is considered very likely that environmental exposures also contribute significantly to human holoprosencephaly, and the chapter describes promising new systems of *in vitro* mouse embryo culture that may aid in identification of teratogens.

Congenital heart defects are the most common class of congenital malformation, and in Chapter 4, Moon provides an overview of cardiac development from a primitive contractile tube into the mature four-chambered mammalian heart. This is followed by a detailed description of how mouse models have been used to understand congenital cardiovascular disease in a variety of human syndromes, including DiGeorge, Holt-Oram, Marfan, Noonan, and others. The use of the mouse as a tool for discovery of additional, novel factors and pathways that regulate development of chambers, tracts, and valves is also discussed; defects in these processes are likely to result in congenital anomalies of the heart.

In Chapter 5, Quinlin *et al.* discuss the ciliopathies, a group of rare genetic diseases that have defects in primary cilia; examples include BBS, Alstrom, Joubert, and Meckel syndromes, and PKD. Primary cilia play key roles in several signaling pathways to regulate the effects of stimuli that range from the mechanosensory to the morphogenetic, and consequently control left–right asymmetry in the embryo, a wide variety of patterning events, planar cell polarity, and development of epithelia. Studies with mice have also illuminated intraflagellar transport, a highly conserved process important for cilia structure and function; mutations in intraflagellar transport proteins underlie many of the ciliopathies.

Wilson covers in detail the diverse PKD and mechanisms of cyst formation in Chapter 6. The genes that underlie autosomal dominant forms of PKD and some of the rare nephronophthisis conditions encode proteins that physically interact not only at the primary cilium but also at adherens junctions and focal adhesions, suggesting involvement in cell–cell and cell–matrix interactions in regulation of epithelial cell proliferation, polarization, and survival.

The volume continues with a chapter on the laminopathies by Cohen and Stewart. The laminopathies comprise a group of inherited diseases and anomalies that result from defects in proteins of the nuclear envelope and lamina. These structures play a key role in integrating myriad cellular processes and link the cytoskeleton to the interior of the nucleus, thus affecting DNA replication, transcription, and nuclear and chromatin organization. Mutations in genes encoding A-type lamins and certain nuclear envelope-associated proteins cause a bewildering array of phenotypes, suggesting that these structures play a variety of cell- and tissue-specific roles.

In Chapter 8, Leibovici *et al.* focus on mouse models of human hereditary deafness. The authors describe the peripheral auditory system, followed by clinical and genetic aspects of human congenital deafness and the limitations of clinical investigations for mechanistic analysis. Mouse models have allowed analysis of development of the cochlea and hair cells, the structures most often defective in congenital deafness. Three different forms of deafness are then discussed as a representative subset of this group of disorders: DFNB9, a form of isolated deafness; several models of Usher syndrome; and DFNB1, the most common form of congenital deafness.

In the final chapter, models of DMD are discussed by Banks and Chamberlain, with particular reference to the development of therapeutic strategies. The severity of this muscle disease is distinct in humans, mice, and dogs that carry mutations in the dystrophin gene. The value of mice versus dogs in assessing gene, cell, and other therapies is compared, and the comparison reveals how important it will be to garner as much information from as many sources as possible prior to treatment of patients.

It is hoped that this volume will shed light not only onto the developmental genetic diseases specifically discussed here, but also provide conceptual insight into congenital anomalies generally. Furthermore, the strengths and limitations of mouse models should be apparent, making it possible to extrapolate these aspects of the system to other anomalies. In the coming years, the continued interplay between clinical observation and basic research is expected to yield deep insight into mammalian developmental processes and the emergence of effective preventive and/ or therapeutic strategies.

Robert S. Krauss, PhD
New York, NY

REFERENCES

Maitra, A., and Kumar, V. (2005). Diseases of infancy and childhood. In Robbins and Cotran PATHOLOGIC BASIS OF DISEASE," (V. Kumar, A. K. Abbas, and N. Fausto, Eds.), 7th Ed. Elsevier Saunders, Philadelphia, PA.

Nguyen, D., and Xu, T., (2008). The expanding role of mouse genetics for understanding human biology and disease. Dis. Model. and Mech. ***1**, 56–66.*

Online Mendelian Inheritance in Man, OMIM (TM). McKusick-Nathans Institute of Genetic Medicine, Johns Hopkins University (Baltimore, MD) and National Center for Biotechnology Information, National Library of Medicine (Bethesda, MD), October 1, 2008. World Wide Web URL:http://www.ncbi.nlm.nih.gov/omim/.

Rosenthal, N., and Brown, S., (2007). The mouse ascending: Perspectives for human-disease models. Nat. Cell Biol. ***9**, 993–999.*

CHAPTER ONE

Modeling Neural Tube Defects in the Mouse

Irene E. Zohn *and* Anjali A. Sarkar

Contents

Center for Neuroscience Research, Children's Research Institute, Children's National Medical Center, Washington, District of Columbia

Current Topics in Developmental Biology, Volume 84
ISSN 0070-2153, DOI: 10.1016/S0070-2153(08)00601-7

Abstract

Neural tube defects (NTDs) are among the most common structural birth defects observed in humans. Mouse models provide an excellent experimental system to study the underlying causes of NTDs. These models not only allow for identification of the genes required for neurulation, they provide tractable systems for uncovering the developmental, pathological and molecular mechanisms underlying NTDs. In addition, mouse models are essential for elucidating the mechanisms of gene–environment and gene–gene interactions that contribute to the multifactorial inheritance of NTDs. In some cases these studies have led to development of approaches to prevent NTDs and provide an understanding of the underlying molecular mechanism of these therapies prevent NTDs.

1. Introduction

1.1. NTDs result from failures in neurulation

Neurulation is a complex morphogenetic process that results in formation of the central nervous system. In the human embryo, neurulation begins around the 17th day following fertilization and is complete before the 30th day, often before many women are aware of the pregnancy. In mouse, neurulation starts at embryonic day 8.5 (E8.5) and is complete by E10.5. This process occurs in two phases termed primary and secondary neurulation (Fig. 1.1). Primary neurulation results in formation of the majority of the central nervous system, while secondary neurulation results in formation of the spine including and caudal to the sacral vertebrate. Both processes involve a series of coordinated morphogenic movements that include regulated changes in cell shape, proliferation, apoptosis, and adhesion in both the neural and surrounding tissues. The pathways regulating these cell behaviors have been recently reviewed (Copp *et al.*, 2003; De Marco *et al.*, 2006). Disruption of any one or combinations of these processes results in neural tube defects (NTDs), an umbrella term used to describe defects in structural formation of the central nervous system. Many different types of NTDs occur and include spina bifida, craniorachischisis, anencephaly, encephalocele, and holoprosencephaly. NTDs represent some of the most common birth defects in humans, affecting approximately one out of one thousand live births.

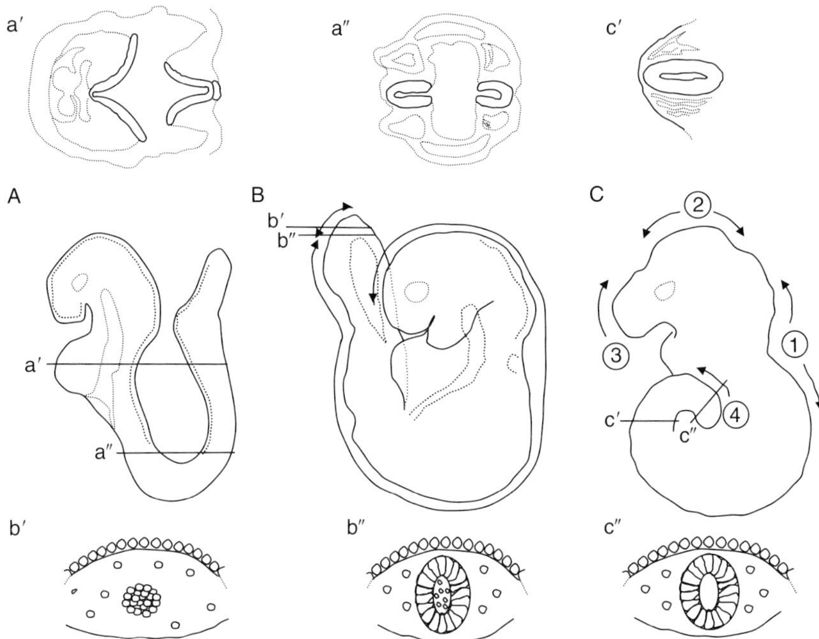

Figure 1.1 Stages of primary and secondary neurulation during development of the central nervous system. Illustration of neurulation in E8.5 (panel A), E9.0 (panel B), and E10.0 (panel C) mouse embryos. Lines represent plane of section in panels a′, a″, b′, b″, c′ and c″. During primary neurulation, neural tube closure is initiated at specific closure points (panel C, 1–3). Closure 1 is located at the junction of the hindbrain and spinal cord, closure 2 at the junction of the forebrain and midbrain, and closure 3 at the extreme rostral end of the forebrain. Primary neurulation begins with the induction of the neural plate from the embryonic ectoderm. Following its formation, the neural plate undergoes convergent extension movements (described in Fig. 1.3), simultaneously with the formation of the neural folds at the lateral edges of the neural plate. Neural fold elevation is aided by the formation of a medial hinge point (MHP, not shown) and paired dorsal–lateral hinge points (DLHPs, not shown). Once the neural folds meet in the dorsal midline, they fuse and the neural and non-neural ectoderm tissues separate to form the dorsal neural tube and overlying ectoderm (c′). In the most caudal portions of the neural tube secondary neurulation predominates (panels b′ and b″). Secondary neurulation involves the aggregation of tailbud cells into a medullary cord (b′). These cells then undergo cavitation to form an epithelial tube without formation of a neural plate intermediate (b″ and c″). The secondary and primary neural tubes fuse to make a seamless neural tube.

Closure of the neural tube does not happen all at once; rather, closure begins at specified closure points, which extend in a zipper-like fashion both cranially and caudally (Fig. 1.1; reviewed in Copp *et al.*, 2003; Detrait *et al.*, 2005). In mouse, three closure points have been described. Closure point 1 is located at the junction of the future hindbrain and spinal cord. Failure to form closure 1 results in craniorachischisis where the entire spinal cord and part of the brain remain open. Closure point 2 is located at the junction of the forebrain and midbrain and closure point 3 at the extreme rostral end of the forebrain. When closure points 2 or 3 fail to form, exencephaly (or anencephaly in humans) results. Other NTDs arise when the neural tube closes in the cranial region but is malformed. For example, encephalocele occurs when the brain protrudes from an opening in the skull and holoprosencephaly when the forebrain fails to divide to form the bilateral cerebral hemispheres.

Multiple forms of spina bifida present with varied degrees of severity depending upon the involvement of the meninges and spinal nerves. The most severe form of spina bifida is meningomyelocele where the meninges and spinal nerves protrude through the open spine. Meningocele and lipomeningocele are less severe and result when the meninges, but not the spinal nerves, protrude through an opening in the spine. Spina bifida occulta and dermal sinus can be asymptomatic and occur when the vertebral bodies do not form properly over the closed spinal cord, or when a channel forms between the skin and spinal cord, respectively.

1.2. Mouse as a model for multifactorial inheritance of NTDs

The majority of NTDs occur sporadically without a family history. However, a number of lines of evidence implicate a genetic component to the etiology of NTDs (reviewed in Detrait *et al.*, 2005; Lynch, 2005). For example, occasionally NTDs present with a family history suggesting that, in some instances, NTDs have a clear genetic etiology. Furthermore, NTDs are often associated with chromosomal abnormalities such as aneuploidies, duplications, and deletions and are noted in many spontaneous abortions with abnormal karyotypes. NTDs are often syndromic, associated with other congenital abnormalities or are part of defined genetic syndromes such as Meckel or Waardenburg syndromes. Finally, twin studies indicate a 5% concordance rate and there is a 50-fold increased risk of recurrence in subsequent affected pregnancies. These types of statistics suggest a multifactorial pattern of inheritance of NTDs. Because of the complex genetics associated with NTDs in humans, the identification of the genes causing NTDs has been difficult.

A multifactorial threshold model has been proposed to account for the pattern of inheritance of NTDs observed in humans (Fig. 1.2; reviewed in Harris and Juriloff, 2007). The multifactorial threshold model postulates that many factors (both genetic and environmental) act in either an additive or synergistic fashion to cause NTDs. Genetic factors include hypomorphic or null mutations in genes required for neurulation. Environmental factors may either positively or negatively influence neurulation. NTDs result when neurulation is significantly disrupted so that this threshold event (e.g., neural tube closure) is not surpassed. Genetic or environmental insults

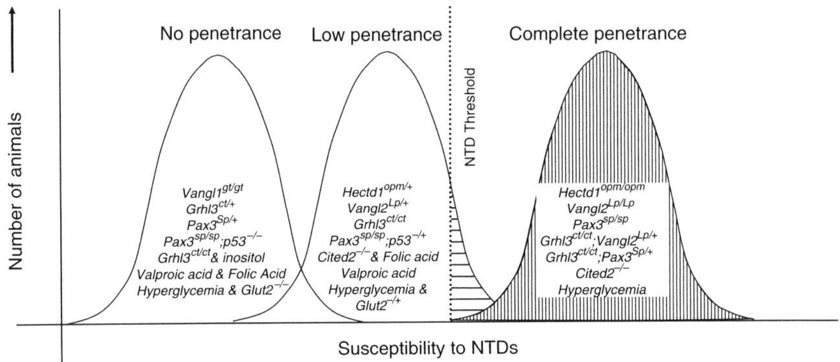

Figure 1.2 Multifactorial threshold model used to explain the complex inheritance of NTDs. Many factors, both genetic and environmental, contribute to the susceptibility for NTDs. In many cases, the effect of an individual factor is not sufficient to cause NTDs. Defects result when neurulation is significantly disrupted so that a threshold event (represented by the dotted line—NTD threshold) is not surpassed. Mouse models fit into one of three groups (no, low, or high penetrance) depending on the severity and number of insults affecting neurulation. Examples given illustrate cases used in the text. For instance, *Grhl3*$^{ct/+}$ mutants exhibit no penetrance of NTDs, mutation of both alleles (*Grhl3*$^{ct/ct}$) results in low penetrance of NTDs and mutation of additional loci (e.g., *Grhl3*$^{ct/ct}$; *Vangl2*$^{Lp/+}$) results in highly penetrant NTDs.

below this threshold will not cause NTDs but when additional susceptibility loci are present or further environmental insults occur, the neural tube fails to close. The genes involved in NTDs may control either a single process required for neurulation such as convergent extension movements, or multiple processes such as convergent extension movements and regulation of proliferation.

Mouse models of NTDs provide important clues as to the genetic causes of NTDs by identifying the genes that are required for neurulation. Since neurulation in the mouse and human are mechanistically similar, the genes required in the mouse provide good candidates for human NTDs. In some instances, these approaches have resulted in identification of mutations associated with NTDs in human patients (see, e.g., Boyles *et al.*, 2005; Hol *et al.*, 1996; Kibar *et al.*, 2007a,b). If the mutation in question does not result in a significant deletion of the gene, the challenge remains to determine if it disrupts gene function significantly to contribute to the defect. One method used to circumvent this issue is to identify major susceptibility genes in humans with the rationale that if a statistically significant proportion of NTDs are associated with a particular genetic variant then it is considered a susceptibility locus. This approach has led to the identification of genetic variants of enzymes in the folate biosynthesis pathway, providing susceptibility loci in some populations (reviewed in Boyles *et al.*, 2005; Kibar *et al.*, 2007a). However, from studies in mouse, we currently know

the identity of at least 200 gene products that are required for neurulation (reviewed in Harris and Juriloff, 2007). The existence of such a large number of potential candidate genes raises the possibility that there will be very few major susceptibility loci and that NTDs may arise as a result of mutations in combinations of the hundreds of the genes required for neurulation. This would have the effect of diluting the statistical significance of the associated mutations identified in these types of genetic association studies. Furthermore, because of the multifactorial nature of the inheritance of NTDs, mutations may be inherited from unaffected parents and result in NTDs only when combined with other genetic or environmental insults. Thus the presence of the variant in a normal individual may not preclude the loci from being causal for the defect. For example, the mutations identified in human patients with NTDs in *PAX1* and *VANGL1* were also present in family members without NTDs (Hol *et al.*, 1996; Kibar *et al.*, 2007b). Yet, further experiments indicate that these mutations significantly disrupt protein function and both genes can contribute to NTDs in mouse models (Helwig *et al.*, 1995; Joosten *et al.*, 1998; Kibar *et al.*, 2007b; Torban *et al.*, 2008). While identification of mutations in particular genes in one or two patients may have little statistical power, this information coupled with the determination that the mutation disrupts the function of a gene required for neural tube closure in mouse, indicates that the mutation may also contribute to the defect in the patients in which it was identified.

In humans, NTDs are mostly nonsyndromic with complex multifactorial inheritance, while in most mouse models, NTDs are mostly syndromic with a recessive pattern of inheritance. Yet there are many examples where a multifactorial threshold pattern of inheritance can be modeled in the mouse. In this chapter, using a few well-characterized mouse models of NTDs, we will illustrate (1) how the multifactorial threshold model of NTDs can be modeled in the mouse, (2) how studies in mouse have helped identify the genes mutated in human NTDs, (3) how studies in mouse have elucidated the molecular basis of gene–environment interactions and (4) how mouse models have provided experiment models for the development of approaches to prevent NTDs. In addition for most of these examples, we will review how studies of these mouse models have uncovered the cellular and molecular mechanisms by which neurulation can fail, resulting in NTDs.

2. Examples of Mouse as a Multifactorial Threshold Model for the Inheritance of NTDs

While the majority of mouse models for NTDs exhibit a Mendelian pattern of inheritance, many examples illustrate that the mutations identified in the mouse can fit the multifactorial threshold model of inheritance observed in human NTDs. These include mutations in genes required for neurulation that cause NTDs with low penetrance either in heterozygous or homozygous mutant embryos and mutations that cause NTDs in multigenic models or in response to environmental insults.

2.1. Mutation of genes required for neurulation can result in low penetrant NTDs in heterozygotes causing apparently sporadic NTDs

Some of the genes required for neurulation cause completely penetrant NTDs in homozygous embryos but result in a low penetrance of NTDs in heterozygous mice. For example, two different alleles of the E3 ubiquitin ligase *Hectd1* (*homologous to the E6-AP carboxyl terminus domain containing-1*) have been reported and both cause completely penetrant NTDs (Zohn *et al.*, 2007). Interestingly, embryos heterozygous for a null mutation in *Hectd1* in the *openmind* (*opm*) mouse line ($Hectd1^{opm/+}$) exhibit a low frequency of NTDs. These results indicate that the level of Hectd1 protein may be critical and that loss of one allele can decrease protein activity below a critical threshold required for neural tube closure. Furthermore, embryos heterozygous for an allele generated by genetrap insertion into the C-terminal catalytic HECT domain ($Hectd1^{XC/+}$) display a slightly higher frequency of NTDs. Since disruption of the HECT domain can create a dominant-negative protein (Huibregtse *et al.*, 1995; Talis *et al.*, 1998), this mutation likely results in dominant-negative activity accounting for the higher frequency of NTDs than observed with the null allele. Other genes required for neural tube closure are embryonic lethal in homozygous mutants but heterozygotes exhibit a low penetrance of NTDs. For example, the chromatin-remodeling complex protein SWI/SNF-related proteins Smarca4 and SmarcaX play essential roles in neural tube closure. Homozygous mutants for $Smarca4^{-/-}$ and $SmarcaX^{-/-}$ are lethal shortly after implantation; however, heterozygous embryos exhibit a low penetrance of exencephaly (Bultman *et al.*, 2000; Kim *et al.*, 2001). Therefore, heterozygous mutations in *HECTD1*, *SMARCA4*, or *SMARCAX* in humans could result in sporadic NTDs as the majority of heterozygous individuals would be phenotypically normal, but sporadically, heterozygotes would exhibit NTDs (Fig. 1.2).

2.2. Mutations of genes required for neurulation can result in low penetrant NTDs in homozygotes

Other examples of a mouse model of the complex genetic inheritance of NTDs are represented by hypomorphic mutations introduced in genes required for neurulation that result in a low penetrance of NTDs in homozygous mutants. One good example of this type of model is the mutations identified in the gene encoding a novel p38-MAPK-interacting protein (*p38IP*; Zohn *et al.*, 2006). A null mutation in *p38IP* results in severe gastrulation defects and embryonic lethality before neural tube closure is complete. In contrast, the *droopy eye* (*drey*) mutation results in aberrant splicing of the *p38IP* transcript with a small amount of normal transcript produced in homozygous mutant embryos. As a consequence, rather than completely penetrant gastrulation defects as observed in the null mutants, $p38IP^{drey/drey}$ mutants exhibit a range of incompletely penetrant phenotypes from gastrulation defects to apparently normal, fertile adults. The most common phenotypes observed in $p38IP^{drey/drey}$ mutants were NTDs such as spina bifida and exencephaly.

2.3. The curly tail mouse and multifactorial NTDs

Another very good example of a hypomorphic mutation in a gene required for neurulation that fits the multifactorial threshold model of inheritance is the *curly tail* (*ct*) mouse model. The *ct* mutation arose spontaneously and has become a classic mouse model of nonsyndromic NTDs inherited in a multifactorial fashion (Gruneberg, 1954; van Straaten and Copp, 2001). Homozygous *ct/ct* mutant embryos exhibit incompletely penetrant exencephaly, spina bifida, and curled tails (Copp *et al.*, 1982; Gruneberg, 1954). As in humans, NTDs in *ct* mutants are multigenic and penetrance is influenced by both genetic and environmental factors (reviewed in van Straaten and Copp, 2001). Because of the similar pattern of inheritance to human NTDs, *ct* has historically been one of the best and most carefully studied animal models for human NTDs.

Despite the long-term efforts of many groups to positionally clone the *ct* mutation, the identity of the gene mutated has only recently been uncovered. Similarities in the phenotypes of *Grainyhead-like-3* ($Grhl3^{-/-}$) and *ct/ct* mutants and the observation that *Grhl3* mapped to the minimal *ct* interval on mouse chromosome 4 (Brouns *et al.*, 2005; Neumann *et al.*, 1994; Ting *et al.*, 2003), suggested that the *Grhl3* was the gene mutated in the *ct* mouse line. This hypothesis was confirmed by a genetic complementation test where spina bifida was observed in over half of the $Grhl3^{-/ct}$ transheterozygotes (Ting *et al.*, 2003). Furthermore, *Grhl3* transcripts were reduced in *ct/ct* mutants due to a mutation in a putative enhancer in the *ct* mutant mouse line (Gustavsson *et al.*, 2007; Ting *et al.*, 2003). Finally, expression of *Grhl3* from a bacterial artificial chromosome transgene rescued spina bifida in *ct* mutants (Gustavsson *et al.*, 2007). Together, these data provide strong evidence that spina bifida in *ct* mutants is due to the reduced expression of *Grhl3*. The differences in severity and penetrance of NTDs in $Grhl3^{-/-}$ and *ct/ct* mutants along with the fact that some *Grhl3* transcript is expressed in *ct/ct* mutants indicates that *ct* is a hypomorphic allele of *Grhl3*.

As in human NTDs, genetic modifiers influence the penetrance of NTDs in *Grhl3*$^{ct/ct}$ mutants. A number of modifier loci for *Grhl3*$^{ct/ct}$ phenotypes have been identified. Three unknown modifier loci have been mapped and other unknown modifiers can cause the *Grhl3*^{ct} mutation to exhibit dominance (Crolla *et al.*, 1990; Letts *et al.*, 1995; Neumann *et al.*, 1994). The penetrance of NTDs in *Grhl3*$^{ct/ct}$ mutants is also influenced by mutations in other genes required for neurulation. For example, while neither *Grhl3*$^{ct/+}$ nor *Pax3*$^{Sp/+}$ heterozygotes exhibit a curly tail phenotype, *Grhl3*$^{ct/+}$;*Pax3*$^{Sp/+}$ compound mutants exhibit this phenotype and *Grhl3*$^{ct/ct}$; *Pax3*$^{Sp/+}$ compound mutants exhibit completely penetrant spina bifida (Fig. 1.2; Estibeiro *et al.*, 1993). Similarly, *Grhl3*^{ct} interacts with *Vangl2*^{Lp} to cause spina bifida in compound mutants (Stiefel *et al.*, 2003). Interestingly, as will be discussed later in this chapter, NTDs in *Grhl3*$^{ct/ct}$, *Pax3*$^{Sp/Sp}$, and *Vangl2*$^{Lp/Lp}$ mutants are due to disruption of very different cellular behaviors, illustrating the idea that disruption of different processes can additively (or synergistically) disrupt the threshold event of neural tube closure resulting in a NTD.

Long before the mutation in the *ct* mouse line was identified, the developmental pathology leading to spina bifida had been extensively characterized. Spina bifida in homozygous *ct*/*ct* mutants results from delayed closure of the posterior neuropore that is caused by an exaggerated ventral curvature of the caudal region of *ct* mutants (Brook *et al.*, 1991; Copp, 1985). This increased curvature imposes a mechanical strain on elevation of the neural folds and delayed posterior neuropore closure. This defect is due to an imbalance in cell proliferation between the neuroepithelium and the underlying ventral tailbud and hindgut endoderm (Copp *et al.*, 1988a). Interestingly, the frequency and severity of spina bifida and curly tail phenotypes can be reduced by treatments that reduce proliferation of the neural tissue, effectively relieving the proliferation imbalance and rescuing the curvature defects (Copp *et al.*, 1988b; Seller, 1983; Seller and Adinolfi, 1981; Seller and Perkins, 1983, 1986).

Studies of *Grhl3*$^{ct/ct}$ mutants have also uncovered the molecular mechanisms that regulate the altered cell proliferation (Fig. 1.3). *Wnt5a* expression was dramatically reduced in the ventral tailbud and hindgut endoderm of *Grhl3*$^{ct/ct}$ mutants. Since *Wnt5a* is required for proliferation of mesodermal progenitor cells (Yamaguchi *et al.*, 1999), defects in cell proliferation in the tailbud of *Grhl3*$^{ct/ct}$ mutants may be due to a decrease in Wnt5a-regulated proliferation. In addition to the altered expression of *Wnt5a*, the expression of retinoic acid receptors (*RARγ* and *RARβ*) is reduced in the tailbud of *Grhl3*$^{ct/ct}$ embryos. Importantly, treatment with retinoic acid reduced the incidence of NTDs in *Grhl3*$^{ct/ct}$ mutants (Chen *et al.*, 1994). Since retinoic acid can regulate the expression of its receptors to control cell proliferation, upregulation of retinoic acid receptors in the tailbud may also serve to rescue the proliferation defect and the spina bifida.

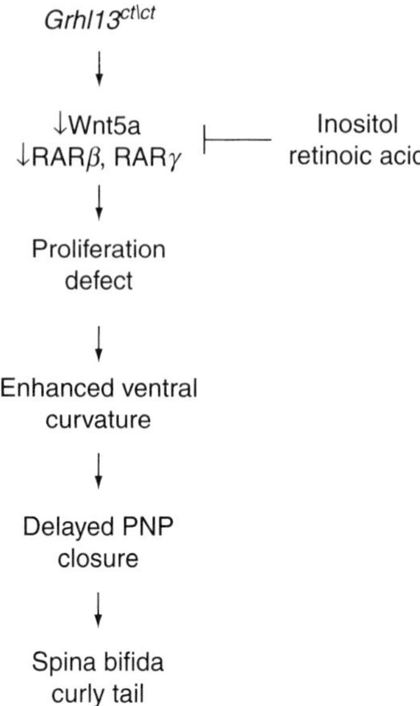

Figure 1.3 Mechanism of NTDs in *Grhl3*$^{ct/ct}$ mutants. *Grhl3*$^{ct/ct}$ exhibit decreased expression of genes regulating proliferation in the tailbud such as *Wnt5a*, *RARβ*, and *RARγ*. These changes result in reduced proliferation and enhanced ventral curvature of the posterior region of the embryo during closure of the posterior neuropore (PNP). This increased curvature results in a delay in closure of the PNP and the characteristic spina bifida or curly tail phenotypes observed in *Grhl3*$^{ct/ct}$ mutants. Factors such as retinoic acid and inositol can rescue the expression of *RARβ*, alleviating the proliferation imbalance and preventing NTDs in *Grhl3*$^{ct/ct}$ mutants. Figure modified from van Straaten and Copp (2001).

Studies of *Grhl3*$^{-/-}$ mutants indicated that *Grhl3* is required for proper formation of the epidermis (Ting *et al.*, 2005; Yu *et al.*, 2006). The first description of the expression pattern of *Grhl3* indicated that it is expressed only in the epithelium and not the tailbud or hindgut endoderm (Ting *et al.*, 2003). However, further studies detected *Grhl3* transcripts in the hindgut endoderm (Gustavsson *et al.*, 2007), one of the tissues with proliferation defects in *Grhl3*$^{ct/ct}$ mutants (Copp *et al.*, 1988a). While the developmental mechanisms leading to spina bifida in *Grhl3*$^{-/-}$ mutants has not been studied in detail, proliferation defects in the ventral tailbud have been documented (Ting *et al.*, 2003).

The examples described in this section illustrate that mouse models of the genes required for neurulation can exhibit patterns of inheritance consistent with the multifactorial threshold model. Mutation of one allele of a gene required for neurulation (e.g., *Hectd1*, *Smarca4*, or *SmarcaX*) can result in a low penetrance of apparently sporadic NTDs. Alternatively, recessive hypomorphic mutations in genes required for neurulation (e.g., *p38IP*^{drey} or *Grhl3*^{ct}) can exhibit a low penetrance of NTDs. Furthermore, genetic and environmental modifiers can influence the penetrance of NTDs in the *Grhl3*^{ct} model. In the next sections, we will discuss other multifactorial mouse models of NTDs. We provide examples that have led to the identification of candidate genes for sequencing in human patients with nonsyndromic NTDs. Furthermore, we provide examples of environmental factors such as hyperglycemia and vitamin supplementation that influence the penetrance of NTDs in mouse models.

3. Mouse Models of NTDs Have Identified Candidate Genes for NTDs in Humans

In addition to providing models of the multifactorial genetics of NTDs, the creation of mouse models of NTDs identifies the genes required for neurulation, providing candidate genes for sequencing in humans. Mouse models have been instrumental for the identification of candidate genes for both nonsyndromic and syndromic NTDs. Furthermore, studies of these mouse models have lead to a mechanistic understanding of the pathology of the human disease. Below we describe multigenic mouse models of NTDs that regulate either convergent extension movements during neurulation or formation of the vertebrae. Disruptions of either process results in NTDs in mouse models and loss-of-function mutations in the genes regulating these processes have been identified in human patients. Additionally, we describe studies of mouse models of NTDs that have been instrumental for the identification of the genes mutated in human diseases such as Meckel, Fraser, and Waardenburg syndromes. Furthermore, these mouse models have been useful for elucidating the underlying cellular and molecular mechanisms responsible for development of these syndromes.

3.1. Mutations in genes regulating planar cell polarity are associated with nonsyndromic NTDs in humans

Mutations in a gene regulating planar cell polarity (PCP) pathways (Fig. 1.4) have recently been identified in patients with NTDs, representing a prime example of how the identification of the genes required for neurulation in the mouse can provide candidate genes for sequencing in humans. The *Drosophila* PCP pathway regulates polarity of a cell within the plane of the epithelium to position asymmetrically localized structures such as the hairs on the wing (reviewed in Adler, 2002). The first realization that this pathway also regulates tissue polarity in vertebrates came from experiments in *Xenopus* where expression of mutant *Disheveled* (*Dsh*) blocked PCP

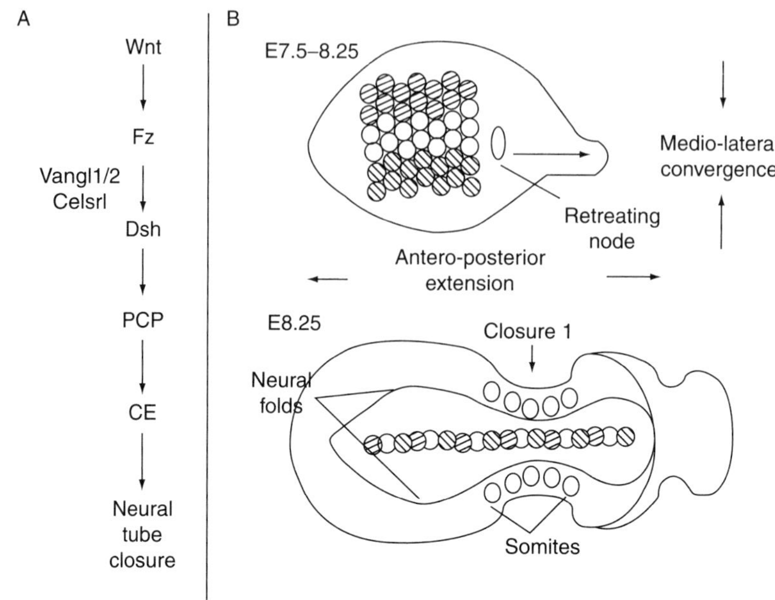

Figure 1.4 Planar cell polarity (PCP) pathways control convergent extension (CE) movements during neurulation. (A) PCP is controlled by a noncanonical Wnt signaling pathway during neurulation. In mouse, mutations in the genes regulating PCP such as *Wnt3a*, *Fz3*, *Fz6*, *Vangl1/2*, *Celsr1* and *Dsh* disrupt CE movements and neural tube closure. (B) CE movements occur in the neural plate and underlying mesoderm during elevation of the neural folds. CE movements involve the mediolateral convergence of cells toward the midline. These cell movements result in lengthening of the embryo along the anterior–posterior axis and narrowing of the embryo. Disruption of CE movements results in the neural folds being too far apart to fuse in the dorsal midline.

signaling and convergent extension movements during gastrulation (Wallingford *et al.*, 2000). Later experiments by the same group demonstrated that PCP signaling is also required for neural tube closure in *Xenopus* (Fig. 1.4; Wallingford and Harland, 2001, 2002). Interestingly, these frog embryos exhibited craniorachischisis, reminiscent of the phenotype observed in the classical mouse mutant *Loop tail* (*Lp*). The *Lp* mutation arose spontaneously causing a characteristic looped-tail phenotype in heterozygous mice and craniorachischisis in homozygous mutant embryos (Strong and Hollander, 1949). Around the same time that Wallingford and colleagues demonstrated that PCP pathways regulate neural tube closure in the frog, positional cloning of *Lp* revealed a missense mutation in *Van Gogh-like-2* (*Vangl2*), a homologue of the *Drosophila* PCP gene *van gogh* (Kibar *et al.*, 2001; Murdoch *et al.*, 2001). Since the identification of the mutation causing the *Lp* phenotype, other PCP genes have been mutated in the mouse and also result in NTDs. Embryos homozygous for mutant *Celsr1*, the vertebrate homologue of *flamingo/starry night*, exhibit craniorachischisis (Curtin *et al.*, 2003). Targeted knockouts of other PCP pathway components such as *Wnt5a*, *Frizzled3* (*Fz3*), *Fz6*, *Dsh1* and *Dsh2* also result in NTDs (Hamblet *et al.*, 2002; Qian *et al.*, 2007; Wang *et al.*, 2006b). Analysis of the phenotypes in these mutants has demonstrated that vertebrate PCP pathways regulate a number of developmental processes in addition to neural tube closure including orientation of stereociliary bundles and eyelid closure (reviewed in Wang and Nathans, 2007).

The realization that mutation of the genes involved in PCP pathways can cause NTDs in mouse models prompted the sequencing of PCP genes in human patients with NTDs (Doudney *et al.*, 2005; Kibar *et al.*, 2007b). In one study, heterozygous missense mutations were identified in *VANGL1* in two familial and one sporadic case of NTDs (Kibar *et al.*, 2007b). In one of these pedigrees, a missense mutation in a conserved region of *VANGL1* (V239I) was found in a girl with severe caudal regression with lipomyeloschisis. The mutation was also present in a brother who had a dermal sinus and the mother who was asymptomatic. In another family, a missense mutation (R274Q) was identified in *VANGL1* in a patient with myelomeningocele. The mother and maternal aunt had less severe NTDs (vertebral schisis) and the mutation was present in the mother (mutation status in aunt was unknown). In addition, a missense mutation (M328T) was identified in *VANGL1* in a sporadic case of myelomeningocele (Kibar *et al.*, 2007b).

Once mutations in *VANGL1* were identified in spina bifida patients, the challenge remained to demonstrate that these mutations actually contribute to the NTDs. First, it was unknown if *Vangl1* was required for neurulation. To address this question, the phenotype of *Vangl1* mutant mice was analyzed (Torban *et al.*, 2008). While homozygous $Vangl1^{gt/gt}$ mutant mice developed normally, a significant percentage of $Vangl2^{Lp/+};Vangl1^{gt/+}$ compound mutants exhibited craniorachischisis, indicating that disruption of the *Vangl1* gene in the mouse can contribute to NTDs. Second, the mutations identified in the human patients were demonstrated to disrupt the function of Vangl1. Mouse Vangl1 and Vangl2 associate with Dsh and the mutations in *Vangl2* identified in two different *Lp* alleles disrupted this interaction (Torban *et al.*, 2004). Significantly, all three of the mutations identified in human patients disrupt the interaction of Vangl1 with Dsh, indicating that these are in fact loss-of-function mutations (Kibar *et al.*, 2007b).

All patients identified in these studies were heterozygous for mutant *VANGL1* (Kibar *et al.*, 2007b). Furthermore, some members of these pedigrees were carriers of the *VANGL1* mutations but had either mild NTDs or were phenotypically normal. Consistent with the multifactorial threshold model for NTDs, while it is likely that the mutations identified in these patients contribute to NTDs, these data suggest that they do so by interacting with other genetic and/or environmental insults. Genetic experiments in the mouse have identified a number of loci that can modify

Vangl2^{Lp} phenotypes. PCP genes can interact with one another to produce severe NTDs in compound mutant embryos. For example, *Vangl2^{Lp}* is known to genetically interact with other PCP genes including *Vangl1*, *Dvl1*, *Dvl2*, *Celsr1*, *Fz3* and *Fz6* to cause NTDs (Curtin *et al.*, 2003; Hamblet *et al.*, 2002; Torban *et al.*, 2008; Wang *et al.*, 2006a,b). Additionally, *Vangl2^{Lp}* can genetically interact with mutations in genes not previously identified as regulating PCP pathways result in to give NTDs including *ct*, *Bardet-Biedl syndrome-1* (*BBS1*), *BBS4*, *BBS6*, *cordon blue* (*cobl*), *protein tyrosine kinase-7* (*PTK7*), and *Scribble* (*Scrbl*; Carroll *et al.*, 2003; Lu *et al.*, 2004; Murdoch *et al.*, 2003; Ross *et al.*, 2005; Stiefel *et al.*, 2003). Recent experiments have uncovered the molecular basis of some of these genetic interactions. For example, the genetic interaction of *Vangl2^{Lp}* with *Scrbl^{Crc}* was surprising since *Scribble* regulates apical–basal polarity (not PCP) in *Drosophila* (Bilder and Perrimon, 2000). Since Scrbi physically interacts with Vangl2 and is required for the asymmetric localization of Vangl2 in the cochlea, the basis for this genetic interaction may be more direct (Montcouquiol *et al.*, 2006).

In addition to identifying key candidate genes for human NTDs, studies of PCP genes in mouse have uncovered the cellular mechanisms responsible for failure of neural tube closure. Remarkably, even before the identification of the genetic lesion in the *Vangl2^{Lp}* mouse line, hints as to the underlying mechanisms of the NTD were realized. A number of studies noted the shorter body axis of *Vangl2^{Lp}* mutants and hypothesized that NTDs may be due to a failure of axial elongation (Gerrelli and Copp, 1997; Smith and Stein, 1962; Wilson and Wyatt, 1994). Others noted a broader notochord and floorplate in *Vangl2^{Lp}* mutants and suggested that failure of neural tube closure may be due to an inability of the neural folds to become apposed in the dorsal midline (Gerrelli and Copp, 1997; Greene *et al.*, 1998). More recent experiments in vertebrates demonstrated that both of these theories of the underlying cause of NTDs in Lp mutants are correct. PCP signaling is necessary for convergent extension movements in the notochord and neural plate (Wallingford and Harland, 2002; Ybot-Gonzalez *et al.*, 2007) and defects in convergent extension result in a wider, shorter midline and floorplate causing the neural folds to be too far apart in the dorsal midline to fuse (Wallingford and Harland, 2002; Wang *et al.*, 2006a; Ybot-Gonzalez *et al.*, 2007). Interestingly, human embryos with craniorachischisis also exhibit a broad floorplate (Kirillova *et al.*, 2000), indicating that similar mechanisms may underlie these types of NTDs in humans.

Mutation of PCP genes can also result in milder forms of NTDs such as spina bifida and exencephaly instead of the more severe craniorachischisis observed in *Vangl2$^{Lp/Lp}$* mutants. For example, *Dvl2$^{-/-}$* or *Vangl2$^{Lp/+}$* mutants exhibit a very low penetrance of spina bifida (Copp *et al.*, 1994; Hamblet *et al.*, 2002). Interestingly, heterozygous *Vangl2$^{Lp/+}$* embryos exhibit a slightly wider and shorter midline and delayed neural tube closure (Wang *et al.*, 2006a), suggesting that even in heterozygous embryos, neural tube closure is compromised. Furthermore, these findings provide a likely mechanism for the development of NTDs in genetic interaction experiments. For example, *Grhl3*$^{ct/ct}$; *Vangl2*$^{Lp/+}$ compound mutants exhibit highly penetrant NTDs (Stiefel *et al.*, 2003). In *Vangl2*$^{Lp/+}$ embryos, neural tube closure is already slightly impaired, priming them for a second insult such as an imbalance in proliferation in the caudal regions of *Grhl3*$^{ct/ct}$ mutants. The combination of these two insults, insufficient on their own to result in a high penetrance of NTDs, is above the threshold required for failure of neural tube closure.

3.2. PAX1 and PDGFRα mutations are associated with nonsyndromic NTDs in humans

Another example of the convergence of studies in mouse and humans to identify NTD-associated mutations in humans is the identification of mutations in *PAX1* and *PDGFRα* genes in human spina bifida patients. Mutations of the *Pax1* gene in mouse cause the *undulated* (*un*) phenotype characterized by a kinky tail and malformations of the vertebrae, but not spina bifida (Balling *et al.*, 1988). The *Patch* (*Ph*) mutant mouse line carries a deletion that includes (among other genes) *Platelet-derived growth factor receptor-alpha* (*Pdgfrα*; Smith *et al.*, 1991; Stephenson *et al.*, 1991). On some genetic backgrounds, homozygous *Ph/Ph* mutants die by E10.5 but do not exhibit NTDs (Morrison-Graham *et al.*, 1992; Orr-Urtreger *et al.*, 1992), while on other genetic backgrounds, embryos survive longer and exhibit spina bifida (Payne *et al.*, 1997). Interestingly, *Pax1*$^{un/un}$; *Ph/+* compound mutant embryos exhibit highly penetrant spina bifida indicating that these mutations can interact in a digenic fashion to cause NTDs (Helwig *et al.*, 1995). Further studies demonstrated that *Pax1* acts as a transcriptional activator of the *Pdgfrα* promoter (Joosten *et al.*, 1998), suggesting that this genetic interaction represents disruption of a linear pathway.

The involvement of *Pax1* in NTDs in a complex genetic fashion in mouse prompted the sequencing of *PAX1* in human patients with nonsyndromic NTDs (Hol *et al.*, 1996). In this study, a point mutation in the DNA-binding domain of *PAX1* was identified in a single fetus that presented with spina bifida. This mutation was not a common polymorphism as it was not present in many control samples. Furthermore, the mutations found either in *un* mutant mice or the human fetus reduced the transcriptional activity of Pax1 (Joosten *et al.*, 1998). These results indicate that the mutation identified in this one case of spina bifida may contribute to the NTD, as it functionally disrupts the activity of Pax1. Since an unaffected mother and grandmother were also heterozygous for mutant *PAX1*, other environmental and/or genetic insults must have also contributed to the NTD in this patient. From the analysis of *un* and *Ph* mutant mouse lines, one genetic risk factor may be *PDGFRα* mutation. Interestingly, polymorphisms in the *PDGFRα* gene have also been associated with NTDs in humans as specific polymorphisms that alter transcription from the *PDGFRα* promoter correlate with a predisposition to spina bifida (Joosten *et al.*, 2001; Zhu *et al.*, 2004). Other potential candidate genes that interact with mutant *Pax1* to cause spina bifida in mouse models include *Transcription factor E2a* (*Tcf2a*) and *Forkhead box C2* (*FoxC2*; Furumoto *et al.*, 1999; Joosten *et al.*, 2005).

Both *Pax1* and *Pdgfrα* transcripts are expressed in the sclerotome portion of the somite and are required for the development of this lineage into the vertebrae (Balling *et al.*, 1988; Deutsch *et al.*, 1988; Orr-Urtreger *et al.*, 1992; Payne *et al.*, 1997). Histological analysis of spina bifida in *Pax1*$^{-/-}$; *Ph*/+ compound mutants or *Ph*/*Ph* homozygotes demonstrates that neural tube closure occurs normally, but the vertebrae do not form properly in the lumbar region of the spine (Helwig *et al.*, 1995; Payne *et al.*, 1997). These findings indicate that spina bifida in these mouse models does not result from defects in primary neurulation; rather, defects in formation of the vertebrae over the closed spinal neural tube.

One of the downstream signaling pathways activated by the PDGF receptor is phosphatidylinositol 3-kinase (PI3K). Mice with a mutation in the PI3K-binding sites of the PDGFRα exhibit loss of PI3K signaling downstream of the receptor and spina bifida (Klinghoffer *et al.*, 2002). These results indicate that spina bifida in *Pdgfrα* mutants is due, at least in part, to loss of PI3K activation. Furthermore, conditional deletion of *Pdgfrα* in sclerotome-derived mesenchyme also resulted in spina bifida (Pickett *et al.*, 2008), demonstrating that the defect leading to spina bifida is in this lineage. Additional experiments indicated that PI3K activation is not necessary for apoptosis, proliferation or chondrogenesis but rather for migration of sclerotome-derived cells to form the dorsal portion of the vertebrae (Pickett *et al.*, 2008).

3.3. Mouse models of NTDs have led to the identification of the genes mutated in Meckel syndrome

Meckel syndrome (MKS) is a rare, autosomal recessive disorder characterized by malformation of multiple organs and accounts for the largest group of syndromic NTDs in humans (reviewed in Smith *et al.*, 2006). The identification of the genes mutated in MKS involved the convergence of studies in both human MKS patients and mouse models (Dawe *et al.*, 2007; Delous *et al.*, 2007; Frank *et al.*, 2008; Kyttala *et al.*, 2006; Smith *et al.*, 2006).

Phenotypes of MKS overlap with ciliopathies such as Bardet-Biedl syndrome (BBS), Joubert syndrome, and polycystic kidney disease, suggesting that defects in cilia formation underlie the pathology of MKS.

Furthermore, mutations in many genetic loci have been associated with these syndromes and the majority of the genes implicated play obvious roles in cilia function. Mutations in the *MKS1* gene were identified by positional cloning strategies in affected pedigrees and the domain structure of MKS1 indicated a role in cilia function (Frank *et al.*, 2008; Kyttala *et al.*, 2006). Therefore, when the *MKS3* locus mapped to a large interval on human chromosome 8q, syntenic with the *wistar polycystic kidneys* (*wpk*) locus in rat, the *wpk* gene became a candidate gene for MKS. Subsequently, the rat mutation was identified in a novel gene implicated in cilia function, followed by the identification of missense mutations in a number of MKS patients (Delous *et al.*, 2007).

Another MKS gene was also first identified in a rodent model. The *Fused toes* (*Ft*) mouse line was created by insertional mutagenesis deleting a number of genes including *retinitis pigmentosa GTPase regulator-interacting protein-like-1* (*Rpgrip1l*; Peters *et al.*, 2002). Homozygous *Ft/Ft* embryos die at midgestation showing phenotypes consistent with defects in cilia function (van der Hoeven *et al.*, 1994) and these phenotypes were recapitulated in targeted knockouts for *Rpgrip1l*, indicating that *Rpgrip1l* is the critical gene responsible for these phenotypes (Vierkotten *et al.*, 2007). The *Ft* locus is in a region syntenic to a MKS interval on human chromosome 16, providing another candidate gene for MKS. Sequencing of *RPGRIP1L* in MKS patients revealed loss-of-function mutations (Delous *et al.*, 2007).

Hints as to how defects in cilia formation may lead to NTDs have also come from studies in the mouse. A slew of recent studies have established an essential role for cilia in the transduction of sonic hedgehog (Shh) signaling (reviewed in Scholey and Anderson, 2006). This novel idea emerged from an ENU mutagenesis screen in mouse conducted by Kathryn Anderson and colleagues at Memorial Sloan Kettering Cancer Center in New York City (Anderson, 2000; Caspary and Anderson, 2006; Garcia-Garcia *et al.*, 2005; Huangfu *et al.*, 2003; Kasarskis *et al.*, 1998; Zohn *et al.*, 2005). From this unbiased genetic screen, a number of mouse lines were established that exhibited syndromic NTDs associated with polydactyly and laterality defects (Caspary *et al.*, 2002, 2007; Eggenschwiler *et al.*, 2001; Garcia-Garcia *et al.*, 2005; Huangfu and Anderson, 2005; Huangfu *et al.*, 2003; Kasarskis *et al.*, 1998; Liu *et al.*, 2005). Further characterization revealed profound defects in dorsal–ventral patterning of the neural tube consistent with defects in Shh signaling in the notochord and floorplate. Positional cloning identified mutations in genes encoding intraflagellar transport (IFT) proteins and other proteins required for the formation and/or maintenance of cilia (Caspary *et al.*, 2002, 2007; Eggenschwiler *et al.*, 2001; Garcia-Garcia *et al.*, 2005; Huangfu and Anderson, 2005; Huangfu *et al.*, 2003; Kasarskis *et al.*, 1998; Liu *et al.*, 2005). IFT proteins are involved in the transport of proteins along the cilia with the aide of motor proteins such as kinesin and dynein. Mouse lines with mutations in IFTs (*IFT52*, *IFT57*,

IFT88, and *IFT72*), *kinesin family member-3a* (*Kif3a*), or *dynein cytoplasmic 2 heavy chain 1* (*Dync2H1*) exhibit disruptions in Shh signaling and NTDs (Houde *et al.*, 2006; Huangfu and Anderson, 2005; Huangfu *et al.*, 2003; Liu *et al.*, 2005; May *et al.*, 2005). Epistasis analysis demonstrated that IFT proteins are required for Shh signaling downstream of the receptor and upstream of the Gli transcription factors (Caspary *et al.*, 2002, 2007; Eggenschwiler *et al.*, 2001; Garcia-Garcia *et al.*, 2005; Huangfu and Anderson, 2005; Huangfu *et al.*, 2003; Kasarskis *et al.*, 1998; Liu *et al.*, 2005). Additionally, Shh pathway components localize to the cilia (Corbit *et al.*, 2005; Haycraft *et al.*, 2005; Rohatgi *et al.*, 2007). Further studies indicate that Shh controls the balance of Gli activator and repressor and IFTs are essential for the processing of Gli proteins (Haycraft *et al.*, 2005; Huangfu and Anderson, 2005; Liu *et al.*, 2005; May *et al.*, 2005), suggesting a model where trafficking of proteins in the cilia promotes the processing of Gli proteins (reviewed in Caspary *et al.*, 2007; Scholey and Anderson, 2006). Interestingly, *Rpgrip1l* mutant mice exhibit defects in Shh signaling affecting the ratio of Gli3 activator and repressor (Vierkotten *et al.*, 2007). These findings suggest that Shh signaling may be disrupted in MKS.

Another ciliopathy, Bardet-Biedl syndrome (BBS), may comprise a spectrum of allelic disorders with MKS as hypomorphic mutations in genes responsible for MKS (*MKS1*, *MKS3*, and *CEP290*) have been associated with BBS and mutations in BBS genes often cause MKS-like phenotypes (Karmous-Benailly *et al.*, 2005; Leitch *et al.*, 2008). While NTDs have not been reported in human patients with BBS, mutation of one of the BBS genes (*Bbs4*) in mouse results in a low penetrance of NTDs (Ross *et al.*, 2005). Furthermore, *Bbs4*, *Bbs1*, and *Bbs6* can genetically interact with $Vangl2^{Lp}$ to cause NTDs (Ross *et al.*, 2005). These mutant embryos also exhibit other aspects of PCP phenotypes such as open eyelids and misorientation of cochlear stereociliary bundles (Ross *et al.*, 2005), suggesting that some BBS proteins may also regulate PCP pathways. The relationship of PCP signaling to cilia assembly and/or maintenance remains to be elucidated and this topic is explored in some recent reviews (Eggenschwiler and Anderson, 2007; Wallingford, 2006).

It remains unclear how loss of IFT proteins and disruption of Shh signal transduction in the neural tube results in NTDs. A number of potential mechanisms have emerged from studies of mouse models. Shh regulates dorsal–ventral patterning of the neural tube and disruption of Shh signaling results in misspecification of cell fates along the dorsal–ventral axis of the neural tube (reviewed in Ulloa and Briscoe, 2007). However, alterations in cell fate are not in themselves sufficient to disrupt closure of the neural tube as $Shh^{-/-}$ mutants exhibit holoprosencephaly but not spina bifida or exencephaly (Chiang *et al.*, 1996). On the other hand, Gli3-mediated repression of Shh target genes is essential for neural tube closure (Hui and Joyner, 1993). Similarly, *patched* (*ptc*), a receptor and negative regulator of Shh signaling, is essential for neural tube closure (Goodrich *et al.*, 1997). Experiments in mouse models have demonstrated that in the spinal cord, Shh signaling suppresses dorsal–lateral hinge point (DLHP) formation and in the absence of Shh, exaggerated DLHPs form (Ybot-Gonzalez *et al.*, 2002). Another mechanism by which IFT mutants may develop NTDs is the altered regulation of cell proliferation. Shh signaling plays an important role in regulation of cell proliferation in the neural tissue (reviewed in Ulloa and Briscoe, 2007). In addition, Shh plays an essential role in the development of the vertebrae (Fan and Tessier-Lavigne, 1994). For example, *Gli2* or *Rab23* mutants (a negative regulator of Shh signaling) exhibit defects in development of vertebrae (Gunther *et al.*, 1994; Mo *et al.*, 1997; Sporle and Schughart, 1998; Sporle *et al.*, 1996).

3.4. Mouse models have led to the identification of the genes mutated in Fraser syndrome

As in MKS, mouse models have played an important role in identification of the genes causing Fraser syndrome (FS) in humans. FS is a rare autosomal recessive disorder resulting in malformation of a variety of organs including limbs, eyes, kidneys, and lungs (reviewed in Smyth and Scambler, 2005). Recently, the convergence of efforts in mouse and humans has led to the identification of the genes mutated in FS (reviewed in Smyth and Scambler, 2005). Mouse models for FS are known as "blebbing" mutants and include *blebbed* (*bl*), *myelencephalic blebs* (*my*), *eye blebs* (*eb*), and *head blebs* (*heb*) (Chapman and Hummel, 1963; Little and Bagg, 1923; Phillips, 1970; Varnum and Fox, 1976). As their names imply, a major phenotypic feature of the blebbing mutants is the development of fluid filled blisters over the limbs, eyes, and ears.

Identification of the genes mutated in blebbing mutants has lead to the discovery of two of the genes causing FS. Targeted deletion of *Fraser syndrome 1 homologue* (*Fras1*) in mouse resulted in blebbing phenotypes and a mutation in *Fras1* was identified in *bl* (McGregor *et al.*, 2003; Vrontou *et al.*, 2003). When the gene causing FS in six unrelated pedigrees were mapped to a region syntenic with the critical region for mouse *bl*, *FRAS1* was sequenced and mutations identified in these FS patients (McGregor *et al.*, 2003). Similarly, the gene mutated in *my* was identified as *Fras1-related extracellular matrix protein-2* (*Frem2*) in three different allelic *my* mutants (*my*, my^{ucl}, and my^{F11}; Jadeja *et al.*, 2005; Little and Bagg, 1923; Timmer *et al.*, 2005). This prompted the sequencing of *FREM2* in FS families not linked to *FRAS1* and the identification of mutations in three of these families (Jadeja *et al.*, 2005). The genes mutated in the two other blebbing mutants have not yet been associated with FS in humans. The *eb* phenotype is caused by mutation in *glutamate receptor-interacting protein-1*

(*Grip1*) and targeted deletion of *Grip1* also results in blebbing phenotypes (Bladt *et al.*, 2002; Takamiya *et al.*, 2004). Similarly, the *heb* phenotype is caused by mutation of *Fras1-related extracellular matrix gene-1* (*Frem1*; Smyth *et al.*, 2004).

Studies of the blebbing mutants have lead to the elucidation of the molecular pathology causing FS-associated malformations (reviewed in Smyth and Scambler, 2005). Blebbing proteins play an essential role in the assembly of structural components of the extracellular matrix during morphogenesis of tissues where a remodeling epidermis interacts with an underlying mesenchyme. Blebbing phenotypes are thought to result from a loss of epithelial adhesion causing the formation of a blister covering the eye or limb. As the lesion heals during subsequent development, malformation of these structures occurs. Alternatively, during formation of organs like the kidneys or lungs, disruption of epithelial–mesenchyme interactions may prevent branching morphogenesis resulting in agenesis or hypoplasia of these organs.

While mutations in FS genes have not been reported in human NTDs, mutation of FS genes in the mouse can cause NTDs depending on the genetic background. *Frem2*$^{my\text{-}F11}$ was isolated in an ENU mutagenesis screen based on the appearance of an exencephalic phenotype (Timmer *et al.*, 2005). This screen was performed on a C57Bl/6 background and outcrossed to a C3H/HeJ background. On the mixed background, *Frem2*$^{my\text{-}F11}$ mutants exhibited exencephaly and died at birth. The NTD was largely suppressed when outcrossed to a castaneus background allowing a small percentage of homozygotes to survive to adulthood that exhibited classical blebbing phenotypes. Another blebbing mutant (*Grip*^{eb}) also exhibited either NTDs or classical blebbed phenotypes (but not both) depending on the genetic background (Swiergiel *et al.*, 2000). Since mutations in the genes causing FS can result in either NTDs or blebbing phenotypes, FS genes may also be good candidates for human NTDs.

3.5. Mouse models of Waardenburg syndrome exhibit NTDs

Another syndrome where NTDs are rarely reported in human patients but represent a predominant phenotype in the mouse model is *PAX3* mutations in Waardenburg syndrome (WS). WS is a rare autosomal dominant inherited disease. NTDs including spina bifida and exencephaly have been associated with WS and *PAX3* mutations in humans (Begleiter and Harris, 1992; Carezani-Gavin *et al.*, 1992; Chatkupt *et al.*, 1993; da-Silva, 1991; de Saxe *et al.*, 1984; Hol *et al.*, 1995; Hoth *et al.*, 1993; Kujat *et al.*, 2007; Moline and Sandlin, 1993; Nye *et al.*, 1998; Pantke and Cohen, 1971; Shim *et al.*, 2004). While the majority of WS patients are heterozygous for mutant *PAX3*, homozygous individuals have been identified. While homozygosity of mutant *PAX3* is associated with NTDs (Ayme and Philip, 1995), not all individuals homozygous for mutant *PAX3* present with NTDs (Wollnik *et al.*, 2003; Zlotogora *et al.*, 1995). These observations suggest that modifiers likely influence the penetrance and expressivity of NTDs associated with *PAX3* mutations and WS.

Mutations in the *Pax3* gene have been identified in the *Splotch* (*Sp*) mutant mouse lines (Bogani *et al.*, 2004; Epstein *et al.*, 1991a,b, 1993; Goulding *et al.*, 1993; Vogan *et al.*, 1993). Homozygous *Sp/Sp* mutant embryos exhibit NTDs such as spina bifida and exencephaly (Auerbach, 1954). *Pax3* is expressed in the dorsal neural tube (Goulding *et al.*, 1991) and *Pax3* deficiency results in excessive apoptosis of the neural tissue (Pani *et al.*, 2002b). This increased apoptosis is likely responsible for NTDs in *Pax3* mutants as inhibiting p53-dependent apoptosis by either chemical inhibitors or genetic mutation of p53, results in a dose-dependent reduction in the penetrance of NTDs (Pani *et al.*, 2002b).

The reason that homozygosity of mutant *Pax3* in the mouse invariantly results in NTDs, while in humans, *PAX3* mutations are only occasionally associated with NTDs remains unknown. One possible explanation is the presence or absence of genetic modifiers or environmental influences. From studies in mouse, a few genetic modifiers of NTDs in *Pax3* mutants have been identified and include *p53*, *Neurofibromin I* (*NfI*), and $Grhl3^{ct}$ (Estibeiro *et al.*, 1993; Lakkis *et al.*, 1999; Pani *et al.*, 2002b). The expressivity of NTDs in *Pax3* mutants is also dependent on environmental influences such as maternal diabetes and folic acid and these are discussed in the next section (Fleming and Copp, 1998; Phelan *et al.*, 1997).

4. Mouse as a Model for the Elucidation of the Molecular Mechanisms of Gene–Environment Interactions Contributing to NTDs

In addition to providing a genetic system to model complex genetic interactions and identification of candidate genes for NTDs in humans, mouse models provide tractable systems to examine the interaction of the genes required for neural tube closure with environmental insults. The study of gene–environment interactions in mouse models has the potential to uncover the molecular basis for these interactions. These findings can then lead to the development of approaches to prevent NTDs. One very good example of this is the work carried out primarily in the laboratory of Mary Loeken at Harvard Medical School in Boston on the molecular mechanisms leading to NTDs in a mouse model of diabetes.

4.1. Studies of mouse models have led to an understanding of the molecular mechanisms underlying NTDs associated with maternal diabetes

Maternal diabetes increases the risk of a number of complications during pregnancy including NTDs (reviewed in Loeken, 2005). Investigations of the molecular mechanisms contributing to NTDs in response to hyperglycemia suggest that exposure of the embryo to elevated glucose alters the expression of genes required for neural tube closure. For example, in a mouse model of diabetic pregnancy, reduced expression levels of *Pax3* transcript were correlated with an increased risk of developing NTDs (Fig. 1.5; Fine *et al.*, 1999; Phelan *et al.*, 1997). Since loss of *Pax3* function is sufficient to cause NTDs, the reduction in *Pax3* expression is considered to be a major mechanism contributing to NTDs in diabetic pregnancies. Hyperglycemia results in the generation of reactive oxygen species (ROS), which are sufficient to reduce expression of *Pax3* and induce NTDs (Chang *et al.*, 2003). Significantly, treatment with antioxidants can prevent changes in *Pax3* expression and NTDs in diabetic pregnancies without affecting serum glucose concentrations (Chang *et al.*, 2003).

Genetic modifiers influence the susceptibility of NTDs associated with maternal diabetes as the genetic background influences the penetrance and

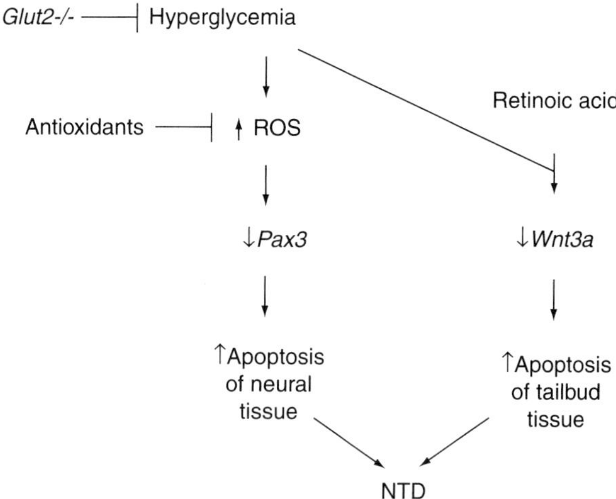

Figure 1.5 Pathways leading to NTDs in diabetes-induced hyperglycemia. Hyperglycemia results in increased levels of reactive oxygen species (ROS), decreased levels of *Pax3* transcripts and apoptosis of neural tissue leading to NTDs. Hyperglycemia and NTDs can be prevented by mutation of the *Glut2* glucose transporter. Similarly, increased ROS and NTDs can be prevented by antioxidants. Hyperglycemia also synergizes with retinoic acid to decrease levels of *Wnt3a* transcripts causing apoptosis of tailbud tissues resulting in caudal agenesis syndrome and spina bifida.

expressivity of diabetes-induced NTDs in rodent models. For example, a rat strain resistant to NTDs in response to hyperglycemia expresses increased levels of transcripts encoding free radical scavenging proteins (Cederberg *et al.*, 2000). Furthermore, *Pax3* expression is not reduced in response to diabetic pregnancy in the C57Bl/6J inbred mouse strain that is resistant to diabetes-associated NTDs (Pani *et al.*, 2002a). While the identity of the genetic modifier in this strain remains unknown, it is a dominant trait as F1 hybrids with a highly susceptible strain are resistant to NTDs (Pani *et al.*, 2002a). One potential genetic modifier of NTDs in diabetic pregnancy is the glucose transporter (*GLUT2*) that transports glucose into embryonic cells under hyperglycemic conditions. The frequency of NTDs were reduced in diabetic pregnancies in $GLUT2^{-/+}$ embryos and completely prevented in $GLUT2^{-/-}$ embryos (Li *et al.*, 2007).

Another complication of diabetic pregnancy is caudal agenesis syndrome which is often associated with some forms of spina bifida such as lipomeningocele (Passarge and Lenz, 1966). Retinoic acid, which by itself can induce caudal agenesis syndrome, can at subthreshold doses increase the susceptibility of embryos from diabetic pregnancies to caudal agenesis syndrome (Chan *et al.*, 2002). Studies in mouse have provided a molecular mechanism for the synergy of exposure to glucose and retinoic acid to cause NTDs. Treatment of mouse embryos with retinoic acid results in extensive apoptosis of the tailbud and the conversion of the remaining tailbud cells to neural tissue (Shum *et al.*, 1999). The excess neural tissue organizes into multiple neural tubes associated with spina bifida. This phenotype is very similar to that observed in *Wnt3a* mutant embryos (Takada *et al.*, 1994; Yoshikawa *et al.*, 1997). A hypomorphic mutation in *Wnt3a* in the *vestigial tail* (*vt*) mouse line results in decreased expression of *Wnt3a* (Greco *et al.*, 1996). $Wnt3a^{vt}$ mutants are more susceptible to caudal agenesis and spina bifida in response to retinoic acid treatment and exhibit more robust reductions in the levels of *Wnt3a* transcripts than wild-type embryos (Chan *et al.*, 2002). Furthermore, hyperglycemia enhances the downregulation of *Wnt3a* expression by retinoic acid (Chan *et al.*, 2002), providing a molecular mechanism for the synergy between retinoic acid and diabetes in the induction of caudal agenesis syndrome and spina bifida.

5. Mouse as a Model for Developing Approaches to Prevent NTDs

Throughout this chapter, we have provided many examples of how mechanistic studies of the underlying causes of NTDs in mouse models have led to the development of approaches to reduce the incidence of NTDs on certain genetic backgrounds. For example, we reviewed the evidence that retinoic acid and treatments that slowed proliferation of the neural tissues can rescue spina bifida in *ct* mutants. Additionally, we discussed how antioxidants reduce the incidence of NTDs in diabetic pregnancy. These preventative approaches may be very specific to the underlying causes of the NTD in these mouse models (e.g., proliferation imbalance in caudal regions and generation of ROS) and may not reduce the incidence of NTDs with other etiologies. In contrast, folic acid and inositol therapies can suppress NTDs in response to a wide variety of genetic and environmental insults and mouse models provide experimental systems for testing their activities and uncovering their mechanism of action.

5.1. Folic acid prevents the majority of NTDs in humans

Epidemiological studies suggested and clinical trials demonstrated that folic acid supplementation before and during the first weeks of pregnancy can significantly reduce the incidence of NTDs (Smithells *et al.*, 1980; Wald *et al.*, 1991). Folic acid supplementation may work in part by compensating for genetic defects in folic acid metabolism. First of all, NTD affected pregnancies are associated with lower levels of maternal plasma folate (see, e.g., Kirke *et al.*, 1993; Smithells *et al.*, 1976; van der Put *et al.*, 1997; Yates *et al.*, 1987). Second, single nucleotide polymorphisms (SNPs) in enzymes involved in folate metabolism have been associated with an increased risk for a NTDs affected pregnancy (reviewed in Blom *et al.*, 2006; Boyles *et al.*, 2005; Kibar *et al.*, 2007a). For example, SNPs in *MTHFR* (*5,10-methylenetetrahydrofolate reductase*) that result in reduced activity of the enzyme are associated with an increased risk of NTDs in some populations and higher levels of folic acid intake are particularly beneficial to these individuals (reviewed in Blom *et al.*, 2006; Boyles *et al.*, 2005).

While the benefits of folic acid supplementation are clear, the mechanism of folic acid suppression of NTDs with many different etiologies still remains a mystery. A number of animal models of folate responsive NTDs have been developed and include *Cart1* (*cartilage homeoprotein-1*), *Cited2* (*CBP/p300-interacting transactivators with glutamic acid/aspartic acid-rich C-terminal domain-2*), $Pax3^{Sp\text{-}2H}$, and *crooked tail* (*cd*) mutants (Barbera *et al.*, 2002; Fleming and Copp, 1998; Zhao *et al.*, 1996). Among the folic acid responsive mouse mutants, some exhibit defects in folic acid metabolism (e.g., $Pax3^{Sp\text{-}2H}$), while others exhibit normal folic acid metabolism (e.g., *Cited2*; Barbera *et al.*, 2002; Fleming and Copp, 1998), indicating that folic acid can suppress NTDs that are not associated with defects in folate metabolism. NTDs in *Cart1*, *Cited2*, and *Pax3* mutants are associated with excessive apoptosis (Bamforth *et al.*, 2001; Barbera *et al.*, 2002; Pani *et al.*, 2002b; Zhao *et al.*, 1996). However, in *Cited2* mutants where NTDs were suppressed with folic acid treatment, excessive apoptosis persisted (Barbera *et al.*, 2002). These results suggest that apoptosis may not be causal for the NTD or that folic acid compensates by other mechanisms. Since folic acid functions as a substrate in a number of metabolic pathways including purine and methionine biosynthesis, folic acid deficiency may affect proliferation by adversely affecting purine biosynthesis or affecting the methylation cycle. Future studies investigating the mechanistic role of folic acid supplementation in folate responsive (and resistant) mouse models promise to provide a greater understanding of the mechanisms by which folic acid suppresses NTDs.

NTDs induced by many teratogens such as valproic acid are suppressed by folic acid supplementation. Valproic acid increases the frequency of NTDs in both mouse models and human patients (reviewed in Cabrera *et al.*, 2004; Finnell *et al.*, 2003; Greene and Copp, 2005). Many possible mechanisms have been proposed to account for the adverse effect of valproic acid on neurulation. Valproic acid causes a reduction in serum folic acid levels in pregnant mothers and valproic acid-induced NTDs are suppressed by treatment with folic acid in mouse models (reviewed in Cabrera *et al.*, 2004; Finnell *et al.*, 2003; Greene and Copp, 2005). Other proposed mechanisms include an affect on proliferation of neural tissue, inhibition of histone deacetylase activity, or Wnt signaling (reviewed in Wiltse, 2005).

Another teratogen that causes NTDs in humans is fumonisin, a mycotoxin produced by a mold that commonly grows on corn. The realization that exposure to fumonisin increases the susceptibility to NTDs came from epidemiological studies of NTD clusters that occurred along the Texas–Mexico border in the early 1990s (reviewed in Cabrera *et al.*, 2004; Marasas *et al.*, 2004). Subsequently, fumonisins were also found to induce folic acid-suppressible NTDs in mouse models (Gelineau-van Waes *et al.*, 2005; Sadler *et al.*, 2002). Other studies have found that treatment with fumonisin results in reduced folate uptake by interfering with the localization of folate-binding protein-1 (folbp1; Stevens and Tang, 1997). These studies provide a molecular mechanism of both the teratogenic action of fumonisin (e.g., reducing the uptake of dietary folate) and how this deficiency can be overcome by folic acid supplementation.

5.2. Inositol prevents folate-resistant NTDs

In clinical trials, folic acid supplementation cannot prevent all NTDs (see, e.g., Wald *et al.*, 1991). Similarly, NTDs in many mouse models, such as *Grhl3*^{ct} mutants, are not suppressed by folic acid treatment (van Straaten *et al.*, 1995). Interestingly, vitamin supplementation with inositol can suppress spina bifida in *Grhl3*$^{ct/ct}$ mutant embryos (Greene and Copp, 1997). Further studies revealed that inositol rescues the proliferation imbalance in the hindgut endoderm of *Grhl3*$^{ct/ct}$ mutants (Cogram *et al.*, 2004). In addition to preventing NTDs in the *ct* mouse model, inositol can reduce the frequency of NTDs in a diabetic rodent model (Khandelwal *et al.*, 1998), suggesting that the ability of inositol to suppress NTDs may be more widely applicable as a preventative therapy. Interestingly, one study in humans found an association of lower levels of inositol in the serum of mothers with NTD affected pregnancies (Groenen *et al.*, 2003). Clinical trials are now underway to evaluate the effectiveness of inositol in reducing the recurrence of folic acid-resistant NTDs in humans. Surprisingly, while the incidence of spina bifida in *Grhl3*$^{ct/ct}$ mutants was reduced by treatment with inositol, NTDs in *Grhl3*$^{-/-}$ mutants were resistant to inositol supplementation (Ting *et al.*, 2003). This suggests that nutritional supplementation may only rescue moderate forms of NTDs or that additional inositol-resistant mechanisms also contribute to spina bifida in *Grhl3*$^{-/-}$ mutants.

6. Conclusions and Future Directions

In humans, NTDs exhibit a complex pattern of inheritance best explained by the multifactorial threshold model. This model postulates that multiple factors (environmental and genetic) contribute to the etiology of NTDs. While most mouse models of NTDs exhibit Mendelian patterns of inheritance, many mouse models of NTDs fit the multifactorial threshold model. Studies of mouse models of NTDs have identified candidate genes and environmental triggers that contribute to NTDs in humans and uncover the cellular and molecular mechanisms underlying these birth defects. However, to date, very few genetic factors contributing to NTDs in humans have been identified. As mouse models uncover additional genes required for neurulation in the mouse, future genetic studies in humans promise to identify mutations in these genes in patients with NTDs. In spite of the significant reduction in the incidence of NTDs due to increased folic acid consumption among women of childbearing age, NTDs still remain one of the most common birth defects in humans. This is due in part to the proportion of NTDs that are not responsive to folic acid. Mouse models have been instrumental in the development of additional preventative approaches to suppress folic acid-resistant NTDs. Furthermore, studies in mouse can identify which NTDs caused by different genetic and environmental insults can be rescued by particular preventative approaches. These studies promise to provide a framework to further reduce the incidence of NTDs in humans.

References

Adler, P. N. (2002). Planar signaling and morphogenesis *in Drosophila. Dev. Cell* **2,** 525–535.
Anderson, K. V. (2000). Finding the genes that direct mammalian development: ENU mutagenesis in the mouse. *Trends Genet.* **16,** 99–102.

Auerbach, R. (1954). Analysis of teh developmental effects of a lethal mutation in the house mouse. *J. Exp. Zool.* **127,** 305–329.
Ayme, S., and Philip, N. (1995). Possible homozygous Waardenburg syndrome in a fetus with exencephaly. *Am. J. Med. Genet.* **59,** 263–265.
Balling, R., *et al.* (1988). Undulated, a mutation affecting the development of the mouse skeleton, has a point mutation in the paired box of Pax 1. *Cell* **55,** 531–535.
Bamforth, S. D., *et al.* (2001). Cardiac malformations, adrenal agenesis, neural crest defects and exencephaly in mice lacking Cited2, a new Tfap2 co-activator. *Nat. Genet.* **29,** 469–474.
Barbera, J. P., *et al.* (2002). Folic acid prevents exencephaly in Cited2 deficient mice. *Hum. Mol. Genet.* **11,** 283–293.
Begleiter, M. L., and Harris, D. J. (1992). Waardenburg syndrome and meningocele. *Am. J. Med. Genet.* **44,** 541.
Bilder, D., and Perrimon, N. (2000). Localization of apical epithelial determinants by the basolateral PDZ protein Scribble. *Nature* **403,** 676–680.
Bladt, F., *et al.* (2002). Epidermolysis bullosa and embryonic lethality in mice lacking the multi-PDZ domain protein GRIP1. *Proc. Natl Acad. Sci. USA* **99,** 6816–6821.
Blom, H. J., *et al.* (2006). Neural tube defects and folate: Case far from closed. *Nat. Rev. Neurosci.* **7,** 724–731.
Bogani, D., *et al.* (2004). New semidominant mutations that affect mouse development. *Genesis* **40,** 109–117.
Boyles, A. L., *et al.* (2005). Candidate gene analysis in human neural tube defects. *Am. J. Med. Genet. C Semin. Med. Genet.* **135C,** 9–23.
Brook, F. A., *et al.* (1991). Curvature of the caudal region is responsible for failure of neural tube closure in the curly tail (ct) mouse embryo. *Development* **113,** 671–678.
Brouns, M. R., *et al.* (2005). Toward positional cloning of the curly tail gene. *Birth Defects Res. A Clin. Mol. Teratol.* **73,** 154–161.
Bultman, S., *et al.* (2000). A Brg1 null mutation in the mouse reveals functional differences among mammalian SWI/SNF complexes. *Mol. Cell* **6,** 1287–1295.
Cabrera, R. M., *et al.* (2004). Investigations into the etiology of neural tube defects. *Birth Defects Res. C Embryo Today* **72,** 330–344.
Carezani-Gavin, M., *et al.* (1992). Waardenburg syndrome associated with meningomyelocele. *Am. J. Med. Genet.* **42,** 135–136.
Carroll, E. A., *et al.* (2003). Cordon-bleu is a conserved gene involved in neural tube formation. *Dev. Biol.* **262,** 16–31.
Caspary, T., and Anderson, K. V. (2006). Uncovering the uncharacterized and unexpected: Unbiased phenotype-driven screens in the mouse. *Dev. Dyn.* **235,** 2412–2423.
Caspary, T., *et al.* (2002). Mouse dispatched homolog1 is required for long-range, but not juxtacrine, Hh signaling. *Curr. Biol.* **12,** 1628–1632.
Caspary, T., *et al.* (2007). The graded response to Sonic Hedgehog depends on cilia architecture. *Dev. Cell* **12,** 767–778.
Cederberg, J., *et al.* (2000). Increased mRNA levels of Mn-SOD and catalase in embryos of diabetic rats from a malformation-resistant strain. *Diabetes* **49,** 101–107.
Chan, B. W., *et al.* (2002). Maternal diabetes increases the risk of caudal regression caused by retinoic acid. *Diabetes* **51,** 2811–2816.
Chang, T. I., *et al.* (2003). Oxidant regulation of gene expression and neural tube development: Insights gained from diabetic pregnancy on molecular causes of neural tube defects. *Diabetologia* **46,** 538–545.
Chapman, D., and Hummel, K. (1963). Eye blebs (eb). *Mouse News Lett.* **28,** 32.
Chatkupt, S., *et al.* (1993). Waardenburg syndrome and myelomeningocele in a family. *J. Med. Genet.* **30,** 83–84.
Chen, W. H., *et al.* (1994). Prevention of spinal neural tube defects in the curly tail mouse mutant by a specific effect of retinoic acid. *Dev. Dyn.* **199,** 93–102.

Chiang, C., *et al.* (1996). Cyclopia and defective axial patterning in mice lacking Sonic hedgehog gene function. *Nature* **383,** 407–413.

Cogram, P., *et al.* (2004). Specific isoforms of protein kinase C are essential for prevention of folate-resistant neural tube defects by inositol. *Hum. Mol. Genet.* **13,** 7–14.

Copp, A. J. (1985). Relationship between timing of posterior neuropore closure and development of spinal neural tube defects in mutant (curly tail) and normal mouse embryos in culture. *J. Embryol. Exp. Morphol.* **88,** 39–54.

Copp, A. J., *et al.* (1982). Neural tube development in mutant (curly tail) and normal mouse embryos: The timing of posterior neuropore closure *in vivo* and *in vitro*. *J. Embryol. Exp. Morphol.* **69,** 151–167.

Copp, A. J., *et al.* (1988a). A cell-type-specific abnormality of cell proliferation in mutant (curly tail) mouse embryos developing spinal neural tube defects. *Development* **104,** 285–295.

Copp, A. J., *et al.* (1988b). Prevention of spinal neural tube defects in the mouse embryo by growth retardation during neurulation. *Development* **104,** 297–303.

Copp, A. J., *et al.* (1994). Developmental basis of severe neural tube defects in the loop-tail (Lp) mutant mouse: Use of microsatellite DNA markers to identify embryonic genotype. *Dev. Biol.* **165,** 20–29.

Copp, A. J., *et al.* (2003). The genetic basis of mammalian neurulation. *Nat. Rev. Genet.* **4,** 784–793.

Corbit, K. C., *et al.* (2005). Vertebrate Smoothened functions at the primary cilium. *Nature* **437,** 1018–1021.

Crolla, J. A., *et al.* (1990). The induction of tail malformations in trisomy 16 mouse fetuses heterozygous for the curly tail recessive gene. *Genet. Res.* **55,** 27–32.

Curtin, J. A., *et al.* (2003). Mutation of Celsr1 disrupts planar polarity of inner ear hair cells and causes severe neural tube defects in the mouse. *Curr. Biol.* **13,** 1129–1133.

da-Silva, E. O. (1991). Waardenburg I syndrome: A clinical and genetic study of two large Brazilian kindreds, and literature review. *Am. J. Med. Genet.* **40,** 65–74.

Dawe, H. R., *et al.* (2007). The Meckel-Gruber syndrome proteins MKS1 and meckelin interact and are required for primary cilium formation. *Hum. Mol. Genet.* **16,** 173–186.

Delous, M., *et al.* (2007). The ciliary gene RPGRIP1L is mutated in cerebello-oculo-renal syndrome (Joubert syndrome type B) and Meckel syndrome. *Nat. Genet.* **39,** 875–881.

De Marco, P., *et al.* (2006). Current perspectives on the genetic causes of neural tube defects. *Neurogenetics* **7,** 201–221.

de Saxe, M., *et al.* (1984). Waardenburg syndrome in South Africa. Part I. An evaluation of the clinical findings in 11 families. *S. Afr. Med. J.* **66,** 256–261.

Detrait, E. R., *et al.* (2005). Human neural tube defects: Developmental biology, epidemiology, and genetics. *Neurotoxicol. Teratol.* **27,** 515–524.

Deutsch, U., *et al.* (1988). Pax 1, a member of a paired box homologous murine gene family, is expressed in segmented structures during development. *Cell* **53,** 617–625.

Doudney, K., *et al.* (2005). Analysis of the planar cell polarity gene Vangl2 and its co-expressed paralogue Vangl1 in neural tube defect patients. *Am. J. Med. Genet. A* **136,** 90–92.

Eggenschwiler, J. T., and Anderson, K. V. (2007). Cilia and developmental signaling. *Annu. Rev. Cell Dev. Biol.* **23,** 345–373.

Eggenschwiler, J. T., *et al.* (2001). Rab23 is an essential negative regulator of the mouse Sonic hedgehog signalling pathway. *Nature* **412,** 194–198.

Epstein, D. J., *et al.* (1991a). Molecular characterization of a deletion encompassing the splotch mutation on mouse chromosome 1. *Genomics* **10,** 89–93.

Epstein, D. J., *et al.* (1991b). Splotch (Sp2H), a mutation affecting development of the mouse neural tube, shows a deletion within the paired homeodomain of Pax-3. *Cell* **67,** 767–774.

Epstein, D. J., *et al.* (1993). A mutation within intron 3 of the Pax-3 gene produces aberrantly spliced mRNA transcripts in the splotch (Sp) mouse mutant. *Proc. Natl Acad. Sci. USA* **90,** 532–536.

Estibeiro, J. P., *et al.* (1993). Interaction between splotch (Sp) and curly tail (ct) mouse mutants in the embryonic development of neural tube defects. *Development* **119,** 113–121.

Fan, C. M., and Tessier-Lavigne, M. (1994). Patterning of mammalian somites by surface ectoderm and notochord: Evidence for sclerotome induction by a hedgehog homolog. *Cell* **79,** 1175–1186.

Fine, E. L., *et al.* (1999). Evidence that elevated glucose causes altered gene expression, apoptosis, and neural tube defects in a mouse model of diabetic pregnancy. *Diabetes* **48,** 2454–2462.

Finnell, R. H., *et al.* (2003). Pathobiology and genetics of neural tube defects. *Epilepsia* **44** (Suppl. 3), 14–23.

Fleming, A., and Copp, A. J. (1998). Embryonic folate metabolism and mouse neural tube defects. *Science* **280,** 2107–2109.

Frank, V., *et al.* (2008). Mutations of the CEP290 gene encoding a centrosomal protein cause Meckel-Gruber syndrome. *Hum. Mutat.* **29,** 45–52.

Furumoto, T. A., *et al.* (1999). Notochord-dependent expression of MFH1 and PAX1 cooperates to maintain the proliferation of sclerotome cells during the vertebral column development. *Dev. Biol.* **210,** 15–29.

Garcia-Garcia, M. J., *et al.* (2005). Analysis of mouse embryonic patterning and morphogenesis by forward genetics. *Proc. Natl Acad. Sci. USA* **102,** 5913–5919.

Gelineau-van Waes, J., *et al.* (2005). Maternal fumonisin exposure and risk for neural tube defects: Mechanisms in an *in vivo* mouse model. *Birth Defects Res. A Clin. Mol. Teratol.* **73,** 487–497.

Gerrelli, D., and Copp, A. J. (1997). Failure of neural tube closure in the loop-tail (Lp) mutant mouse: Analysis of the embryonic mechanism. *Brain Res. Dev. Brain Res.* **102,** 217–224.

Goodrich, L. V., *et al.* (1997). Altered neural cell fates and medulloblastoma in mouse patched mutants. *Science* **277,** 1109–1113.

Goulding, M. D., *et al.* (1991). Pax-3, a novel murine DNA binding protein expressed during early neurogenesis. *EMBO J.* **10,** 1135–1147.

Goulding, M., *et al.* (1993). Analysis of the Pax-3 gene in the mouse mutant splotch. *Genomics* **17,** 355–363.

Greco, T. L., *et al.* (1996). Analysis of the vestigial tail mutation demonstrates that Wnt-3a gene dosage regulates mouse axial development. *Genes Dev.* **10,** 313–324.

Greene, N. D., and Copp, A. J. (1997). Inositol prevents folate-resistant neural tube defects in the mouse. *Nat. Med.* **3,** 60–66.

Greene, N. D., and Copp, A. J. (2005). Mouse models of neural tube defects: Investigating preventive mechanisms. *Am. J. Med. Genet. C Semin. Med. Genet.* **135C,** 31–41.

Greene, N. D., *et al.* (1998). Abnormalities of floor plate, notochord and somite differentiation in the loop-tail (Lp) mouse: A model of severe neural tube defects. *Mech. Dev.* **73,** 59–72.

Groenen, P. M., *et al.* (2003). Maternal myo-inositol, glucose, and zinc status is associated with the risk of offspring with spina bifida. *Am. J. Obstet. Gynecol.* **189,** 1713–1719.

Gruneberg, H. (1954). Genetical studies on the skeleton of the mouse. VIII. Curly tail. *J. Genet.* **52,** 52–67.

Gunther, T., *et al.* (1994). Open brain, a new mouse mutant with severe neural tube defects, shows altered gene expression patterns in the developing spinal cord. *Development* **120,** 3119–3130.

Gustavsson, P., *et al.* (2007). Increased expression of Grainyhead-like-3 rescues spina bifida in a folate-resistant mouse model. *Hum. Mol. Genet.* **16,** 2640–2646.

Hamblet, N. S., *et al.* (2002). Dishevelled 2 is essential for cardiac outflow tract development, somite segmentation and neural tube closure. *Development* **129,** 5827–5838.
Harris, M. J., and Juriloff, D. M. (2007). Mouse mutants with neural tube closure defects and their role in understanding human neural tube defects. *Birth Defects Res. A Clin. Mol. Teratol.* **79,** 187–210.
Haycraft, C. J., *et al.* (2005). Gli2 and Gli3 localize to cilia and require the intraflagellar transport protein polaris for processing and function. *PLoS Genet.* **1,** e53.
Helwig, U., *et al.* (1995). Interaction between undulated and Patch leads to an extreme form of spina bifida in double-mutant mice. *Nat. Genet.* **11,** 60–63.
Hol, F. A., *et al.* (1995). A frameshift mutation in the gene for PAX3 in a girl with spina bifida and mild signs of Waardenburg syndrome. *J. Med. Genet.* **32,** 52–56.
Hol, F. A., *et al.* (1996). PAX genes and human neural tube defects: An amino acid substitution in PAX1 in a patient with spina bifida. *J. Med. Genet.* **33,** 655–660.
Hoth, C. F., *et al.* (1993). Mutations in the paired domain of the human PAX3 gene cause Klein-Waardenburg syndrome (WS-III) as well as Waardenburg syndrome type I (WS-I). *Am. J. Hum. Genet.* **52,** 455–462.
Houde, C., *et al.* (2006). Hippi is essential for node cilia assembly and Sonic hedgehog signaling. *Dev. Biol.* **300,** 523–533.
Huangfu, D., and Anderson, K. V. (2005). Cilia and Hedgehog responsiveness in the mouse. *Proc. Natl Acad. Sci. USA* **102,** 11325–11330.
Huangfu, D., *et al.* (2003). Hedgehog signalling in the mouse requires intraflagellar transport proteins. *Nature* **426,** 83–87.
Hui, C. C., and Joyner, A. L. (1993). A mouse model of greig cephalopolysyndactyly syndrome: The extra-toesJ mutation contains an intragenic deletion of the Gli3 gene. *Nat. Genet.* **3,** 241–246.
Huibregtse, J. M., *et al.* (1995). A family of proteins structurally and functionally related to the E6-AP ubiquitin-protein ligase. *Proc. Natl Acad. Sci. USA* **92,** 5249.
Jadeja, S., *et al.* (2005). Identification of a new gene mutated in Fraser syndrome and mouse myelencephalic blebs. *Nat. Genet.* **37,** 520–525.
Joosten, P. H., *et al.* (1998). Altered regulation of platelet-derived growth factor receptor-alpha gene-transcription in vitro by spina bifida-associated mutant Pax1 proteins. *Proc. Natl Acad. Sci. USA* **95,** 14459–14463.
Joosten, P. H., *et al.* (2001). Promoter haplotype combinations of the platelet-derived growth factor alpha-receptor gene predispose to human neural tube defects. *Nat. Genet.* **27,** 215–217.
Joosten, P. H., *et al.* (2005). Pax1/E2a double-mutant mice develop non-lethal neural tube defects that resemble human malformations. *Transgenic Res.* **14,** 983–987.
Karmous-Benailly, H., *et al.* (2005). Antenatal presentation of Bardet-Biedl syndrome may mimic Meckel syndrome. *Am. J. Hum. Genet.* **76,** 493–504.
Kasarskis, A., *et al.* (1998). A phenotype-based screen for embryonic lethal mutations in the mouse. *Proc. Natl Acad. Sci. USA* **95,** 7485–7490.
Khandelwal, M., *et al.* (1998). Dietary myo-inositol therapy in hyperglycemia-induced embryopathy. *Teratology* **57,** 79–84.
Kibar, Z., *et al.* (2001). Ltap, a mammalian homolog of *Drosophila* Strabismus/Van Gogh, is altered in the mouse neural tube mutant Loop-tail. *Nat. Genet.* **28,** 251–255.
Kibar, Z., *et al.* (2007a). Toward understanding the genetic basis of neural tube defects. *Clin. Genet.* **71,** 295–310.
Kibar, Z., *et al.* (2007b). Mutations in VANGL1 associated with neural-tube defects. *N. Engl. J. Med.* **356,** 1432–1437.
Kim, J. K., *et al.* (2001). Srg3, a mouse homolog of yeast SWI3, is essential for early embryogenesis and involved in brain development. *Mol. Cell. Biol.* **21,** 7787–7795.

Kirillova, I., *et al.* (2000). Expression of the sonic hedgehog gene in human embryos with neural tube defects. *Teratology* **61,** 347–354.
Kirke, P. N., *et al.* (1993). Maternal plasma folate and vitamin B12 are independent risk factors for neural tube defects. *Q. J. Med.* **86,** 703–708.
Klinghoffer, R. A., *et al.* (2002). An allelic series at the PDGFalphaR locus indicates unequal contributions of distinct signaling pathways during development. *Dev. Cell* **2,** 103–113.
Kujat, A., *et al.* (2007). Prenatal diagnosis and genetic counseling in a case of spina bifida in a family with Waardenburg syndrome type I. *Fetal Diagn. Ther.* **22,** 155–158.
Kyttala, M., *et al.* (2006). MKS1, encoding a component of the flagellar apparatus basal body proteome, is mutated in Meckel syndrome. *Nat. Genet.* **38,** 155–157.
Lakkis, M. M., *et al.* (1999). Neurofibromin deficiency in mice causes exencephaly and is a modifier for Splotch neural tube defects. *Dev. Biol.* **212,** 80–92.
Leitch, C. C., *et al.* (2008). Hypomorphic mutations in syndromic encephalocele genes are associated with Bardet-Biedl syndrome. *Nat. Genet.* **40,** 443–448.
Letts, V. A., *et al.* (1995). A curly-tail modifier locus, mct1, on mouse chromosome 17. *Genomics* **29,** 719–724.
Li, R., *et al.* (2007). Expression of the gene encoding the high-Km glucose transporter 2 by the early postimplantation mouse embryo is essential for neural tube defects associated with diabetic embryopathy. *Diabetologia* **50,** 682–689.
Little, C., and Bagg, H. (1923). The occurrence of two heritable types of abnormality among descendants of X-rayed mice. *Am. J. Roentgenol.* **10,** 975–989.
Liu, A., *et al.* (2005). Mouse intraflagellar transport proteins regulate both the activator and repressor functions of Gli transcription factors. *Development* **132,** 3103–3111.
Loeken, M. R. (2005). Current perspectives on the causes of neural tube defects resulting from diabetic pregnancy. *Am. J. Med. Genet. C Semin. Med. Genet.* **135,** 77–87.
Lu, X., *et al.* (2004). PTK7/CCK-4 is a novel regulator of planar cell polarity in vertebrates. *Nature* **430,** 93–98.
Lynch, S. A. (2005). Non-multifactorial neural tube defects. *Am. J. Med. Genet. C Semin. Med. Genet.* **135,** 69–76.
Marasas, W. F., *et al.* (2004). Fumonisins disrupt sphingolipid metabolism, folate transport, and neural tube development in embryo culture and *in vivo*: A potential risk factor for human neural tube defects among populations consuming fumonisin-contaminated maize. *J. Nutr.* **134,** 711–716.
May, S. R., *et al.* (2005). Loss of the retrograde motor for IFT disrupts localization of Smo to cilia and prevents the expression of both activator and repressor functions of Gli. *Dev. Biol.* **287,** 378–389.
McGregor, L., *et al.* (2003). Fraser syndrome and mouse blebbed phenotype caused by mutations in FRAS1/Fras1 encoding a putative extracellular matrix protein. *Nat. Genet.* **34,** 203–208.
Mo, R., *et al.* (1997). Specific and redundant functions of Gli2 and Gli3 zinc finger genes in skeletal patterning and development. *Development* **124,** 113–123.
Moline, M. L., and Sandlin, C. (1993). Waardenburg syndrome and meningomyelocele. *Am. J. Med. Genet.* **47,** 126.
Montcouquiol, M., *et al.* (2006). Asymmetric localization of Vangl2 and Fz3 indicate novel mechanisms for planar cell polarity in mammals. *J. Neurosci.* **26,** 5265–5275.
Morrison-Graham, K., *et al.* (1992). A PDGF receptor mutation in the mouse (Patch) perturbs the development of a non-neuronal subset of neural crest-derived cells. *Development* **115,** 133–142.
Murdoch, J. N., *et al.* (2001). Severe neural tube defects in the loop-tail mouse result from mutation of Lpp1, a novel gene involved in floor plate specification. *Hum. Mol. Genet.* **10,** 2593–2601.

Murdoch, J. N., *et al.* (2003). Disruption of scribble (Scrb1) causes severe neural tube defects in the circletail mouse. *Hum. Mol. Genet.* **12,** 87–98.

Neumann, P. E., *et al.* (1994). Multifactorial inheritance of neural tube defects: Localization of the major gene and recognition of modifiers in ct mutant mice. *Nat. Genet.* **6,** 357–362.

Nye, J. S., *et al.* (1998). Myelomeningocele and Waardenburg syndrome (type 3) in patients with interstitial deletions of 2q35 and the PAX3 gene: Possible digenic inheritance of a neural tube defect. *Am. J. Med. Genet.* **75,** 401–408.

Orr-Urtreger, A., *et al.* (1992). Developmental expression of the alpha receptor for platelet-derived growth factor, which is deleted in the embryonic lethal Patch mutation. *Development* **115,** 289–303.

Pani, L., *et al.* (2002a). Polymorphic susceptibility to the molecular causes of neural tube defects during diabetic embryopathy. *Diabetes* **51,** 2871–2874.

Pani, L., *et al.* (2002b). Rescue of neural tube defects in Pax-3-deficient embryos by p53 loss of function: Implications for Pax-3-dependent development and tumorigenesis. *Genes Dev.* **16,** 676–680.

Pantke, O. A., and Cohen, M. M., Jr. (1971). The Waardenburg syndrome. *Birth Defects Orig. Artic. Ser.* **7,** 147–152.

Passarge, E., and Lenz, W. (1966). Syndrome of caudal regression in infants of diabetic mothers: Observations of further cases. *Pediatrics* **37,** 672–675.

Payne, J., *et al.* (1997). Spina bifida occulta in homozygous Patch mouse embryos. *Dev. Dyn.* **209,** 105–116.

Peters, T., *et al.* (2002). The mouse Fused toes (Ft) mutation is the result of a 1.6-Mb deletion including the entire Iroquois B gene cluster. *Mamm. Genome* **13,** 186–188.

Phelan, S. A., *et al.* (1997). Neural tube defects in embryos of diabetic mice: Role of the Pax-3 gene and apoptosis. *Diabetes* **46,** 1189–1197.

Phillips, R. (1970). Blebbed, bl. *Mouse News Lett.* **42,** 26.

Pickett, E. A., *et al.* (2008). Disruption of PDGFRalpha-initiated PI3K activation and migration of somite derivatives leads to spina bifida. *Development* **135,** 589–598.

Qian, D., *et al.* (2007). Wnt5a functions in planar cell polarity regulation in mice. *Dev. Biol.* **306,** 121–133.

Rohatgi, R., *et al.* (2007). Patched1 regulates hedgehog signaling at the primary cilium. *Science* **317,** 372–376.

Ross, A. J., *et al.* (2005). Disruption of Bardet-Biedl syndrome ciliary proteins perturbs planar cell polarity in vertebrates. *Nat. Genet.* **37,** 1135–1140.

Sadler, T. W., *et al.* (2002). Prevention of fumonisin B1-induced neural tube defects by folic acid. *Teratology* **66,** 169–176.

Scholey, J. M., and Anderson, K. V. (2006). Intraflagellar transport and cilium-based signaling. *Cell* **125,** 439–442.

Seller, M. J. (1983). The cause of neural tube defects: Some experiments and a hypothesis. *J. Med. Genet.* **20,** 164–168.

Seller, M. J., and Adinolfi, M. (1981). The curly-tail mouse: An experimental model for human neural tube defects. *Life Sci.* **29,** 1607–1615.

Seller, M. J., and Perkins, K. J. (1983). Effect of hydroxyurea on neural tube defects in the curly-tail mouse. *J. Craniofac. Genet. Dev. Biol.* **3,** 11–17.

Seller, M. J., and Perkins, K. J. (1986). Effect of mitomycin C on the neural tube defects of the curly-tail mouse. *Teratology* **33,** 305–309.

Shim, S. H., *et al.* (2004). Molecular cytogenetic characterization of multiple intrachromosomal rearrangements of chromosome 2q in a patient with Waardenburg's syndrome and other congenital defects. *Clin. Genet.* **66,** 46–52.

Shum, A. S., *et al.* (1999). Retinoic acid induces down-regulation of Wnt-3a, apoptosis and diversion of tail bud cells to a neural fate in the mouse embryo. *Mech. Dev.* **84,** 17–30.

Smith, L. J., and Stein, K. F. (1962). Axial elongation in the mouse and its retardation in homozygous looptail mice. *J. Embryol. Exp. Morphol.* **10,** 73–87.
Smith, E. A., *et al.* (1991). Mouse platelet-derived growth factor receptor alpha gene is deleted in W19H and patch mutations on chromosome 5. *Proc. Natl Acad. Sci. USA* **88,** 4811–4815.
Smith, U. M., *et al.* (2006). The transmembrane protein meckelin (MKS3) is mutated in Meckel-Gruber syndrome and the wpk rat. *Nat. Genet.* **38,** 191–196.
Smithells, R. W., *et al.* (1976). Vitamin deficiencies and neural tube defects. *Arch. Dis. Child.* **51,** 944–950.
Smithells, R. W., *et al.* (1980). Possible prevention of neural-tube defects by periconceptional vitamin supplementation. *Lancet* **1,** 339–340.
Smyth, I., and Scambler, P. (2005). The genetics of Fraser syndrome and the blebs mouse mutants. *Hum. Mol. Genet.* **14**(Spec No. 2), R269–R274.
Smyth, I., *et al.* (2004). The extracellular matrix gene Frem1 is essential for the normal adhesion of the embryonic epidermis. *Proc. Natl Acad. Sci. USA* **101,** 13560–13565.
Sporle, R., and Schughart, K. (1998). Paradox segmentation along inter- and intrasomitic borderlines is followed by dysmorphology of the axial skeleton in the open brain (opb) mouse mutant. *Dev. Genet.* **22,** 359–373.
Sporle, R., *et al.* (1996). Severe defects in the formation of epaxial musculature in open brain (opb) mutant mouse embryos. *Development* **122,** 79–86.
Stephenson, D. A., *et al.* (1991). Platelet-derived growth factor receptor alpha-subunit gene (Pdgfra) is deleted in the mouse patch (Ph) mutation. *Proc. Natl Acad. Sci. USA* **88,** 6–10.
Stevens, V. L., and Tang, J. (1997). Fumonisin B1-induced sphingolipid depletion inhibits vitamin uptake via the glycosylphosphatidylinositol-anchored folate receptor. *J. Biol. Chem.* **272,** 18020–18025.
Stiefel, D., *et al.* (2003). Tethering of the spinal cord in mouse fetuses and neonates with spina bifida. *J. Neurosurg.* **99,** 206–213.
Strong, L., and Hollander, W. (1949). Hereditary loop-tail in the house mouse accompanied by imperforate vagina and craniorachischisis when homozygous. *J. Hered.* **40,** 329–334.
Swiergiel, J. J., *et al.* (2000). Developmental eye and neural tube defects in the eye blebs mouse. *Dev. Dyn.* **219,** 21–27.
Takada, S., *et al.* (1994). Wnt-3a regulates somite and tailbud formation in the mouse embryo. *Genes Dev.* **8,** 174–189.
Takamiya, K., *et al.* (2004). A direct functional link between the multi-PDZ domain protein GRIP1 and the Fraser syndrome protein Fras1. *Nat. Genet.* **36,** 172–177.
Talis, A. L., *et al.* (1998). The role of E6AP in the regulation of p53 protein levels in human papillomavirus (HPV)-positive and HPV-negative cells. *J. Biol. Chem.* **273,** 6439–6445.
Timmer, J. R., *et al.* (2005). Tissue morphogenesis and vascular stability require the Frem2 protein, product of the mouse myelencephalic blebs gene. *Proc. Natl Acad. Sci. USA* **102,** 11746–11750.
Ting, S. B., *et al.* (2003). Inositol- and folate-resistant neural tube defects in mice lacking the epithelial-specific factor Grhl-3. *Nat. Med.* **9,** 1513–1519.
Ting, S. B., *et al.* (2005). A homolog of Drosophila grainy head is essential for epidermal integrity in mice. *Science* **308,** 411–413.
Torban, E., *et al.* (2004). Independent mutations in mouse Vangl2 that cause neural tube defects in looptail mice impair interaction with members of the Dishevelled family. *J. Biol. Chem.* **279,** 52703–52713.
Torban, E., *et al.* (2008). Genetic interaction between members of the Vangl family causes neural tube defects in mice. *Proc. Natl Acad. Sci. USA* **105,** 3449–3454.
Ulloa, F., and Briscoe, J. (2007). Morphogens and the control of cell proliferation and patterning in the spinal cord. *Cell Cycle* **6,** 2640–2649.

van der Hoeven, F., *et al.* (1994). Programmed cell death is affected in the novel mouse mutant Fused toes (Ft). *Development* **120,** 2601–2607.

van der Put, N. M., *et al.* (1997). Altered folate and vitamin B12 metabolism in families with spina bifida offspring. *Q. J. Med.* **90,** 505–510.

van Straaten, H. W., and Copp, A. J. (2001). Curly tail: A 50-year history of the mouse spina bifida model. *Anat. Embryol. (Berl.)* **203,** 225–237.

van Straaten, H. W., *et al.* (1995). Dietary methionine does not reduce penetrance in curly tail mice but causes a phenotype-specific decrease in embryonic growth. *J. Nutr.* **125,** 2733–2740.

Varnum, D., and Fox, S. (1976). Heb—head bleb. *Mouse News Lett.* **55,** 16.

Vierkotten, J., *et al.* (2007). Ftm is a novel basal body protein of cilia involved in Shh signalling. *Development* **134,** 2569–2577.

Vogan, K. J., *et al.* (1993). The splotch-delayed (Spd) mouse mutant carries a point mutation within the paired box of the Pax-3 gene. *Genomics* **17,** 364–369.

Vrontou, S., *et al.* (2003). Fras1 deficiency results in cryptophthalmos, renal agenesis and blebbed phenotype in mice. *Nat. Genet.* **34,** 209–214.

Wald, N., *et al.* (1991). Prevention of neural tube defects: Results of the Medical Research Council Vitamin Study. MRC Vitamin Study Research Group. *Lancet* **338,** 131–137.

Wallingford, J. B. (2006). Planar cell polarity, ciliogenesis and neural tube defects. *Hum. Mol. Genet.* **15**(Spec No. 2), R227–R234.

Wallingford, J. B., and Harland, R. M. (2001). Xenopus Dishevelled signaling regulates both neural and mesodermal convergent extension: Parallel forces elongating the body axis. *Development* **128,** 2581–2592.

Wallingford, J. B., and Harland, R. M. (2002). Neural tube closure requires Dishevelled-dependent convergent extension of the midline. *Development* **129,** 5815–5825.

Wallingford, J. B., *et al.* (2000). Dishevelled controls cell polarity during Xenopus gastrulation. *Nature* **405,** 81–85.

Wang, Y., and Nathans, J. (2007). Tissue/planar cell polarity in vertebrates: New insights and new questions. *Development* **134,** 647–658.

Wang, J., *et al.* (2006a). Dishevelled genes mediate a conserved mammalian PCP pathway to regulate convergent extension during neurulation. *Development* **133,** 1767–1778.

Wang, Y., *et al.* (2006b). The role of Frizzled3 and Frizzled6 in neural tube closure and in the planar polarity of inner-ear sensory hair cells. *J. Neurosci.* **26,** 2147–2156.

Wilson, D. B., and Wyatt, D. P. (1994). Analysis of neurulation in a mouse model for neural dysraphism. *Exp. Neurol.* **127,** 154–158.

Wiltse, J. (2005). Mode of action: Inhibition of histone deacetylase, altering WNT-dependent gene expression, and regulation of beta-catenin—Developmental effects of valproic acid. *Crit. Rev. Toxicol.* **35,** 727–738.

Wollnik, B., *et al.* (2003). Homozygous and heterozygous inheritance of PAX3 mutations causes different types of Waardenburg syndrome. *Am. J. Med. Genet. A* **122,** 42–45.

Yamaguchi, T. P., *et al.* (1999). A Wnt5a pathway underlies outgrowth of multiple structures in the vertebrate embryo. *Development* **126,** 1211–1223.

Yates, J. R., *et al.* (1987). Is disordered folate metabolism the basis for the genetic predisposition to neural tube defects? *Clin. Genet.* **31,** 279–287.

Ybot-Gonzalez, P., *et al.* (2002). Sonic hedgehog and the molecular regulation of mouse neural tube closure. *Development* **129,** 2507–2517.

Ybot-Gonzalez, P., *et al.* (2007). Convergent extension, planar-cell-polarity signalling and initiation of mouse neural tube closure. *Development* **134,** 789–799.

Yoshikawa, Y., *et al.* (1997). Evidence that absence of Wnt-3a signaling promotes neuralization instead of paraxial mesoderm development in the mouse. *Dev. Biol.* **183,** 234–242.

Yu, Z., *et al.* (2006). The Grainyhead-like epithelial transactivator Get-1/Grhl3 regulates epidermal terminal differentiation and interacts functionally with LMO4. *Dev. Biol.* **299,** 122–136.

Zhao, Q., *et al.* (1996). Prenatal folic acid treatment suppresses acrania and meroanencephaly in mice mutant for the Cart1 homeobox gene. *Nat. Genet.* **13,** 275–283.

Zhu, H., *et al.* (2004). Promoter haplotype combinations for the human PDGFRA gene are associated with risk of neural tube defects. *Mol. Genet. Metab.* **81,** 127–132.

Zlotogora, J., *et al.* (1995). Homozygosity for Waardenburg syndrome. *Am. J. Hum. Genet.* **56,** 1173–1178.

Zohn, I. E., *et al.* (2005). Using genomewide mutagenesis screens to identify the genes required for neural tube closure in the mouse. *Birth Defects Res. A Clin. Mol. Teratol.* **73,** 583–590.

Zohn, I. E., *et al.* (2006). P38 and a p38-interacting protein are critical for downregulation of E-cadherin during mouse gastrulation. *Cell* **125,** 957–969.

Zohn, I. E., *et al.* (2007). The Hectd1 ubiquitin ligase is required for development of the head mesenchyme and neural tube closure. *Dev. Biol.* **306,** 208–221.

CHAPTER TWO

The Etiopathogenesis of Cleft Lip and Cleft Palate: Usefulness and Caveats of Mouse Models

Amel Gritli-Linde

Contents

Abstract

Cleft lip and cleft palate are frequent human congenital malformations with a complex multifactorial etiology. These orofacial clefts can occur as part of a syndrome involving multiple organs or as isolated clefts without other detectable defects. Both forms of clefting constitute a heavy burden to the affected individuals and their next of kin. Human and mouse facial traits are utterly dissimilar. However, embryonic development of the lip and palate are strikingly

Department of Oral Biochemistry, Sahlgrenska Academy at the University of Gothenburg, Medicinaregatan 12F, Göteborg, Sweden

Current Topics in Developmental Biology, Volume 84
ISSN 0070-2153, DOI: 10.1016/S0070-2153(08)00602-9

similar in both species, making the mouse a model of choice to study their normal and abnormal development. Human epidemiological and genetic studies are clearly important for understanding the etiology of lip and palate clefting. However, our current knowledge about the etiopathogenesis of these malformations has mainly been gathered throughout the years from mouse models, including those with mutagen-, teratogen- and targeted mutation-induced clefts as well as from mice with spontaneous clefts. This review provides a comprehensive description of the numerous mouse models for cleft lip and/or cleft palate. Despite a few weak points, these models have revealed a high order of molecular complexity as well as the stringent spatio-temporal regulations and interactions between key factors which govern the development of these orofacial structures.

1. Introduction

Over the past 100 years, the mouse has been feeding us with important biomedical insights. Compared to other mammals, mice are small, prolific and amenable to experimental and genetic manipulation *in vitro* or *in vivo*. Mouse models for various human ailments, including metabolic, developmental, neoplastic, neurological, and age-related diseases, have been and continue to be created. These achievements epitomize the prowesses of biotechnology. The mouse models provide us with an unfathomable wealth of information about the etiopathogenesis of diseases and set the stage for tailoring methods for improved diagnosis, prophylaxis, drug testing, and therapy.

Despite striking external differences humans and mice share ≈99% of their genes (Waterston *et al.*, 2002). In addition, during embryonic development, especially during early craniofacial morphogenesis (at days 32 and 10 of human and mouse embryogenesis, respectively) the human and mouse embryos are essentially alike and are approximately of similar sizes. Furthermore, orofacial development is basically the same in mice and humans. Most importantly, a score of spontaneous and induced mutations in mice replicate several human congenital craniofacial malformations, such as cleft lip and cleft palate.

Clinically, cleft lip (CL) and/or cleft palate (CP) is a spectrum of anomalies (Fig. 2.1) including unilateral or bilateral CL; unilateral or bilateral CL combined with dento-alveolar ridge and primary palate clefting; complete unilateral or bilateral CL combined with cleft of the dento-alevolar ridge and CP (where both the primary and secondary palates are affected in addition to the lip and alveolar ridge); as well as isolated cleft palate (cleft palate only, CPO) in which either the entire or part of the secondary palate is cleft (Cobourne, 2004; Muenke, 2002). Median cleft lip and midline facial clefting are different entities that may be found in

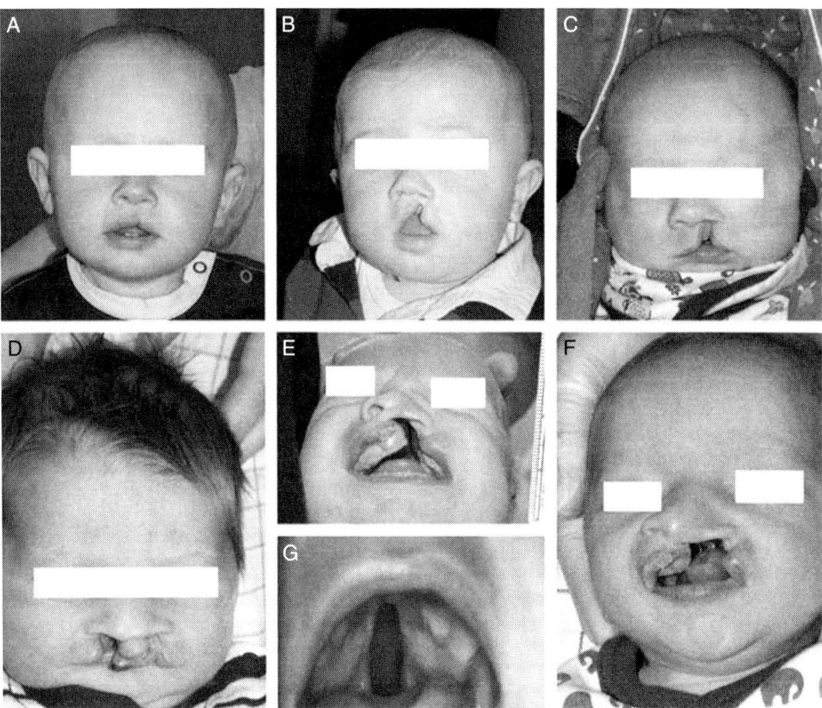

Figure 2.1 Clinical appearance of orofacial clefts. Partial unilateral cleft lip (A). Unilateral total cleft lip (B). Unilateral cleft lip and cleft of the alveolar ridge (C). Bilateral cleft of the lip, alveolar ridge and primary palate (D). Narrow (E) and wide (F) unilateral total clefts involving the lip, alveolus, primary palate and secondary palate. Cleft of the secondary palate (G). Photographs kindly provided by Drs. Sara Rizell (A–F) and Martyn T. Cobourne (G).

conjunction with syndromes such as holoprosencephaly (HPE) (Muenke, 2002). These can also be caused by defects in the neural crest-derived mesenchyme of the frontonasal process and/or by lack of growth and fusion of its derivatives, the medial nasal processes (Johnston and Sulik, 1979).

CL with or without CP (CL/P) and CPO are birth defects that affect not only humans and mice but also other mammals. While these orofacial clefts are not encountered in the wilderness because of early lethality of the affected newborns, they have been documented in a number of farm (bovine, ovine, and caprine) and zoo (tigers, lions, jaguars) animals as well as pets (dogs and cats) (Johnston and Bronsky, 1995; Loevy and Fenyes, 1968; Mulvihill *et al.*, 1980; Shupe *et al.*, 1968). Compelling evidence from human, laboratory mice, and domestic animal studies indicates that the etiology of these malformations is complex, where interacting genetic and environmental factors are part of the equation.

One of the first examples of environmental inputs in a malformation that has a genetic component, and thus emphasizing the interplay between genes and exogenous factors, is illustrated by previous work on cortisone-induced CP in different strains of mice. When pregnant females of the A/J strain were fed cortisone, they generated 100% of offsprings with CP as compared to 17% of cortisone-induced CP in other strains. Thus, in the A/J strain, neither genetic nor environmental factors are the sole cause of CP, rather the interaction of the two determines the degree of vulnerability. Furthermore, the time of palatal closure appears to be an important determinant of liability (sensitivity) to CP, since the liability of the A/J and the SW/Fr strains to cortisone has been found to be related to their normally delayed palatal closure time as compared to the low-incidence strains (Biddle and Fraser, 1976; Vekemans and Fraser, 1979; Walker and Fraser, 1956, 1957).

CL/P and CPO are the most frequent human craniofacial birth defects. They affect 1 in 500 to 1000 newborns worldwide, with CL/P being more frequent than CPO (Marazita, 2002). These orofacial clefts occur either as an isolated, nonsyndromic condition with a genetic contribution of 20–50%, or as part of a syndrome in association with other congenital defects (Marazita, 2002; Schutte and Murray, 1999). Orofacial cleftings have been documented in over 300 syndromes, in which they occur as part of a Mendelian disorder at a single genetic locus, or are caused by chromosomal abnormalities and teratogens. Specific genes within cytogenetically visible chromosomal anomalies such as deletions and translocations are being identified, making less obvious the distinction between single gene alterations and chromosomal abnormalities (FitzPatrick *et al.*, 2003).

The majority of CL/P or CPO conditions are, however, nonsyndromic and nonMendelian (Calzolari *et al.*, 2007; Marazita, 2002; Muenke, 2002). CP associated with CL is generally considered as a consequence of lip clefting, and family-based studies suggest that nonsyndromic CPO is genetically different from CL/P (Curtis *et al.*, 1961; Fraser, 1970; Woolf *et al.*, 1963).

Individuals with CL/P and CPO necessitate a lengthy multidisciplinary treatment, involving surgery, orthodontic interventions, and speech and psychological therapies. Despite palpable progress in their treatment, these orofacial clefts may lead to long-term anatomical, functional, and psychological sequelae (Christensen and Mortensen, 2002; Christensen *et al.*, 2004). Unfortunately, recent evidence from a Danish study suggests an increased mortality (during infancy, childhood, and adulthood) of humans affected with orofacial clefts as compared to individuals born without clefts. Causes of mortality include pneumonia, aspiration, sepsis, suicide, epilepsy, cardiovascular diseases, and cancer (Christensen *et al.*, 2004).

Since the publication of recent reviews discussing what was then the current knowledge about the genetics of orofacial clefting and the molecular and cellular mechanisms underlying normal and abnormal development of the lip and palate (Chai and Maxon, 2006; Gritli-Linde, 2007; Jiang *et al.*, 2006;

Juriloff and Harris, 2008; Vieira, 2008), a considerable amount of new and thrilling insights from human genetic studies and mouse models have been gained. This undoubtedly portrays the rapid progress of the field, but also a vivid interest and tenacity of the scientific community to reveal the secrets of lip and palate development and identify the molecular and cellular alterations that lead to clefting, with the ultimate goal for improving diagnosis, prophylaxis, genetic counseling, and care for the affected and their relatives.

In addition to bringing to light factors involved in orofacial clefting that have been left aside in previous reviews, this review thus intends to highlight the many recent discoveries in cleft lip and palate research and integrate them into previous (old and recent) paradigms. The importance of mouse models and what they have taught us so far, as well as some caveats on relying only on mice and thence falling into the trap are also discussed.

2. The Genetic Etiology of CL/P and CPO

The extensive endeavour to identify genes that cause CL/P and CPO has been most fruitful for syndromic clefts at a single genetic locus (Cobourne, 2004; Jugessur and Murray, 2005; Rice, 2005; Stanier and Moore, 2004).

Genes identified so far (Table 2.1) include *PVRL1* (*Poliovirus receptor-related1*) encoding nectin-1 (Margarita Island CL/P-ectodermal dysplasia1, CLPED1; Suzuki *et al.*, 2000); *SHH* encoding the signaling protein Sonic hedgehog (Holoprosencephaly 3; Muenke, 2002); *PTCH* encoding the Hedgehog receptor Patched1 (Ptc1) (Holoprosencephaly 7; Ming *et al.*, 2002; Ribeiro *et al.*, 2006); *FGFR1* and *FGFR2* which encode the fibroblast growth factor receptor1 (Kallmann syndrome 2; Dodé *et al.*, 2003, 2007), and the fibroblast growth factor receptor2 (Apert syndrome, Kreiborg and Cohen, 1992; Moloney *et al.*, 1996; Park *et al.*, 1995; Wilkie *et al.*, 1995); *CDH1* encoding E-cadherin (Frebourg *et al.*, 2006); *CHD7*, encoding the chromodomain helicase DNA-binding protein 7 (CHARGE syndrome; Sanlaville *et al.*, 2006; Vissers *et al.*, 2004); *PHF8*, encoding a transcription factor (Abidi *et al.*, 2007; Laumonnier *et al.*, 2005) as well as *DHCR7*, the gene encoding 3-β-hydroxysterol-Delta (7)-reductase (Muenke, 2002; Wassif *et al.*, 1998). Other genes include those coding for transcription factors such as the interferon regulatory factor 6 (*IRF6*; van der Woude and popliteal pterygium syndromes; Kondo *et al.*, 2002), Msx1 (CL/P and oligodontia syndrome; van den Boogaard *et al.*, 2000), Tbx1 (Yagi *et al.*, 2003), Tbx22 (X-linked CP and ankyloglossia; Braybrook *et al.*, 2001, 2002; Marçano *et al.*, 2004), Pax9 (Schuffenhauer *et al.*, 1999), Foxel (Castanet *et al.*, 2002; Clifton-Blight *et al.*, 1998), Gli2 (Roessler *et al.*, 2003) and p63, a p53 homolog (*P63*; ectrodactyly, ectodermal dysplasia, clefting syndrome 3; EED3; Celli *et al.*, 1999; Ianakiev *et al.*, 2000). *P63* also seems to be involved

Table 2.1 Genes implicated in human syndromic orofacial clefting based on evidence from human genetic studies, mouse models, and expression data in orofacial primordia

Gene	Phenotype	Gene product	References	Non - syndromic	Mouse model	Expression
CDH1	CL/P and hereditary diffuse gastric cancer	CAM (E-cadherin)	Frebourg *et al.*, 2006	No	KO, early lethal; conditional KO, no clefting	E
CHD7	CL/P in CHARGE syndrome	Chromodomain helicase DNA- binding protein	Vissers *et al.*, 2004; Sanlaville *et al.*, 2006	Yes	Yes, CPO	Mainly in E
DHCR	Smith–Lemli–Opitz syndrome: defects in cholesterol biosynthesis, multiple organ malformations, growth retardation, dysmorphic facial features including CL/P or CPO, postaxial polydactyly	3-β-hydroxysterol-Delta 7-reductase	Wassif *et al.*, 1998; Muenke, 2002	No	Yes, CPO	NA
FGFR1	Kallmann syndrome (KAL2): CL/P or CPO, anosmia, hypogonadotrophic hypogonadism.	TK receptor	Dodé *et al.*, 2003, 2007	Yes	Yes, CPO	E + M

FGFR2	Apert syndrome: CPO (in 44% of cases), craniosynostosis and syndactyly	TK receptor	Kreiborg and Cohen, 1992; Park *et al.*, 1995; Wilkie *et al.*, 1995; Moloney *et al.*, 1996	Yes	Yes, no clefting	E + M
FOXE1	Bamforth–Lazarus syndrome: hypothyroidism, athyroidal, CPO, choanal atresia, spiky hair	TF	Clifton-Blight *et al.*, 1998; Castanet *et al.*, 2002	Yes	Yes, CPO	E
GLI2	CL and CP, pituitary anomalies and holoprosencephaly-like features	TF	Roessler *et al.*, 2003	Yes	Yes, CPO	E + M
IRF6	CL/P or CPO, pits of the lower lip and hypodontia in VWS. CL/P, ankyloplepharon, oral adhesions, syndactyly, and pterygia in PPS	TF	Kondo *et al.*, 2002	Yes	Yes, CPO	E
MSX1	CL/P or CPO/selective tooth agenesis	TF	van den Boogaard *et al.*, 2000	Yes	Yes, CPO	M

(*continued*)

Table 2.1 (*continued*)

Gene	Phenotype	Gene product	References	Non - syndromic	Mouse model	Expression
PAX9	Bilateral CL and CP, colobomas of the optic nerve and retina, agenesis of the corpus callosum, dysphagia, reduced esophageal peristalsis, pes calcaneovarus.	TF	Schuffenhauer *et al.*, 1999	Yes	Yes, CPO	M
PHF8	X-linked mental retardation and CL/P	TF	Laumonnier *et al.*, 2005; Abidi *et al.*, 2007	No	No	Ubiquitous
PTCH	Holoprosencephaly 7, a spectrum of forebrain and midline anomalies and midline CL	Hedgehog receptor	Ming *et al.*, 2002; Muenke 2002; Ribeiro *et al.*, 2006	Yes	Early embryonic lethal	E + M
PVRL1	Autosomal recessive CLPED1: CL/P, hair and tooth anomalies	Ig-like CAM/ viral receptor (nectin-1)	Suzuki *et al.*, 2000	Yes	Yes, no clefting	E
P63	Autosomal dominant EEC syndrome: hand and feet defects, skin and adnexa	TF	Celli *et al.*, 1999; McGrath *et al.*, 2001;	Yes	Yes, truncated palate	E

	anomalies, hypolastic teeth, CL/P. Hay-Wells syndrome: fused eyelids, CL/P skin, adnexa and tooth anomalies. LMS: ectrodactyly, mammary gland/nipple hypoplasia, CPO.		van Bokhoven *et al.*, 2001			
SATB2	CPO, craniofacial anomalies, osteoporosis, and cognitive defects	TF	Leoyklang *et al.*, 2007	Yes	Yes, CPO	M
SHH	Holoprosencephaly, a spectrum of anomalies ranging from severe (cyclopia) to subtle midline asymmetries. CL/P are part of the spectrum	SP	Muenke, 2002	No	Yes, CPO	E
TBX1	Anomalies encompassing most of the features of DiGeorge/velocardiofacial	T-box TF	Yagi *et al.*, 2003	No	Yes, CPO	E

(*continued*)

Table 2.1 *(continued)*

Gene	Phenotype	Gene product	References	Non - syndromic	Mouse model	Expression
	syndromes: CPO, thymus and parathyroid gland hypoplasia, vertebra, facial, and cardiac outflow anomalies.					
TBX22	X-linked CPO and ankyloglossia	T-box TF	Braybrook *et al.*, 2001, 2002; Marçano *et al.*, 2004	Yes	No	M
TCOF1	Treacher–Collins syndrome: Hypoplasia of the maxilla and mandible, ear anomalies, and CPO	Treacle (nucleolar protein)	Edwards *et al.*, 1997	No	Yes, CPO	M
TFAP2A	Branchio-oculo-facial syndrome: CL and/or CP, ocular,	TF	Milunsky *et al.*, 2008	No	Yes, midline clefting	NCC + M

	cutaneous, renal, dental, and hair anomalies					
TGFBR1 or *TGFBR2* (activating mutations)	Cardiovascular, craniofacial, skeletal, and cognitive alterations, and bifid uvula and/or CPO	Tgf β receptor	Loeys *et al.*, 2005	No	No	E + M

CAM, cell adhesion molecule; CL/P, cleft lip with or without cleft palate; CLPED1, cleft lip-palate -ectodermal dysplasia1; CP, cleft palate; CPO, cleft palate only; E, epithelium of orofacial primordia; NCC, neural crest cells; EEC, ectrodactyly–ectodermal dysplasia–clefting syndrome; LMS, limb–mammary syndrome; M, mesenchyme of orofacial primordia; NA, not available; PPS, popliteal pterygium syndrome; SP, signaling protein; TF, transcription factor, TK, tyrosine kinase, VWS, van der Woude syndrome.

in other EED-like syndromes, including ankyloblepharon, ectodermal dysplasia, clefting syndrome (or Hay–Wells syndrome) characterized by fused eye lids (ankyloblepharon) in addition to CL/P and defects related to ectodermal dysplasia (McGrath *et al.*, 2001), and the limb-mammary syndrome (LMS, van Bokhoven *et al.*, 2001). Interestingly, in contrast to EEC syndrome in which individuals show CL/P and never CPO, individuals with LMS display CPO (van Bokhoven *et al.*, 2001). Activating mutations in the genes encoding the transforming growth factor-β receptors have been found in a new syndrome characterized by multiple anomalies, including CPO and/or bifid uvula (Loeys *et al.*, 2005). Recently, a nonsense heterozygous mutation in the gene coding for the transcription factor Satb2 has been associated with a new syndrome that includes CPO, osteoporosis and cognitive defects (Leoyklang *et al.*, 2007). Nearly all of the above genes have also been implicated in nonsyndromic (isolated) clefts (Table 2.2).

Identification of causal genes for nonsyndromic or isolated CL/P or CPO is a more difficult task, given the nature of this malformation that occurs as a complex multifactorial trait with varying levels of penetrance and environmental inputs. About 2–14 interacting loci have been suggested to be involved in this form of clefting (Schliekelman and Slatkin, 2002). Fortunately, recent publications have pointed to a number of causal or potentially causal genes for isolated clefting. Emphasis was placed on the necessity to use large sample sizes of controls to distinguish rare polymorphic variants from etiologic mutations and prioritize functional studies for rare point mutations (Vieira *et al.*, 2005). Candidate genes (Table 2.2) are selected on the grounds of their expression patterns during lip and palate development in mice, the occurrence of CL/P or CPO in mouse models, and/or the chromosomal location of a newly cloned gene.

TGFβ3 encoding the secreted protein transforming growth factor β3 (Tgfβ3) and *MSX1* have been associated with nonsyndromic clefting (Ichikawa *et al.*, 2006; Jezewski *et al.*, 2003; Lidral *et al.*, 1998; Suzuki *et al.*, 2004; Tongkobpetch *et al.*, 2006; Vieira *et al.*, 2005). Other genes including *CHD7* (Felix *et al.*, 2006); *IRF6* (Scapoli *et al.*, 2005; Zucchero *et al.*, 2004); *P63* (Leoyklang *et al.*, 2006); *GAD1* encoding glutamic acid decarboxylase 67 (Gad67), a γ-aminobutyric acid (GABA) biosynthetic enzyme (Kanno *et al.*, 2004); *GABRB3*, coding for the β3 subunit of $GABA_A$ receptor (Inoue *et al.*, 2008; Scapoli *et al.*, 2002); the genes coding for Tbx22 (Braybrook *et al.*, 2002; Marçano *et al.*, 2004); nectin-1 (Avila *et al.*, 2006; Scapoli *et al.*, 2006; Sözen *et al.*, 2001); Pax9 (Ichikawa *et al.*, 2006); Ptc1 (Mansilla *et al.*, 2006); Pvr12/nectin-2, a receptor likely involved in cell–cell adhesion (Warrington *et al.*, 2006); Ryk, a receptor related to tyrosine kinases and devoid of a catalytic activity (Watanabe *et al.*, 2006); Fgfs and their receptors; as well as the gene encoding Estrogen receptor 1 (Osoegawa *et al.*, 2007a,b; Riley *et al.*, 2008) have been incriminated or suggested as candidate causal factors for isolated cleft lip and/or

Table 2.2 Genes and candidate genes implicated in nonsyndromic human orofacial clefting as suggested by human genetics studies, mouse models, and expression analyses

Gene	Phenotype	Gene product	References	Mouse model	Expression
CHD7	CL/P	Chromodomain helicase DNA-binding protein	Félix *et al.*, 2006	Yes, CPO	Mainly in E
ESR1	CL or CPO	Ligand-activated TF (estrogen receptor)	Osoegawa *et al.*, 2008	Yes, no clefting defect	NA
FGF8	CL and CP	SP	Riley *et al.*, 2007b	CPO or midfacial cleft	E
FGF3	CL and CP	SP	Riley *et al.*, 2007b	Yes, no clefting	M
FGF10	CL and CP	SP	Riley *et al.*, 2007b	Yes, CPO	M
FGF18	CL and CP	SP	Riley *et al.*, 2007b	Yes, CPO	E
FGFR1	CPO or CL and CP	TK receptor	Riley *et al.*, 2007a,b	Yes, CPO	E + M
FGFR2	CL and CP or CPO	TK receptor	Riley *et al.*, 2007a,b; Osoegawa *et al.*, 2008	Yes, CPO	E + M
FGFR3	CL or CL and CP	TK receptor	Riley *et al.*, 2007b	Yes, no clefting	M
FOXE1	CL/P	TF	Vieira *et al.*, 2005	Yes, CPO	E
GABRB3	CL/P	$\beta 3$ subunit of $GABA_A$ receptor	Scapoli *et al.*, 2002; Inoue *et al.*, 2008	Yes, CPO	E + M

(continued)

Table 2.2 (*continued*)

Gene	Phenotype	Gene product	References	Mouse model	Expression
GAD1	CL/P	Gad67, enzyme (GABA synthesis)	Kanno *et al.*, 2004	Yes, CPO	E
GLI2	CL/P	TF	Vieira *et al.*, 2005	Yes, CPO	E + M
IRF6	CL or CPO	TF	Zucchero *et al.*, 2004; Scapoli *et al.*, 2005	Yes, CPO	E
JAG2	CL/P	Cell surface ligand for Notch receptors	Vieira *et al.*, 2005	Yes, CPO	E
LHX8	CL/P	LIM-homeodomain TF	Vieira *et al.*, 2005	Yes, CPO	Mainly in M
MSX1	CL or CPO	TF	Jezewski *et al.*, 2003; Suzuki *et al.*, 2004	Yes, CPO	M
MSX2	CL/P	TF	Vieira *et al.*, 2005	Yes, no clefting	E + M
MYH9	CL/P	Heavy chain of nonmuscle myosin IIA	Martinelli *et al.*, 2007	No	E + endothelia
PTCH	CL/P or CPO	Hedgehog receptor	Mansilla *et al.*, 2006	Early embryonic lethality	
PAX9	CL/P	TF	Ichikawa *et al.*, 2006	Yes, CPO	M
P63	CL	TF	Leoyklang *et al.*, 2006	Yes, truncated palate	E
PVR	CL/P or CPO	Poliovirus receptor	Warrington *et al.*, 2006	No	NA

PVRL1	CL and CP	Ig-like CAM/viral receptor (nectin-1)	Sözen *et al.*, 2001; Scapoli *et al.*, 2006; Avila *et al.*, 2006	Yes, no clefting	E
PVRL2	CL and CP	Ig-like CAM/viral receptor (nectin-2)	Warrington *et al.*, 2006	Yes, no clefting	NA
RUNX2	CL/P	TF	Sull *et al.*, 2008	Yes, CPO	E + M
RYK	CL and CP	TK-related	Watanabe *et al.*, 2006	Yes, CPO	E + M
SATB2	CPO or CL/P	TF	Brewer *et al.*, 1999; FitzPatrick *et al.*, 2003; Veira *et al.*, 2005	Yes, CPO	M
SKI	CL/P	TF	Vieira *et al.*, 2005	Yes, median CL/P	E + M
SPRY2	median CL/P	Fgf signaling antagonist	Vieira *et al.*, 2005	Yes, CPO	Early development (E + M), later development (E)
SUMO1	CL and CP	Small ubiquitin-related modifier	Alkuraya *et al.*, 2006	Yes, CPO or oblique facial clefting	E + M
TBX10	CL/P	T-box TF	Vieira *et al.*, 2005	Yes, CL and CP	No expression in lip and palate primordia
TBX22	CPO	T-box TF	Braybrook *et al.*, 2002; Marcano *et al.*, 2004	No	M

(*continued*)

Table 2.2 (*continued*)

Gene	Phenotype	Gene product	References	Mouse model	Expression
TGFβ3	CPO	SP	Lidral *et al.*, 1998; Ichikawa *et al.*, 2006	Yes, CPO	E
WNT3A	CL/P	SP	Chiquet *et al.*, 2008	Yes, early embryonic lethality	M (frontonasal process)
WNT5A	CL/P	SP	Chiquet *et al.*, 2008	Yes, CPO	M
WNT11	CL/P	SP	Chiquet *et al.*, 2008	Yes, clefting not known	E

CAM, cell adhesion molecule; CL, cleft lip; CL/P, cleft lip with or without cleft palate; CPO, cleft palate only; E, epithelium of orofacial primordia; GABA, γ-aminobutyric acid; M, mesenchyme of orofacial primordia; NA, not available; SP, signaling protein, TF, transcription factor; TK, tyrosine kinase.

cleft palate. Data from sequence analyses alone incriminated point mutations in genes encoding the transcription factors Msx1, Msx2, Foxe1, Lhx8, Satb2, Tbx10, Gli2, the proto-oncogene Ski, the Fgf antagonist Sprouty2 (Spry2), and Jagged 2 (Jag2), a ligand for the Notch family receptors, as rare causes for isolated CL/P. However, tests in a larger control group disclosed variants in the *TBX10*, *LHX8*, *SKI*, and *SPRY2* mutations (Vieira *et al.*, 2005). Interestingly, linkage disequilibrium data endorse a role for variants in or near *MSX2*, *JAG2*, and *SKI*. It would be valuable if future studies investigated whether mutations in the above candidate genes affect their expression levels, and if some of the rare variants observed are sufficient to cause isolated cleftings. The role of *TGFA*, encoding transforming growth factor α, in CL/P is not clear as conflicting results were derived from association and linkage studies (Marazita *et al.*, 2004; Mitchell, 1997). Recently, direct sequencing disclosed nine previously unreported noncoding rare variants in single individuals (Vieira *et al.*, 2005).

Breakpoint mapping techniques and expression data identified the *SATB2* gene as a candidate gene for craniofacial malformations associated with deletion and translocation at a chromosomal region (2q32–q33) which is one of only three genomic regions for which haploinsufficiency has been associated with isolated CPO (Brewer *et al.*, 1999; Fitzpatrick *et al.*, 2003).

The list of genes shown to be responsible or suggested as potential causal factors for CL/P and CPO has been further extended by more recent studies. Following a family-based association analysis and based on its expression patterns, the gene encoding the heavy chain of nonmuscle myosin IIA (*MYH9*) has recently been implicated as a causal factor for isolated CL/P (Martinelli *et al.*, 2007). Association between markers in the *RUNX2* locus and risk for isolated CL/P have been suggested in a case-parent trio study from four populations (Sull *et al.*, 2008). *RUNX2* encodes the Runt-related transcription factor2 (Runx2) involved in tooth and bone development (Åberg *et al.*, 2004a; Ducy *et al.*, 1999), and mutations in *RUNX2* have been associated with cleidocranial dysplasia (Sull *et al.*, 2008 and references therein). *SUMO1*, a gene encoding the small ubiquitin-like modifier protein, has been found to be interrupted by a 2q breakpoint leading to haploinsufficiency in a case of isolated cleft lip and palate (Alkuraya *et al.*, 2006).

3. Embryonic Development of the Upper Lip, Primary Palate and Secondary Palate

Facial development in the mouse is heralded by the appearance, at embryonic day 9.5 (E9.5; corresponding to the 4th week of gestation in humans), of five facial prominences/processes: the impaired frontonasal process, and a pair each of maxillary (MxP) and mandibular processes. These tissue swellings consist of an ectodermally derived epithelium externally and a core of cranial neural crest-derived mesenchyme. Facial and lingual muscles are mesodermal derivatives. Fusion of the bilateral mandibular processes create the lower lip and mandible as well as the anterior portion of the tongue. Development of the upper lip has been described in detail recently (Jiang *et al.*, 2006). Briefly, following their formation at the ventral part of the frontonasal process at E10 (late 4th week of gestation in humans), the nasal placodes invaginate, and further morphogenetic movements and growth create the paired medial (MNP) and lateral nasal (LNP) processes. Growth and merger of the MNP with each other and with the MxP create the intermaxillary segment, which consists of the upper lip, the upper jaw that bears the two upper incisors (four incisors in humans), as well as the primary palate. By E12.5 (7th week of human gestation), the upper lip and primary palate have formed. The LNP contribute to the alae of the nose, whereas the nostrils are the product of a merger between the MxP, MNP, and LNP. Failure of adequate growth or fusion between the MNP and MxP generates a spectrum of orofacial clefting involving the upper lip, the alveolus bearing the incisors, and/or the primary palate.

Secondary palate development (Gritli-Linde, 2007) initiates at E11.5 (early week 7 of gestation in humans) by the appearance of primordia of the palatal shelves (PS), which emerge as bilateral outgrowths from the inner side of the MxP and extend antero-posteriorly along the lateral walls of the oropharynx (Fig. 2.2A). During their active growth phase, from E12 to E14 (gestation weeks 7–8 in humans), the PS are oriented vertically in the oral cavity and are sandwitched between the cheeks and the lateral sides of the elevated tongue (Figs. 2.2B and C). At E14.5–E15 (depending on mouse strains; gestation week 9 in humans), the PS elevate into a horizontal posititition above the dorsum of the tongue (Fig. 2.2D). Further polarized growth leads to approximation of the opposing PS, which then adhere along their medial edge epithelia (MEE) creating the transient multilayered medial epithelial seam (MES) (Figs. 2.2F–G). Anteriorly, the PS fuse with the primary palate. The PS also fuse dorsally with the nasal septum following contact between the PS epithelium and the vomerine epithelium that covers the ventral part of the nasal septum (Figs. 2.2H–J). Progressive disintegration of the MES (Fig. 2.2H) as well as removal of the transient epithelial seams generated following contact of the PS with the primary palate and vomerine epithelia allow successful fusions of these primordia and separation of the nasal and oral cavities, a condition required for simultaneous breathing and feeding. By E16.5 (week 10 of gestation in humans), secondary palate formation is completed (Figs. 2.2K and L). The hard palate consists of palatal processes of the maxillary and palatine bones which form following differentiation of PS mesenchymal cells into osteoblasts. The posterior-most portion of the secondary palate, the soft palate (a complex muscular organ), forms the velum and uvula.

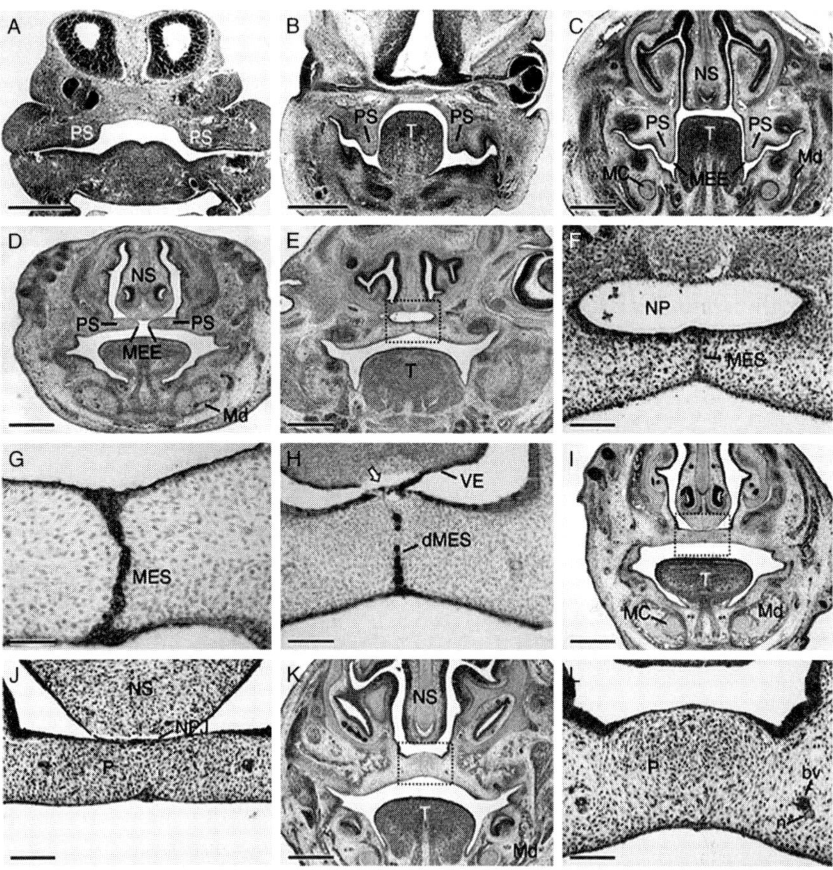

Figure 2.2 Histological sections showing the different steps of development of the murine secondary palate. At E11.5, the palatal shelves (PS) appear as outgrowths from the internal side of the maxillary processes (MxP) (A). During their growth phase at E13.5 (B) and E14.5 (C), the PS are vertical. The tongue (T) is elevated (B and C). At E14.5–E15, the PS have elevated above the tongue and are oriented horizontally (D). At E15–E15.5 (E and F), adhesion of the opposing medial edge epithelia (MEE) following further extension of the PS forms the transient medial epithelial seam (MES). Panel (F) is a high magnification view of the area indicated in (E). Sections immunostained with an anti-E-cadherin antibody which highlights the MES before (G) and during (H) its progressive regression. Note the epithelial islands, transient remnants of the degenerating MES (dMES) and the site of adhesion between the vomerine epithelium (VE) and palate (arrow in H). At E16–E16.5 (I–L), disappearance of the MES allows mesenchymal confluence and successful palate (P) fusion (I–L). The epithelial seam along the nasoplatine junction (NPS) is a result of adhesion between the vomerine epithelium (VE) and PS epithelium. This seam will eventually degenerate allowing successful fusion of the palate with the nasal septum (NS). Additional abbreviations: bv, blood vessel; MC, Meckel's cartilage; Md, mandibular bone; n, nerve. Scale bars: 500 μm (A–E, I, K), 100 μm (F, H, J, L), 50 μm (G). (See Color Insert.)

As secondary palate development proceeds, proliferation within the frontonasal process and MNP generates the nasal septum. This grows ventrally from the roof of the nasal cavity and merges along the midline with the primary palate and the anterior part of the secondary palate, thus dividing the nasal cavity into two chambers. The nasal cavities communicate with the pharynx behind the secondary palate through the definitive choanae.

In sum, colonization by neural crest cells, growth, morphogenetic movements, epithelial adhesion, and degeneration, as well as epithelial and mesenchymal differentiation are key steps underlying lip and palate development. Defects in these steps singly or in combination, as a result of a genetic anomaly or environmental insult or both, engender congenital orofacial clefts. Although the development of the upper lip/primary palate and secondary palate depends largely on cranial neural crest cells (CNCC), the secondary palate is further indirectly dependent on normal development of other craniofacial structures such as the craniofacial skeleton and tongue. Therefore, morphological or functional anomalies in those structures can generate a CPO as well.

4. Cellular and Molecular Mechanisms Governing Lip and Palate Development: Insights from Mouse Models for CL/P and CPO

Mouse models for orofacial clefting provide an invaluable amount of information to delineate the etiopathogenesis of clefting, gene–gene and gene–environment interactions. Good candidate genes for human nonsyndromic clefting are those that not only are expressed during the critical stages of lip and palate development, but also cause a clefting phenotype when mutated or removed in mice. For obvious reasons, it is not feasible to determine the early events of orofacial clefting in humans, even in cases where the gene responsible has been identified. Therefore, whether human cleftings are due to lack of growth, abnormal apoptosis, abnormal morphogenetic and patterning events, or failure of fusion of lip and palate primordia would remain largely unknown. However, mouse models allow us to get around this impasse and study the cellular, morphogenetic, and molecular changes that take place during the genesis of CL/P and CPO.

4.1. Mouse models for CL/P

As is evident from Table 2.3, mice exhibiting CL or CL and CP are scarce as compared to models with CPO. Mouse models for CL/P have been comprehensively reviewed recently (Jiang *et al.*, 2006; Juriloff and Harris, 2008). Therefore, only a brief account and update will be given here.

Table 2.3 Mouse models for cleft lip and/or cleft palate

Genetic loss-of-function Δ	Phenotype and causes of clefting	References	Gene/protein expression
Signaling proteins and receptors			
Activinb-A	CPO(★), lack of whiskers and lower incisors	[j,k]Matzuk *et al.*, 1995a,b	M
ActRcII	CPO(★), hypotrophic mandible, abnormal Meckel's cartilage, and craniofacial skeletal anomalies	[j]Matzuk *et al.*, 1995a	Weak expression in E
Bmp4 (Nestin-Cre-mediated inactivation)	CL	Liu *et al.*, 2005	E
Bmpr1a (Alk3) (Nestin-Cre-mediated inactivation)	Bilateral CL and CP: The CL is due to enhanced apoptosis in the fusing lip primordia. The CP is caused by cell proliferation defects and altered anterior posterior patterning. Tooth developmental arrest	Liu *et al.*, 2005	E + M
Bmp type I receptor (Alk2) (Wnt1-Cre-mediated inactivation)	CPO(★), delayed elevation of PS, hypotrophic mandible, and multiple craniofacial skeletal defects	Dudas *et al.*, 2004b	E + M
Egfr	CPO: Failure of fusion of the PS (persistence of the MEE). Hypotrophic mandible	Miettinen *et al.*, 1999	E
Ephb2; *Ephb3*	CPO: Hypoplastic PS. Abnormal corpus callosum and axon tract defects	Orioli *et al.*, 1996	*Ephb2* (E + M), *Ephb3* (M)
Et1	CPO(★) and elevated blood pressure	[f]Kurihara *et al.*, 1994	E + M (BA)
Fgf8 (hypomorphic)	CPO (★) or absent or reduced palatine bones. Craniofacial skeletal defects including a	Abu-Issa *et al.*, 2002; Frank *et al.*, 2002	E

(continued)

Table 2.3 (*continued*)

Genetic loss-of-function Δ	Phenotype and causes of clefting	References	Gene/protein expression
	hypomorphic mandible. Cardiovascular defects. Anomalies caused by increased apoptosis in NCC progeny		
Fgf9	CPO in 40% of mutants and lung hypoplasia	Colvin *et al.*, 2001	E
Fgf10	CPO: Proliferation defects and increased apoptosis in PS, loss of Shh expression and aberrant adhesion of PS with other oral epithelia	Rice *et al.*, 2004; Alappat *et al.*, 2005	M
Fgf18	CPO(*), craniofacial and other skeletal anomalies	Liu *et al.*, 2002; Ohbayashi *et al.*, 2002	E
Fgfr1 (hypomorphic)	CPO(*): PS fail to elevate. Craniofacial skeletal defects and BA patterning anomalies	Trokovic *et al.*, 2003	E + M
Fgfr2b	CPO: Altered proliferation in PS. Other anomalies include tooth, salivary gland, skin, limb, lung, and pituitary defects.	De Moerlooze *et al.*, 2000; Rice *et al.*, 2004	E + M
Fgfr2c (gain-of-function)	CPO(*), multiple joint fusions, lung, and tracheal defects	Eswarakumar *et al.*, 2004	M
Follistatin	CPO(*), tooth, whisker, skin, and rib anomalies	[1]Matzuk *et al.*, 1995c	M
Gabrb3	CPO: PS elevate but fail to make contact. Neuronal defects	Homanics *et al.*, 1997; Culiat *et al.*, 1993, 1995	E + M
Gabrb3 (loss-of-function in the palate)	CPO: PS elevate but fail to abut	Hagiwara *et al.*, 2003	E + M

Itgav	CPO: PS elevate but fail to make contact. Intracerebral and intestinal hemorrhages	Bader *et al.*, 1998	NA
Jagged2	CPO: Aberrant adhesion between PS and oral epithelia secondary to altered differentiation of the epithelium of the tongue and mandible. Limb and thymic anomalies	Jiang *et al.*, 1998; Casey *et al.*, 2006	E
Pdgfc	CPO: PS are hypotrophic, show delayed lifting and fail to fuse. Subcutaneous edema	Ding *et al.*, 2004	E
Pdgfc (*EII-Cre*-mediated ablation)	Similar phenotype to that of *Pdgfc*mutants	Wu and Ding, 2007	E
Pdgfc; *Pdgfa* compound mutants	Midfacial cleft (*), subepidermal blistering, renal, skeletal and vascular defects. Phenocopies the defects in *Pdgfra* mutants	Ding *et al.*, 2004	E
Pdgfra (*Wnt1-Cre*-mediated inactivation	CP (*) and a range of midfacial clefting, subepidermal blistering, renal, vascular and skeletal anomalies	Tallquist and Soriano, 2003	M
Pdgfra $^{+/-}$; *Plekha1* $^{-/-}$ compound mutants	Midfacial clefting similar to the *Pdgfra* phenotype or CPO	Schmahl *et al.*, 2007	M
RARa; *RARg* compound mutants	CP (*) and midfacial clefting. Other skeletal anomalies	Lohnes *et al.*, 1994	E + M
Rspo2	CPO, craniofacial skeletal and limb defects, sporadic kidney agenesis, lung immaturity, and pulmonary vascular anomalies	Nam *et al.*, 2007	E
Ryk	CPO (*): Delayed PS elevation but normal fusion *in vitro*. Skeletal anomalies	Halford *et al.*, 2000	E + M

(*continued*)

Table 2.3 *(continued)*

Genetic loss-of-function Δ	Phenotype and causes of clefting	References	Gene/protein expression
Shh (*K14-Cre*-mediated inactivation)	CPO, tooth and skin anomalies	Rice *et al.*, 2004	E
Tgfb2	CPO (★). Skeletal anomalies and cardiac, ocular, urogenital, and inner ear defects	[p]Sanford *et al.*, 1997	E + M
Tgfβ3	CPO: Failure of fusion of PS. Partial rescue of CP following expression of *Tgfb1* in the *Tgfβ3* locus. Lung defects	Kaartinen *et al.*, 1995; Proetzel *et al.*, 1995; Yang and Kaartinen, 2007	E
Tgfbr1 (*Alk5*) (*K14-Cre*-mediated inactivation)	CPO: impaired PS adhesion and fusion (partial) due to decreased MEE filopodia and to lack of apoptosis of the MES	Dudas *et al.*, 2006	E + M
Tgfbr1 (*Alk5*) (*Wnt1-Cre*-mediated inactivation)	CPO (★): Increased apoptosis and cell proliferation in the PS. Anomalies in other skeletal craniofacial structures may also contribute to CP	Dudas *et al.*, 2006	E + M
Tgfbr1 (*Alk5*) (*Tgfβ3-Cre*-mediated inactivation)	CPO similar to that in *Tgfβ3* mutants. Hydrocephalus and intracranial hemorrhages	Yang *et al.*, 2008	E + M
Tgfbr2 (*Wnt1-Cre*-mediated inactivation)	CPO: Reduced proliferation in PS, PS fuse normally *in vitro*. Dura mater and craniofacial skeletal anomalies	Ito *et al.*, 2003	E + M
Tgfbr2 (*K14-Cre*-mediated inactivation)	CPO: Impaired PS fusion (partial) due to lack of apoptosis and persistent proliferation of the MEE/MES	Xu *et al.*, 2006	E + M
Wnt5a	CPO (★) and skeletal anomalies	Yang *et al.*, 2003	M

clf1, *clf2* spontaneous mutant (*clf1*, a spontaneous hypomorph of *Wnt9b*; *clf2*, unknown)	nonsyndromic CL/P	Juriloff *et al.*, 2006	E (*Wnt9b*)
Wnt9b	Incompletely penetrant CL/P and urogential defects	Carroll *et al.*, 2005	E
Transcription factors and nuclear proteins			
Different compound mutants of *Alx4* and *Cart1*	CP (*) and median cleft lip and nose. Polydactyly and other skeletal anomalies	Qu *et al.*, 1999	M
Alx3; *Alx4* compound mutants	CP (*) and median cleft lip and nose due to abnormal lateral position of the nasal processes and apoptosis. Other craniofacial skeletal anomalies	Beverdam *et al.*, 2001	M
Arid5b	Submucous CPO: reduced palatal processes of the palatine and presphenoid bones	Schmahl *et al.*, 2007	M
BC055757	Similar phenotype to *Arid5b*	Schmahl *et al.*, 2007	M
Arid5b; *BC055757* compound mutants	More severe palatal bone reduction than in single mutants	Schmahl *et al.*, 2007	M
Arid5b$^{-/-}$; *Pdgfra*$^{+/-}$ compound mutants	Complete CPO or reduced palatal bones.	Schmahl *et al.*, 2007	M
Chd7 (*Whirligig* heterozygotes)	Complete or partial CPO due to delayed PS elevation or failure of their fusion, edema, choanal, ocular, genital, inner ear, and cardiovascular defects in heterozygotes replicating the clinical features of CHARGE syndrome	Bosman *et al.*, 2005	Mainly in E

(continued)

Table 2.3 *(continued)*

Genetic loss-of-function Δ	Phenotype and causes of clefting	References	Gene/protein expression
Dlx1; *Dlx2* compound mutants	CPO (*): Delayed PS elevation. Other craniofacial skeletal anomalies	[n]Qiu *et al.*, 1997	No expression in MxP
Dlx5	CPO (*): Delayed PS elevation. Other craniofacial skeletal anomalies	[a]Acampora *et al.*, 1999;[e]Depew *et al.*, 1999	No expression in MxP
dEF1 (aliases *Zfhep and Zfhx1a*)	CPO (*). T-cell defects, craniofacial and other skeletal anomalies	[r]Takagi *et al.*, 1998	M
Eya1	CPO (129/Sv and Balb/C strains). PS fuse abnormally with the nasal septum (in the C57BL/6J strain). Failure of fusion of eye lids, skeletal, inner ear and renal defects	Xu *et al.*, 1999	E + M
Foxc2 (previously *Mfh1*)	CPO (*): Delayed PS elevation. Aortic arch and craniofacial skeletal anomalies	[g]Lida *et al.*, 1997	M
Foxe1 (previously *Titf2*)	CPO: PS elevate but fail to fuse with each other. Sublingual or complete agenesis of thyroid gland	De Felice *et al.*, 1998	E
Foxf2	CPO (*)?	Wang *et al.*, 2003	M
Gli2	CPO (*): Delayed elevation of PS. Craniofacial skeletal anomalies	Mo *et al.*, 1997	E + M
Gli3^{xtJ}	CPO (*): Delayed PS elevation. Craniofacilal skeletal anomalies	Mo *et al.*, 1997	E + M
Hic1	CPO (*) and a spectrum of craniofacial anomalies including, acrania, exencephaly and cyclopia. Also omphalocele, edema and reduced body size. These defects are found in MDS individuals	Carter *et al.*, 2000	M

Hoxa2	CPO (⋆)	Gendron-Maguire *et al.*, 1993; Rijli *et al.*, 1993; Barrow and Capecchi, 1999	No expression in lip + palate
$Irf6^{R84C/R84C}$ (missense mutation)	CPO due to fusion between epithelia of PS and tongue. Other intraoral adhesions. Adhesions of tail and hindlimbs with the body wall, skeletal anomalies, obliterated esophagus, lack of mature hair and whisker follicles, skin barrier defects. These alterations are likely caused by failure of cell cycle exit and terminal differentiation of keratinocytes. $Irf6^{+/R84C}$ heterozygotes show intraoral adhesions but no clefting	Richardson *et al.*, 2006	E
$Irf6^{+/R84C}$; $Sfn^{+/Er}$ compound mutants	CPO due to aberrant intraoral adhesions, obliterated esophagus, partial adhesions of tail and hindlimbs with body wall, syndactyly	Richardson *et al.*, 2006	E
$Irf6^{gtl/gtl}$ (null allele)	CPO due to aberrant intraoral epithelial adhesions. Adhesions between the tail and hindlimbs, esophagal obliteration. Skin anomalies caused by failure of differentiation of keratinocytes which remain proliferative	Ingraham *et al.*, 2006	E
Lhx8	CPO: PS elevate but fail to make contact	Zhao *et al.*, 1999	E + M

(continued)

Table 2.3 (*continued*)

Genetic loss-of-function Δ	Phenotype and causes of clefting	References	Gene/protein expression
Menin (*Pax3*-or *Wnt1-Cre*-mediated inactivation	Complete CPO or deficient soft palate: PS elevate but remain apart. Reduced $p27^{kip1}$ expression, hyperplastic but hypotrophic PS. Decreased extracellular matrix in PS. Minor defects in basisphenoid. Rib defects following *Pax3-Cre*-mediated inactivation.	Engleka *et al.*, 2007	E + M
Meox2 homozygous or heterozygous inactivation	CPO (*): post-fusion cleft	Jin and Ding, 2006a	M
Mkx (previously *Irxl1*), a candidate gene for the *Twirler* mutation	CL/P or CPO. Hypotrophic PS, delayed PS elevation. Inner ear defects and obesity	Gong *et al.*, 2000; Gong and Eulenberg, 2001; Liu *et al.*, 2006	M
Mnt	CPO (varying degrees of clefting, from complete to localized small clefts), reduced body size, hypotrophic mandible. MNT/*Mnt* is another MDS region gene.	Toyo-Oka *et al.*, 2004	Ubiquitous
Msx1	CPO: Altered proliferation in PS. Tooth developmental arrest and other craniofacial skeletal anomalies	Satokata and Maas, 1994; Zhang *et al.*, 2002	M
Myf5; *MyoD*	Primary palate and secondary palate do not fuse with each other. Skeletal anomalies and lack of skeletal muscle formation	[o]Rot-Nikcevic *et al.*, 2005	Skeletal muscle

Osr2	CPO: Impaired proliferation and medio-lateral patterning in PS and delayed PS elevation. Altered expression of *Osr1*, *Pax9*, and *Tgfβ3*. Failure of fusion of eye lids	Lan *et al.*, 2004	M
Ovca1 or *Ovca1–2*	CPO, lung immaturity, reduced body size, fetal liver degeneration, preaxial polydactyly of right hind limb. Reduced proliferation. Rescue of CPO and embryo size in some *Ovca1–2*; *p53* compound mutants. Phenotype encompasses some features in MDS	Chen and Behringer, 2004	Ubiquitous
p63	Truncated palate	Mills *et al.*, 1999; Yang *et al.*, 1999	E
Pax9	CPO (★), tooth and craniofacial skeletal anomalies	Peters *et al.*, 1998	M
Pds5B (*Aprin*)	CPO (★): Hypotrophic PS, delayed elevation of PS. Short snout, thin upper lip, hypotrophic mandible. Other skeletal anomalies, distal colon aganglionosis, cardiac defects and abnormal migration and projections of sympathetic neurons. The model phenocopies defects in Cornelia de Lange syndrome	Zhang *et al.*, 2007	E + M
Pitx1	CPO (★), hindlimb and mandible growth defects	Lanctôt *et al.*, 1999; Szeto *et al.*, 1999	E or M

(*continued*)

Table 2.3 (*continued*)

Genetic loss-of-function Δ	Phenotype and causes of clefting	References	Gene/protein expression
Pitx2	CPO: PS elevate but are hypotrophic. Abnormal cardiac morphogenesis, abnormal maxillary and mandibular prominences and tooth developmental arrest	Lu *et al.*, 1999a	E or M
Prx1 (previously *Mhox*)	CPO (⋆) and skeletal defects	[i]Martin *et al.*, 1995	M
Prx1; *Prx2*	CPO (⋆): PS fail to elevate. Skeletal defects, cleft mandible, developmental arrest of the lower incisors, polydactyly, and inner, middle and external ear defects	[t]ten Berge *et al.*, 1998; [h]Lu *et al.*, 1999b	M
Pygo2	CPO. Exencephaly, renal defects, and lens agenesis	Schwab *et al.*, 2007	NA
Rae28	CPO (⋆), posterior skeletal transformation and NCC defects correlating with altered Hox codes. Ocular, cardiac, parathyroid and thymus anomalies	[s]Takihara *et al.*, 1997	NA
Recq14	CPO, preaxial polydactyly of the hind limb, skin anomalies and increased cancer susceptibility	Mann *et al.*, 2005	Ubiquitous
Runx2	CPO (⋆): PS elevate but do not abut. Skeletal anomalies, tooth defects and failure of eye lid fusion.	Åberg *et al.*, 2004a	E + M
Sall3	Hypoplastic soft palate and epiglottis. Cranial nerve deficiencies	Parrish *et al.*, 2004	E + M (PS)

Satb2	CPO: Delayed elevation of PS. PS with peculiar bulges and reduced *Lhx8* expression. Skeletal anomalies	Dobreva *et al.*, 2006	M
Satb2 haploinsufficiency	CPO: PS elevate and are hyptrophic. Craniofacial skeletal anomalies	Britanova *et al.*, 2006	M
Satb2	CPO: Altered expression patterns of *Alx4* and *Msx1* in PS. Reduced *Pax9* expression in BA1. Abnormal apoptosis in *Satb2*expression territories. Craniofacial skeletal and tongue anomalies	Britanova *et al.*, 2006	M
Shox2	Cleft of the anterior portion of the secondary palate (hard palate) due to abnormal proliferation and apoptosis	Yu *et al.*, 2005	M
Shox2 (*Wnt1-Cre*-mediated inactivation)	CPO confined anteriorly: Delayed PS fusion but failure of fusion between the PP and SP. Reduced osteogenesis within the hard palate secondary to reduced expression of *Runx2* and *Osterix*. Increased *Bmp2* expression	Gu *et al.*, 2008	M
Sim2	CPO: PS are hypocellular and exhibit increased extracellular glycosaminoglycans	Shamblott *et al.*,, 2002	NA
Ski	Median CL/PP. Digit and ocular anomalies and skeletal muscle defects	Berk *et al.*, 1997; Colmenares *et al.*, 2002	E + M
Snai2 mutants or *Snai2*$^{-/-}$; *Snai1*$^{+/-}$ compound mutants	CPO (*): Failure of MES formation secondary to delayed PS elevation and head growth or due to lack of apoptosis and failure of periderm migration	Murray *et al.*, 2007	*snail* (M); *Snai2* (E + M)

(continued)

Table 2.3 (*continued*)

Genetic loss-of-function Δ	Phenotype and causes of clefting	References	Gene/protein expression
Snai2; *Snai1* compound mutants (*Wnt1-Cre*-mediated inactivation of *Snai1*)	CPO (★): Failure of PS elevation. Skull, mandible and Meckel's cartilage defects	Murray *et al.*, 2007	
Sp8	Midfacial clefting, neural tube defects, limb truncation.	Bell *et al.*, 2003	E (MNP + LNP)
Sox5	CPO (★) and skeletal anomalies	[q]Smits *et al.*, 2001	M, brain and cartilage
Sox9 haploinsufficiency	CPO (★), craniofacial skeletal anomalies	Bi *et al.*, 2001	M
Sox9 (*Wnt1-Cre*-mediated inactivation)	CPO (★), craniofacial skeletal anomalies	Mori-Akiyama *et al.*, 2003	M
Sox11	Unilateral or bilateral CL and CP or anterior cleft due to failure of fusion of the PS with each other and with the PP. Failure of eye lid fusion, lung hypoplasia, cardiac defects and other skeletal anomalies	Sock *et al.*, 2004	E + M
Tbx1	CPO (★). Developmental anomalies encompassing the quasitotality of DGS/VCFS features	Jerome and Papaioannou, 2001	E (lip + palate primordia). M (tongue)
Tcfap2a	Midline cleft face and mandible, exencephaly, and other severe anomalies	Schorle *et al.*, 1996; Zhang *et al.*,, 1996	NCC + M
Tcfap2a chimeras	Isolated CL/P	Nottoli *et al.*, 1998	NCC + M
Tcfap2a (*Wnt1-Cre*-mediated inactivation)	CPO (★): PS elevate but fail to abut. Other craniofacial skeletal anomalies	Brewer *et al.*, 2004	NCC + M

Tcof1 haploinsufficiency	CPO (*) and craniofacial skeletal malformations phenocopying those in TCS. Rescue of the malformations following pharmacological or genetic inactivation of p53/*Trp53*	Dixon *et al.*, 2006; Jones *et al.*, 2008	NCC + M
Tshz1	Cleft of the soft palate (velum). PS elevate but fail to fuse. Some abnormal fusion occurs in the same palatal specimen. Middle ear anomalies and homeotic transformations of axial skeleton	Coré *et al.*, 2007	Not expressed in the palate at E12.5–E14.5
Vax1	Fully penetrant CPO (*) with varying degrees of severity, coloboma and axon guidance defects	[d]Bertuzzi *et al.*, 1999	CNS
Cytoplasmic and membrane-bound proteins			
Apaf1	CPO: Failure of fusion of PS owing to failure of apoptosis. Brain overgrowth, persistent interdigital web, and ocular anomalies	Cecconi *et al.*, 1998	E + M
Cacnals (*mdg* mutants)	CPO (*) due to paralysis of the tongue	Pai, 1965	Skeletal muscle
CASK	CPO: PS elevate but fail to abut	Atasoy *et al.*, 2007	Epithelia and brain
Crk	CPO: PS elevate but remain apart or presence of edema at the nasal side of the medial epithelia seam. Edema in the nasal septum, midline focal hemorrhagic edema on the snout, edema on the neck and back and cardiovascular defect. Some of the defects are features of MDS	Park *et al.*, 2006	Ubiquitous expression

(continued)

Table 2.3 *(continued)*

Genetic loss-of-function Δ	Phenotype and causes of clefting	References	Gene/protein expression
	Defects caused by abnormal activation of Wnt signaling and apoptosis in NCC-derived tissues. Also kidney agenesis and rupture of the gut due to degeneration of the NCC-derived enteric nervous system		
IKKa (IKK1, Chuk)	CPO (*), skin, whisker, limb, and gastrointestinal tract anomalies	Li *et al.*, 1999	NA
Insig1; *Insig2* compound mutants	CPO or midfacial clefting (CL, CP, and split nose): Hypotrophic PS. Some cleft face embryos show exencephaly. Poor development of the basisphenoid and hyoid bones and Meckel's cartilage in cleft face embryos. Middle and inner ear anomalies. Reduced body size. Accumulation of sterol intermediates in liver and head structures. Rescue of craniofacial malformations by lovastatin	Engelking *et al.*, 2006	NA
Kcnj2 (Kir2.1)	Isolated fully penetrant CPO. No other morphological anomalies	Zaritsky *et al.*, 2000	Arterial smooth muscle cells
Nf2 mosaics (*NesCre1*-mediated inactivation	CPO: PS elevate but do not fuse. Retinal coloboma, lens herniation and failure of eye lid fusion. A spectrum of neural tube defects, omphalocele and cardiac	McLaughlin *et al.*, 2007	E + Schwann cells and neurons

	ventricular septal defects. Also skeletal and other visceral defects		
p57^{kip2} (homozygotes and targeted maternal allele in heterozygotes)	CPO due to increased apoptosis in PS epithelium and mesenchyme. Renal, placental, intestinal, and skeletal anomalies. Macroglossia and omphalocele	Yan *et al.*, 1997; Zhang *et al.*, 1997; Caspary *et al.*, 1999	Expression in cells that exit the cell cycle
Sc5d	CPO, micrognathia, narrow frontonasal process, calvarial defects, kinked tail, short limbs, and limb patterning defects (postaxial polydactyly). Elevated lathosterol and reduced cholesterol levels. Mouse model for human lathosterolosis	Krakowiak *et al.*, 2003	NA
Smo (*Wnt1-Cre*-mediated inactivation)	CPO (*), extensive loss of craniofacial skeletal structures and agenesis of the tongue and lower incisors	Jeong *et al.*, 2004	E + M
Smo activation (*Wnt1-Cre*; *R26SmoM2*)	Midfacial clefting, exencephaly, and quasitotal absence of head skeleton	Jeong *et al.*, 2004	E + M
Spry2 (*Spry2* locus removal by the *36Pub* deletion)	CL (low incidence) or CPO (high incidence): Failure of PS to elevate, elevated and ectopic expression of FGF-responsive genes, failure of *Tbx22* expression in the posterior palate and increased proliferation of the PS epithelium and mesenchyme	Welsh *et al.*, 2007	Early development (E + M), later development (E)
Sumo1$^{Gt/+}$	CL or CP. Embryonic lethality suggests the occurrence of developmental anomalies in other organs	Alkuraya *et al.*, 2006	E + M

(continued)

Table 2.3 (*continued*)

Genetic loss-of-function Δ	Phenotype and causes of clefting	References	Gene/protein expression
Sumo1$^{Gt/+}$; *Eyal* $^{+/-}$ *compound mutants*	CL or CP. Note: *Eya1* encodes a TF	Alkuraya *et al.*, 2006	See above for *Sumo1* and *Eya1*
Viaat	CPO (*), reduced neurotransmitter release in some neurons	Wojcik *et al.*, 2006	CNS
Extracellular matrix components			
Col2a1	CPO (*) and skeletal anomalies	[m]Pace *et al.*, 1997	Cartilage
Perlecan	CPO (*) and skeletal dysplasia	[b]Arikawa-Hirasawa *et al.*, 1999	Basement membranes, cartilage
Insertional mutations			
CASK (loss-of-function)	CPO (*), retrognathia and spinal kinks	[u]Wilson *et al.*, 1993; Laverty and Wilson, 1998	Epithelia and brain
Sp8 hypomorphic allele in the *legless mutation*	Neural tube, limb and craniofacial defects similar to those in *Sp8* mutants	Bell *et al.*, 2003	E (MNP + LNP)
Tbx10 (gain-of-function). *Dancer* mutation	CL and CP due to ectopic expression of *Tbx10*	Bush *et al.*, 2004	No expression in lip + palate
p23-Tbx10 transgenic mice	CL and CP similar to that of *Dancer* mice	Bush *et al.*, 2004	No expression in lip + palate

Δ genetic loss-of-function except where indicated; (*) indicates cleft palate conditions that are or may be secondary to other craniofacial skeletal defects and/or steric hindrance by the tongue; BA, branchial arch; CL, cleft lip; CP, cleft of the secondary palate; CPO, only cleft of the secondary palate; CL/P, cleft lip with or without cleft palate; CNS, central nervous system; DGS/VCFS, DiGeorge syndrome/velocardiofacial syndrome; E, epithelium; LNP, lateral nasal process; M, mesenchyme; MDS, Miller–Dieker syndrome; MNP, median nasal process; MxP, Maxillary process; MES, medial epithelial seam; NA, not available; NCC, neural crest cells; PP, primary palate; PS, palatal shelf (ves); SLOS, Smith–Lemli–Opitz syndrome; SP, secondary palate; TCS, Treacher–Collins syndrome.

[a] Acampora, D., Merlo, G. R., Paleari, L., Zenega, B., Postiglione, M. P., Mantero, S., Bober, E., Barbieri, O., Simeoni, A., and Levi, G. (1999). Craniofacial, vestibular and bone defects in mice lacking the distal-less-related gene *Dlx5*. *Development* **126,** 3795–3809.

[b] Arikawa-Hirasawa, E., Watanabe, H., Takami, H., Hassell, J. R., and Yamada, Y. (1999). Perlecan is essential for cartilage and cephalic development. *Nat. Genet.* **23,** 354–358.

[c] Bader, B. L., Rayburn, H., Crowley, D., and Hynes, R. O. (1998). Extensive vasculogenesis, angiogenesis and organogenesis precede lethality in mice lacking all αv integrins. *Cell* **95,** 507–519.

[d] Bertuzzi, S., Hindges, R., Mui, S. H., O'Leary, D. D., and Lemke, G. (1999). The homeodomain protein vax1 is required for axon guidance and major tract formation in the developing forebrain. *Genes Dev.* **13,** 3092–3105.

[e] Depew, M. J., Liu, J. K., Long, J. E., Presley, R., Meneses, J. J., Pedersen, R. A., and Rubenstein, J. L. (1999). Dlx5 regulates regional development of the branchial arches and sensory capsules. *Development* **126,** 3831–3846.

[f] Kurihara, Y., Kurihara, H., Suzuki, H., Kodama, T., Maemura, K., Nagai, R., Oda, H., Kuwaki, T., Cao, W.H., Kamada, N., Jishage, K., Ouchi, Y., *et al.* (1994). Elevated blood pressure and craniofacial anomalies in mice deficient in endothelin-1. *Nature* **368,** 703–710.

[g] Lida, K., Koseki, H., Kakinuma, H., Kato, N., Mizutani-Koseki, Y., Ohushi, H., Yoshioka, H., Noji, S., Kawamura, K., Kataoka, Y., Ueno, F., Tanigishi, M., *et al.* (1997). Essential roles of the winged helix transcription factor MFH-1 in aortic arch patterning and skeletogenesis. *Development* **124,** 4627–4638.

[h] Lu, M. F., Cheng, H. T., Kern, M. J., Potter, S. S., Tran, B., Diekwisch, T. G., and Martin, J. F. (1999b). Prx-1 functions cooperatively with another paired-related homeobox gene, Prx-2, to maintain cell fates within the craniofacial mesenchyme. *Development* **126,** 495–504.

[i] Martin, J. F., Bradley, A., and Olson, E. N. (1995). The paired-like homeobox gene MHox is required for early events of skeletogenesis in multiple lineages. *Genes Dev.* **9,** 1237–1249.

[j] Matzuk, M. M., Kumar, T. R., and Bradley, A. (1995a). Different phenotypes for mice deficient in either activins or activin receptor type II. *Nature* **374,** 356–360.

[k] Matzuk, M. M., Kumar, T. R., Vassali, A., Bickenbach, J. R., Roop, D. R., Jaenish, R., and Bradley, A. (1995b). Functional analysis of activins during mammalian development. *Nature* **374,** 354–356.

[l] Matzuk, M. M., Lu, N., Vogel, H., Sellheyer, K., Roop, D. R., and Bradley, A. (1995c). Multiple defects and perinatal death in mice deficient in follistatin. *Nature* **374,** 360–363.

[m] Pace, J. M., Li, Y., Seegmiller, R. E., Teuscher, C., Taylor, B. A., and Olsen, B. R. (1997). Disproportionate micromelia (Dmm) in mice caused by a mutation in the C-propeptide coding region of Co12a1. *Dev. Dyn.* **208,** 25–33.

[n] Qiu, M., Bulfone, A., Ghattas, I., Meneses, J. J., Christensen, L., Sharpe, P. T., Presley, R., Pedersen, R. A., and Rubenstein, L. R. (1997). Role of the Dlx homeobox genes in proximodistal patterning of the branchial arches: Mutations of Dlx1, Dlx2 and Dlx1 and 2 alter morphogenesis of proximal skeletal and soft tissue structures derived from the first and second arches. *Dev. Biol.* **185,** 165–184.

[o] Rot-Nikcevic, I., Reddy, T., Doening, K. J., Belliveau, A. C., Hallgrimsson, B., Hall, B. K., and Kablar, B. (2005). $Myf5^{-/-}$: $MyoD^{-/-}$ amyogenic fetuses reveal the importance of early contraction and static loading by striated muscle in mouse skeletogenesis. *Dev. Genes Evol.* **216,** 1–9.

[p] Sanford, P. L., Ormsby, I., Gittenberger, A. C., Sariola, H., Friedman, R., Boivin, G. P., Cardell, E. L., and Doetschman, T. (1997). TGFβ2 knockout mice have multiple developmental defects that are non-overlapping with other TGFβ knockout phenotypes. *Development* **124,** 2659–2670.

[q] Smits, P., Li, P., Mandel, J., Zhang, Z., Deng, J. M., Behringer, R. R, de Crombrugghe, B., and Lefebvre, V. (2001). The transcription factors L-Sox5 and Sox6 are essential for cartilage formation. *Dev Cell* **1,** 277–290.

[r] Takagi, T., Moribe, H., Kondoh, H., and Higashi, Y. (1998). *δEF1*, a zinc finger and homeodomain transcription factor, is required for skeleton patterning in multiple lineages. *Development* **125,** 21–31.

[s] Takihara, Y., Tomotsune, D., Shirai, M., Katoh-Fukui, Y., Motaleb, M. A., Nomura, M., Tsuchiva, R., Fujita, Y., Shibata, Y., Higashinakagawa, T., and Shimada, K. (1997). Targeted disruption of the mouse homologue of the Drosophila polyhomeotic gene leads to altered anteroposterior patterning and neural crest defects. *Development* **124,** 3673–3682.

[t] ten Berge, D., Brouwer, A., Korving, J., Martin, J. F., and Meijlink, F. (1998). Prx1 and Prx2 in skeletogenesis: roles in the craniofacial region, inner ear and limbs. *Development* **125,** 3831–3842.

[u] Wilson, J. B., Ferguson, M. W. J., Jenkins, N. A., Lock, L. F., Copeland, N. G., and Levine, A. J. (1993). Transgenic mouse model of X-linked cleft palate. *Cell Growth Differ.* **4,** 67–76.

[v] Wu, X., and Ding, H. (2007). Generation of conditional knockout alleles for PDGF-C. *Genesis* **45,** 653–657.

The first models for CL/P were mice of the A/- strains and related strains which display susceptibity to spontaneous CL/P. These strains are the best in mimicking human nonsyndromic CL/P in terms of genetic causes and environmental sensitivity (Juriloff and Harris, 2008). Two loci underlie the genetic etiology of CL/P in the A/- strains and their related strains, the *clf1* and *clf2* recessive mutation and semidominant polymorphic variant, respectively. Interestingly, complementation crosses indicated that the *clf1* mutation is a hypomorphic defect of the *Wnt9b* gene, a member of the *Wnt* family of genes encoding secreted molecules that play key roles during embryogenesis and tissue homeostasis and are implicated in cancer (Juriloff *et al.*, 2006). Lack of function of two alleles of *Wnt9b* generates low-penetrance CL/P and urogenital defects (Carroll *et al.*, 2005). The involvement of Wnt9b signaling, likely via the Wnt canonical pathway, during lip formation is further supported by its high expression in epithelia of facial prominences. Another Wnt family member, *Wnt3*, is partly coexpressed with *Wnt9b* during lip formation and may participate in its development (Lan *et al.*, 2006). This is further indicated by a human lethal syndrome that includes CL/P among other malformations (Juriloff and Harris, 2008 and references therein). Importantly, several WNT genes have been associated with human nonsyndromic CL/P (Chiquet *et al.*, 2008).

As mentioned above, *SUMO1* haploinsufficiency causes CL and CP. The causative role for *SUMO1* haploinsufficiency in human clefting was further firmly established by the demonstration of *Sumo1* expression in the epithelia and mesenchyme of the murine developing palate and upper lip and by the occurrence of cleft palate or oblique facial cleft in *Sumo1* heterozygous mice (Alkuraya *et al.*, 2006). Several key players in lip and palate development have been shown to be modified by sumoylation via SUMO1 (discussed in Section 4.2.1.6), endowing SUMO1 with an important task in the multimolecular events governing lip and palate formation.

CNCC contribute to the formation of an important part of the craniofacial mesenchyme, including that of lip and palate primordia. Therefore, molecular changes that disrupt the specification, migration, survival, patterning or proliferation of CNCC usually have a severe impact on orofacial and cranial structures. Neural crest cell formation and survival requires the cooperation of a number of signaling cascades, including the Fgf, Wnt, and Bmp pathways (Nie *et al.*, 2006a,b). The Bmp and Fgf signaling pathways play a key role in the formation of orofacial primordia. Several Fgf family members and Fgfrs are expressed during early orofacial development where they mediate epithelial–mesenchymal interactions (Nie *et al.*, 2006a). Despite the early lethality of mouse embryos following removal of *Fgf* genes and the presence of functional redundancy amongst Fgf family members, some information exists and points to their important role for the development and outgrowth of orofacial primordia. *Fgf3*, *Fgf8*, *Fgf10*, *Fgfr1*, and *Fgfr2* are highly expressed in the medial and/or LNP (Bachler and

Neubüser, 2001), and their implication in human orofacial clefting indicates a role in lip formation (Osoegawa *et al.*, 2007a,b; Riley *et al.*, 2008). However, mouse mutations of *Fgfs* and their receptors generate CPO. High incidence of CPO and low incidence of CL have been reported to occur in mice lacking the function of Spry2 (Welsh *et al.*, 2007), consistent with recent human studies implicating this gene in nonsyndromic CL/P.

Bmps belong to the Tgfβ superfamily, the members of which play key roles during embryogenesis, organogenesis, and tissue homeostasis by controlling several cellular activites, including fate specification, proliferation, apoptosis, and differentiation. Because of their similarity with the ancestral *decapentaplegic* (*dpp*) of *Drosophila*, Bmp2 and Bmp4 are grouped together as the dpp subfamily. Other family members, including Bmp5, -6, -7, and -8 constitute the 60A subfamily. Bmps signal through type I and type II receptors. Type I receptors include Alk2, Alk3 or Bmpr1a, and ALK6 or Bmpr1b, and Type II receptors comprise BRII, ActRIIA, and ActRIIB (Nie *et al.*, 2006b). Activation of the Bmp signaling cascade is initiated by ligand binding leading to association of two receptors, one of each type, which results in phosphorylation of the type I receptor by the type II receptor and phosphorylation of downstream effectors, including Smads1, 5, and 8, which thus provide a readout of Bmp activity in a tissue.

The closely related Bmp2 and Bmp4 share only 60% identity at the amino acid level with Bmp7, and this translates into different responses of specific embryonic cells to Bmp2/Bmp4 activity as compared to Bmp7 (song *et al.*, 1998). However, in other embryonic cell types Bmp2, Bmp4, and Bmp7 elicit the same response, such as induction of *Msx1* in the neural plate (Furuta *et al.*, 1997; Shimamura and Robenstein, 1997). *Bmp7* (Fig. 2.3A–E), *Bmp4* (Gong and Guo, 2003; Nie *et al.*, 2006b), and *Bmp2* (Nie *et al.*, 2006b) are highly expressed in early orofacial primordia in areas that are adjacent or complementary to the expression domains of *Msx1* and *Msx2*, two established transcriptional targets of Bmp signaling in several developmental settings. However, early embryonic lethality or functional redundancy among Bmps has hindered investigations of their role during craniofacial development (Nie *et al.*, 2006b). Loss of function of Alk3 leads to failure of mesoderm formation, a phenotype that is more severe than the malformations generated by ablation of either *Bmp2*, *Bmp4*, or *Bmp7* (Nie *et al.*, 2006b). *Bmp5*/*Bmp7* double mutant embryos die at E10.5 and display a range of anomalies that reflect their sites of expression, including heart, branchial arch, allantois, forebrain, and somite defects (Solloway and Robertson, 1999).

While the role of Bmp5 and Bmp7 during lip and palate development is not accessible to study in the above double mutants, the fact that *Bmp7* (Fig. 2.3A–E) is expressed during development of these primordia suggests a role that could be similar to or different from that of Bmp2/Bmp4. That Bmp signaling plays a key role during lip and palate formation was demonstrated in mutants with partial loss-of-function of Bmp4 in the epithelium of

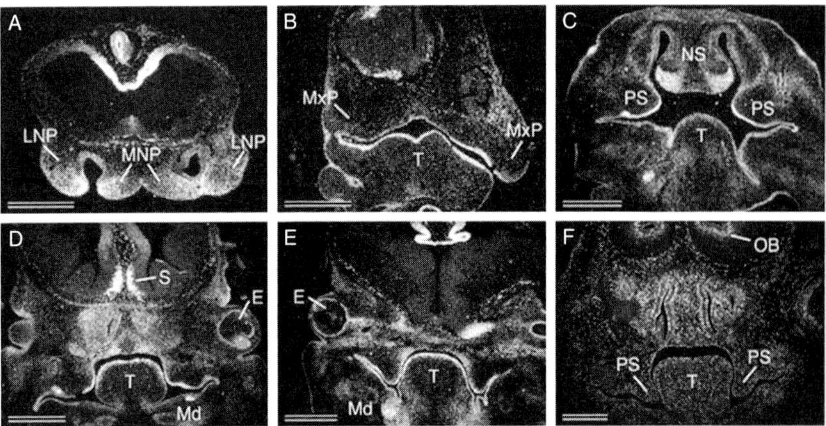

Figure 2.3 Dark-field images of sections from mouse heads. Sections hybridized with a *Bmp7* probe showing the patterns of expression of *Bmp7* mRNA (shiny areas) in lip and palate primordia (A–E). At E11 (A) and E11.5 (B), *Bmp7* is expressed in both the epithelium and mesenchyme of the medial (MNP) and lateral (LNP) nasal processes as well as in the MxP. At E12.5 (C–D), *Bmp7* is expressed in the palatal shelf (PS) and nasal septum (NS) epithelium and mesenchyme, in the anterior (C), middle (D) and posterior (E) palatal regions. Section showing tritiated thymidine labeling (shiny dots) indicating proliferating cells at E14. The PS contain numerous cycling cells during this vertical growth phase. Additional abbreviations: E, eye; Md, mandible; OB, olfactory bulb; S, telencephalic septum; T, tongue. Scale bars: 500 μm.

lip primordia and in mice deficient in *Bmprla* (*Alk3*) in the epithelium of lip primordia and in PS epithelium and mesenchyme. Partial epithelial loss of *Bmp4* generates CL which seems to be spontaneously repaired in some of the mutants (Liu *et al.*, 2005). Conditional *Alk3* mutants exhibit CL and CP, and the CL has been shown to be caused by increased apoptosis in the MNP (Liu *et al.*, 2005). While the *BMP4* gene in humans maps to chromosome 14q22–q23, a region implicated in nonsyndromic CL/P by linkage studies (Marazita *et al.*, 2004), *BMPR1A* and *BMP7* which map to human chromosomes 10q22 and 20q13, respectively, are not implicated in CL/P (Marazita *et al.*, 2004). In mice, *Bmpr1a* and *Bmpr1b* are coexpressed, however mutants lacking *Bmpr1b* are exempt of orofacial clefting (Juriloff and Harris, 2008 and references therein). Interestingly, deletion and linkage studies have implicated human *BMPR1B* in CL/P (Juriloff and Harris, 2008 and references therein). Altogether, these findings suggest an important role for Bmp signaling during normal and cleft lip formation.

Interpretation of developmental signals and their transduction into cellular responses involves a range of cellular effectors and transcription factors. In addition to Msx1 and Msx2, several transcription factors are expressed in the developing orofacial structures where they play a determinant role in their formation. Genetic removal of *Tcfap2a* a gene encoding different isoforms of AP2-α, a transcription factor highly expressed in CNCC, leads to severe malformations, including midline clefts, exencephaly, and other aberrations (Schorle *et al.*, 1996; Zhang *et al.*, 1996). Interestingly, some chimeric embryos harboring a mix of *Tcfap2a* mutant and wild type cells show isolated CL/P (Nottoli *et al.*, 1998). On the other hand, *Wnt1-Cre*-mediated deletion of *Tcfap2a* in neural crest cells and their derivatives generates CPO and skeletal anomalies (Brewer *et al.*, 2004). Given its chromosomal location, defects in human *TFAP2A* gene could be implicated in nonsyndromic CL/P (Juriloff and Harris, 2008). Interestingly, a recent study showed that deletions or mutations in the *TFAP2A* (*TCFAP2A*) gene cause the clefting disorder Branchio-Oculo-Facial syndrome, which is characterized by skin regional defects, ocular anomalies, ectodermal defects affecting hair and teeth, and frequent CL and CP (Milunsky *et al.*, 2008).

In addition to multiple anomalies, unilateral or bilateral CL and CP or clefting restricted anteriorly (as a result of lack of fusion of the PS with each other and with the primary palate) have been reported in mutants lacking the function of Sox11, a high-mobility-group domain containing transcription factor expressed in neural crest derivatives as well as in other tissues, including the nervous system, facial primordia and limbs (Sock *et al.*, 2004). Although the mechanisms leading to clefting in the *Tcfap2a* and *Sox11* mutants are not clear, their effects might be connected because of the presence of Sox binding sites within a *Tcfap2a* enhancer that drives expression of this gene in the orofacial region (Donner and Williams, 2006). It would be interesting to unveil the identity of the regulators and downstream targets of *Sox11* during lip and palate formation.

Other models for CL/P include mutants lacking *Folr1* (*Folbp1*) and the *Dancer* and *Twirler* mice. Mice with loss-of-function of *Folr1*, the gene encoding a folate binding membrane-bound protein have multiple malformations including median clefting, unilateral or bilatreal CL/P, or combinations of the above (Piedrahita *et al.*, 1999; Spiegelstein *et al.*, 2004; Tang and Finnel, 2003). The semidominant spontaneous mutations, *Dancer* and *Twirler*, generate nearly completely penetrant CL/P in homozygotes. Interestingly, lip and palate clefting in *Dancer* mice has been shown to result from ectopic expression of *Tbx10* subsequent to insertion of a heterologous promoter (Bush *et al.*, 2004). As for the *Twirler* mutation, which generates CL/P or CPO in homozygous mice (Gong and Eulenberg, 2001; Gong *et al.*, 2000), recent studies have suggested *Mkx* (previousely *Irxl1*, *Iroquois-related homeobox like1*) as a candidate gene (Liu *et al.*, 2006). These findings pave the way for future research to determine the cellular and molecular alterations underlying the clefting in *Twirler* and *Dancer* mice.

Midline facial and median lip/palate cleftings are encountered in a number of mutants and, as stated above, these are usually associated with severe defects in CNCC and/or their progeny. These include loss-of-function of genes encoding members of the Aristaless-like homeobox transcription factors in *Alx4; Cart1* (Qu *et al.*, 1999), or *Alx3; Alx4* (Beverdam *et al.*, 2001) compound mutants, platelet-derived growth factor (*Pdgfc;Pdgfa*) (Ding *et al.*, 2004), and *Plekhal*$^{-/-}$; *Pdgfra*$^{+/-}$ compound mutants (Schmahl *et al.*, 2007) which replicate loss-of-function of Pdgfr-α (Tallquist and Soriano, 2003), *Insig1; Insig2* compound mutants which accumulate sterols in their tissues (Engelking *et al.*, 2006), as well as retinoic acid receptor genes (Lohnes *et al.*, 1994).

Mutants lacking the function of the proto-oncogene Ski develop mild to severe midfacial clefting involving the lip, primary palate, and nose (Berk *et al.*, 1997; Colmenares *et al.*, 2002). This factor is known to act as a repressor of Tgfβ and Bmp signaling (Sun *et al.*, 1999; Wang *et al.*, 2000; Xu *et al.*, 2000) and has been shown to negatively regulate the activities of both the repressor and activator forms of Gli3, a member of the Hedeghog signaling pathway (Dai *et al.*, 2002). These observations incite the question of whether the clefting of the *Ski* mutant mice is caused by an abnormal increase in Tgfβ or Bmp signaling or is engendered by altered functions of Gli3 activator or repressor forms, or all/none of the above. Human *SKI* has been implicated in nonsyndromic CL/P (Table 2.2) and has been suggested as a candidate gene for the clefting defects in humans with monosomy 1p36 (Colmenares *et al.*, 2002).

The lethal *legless* (*lgl*) mutation causes multiple developmental anomalies with low frequency of CL/P and involves three contiguous genes, *Sp4*, *lrd*, and *Sp8*, with the first two being deleted and the third one being disrupted by insertion of a transgene. As indicated by the phenotype of *Sp8* mutants generated by targeted gene ablation and data from a complementation test, the limb and craniofacial defects of *lgl* mutants are secondary to a transgenic insertion generating a hypomorphic allele of the *Sp8* gene (Bell *et al.*, 2003). Both the *lgl* and *Sp8* mutants have midfacial clefts involving the lip and palate as well as forebrain anomalies in addition to other malformations. The midfacial clefting has been suggested to be secondary to neural tube defects. *Sp8* is expressed in the MNP and LNP but not in the MxP and, might thus, have a minor role in lip formation. The cleft palate of *lgl* and *Sp8* mutants might also be secondary to craniofacial skeletal defects. *Sp8* has been shown to be a target of Wnt/β-catenin and Bmpr1a signaling that mediate apical ectodermal ridge formation in the developing limb (Bell *et al.*, 2003). Given the involvement of Wnt and Bmpr1a in cleft lip, similar regulation of *Sp8* by these factors might occur in the developing lip as well.

The amplitude of signalings is tightly regulated, and in several instances too much or too little of signaling generates congenital malformations. This is well illustrated in mutants lacking the function of inhibitor of β-catenin and T-cell factor (ICAT), which interferes with the interaction between β-catenin and T-cell factor (TCF). *Icat* mutants display either a severe midfacial clefting and truncation of the rostral head or CP and a minor median CL. The craniofacial defects are caused by apoptosis subsequent to aberrant activation of Wnt signaling in CNCC derivatives (Satoh *et al.*, 2004).

The above models displaying median cleft lip/palate are similar to midline facial clefting, that are found in some human syndromes.

4.2. Mouse models for cleft of the secondary palate

The sequences of development of the mammalian secondary palate were already established at the time the subject was reviewed by Peter in 1924. Although further research is required for better understanding of the mechanisms behind palate formation and the etiopathogenesis of CPO, a substantial amount of new insights into these issues have been gathered during the last two decades as a consequence of the availability of mouse models and use of new investigative technologies.

Palate development is a multistep process requiring adequate growth of the PS, their elevation on schedule, adhesion of the MEE of the bilateral PS, formation and subsequent degeneration of the MES, and finally differentiation of a subset of mesenchymal cells into osteoblasts or muscle cells. Alterations in these developmental events generate a patent CPO or a submucous cleft. The latter consist of a reduction/absence of bone or muscles in the hard and soft palates, respectively. A constellation of genes encoding signaling molecules, receptors, downstream effectors, transcription factors, scaffolding proteins, protein modifiers, extracellular matrix (ECM) components, and other cellular factors have been shown to generate CPO in mice (Table 2.3) following loss-of-function, spatiotemporal misexpression or overexpression. However, some genes are not expressed during palate development, and thus their altered function generates CPO as a consequence of craniofacial skeletal malformations or tongue anomalies. Others cause CPO subsequent to intrinsic cellular and molecular disruptions within the PS which impede their growth, elevation or fusion.

4.2.1. Cellular and molecular mechanisms of palatal shelf growth

Embryogenesis and organogenesis are the products of interactions between adjacent tissue layers leading to specification, migration, proliferation, death or survival, and differentiation of cells. The vertical growth phase of the PS implies cell survival and active cell proliferation (Fig. 2.3F). These are achieved as a result of reciprocal interactions between the PS epithelium and mesenchyme. Epithelial–mesenchymal interactions are instrumental during development of several organs as diverse as hair follicles, teeth, glands, limbs, etc., and molecular and structural alterations in the epithelium affect the behavior of the mesenchyme and vice versa.

Decreased cell proliferation or increased cell death in the PS generate hypotrophic PS that are either unable to elevate as a consequence of a severe size reduction or elevate but are unable to contact each other, thus resulting in CPO. Several mouse models have unveiled key players during PS growth phase (Table 2.3).

4.2.1.1. Cell cycle regulation, genome stability and ribosome biogenesis Genomic instability generates not only cancer but also birth defects. This is illustrated by clefting following exposure of experimental animals to ionizing radiation as well as humans with chromosomal instability caused by genetic disorders, such as Fanconi anemia and immunodeficiency-centromeric instability-facial anomalies syndrome. The autosomal recessive type II Rothmund–Thomson syndrome (RTS) is another genetic disorder due to chromosomal instability. It is characterized by congenital skin and skeletal anomalies, genomic instability, and increased susceptibility to neoplasia (Mann *et al.*, 2005 and references therein). Mice lacking the function of a RecQ DNA helicase, Recq14, replicate the phenotype of humans with type II RTS (Mann *et al.*, 2005). Interestingly, *Recq14* mutants display a spectrum of palatal defects ranging from a complete to minor palatal clefting and patterning aberrations of rugae palatinae. The mutants have no other craniofacial anomalies, implying that CPO is caused by defects within the PS as further indicated by the abnormal rugae. Cells from *Recq14* mutants show increased frequencies of premature centromere separation and aneuploidy, suggesting a crucial role for Recq14 in sister chromatid cohesion, and that chromosomal instability may be the cause for congenital malformations and cancer predisposition in these mutants (Mann *et al.*, 2005). In which way loss-of-function of Recq14 affects palatogenesis is at present not known.

Recently, mice lacking the sister chromatid cohesion protein PDS5B have been found to display CPO, micrognathia, and a range of anomalies reminiscent of Cornelia de Lange syndrome (CLS) (Zhang *et al.*, 2007). However, like humans with CLS, no defects in sister chromatid cohesion were detected in chromosomes from *Pds5B* mutant mice. This could be due to functional redundancy between PDS5B and its homolog, PDS5A. At present the exact function of PDS5B is not known. However, PDS5B harbors a DNA-binding domain, localizes to the nucleolus, and is expressed in post-mitotic neurons. Together with the multiple anomalies in this mouse model and in humans with CLS, these observations point to a role for PDS5B and the cohesin complex beyond that in chromosomal segregation. The CPO in *Pds5B* mutants may be secondary to abnormal growth of the PS, as they appeared smaller than those of control mice. Together with the micrognathia, the small size of the PS may have contributed to their delayed elevation. The localization of PDS5B in the nucleolus has been suggested to be linked to a function in ribosome biogenesis (Zhang *et al.*, 2007).

TCOF1 encoding the nucleolar phosphoprotein Treacle, implicated in ribosomal biogenesis, is mutated in the autosomal dominant Treacher–Collins syndrome (TCS) (Edwards *et al.*, 1997; Valdez *et al.*, 2004).

Haploinsufficiency of *Tcof1* in mice reproduces the craniofacial anomalies in TCS, including CPO. These have been shown to be secondary to defects in formation, proliferation and survival of CNCC (Dixon *et al.*, 2006). These defects have previously been suggested to be secondary to altered ribosomal biogenesis (Dixon *et al.*, 2006). However, recent evidence from studies of *Tcof1*$^{+/-}$ mice demonstrated that apoptosis and altered proliferation of CNCC was caused by stabilization of p53, a tumor suppressor and a cell cycle checkpoint protein, and overexpression of p53-responsive genes, including among others, *Ccng1* encoding cyclin G1 which arrests cells at the G1 phase of the cell cycle. This activation correlated with cell cycle arrest and caspase3-mediated apoptosis in CNCC progenitors, thus accounting for the CNCC deficiencies that characterize TCS (Jones *et al.*, 2008). Importantly, pharmacological inhibition of p53 as well as genetic inactivation of *Trp53* rescued the craniofacial defects of *Tcof1*$^{+/-}$ mice. This occurred despite decreased ribosomal biogenesis, implying that p53-dependent apoptosis, rather than abnormal ribosomal biogenesis, is behind the craniofacial malformation (Jones *et al.*, 2008). A model has been suggested in which nucleolar stress subsequent to reduced ribosomal biogenesis caused by *Tcof1*$^{+/-}$ deficiency leads to p53 stabilization followed by cell cycle arrest and apoptosis of CNCC and their progeny. Although growth deficiency might underlie palatal clefting in the *Tcof1*$^{+/-}$ mutants, there is also a possibility that clefting is the result of craniofacial skeletal malformations. Thus, rescue of the CPO by inactivation of p53/*Trp53* might result from the general rescue of the other craniofacial skeletal defects.

Another model in which genetic removal of *Trp53* rescues the CPO is that of mice lacking *Ovca1*, the ortholog of human *OVCA1* which encodes a tumor suppressor (Chen and Behringer, 2004). While *Ovca1* heterozygotes spontaneously develop cancer, the *Ovca1* or *Ovca1–2* homozygous mutants show a range of anomalies, including CPO, reduced body size, limb defects, lung immaturity, and liver degeneration. These malformations are predominantly due to loss of function of *Ovca1*, since *Ovca1–2* mutants show an identical phenotype to *Ovca1* mice. Data from mouse embryo fibroblasts (MEF) from *Ovca1* mutants demonstrated that *Ovca1* is necessary for cell cycle progression. In addition to lethality at birth, some mutants died before or after E13.5, depending on the genetic background, but the cause of death is not known. Importantly, CPO was rescued only in *Ovca1*; *p53* double homozygous mutants that exhibited normal body size. This suggests that growth and proliferation defects are the cause of CPO following loss of *Ovca1*. However, further study of palate development both before and after *Ovca1* mutant rescue is warranted. Interestingly, CPO, limb anomalies, and reduced body size found in *Ovca1* mice are also part of the phenotype of mutants lacking *Hic1* (*Hypermethylated in cancer1*), encoding a transcription factor that interacts with C-terminal binding protein and recruits histone deacetylases to repress transcription (Carter *et al.*, 2000).

HIC1, *OVCA1*, *CRK* which encodes an adaptor protein through two splice variants, *CRKI* and *CRKII*, and *MNT* encoding a Mad-family bHLH transcription factor regulating cell proliferation, differentiation, and survival, are amongst genes at 17p13.3 that have been identified or proposed to be deleted in a heterozygous fashion in Miller–Dieker syndrome (MDS). This syndrome is characterized by brain, body wall closure, limb, and craniofacial anomalies (Carter *et al.*, 2000; Chen and Behringer, 2004; Park *et al.*, 2006; Toyo-Oka *et al.*, 2004; Yingling *et al.*, 2003). Importantly, the mouse counterparts of most of the known or predicted human genes of the MDS critical region are situated in a similar order on mouse chromosome 11. Like *Hicl* and *Ovca1* mutants, mice lacking *Crk* (Park *et al.*, 2006) or *Mnt* (Toyo-Oka *et al.*, 2004) display craniofacial anomalies, including CPO. Mnt/Max heterodimers are found in proliferating cells containing Myc/Max heterodimers, and Mnt has been suggested as a putative antagonist of Myc. The causes of growth defects in *Mnt* mutants are at present unclear. Nevertheless, the PS in *Mnt* mutants are hypotrophic and show delayed elevation, suggesting that CPO is secondary to either PS growth defect or their delayed elevation caused by mandible hypolasia or both.

In addition to palatal defects, *Crk* mutants exhibited focal edema, hemorrhages, and cardiovascular defects (Park *et al.*, 2006). The palatal defects in *Crk* mutants ranged from a complete cleft to delayed removal of the MES and development of edema on the nasal surface of the MES. In addition, huge edema formed in the nasal septum. The defects in *Crk* mutants are likely caused by altered vascular integrity, resulting in dilatation and rupture of blood vessels (Park *et al.*, 2006). The developmental phenotype in *Crk* mutants has been suggested to be predominantly due to loss of function of *CrkI* (Park *et al.*, 2006), since removal of *CrkII* in a gene trap mutant with intact *CrkI* has no effects (Park *et al.*, 2006 and references therein). The Crk family plays a key role in regulating cell shape, in reorganizing the cytoskeleton during cell migration, and in integrating growth and adhesion signals at focal adhesions (Yingling *et al.*, 2003 and references therein). Although the exact role of *Crk* during palatogenesis remains to be defined, the anomalies in the *Crk* mutants point to an important function for this gene in this process.

In addition to targeted mutations, mouse *N*-ethyl-*N*-nitrosourea (ENU) mutagenesis programmes yielded a large number of mutants to study gene function. Among these, mutations of *Chd7*, the mouse orthologue of the human *CHD7* gene implicated in CHARGE syndrome, generated abnormalities found in CHARGE syndrome patients (Bosman *et al.*, 2005). Heterozygous *Wirligig* (*Whi*/+) mutants exhibit CPO (in 35% of mutants), choanal atresia, failure of closure of the interventricular septum, inner ear, ocular and genital defects (Bosman *et al.*, 2005). Some *Whi*/+ mutants display complete CPO with the bilateral PS either widely spaced or unfused though touching one another, and others display a minor localized cleft (Bosman *et al.*, 2005), indicating growth defects, delayed elevation or failure of fusion of the PS. Unfortunately, it is not possible to distinguish between the two mechanisms, as only scanning electron microscopy data were shown in this study. Therefore, histological sections combined with molecular analyses are required to unravel the mechanisms behind CPO in these mutants. *Chd7* encoding the chromodomain helicase DNA-binding protein 7 is expressed in the epithelium of the developing palate (Bosman *et al.*, 2005) implying a direct role in its formation, and human studies suggest that *CHD7* may act as a modifier for nonsyndromic clefting (Felix *et al.*, 2006). Thus, further studies of this gene during palatogenesis are worthwhile.

Failure of cells to exit the cell cycle generates neoplasia or developmental defects. In the latter, when the cell cycle goes awry, the apoptotic machinery is activated. This is exemplified in the retina and pancreas of mice lacking $p57^{kip2}$ where failure of cell cycle exit leads to apoptosis (Dyer and Cepko, 2000; Georgia *et al.*, 2006). $p57^{kip2}$ is a member of the Cip/Kip family of mammalian cyclin dependent kinase (Cdk) inhibitors, which inhibit G1-cyclin-cdk complexes by blocking the activity of the cdk subunit. Mice lacking the function of $p57^{kip2}$ develop a CPO in addition to other defects (Caspary *et al.*, 1999; Yan *et al.*, 1997), and this anomaly has been suggested to be a result of increased apoptosis of PS epithelial and mesenchymal cells (Yan *et al.*, 1997). However, apoptosis in this mouse model might be secondary to cellular hyperproliferation and may not be the primary cause for palatal clefting, since increased cell proliferation can generate CPO as shown in *Spry2* mutants (Welsh *et al.*, 2007; discussed in Section 4.2.2.2) and in *Menin*-deficient mice (Engleka *et al.*, 2007). In fact, sustained cell proliferation and apoptosis was found in the developing lens of $p57^{kip2}$ mutants (Zhang *et al.*, 1997). $p57^{kip2}$ deficient mice show several defects that characterize human Beckwith–Wiedemann syndrome (BWS), including CPO, indicating the involvement of $p57^{kip2}$ in BWS and a function for this factor during palatogenesis.

Menin is a nuclear tumor suppressor expressed ubiquitously. Loss-of-function of *MEN1* causes multiple endocrine neoplasia type I syndrome, characterized by tumors affecting endocrine tissues (Marx, 2005). Mice heterozygous for *Menin* deletion develop endocrine tumors similar to human *MEN1* tumors (Argawal *et al.*, 2005). Menin is involved in the regulation of an array of cellular activities, including gene expression, cell cycle, and apoptotic events as well as genome stability (Argawal *et al.*, 2005). Loss-of-function of Menin in murine neural crest cells and their derivatives following *Pax3-Cre-* or *Wnt1-Cre*-mediated inactivation causes complete CPO or deficiency of the soft palate. Anomalies in the basisphenoid bone were detected in both cleft and noncleft mutants (Engleka *et al.*, 2007). The PS elevated on schedule, however they were unable to abut because of their reduced size. Analyses disclosed a net reduction in the expression of $p27^{kip1}$, a Menin transcriptional target and a cdk inhibitor, which likely led to hyperproliferation within the PS mesenchyme. As a consequence, the PS became hyperplastic and displayed a reduction in the extracellular space (Engleka *et al.*, 2007). Growth of the PS results from both cell proliferation/survival and expansion of the ECM. Therefore, the decreased size (hypotrophia) of the PS of *Menin*-deficient mice could be a result of abnormal ECM deposition by cells that are busy proliferating. Other factors besides $p27^{kip1}$ may also be deregulated following Menin loss-of-function since tissue hyperplasia but not CPO occurs in mutants lacking $p27^{kip1}$ function.

4.2.1.2. The cytoskeleton The X-linked gene *FLNA* encodes filamin A, a cytoskeletal protein involved in crosslinking actin filaments into orthogonal networks. Filamin A interacts with a range of proteins, including transmembrane receptor complexes, integrins and ion channels. Mutations in *FLNA* give rise to a range of defects in humans (Robertson, 2005). The ENU-induced mouse mutation *Dilp2* has recently been shown to involve a nonsense mutation in exon 44 of the *Flna* gene resulting in a truncated protein, and the *Dilp2* phenotype has been suggested to be due solely to the mutation in the *Flna* gene (Hart *et al.*, 2006). Mutant males and heterozygous females show low levels or absent mutant transcripts, respectively, as a consequence of nonsense-mediated decay (Hart *et al.*, 2006). Mutant males survive until E15.5 and display a range of cardiovascular defects, whereas some of the carrier females have cardiac defects. In addition, mutant males and carrier females exhibit incomplete sternal fusion and CPO. In carrier females, the PS elevate but are unable to abut, whereas in mutant males the hypotrophic PS remain vertically oriented.

Filamin A is involved in cell migration, however the phenotype of the mutant mice as well as the behavior of *Flna*-deficient MEF rules out migration defects (Hart *et al.*, 2006). Given its role in cytoskeletal rearrangement, Filamin A may be crucial for cell polarization and integration of external stimuli, these issues can be explored with this interesting mouse model.

4.2.1.3. Fgf, Bmp and Tgfβ signaling pathways The requirement of Fgf signaling for craniofacial and palate formation is well illustrated in mice lacking the function of Fgf8 (Abu-Issa *et al.*, 2002; Frank *et al.*, 2002), Fgf9 (Colvin *et al.*, 2001), Fgf10 (Alappat *et al.*, 2005; Rice *et al.*, 2004), Fgf18 (Liu *et al.*, 2002; Ohbayashi *et al.*, 2002), Fgfr1 (Trokovic *et al.*, 2003), Fgfr2b (De Moerlooze *et al.*, 2000; Rice *et al.*, 2004), and in mutants with gain-of function of Fgfr2c (Eswarakumar *et al.*, 2004). However, several of these mutants have major craniofacial skeletal anomalies, and CPO may thus be a secondary event. These include mutants lacking *Fgf8, Fgf18, Fgfr1,* and *Fgfr2c.*

The mechanism of action of Fgf signaling during palate growth are better understood in mice harboring null mutations of *Fgf10* and *Fgfr2b* (Alappat *et al.*, 2005; Rice *et al.*, 2004). In these, CPO is a result of lack of growth of the PS subsequent to reduced cell proliferation of the PS epithelium and mesenchyme and epithelial apoptosis. These mouse models revealed that Fgf10–Fgfr2b interaction coordinates an epithelial–mesenchymal signaling loop, in which Fgf10 emanating from the mesenchyme binds to Fgfr2b in the PS epithelium. This interaction is crucial not only for epithelial proliferation and survival, but it is also necessary for expression of *Shh*. In turn, epithelial Shh activates PS mesenchyme proliferation (Rice *et al.*, 2004).

However, increased Fgf signaling is also detrimental to palate development as revealed in mice lacking the function of *Spry2*, an Fgf antagonist (Welsh *et al.*, 2007; discussed in Section 4.2.2.2) and following loss-of-function of the Short stature homeobox2 transcription factor (Shox2) (Yu *et al.*, 2005). *Shox2* null mice develop CPO, but the clefting is confined to the anterior hard palate, whereas the soft palate is intact (Yu *et al.*, 2005). This form of clefting is found in both humans and animals (Schüpbach, 1983). The CPO in *Shox2* mutants is caused by increased apoptosis in the epithelium and reduced cell proliferation in the epithelium and mesenchyme of the anterior palate (Yu *et al.*, 2005), and these cellular defects are likely consequent upon abnormal expression of *Fgf10* and *Fgfr2c* in this developing palatal region. Thus, both increased and decreased Fgf signaling leads to altered cell proliferation and apoptosis.

It seems that Bmp inputs are necessary but not sufficient for the confined expression of *Shox2* in the anterior part of palate, and thus *Shox2* induction requires multiple upstream inducers (Yu *et al.*, 2005). Additional mechanistic insights into the function of Shox2 were gathered from the study of mice lacking *Shox2* in CNCC derivatives following *Wnt1-Cre*-mediated recombination (Gu *et al.*, 2008). The *Shox2* conditional mutants developed a cleft due to failure of fusion of the primary palate with the secondary palate, but unlike the *Shox2* null mice (Yu *et al.*, 2005) they survived for several days, although they eventually succumbed, probably due to malnutrition (Gu *et al.*, 2008). The PS of the secondary palate showed delayed closure due to reduced cell proliferation at E13.5 when they were vertically oriented. However, by E15.5 they recovered cell proliferation profiles similar to control mice at E13.5. The *Shox2* mutant PS mesenchyme exhibited reduced expression of *Sox9*, a transcription factor that seems to be required for palatogenesis, as indicated by *Sox9* mutant mice (Bi *et al.*, 2001; Mori-Akiyama, *et al.*, 2003). Interestingly, although the hard palate eventually managed to close, it exhibited reduced osteogenesis, probably subsequent to reduced production of the osteogenetic transcription factors Runx2 and its target Osterix. Further *in vivo* and *in vitro* analyses of *Shox2* conditional mutants suggested that reduced expression of *Runx2* and *Osterix* was secondary to increased Bmp2 signaling, thus echoing findings with Bmp4 in the developing limb (Yu *et al.*, 2007). To conclude, Shox2 requires Bmp2 inputs to be expressed, and it seems to regulate Bmp and Fgf signaling. Given the connections of these signaling pathways with other key players in palatogenesis, such as Msx1 and Shh, Shox2 emerges as a factor exerting a central role in the developing hard palate.

Runx2 deficient mice show CPO, failure of eye lid fusion, tooth and skeletal anomalies (Åberg *et al.*, 2004a; Ducy *et al.*, 1999), and this gene has been implicated in nonsyndromic CL/P (Table 2.2). Although the CPO in *Runx2* mutants may be secondary to craniofacial skeletal defects, this gene is indeed expressed in the PS (Figs. 2.5G–I) which indicates some role in the palate proper (see also Section 4.2.2.2). In the developing tooth, Runx2 mediates Fgf signaling from epithelium to mesenchyme, and expression of *Runx2* necessitates the activity of Msx1 (Åberg *et al.*, 2004b). At least in osteoblast progenitors, Satb2, an important player in palate formation, has been shown to interact with Runx2 and to enhance its activity (Dobreva *et al.*, 2006).

The Tgf*β* superfamily is highly implicated in palate development in both humans and mice. Nestin-Cre-mediated inactivation of *Bmpr1a* (*Alk3*) in both the epithelium and mesenchyme of the developing lip and palate generates bilateral CL and CP (Liu *et al.*, 2005), and the clefting of the palate has been suggested to be secondary to altered cell proliferation and misexpression of the transcription factors *Barx1* and *Pax9*. By contrast, inactivation of *Alk3* in CNCC derivatives, including the PS mesenchyme generates a CPO, possibly as a result of skeletal anomalies (Dudas *et al.*, 2004b). Inactivation of *Bmpr1a* in ectodermal derivatives such as teeth, palate, and hair follicles severely impairs hair and tooth development, whereas the palate seems to be intact (Gritli-Linde, 2007 and references therein), indicating that the main function of Bmpr1a resides within the PS mesenchyme. The PS mesenchyme expresses *Msx1*, a target of Bmp signaling, and null mutants for *Msx1* display CPO owing to reduced cell proliferation (Satokata and Maas, 1994; Zhang *et al.*, 2002). Interestingly, Msx1 has been found to be necessary for the expression of *Bmp4* and/or *Bmp2* in both the developmentally compromised tooth and palate primordia of *Msx1* mutants (Zhang *et al.*, 2000, 2002). Moreover, recombinant Bmp4 or a *Bmp4* transgene expressed in palate mesenchyme have been shown to rescue the tooth anomalies and CPO, respectively, in *Msx1* mutants (Bei *et al.*, 2000; Zhang *et al.*, 2002).

The role of Tgf*β* signaling during the growth phase of the PS has recently been unveiled in mice with conditional inactivation of Tgf*β* receptors (*Tgfbr*) in CNCC derivatives (Dudas *et al.*, 2006; Ito *et al.*, 2003). The secreted Tgf*β* peptides signal by activating the serine/threonine kinase quaternary complex consisting of two type I and two type II receptors. Inactivations of the type I receptor (*Alk5*) and the type II receptor

(*Tgfbr2*) following *Wnt1-Cre*-mediated recombination generate craniofacial anomalies that are more severe in the *Alk5*-deficient mice (Dudas *et al.*, 2006) than in *Tgfbr2* mutants (Ito *et al.*, 2003). This suggests that Alk5 is involved in mediating signaling by other ligands in addition to Tgfβ1–3, and that Alk5 is capable of functioning with type II receptors other than Tgfβr type II (Dudas *et al.*, 2006). The differences between the signaling activities of these two receptors are further revealed in the PS mesenchyme which was hyper-proliferative and underwent a massive apoptosis in *Alk5* mutants (Dudas *et al.*, 2006), while in *Tgfbr2* mutants, it displayed reduced cell proliferation (Ito *et al.*, 2003). Over-activation of Tgfβ signaling, as in humans carrying mutations in *TGFBR1* and *TGFBR2*, generates CPO (Loeys *et al.*, 2005). Thus, both loss-of-function and gain-of-function of Tgfβ signaling impairs palatogenesis, echoing situations in other signaling pathways and reiterating the requirement for a delicate balance in signaling activities for normal organogenesis.

*4.2.1.4. **Shh signaling and cholesterol metabolism*** Shh is a member of the Hedgehog family of signaling molecules. The Hedgehog pathway is complex and involves a multitude of components that function in processing of the ligands, their release, trafficking, reception, range of activity as well as in signal transduction (Wang *et al.*, 2007). As indicated by analysis of the developing palate in *Fgf10* and *Fgfr2b* mutant mice, Shh appears to be a necessary mitogen for the PS mesenchyme (Rice *et al.*, 2004). Furthermore, *in vitro* manipulation has shown that Bmp2 is a mesenchymal mitogenic factor that is induced and maintained by Shh (Zhang *et al.*, 2002). Shh is a powerful mitogen, but it is also crucial for cell fate specification, survival and differentiation in several developing organs. Unfortunately, sustained activation of the Hedgehog pathway leads to neoplasia (McMahon *et al.*, 2003, Wang *et al.*, 2007). As an indication of the importance of Shh signaling during development, murine null mutation of *Shh* causes the most severe form of HPE as well as defects in multiple other organs (Chiang *et al.*, 1996). The severity of the craniofacial defects (Fig. 2.4B) hinders study of Shh during palate development. However, inactivation of *Shh* specifically in epithelia using the *Cre/loxP* system allows fairly normal craniofacial development (Rice *et al.*, 2004). Mice lacking Shh in the PS epithelium display altered Shh signaling in both the epithelium and mesenchyme, as Shh is known to signal both short and long range (Gritli-Linde *et al.*, 2001) and components of the Shh signaling pathway are expressed in both the PS epithelium and mesenchyme (Rice *et al.*, 2006). Importantly, conditional *Shh* mutants have a wide CPO (Figs. 2.4D and F; Rice *et al.*, 2004). The clefting appears to result from impaired Shh signaling in the mesenchyme, since mutants harboring a nonfunctional Smoothened (Smo), an obligate transducer of all Hedgehog signaling, in the PS epithelium have normal

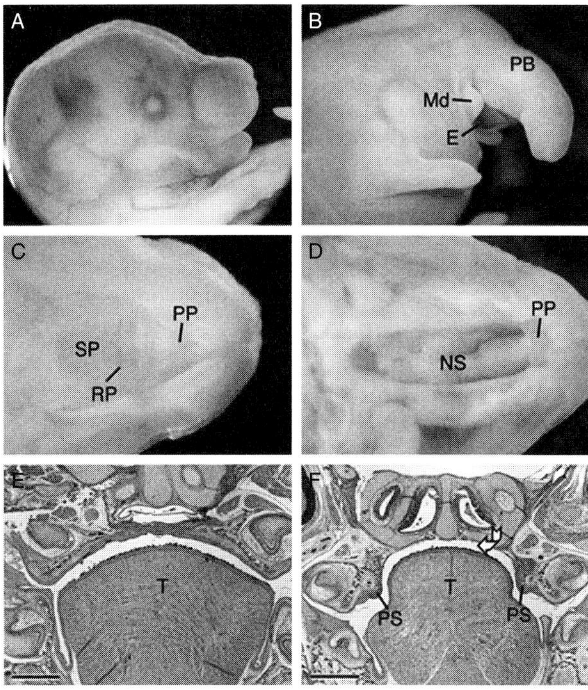

Figure 2.4 Craniofacial malformations following loss-of-function of Shh in the mouse. Wild type (A) and *Shh*$^{-/-}$ mutant (B) embryos at E18.5. The mutant displays a severe form of holoprosencephaly. Note the absence of maxilla, the formation of a rudimentary mandible (Md) and formation of a proboscis (PB) as a result to defects of the frontonasal process. The fused eyes (E) are located under the proboscis. Views of the palates of E18.5 wild type (C) and *K14-Cre*; *Shh*$^{n/c}$ mutant (D). Conditional loss-of-function of Shh in the palate generates a wide cleft of the secondary palate (D). Sections from mouse heads at E18.5 showing a fused secondary palate in a wild type embryo (E) and a wide cleft palate (arrow) in a *K14-CreShh*$^{n/c}$ mutant. The mutant displays severely hypoplastic PS that remained in a vertical position. Abbreviations: NS, nasal septum; PP, primary palate; RP, rugae palatinae; SP, secondary palate; T, tongue. Scale bars: 500 μm. (See Color Insert.)

palate development. Supporting these suggestions, Shh has been shown to stimulate PS mesenchyme explants *in vitro* (Rice *et al.*, 2004).

Fidget mice have defects in *Fign*, the gene encoding fidgetin, as a result of retrotransposon-insertion mutations that interfere with RNA processing (Cox *et al.*, 2000). Fidgetin is a member of the AAA proteins (ATPases associated with diverse cellular activities) that function as chaperones involved in multiple cellular activities, including endosome sorting, vesicle-mediated transport, proteasome function, peroxisome biogenesis, and microtubule regulation (Lupas and Martin, 2002). Fidgetin has been shown to physically interact and colocalize in the nuclear matrix with Akap95, a c-AMP-dependent protein kinase A (PKA) anchoring protein

(Yang *et al.*, 2006). A subset of *Fign* mutants develop CPO in addition to a circling behavior and microphtalmia. The number of embryos with CPO increased in *Fign* mutants harboring loss-of-function of *Akap95* (Yang *et al.*, 2006), providing a biological significance for the physical interaction between fidgetin and Akap95. The palatal defect in the above mutants has been suggested to be linked to altered function/subcellular localization of PKA and that these defects may lead to altered Shh signaling (Yang *et al.*, 2006). These assumptions are sound given the known function of PKA in inhibiting the Hedgehog signaling pathway (McMahon *et al.*, 2003) but warrant further studies.

In addition to the above mouse models, the importance of Shh signaling during craniofacial development is further well reflected in mutations of *SHH*, which cause HPE in humans (Muenke and Beachy, 2000). HPE is a spectrum of anomalies involving the forebrain and midline craniofacial structures with a varying severity even within members of the same pedigree. In the most severe forms, inflicted individuals display a small forebrain (prosencephalon) that fails to develop into two halves, defects in the eye field generating cyclopia, as well as midline facial clefting. The less severe forms, microform HPE, manifest as milder craniofacial malformations such as a close-set eyes (hypotelorism), premaxillary defects, and/or as a single median maxillary central incisor (Muenke, 1995; Muenke and Beachy, 2000). HPE is a complex disorder where genetic and environmental factors or both are implicated in its genesis. Recently, growth arrest-specific 1 (Gas1), a membrane glycoprotein, has been shown to play a positive and crucial role within the Shh pathway during craniofacial development. Mice harboring a targeted null mutation of *Gas1* have severe ear anomalies and show features of microform HPE, including premaxillary defects, fused upper incisors, and partial or complete CPO (Seppala *et al.*, 2007). In the developing palate and other affected craniofacial structures of *Gas1* mutants, cells situated at a distance from the source of Shh production showed dwindling responses to Shh signaling, indicating the ability of Gas1 to potentiate Shh activity. Interestingly, *Gas1* null mutants harboring one single functional *Shh* allele show exacerbation of their craniofacial defects, where among others the CPO phenotype became fully penetrant and the clefting became complete (Seppala *et al.*, 2007).

These findings, together with the fact that human *GAS1* maps to chromosome 9q21.3-q22, a region associated with nonsyndromic cleft palate (Marazita *et al.*, 2004), make *GAS1* a good candidate gene for nonsyndromic palatal clefting and other anomalies associated with this chromosomal region (Seppala *et al.*, 2007 and references therein).

HPE can occur in Pallister–Hall syndrome (PHS) and Smith–Lemli–Opitz syndrome (SLOS) (Muenke and Beachy, 2000). PHS is caused by mutations in another Hedgehog pathway component, GLI3, which are thought to lead to inhibition of the Shh pathway (Muenke *et al.*, 2000).

SLOS is one of eight human inborn disorders of cholesterol synthesis. SLOS is caused by mutations in the gene encoding 7-dehydrocholesterol (7-Dhc) reductase (Dhcr7) leading to cholesterol deficiency and accumulation of 7-Dhc (Porter, 2006). Interestingly, mice lacking a functional *Dhcr7* gene phenocopy the defects in SLOS patients, including CPO (Fitzky *et al.*, 2001; Wassif *et al.*, 2001). Other mouse models for human cholesterol synthesis pathway disorders include mice with a null mutation of the gene encoding lathosterol 5-desaturase (*Sc5d*) and the X-linked semidominant mouse mutation *Tattered* (*Td*), models for lathosterolosis and X-linked dominant chondrodysplasia punctata type 2 (CDPX2), respectively (Derry *et al.*, 1999; Krakowiak *et al.*, 2003). The *Td* mutation and the human CDPX2 are caused by mutations in the Δ^8–Δ^7 sterol isomerase emopamil binding protein (Ebp, encoded by *Ebp* in mouse) (Derry *et al.*, 1999). Similar to SLOS, defects in each of these two enzymes lead to reduced cholesterol synthesis and an accumulation of intermediate sterols upstream of the deficient enzyme (Porter, 2006). Furthermore, the hemizygous *Td* males and *Sc5d* mutant embryos display CPO in addition to other anomalies (Derry *et al.*, 1999; Krakowiak *et al.*, 2003, Table 2.3). The PS in *Dhcr7* mutants elevate but fail to abut, possibly as a result of their abnormal small size, because the mutants do not show detectable craniofacial skeletal anomalies (Wassif *et al.*, 2001).

Many organs affected in SLOS infants are those which require Hedgehog signaling, and specific inhibition of Dhcr7 causes HPE in mice and impairs induction of Shh target genes in neural plate explants (Incardona *et al.*, 1998). Interestingly, there is precedent showing that cells from mouse models of SLOS and lathosterolosis as well as normal cells pharmacologically depleted of sterols are unable to respond to Shh signaling as a consequence of altered Smo activity (Cooper *et al.*, 2003). Recently, cholesterol and specific cholesterol derivatives, oxysterols, have been shown to stimulate Smo activity. How this activation is triggered is presently unclear (Corcoran and Scott, 2006). A careful study of the spatiotemporal patterns of expression of Shh target genes in the developing palate of the above mouse models would certainly reveal whether the clefting is due to altered Shh signaling or not.

The malformations in humans and mouse models with defects in the cholesterol biosynthetic pathways have been suggested to be caused either by cholesterol deficiency or the accumulation of precholesterol sterol intermediates or both. Recently, this issue has been clarified, at least in one mouse model lacking the function of *Insig1* and *Insig2* (*Insig*-DKO) (Engelking *et al.*, 2006). Insig-1 and Insig-2 are membrane proteins resident in the endoplasmic reticulum (ER) membrane that are essential for feedback inhibition of cholesterol synthesis. Transcription of all enzymes in the cholesterol synthetic pathway is activated by the sterol regulatory element-binding proteins (SREBs). Transport of SREBs from the ER to the Golgi, where they are processed into active fragments able to enter the nucleus, is inhibited by cholesterol or oxysterols, thus blocking cholesterol synthesis. SREBs are transported from the ER to the Golgi by an escort protein called Scap. When cholesterol and oxysterol levels rise, Scap is retained in the ER by binding to Insig-1 or-2. Cholesterol acts by binding to Scap, thereby facilitating its binding to Insigs, whereas oxysterols bind to Insigs causing them to bind Scap (Espenshade and Hughes, 2007; Radhakrishnan *et al.*, 2007). *Insig*-DKO mice exhibit either a CPO or a midline cleft face involving the lip, premaxilla, palate, and nasal septum. Some of the facial cleft mutants show exencephaly. The PS in the mutants with CPO are hypotrophic and either fail to elevate or are delayed to do so (Engelking *et al.*, 2006). Importantly, maternal treatment with the HMG–CoA reductase (an early enzyme in the cholesterol synthetic pathway) inhibitor lovastatin, not only reduced sterol synthesis and decreased pre-cholesterol intermediates but also rescued the clefting in *Insig*-DKO fetuses (Engelking *et al.*, 2006). This indicates that the clefting is an indirect consequence of sterol accumulation rather than directly caused by Insig deficiency. Both abnormally high and reduced Shh signaling generate a range of orofacial clefting in humans and mice (Jeon *et al.*, 2004; Mo *et al.*, 1997; Muenke, 2002; Rice *et al.*, 2004; Seppala *et al.*, 2007). These observations, together with the role of specific oxysterols in Hh signaling, raise the question whether the clefting in *Insig*-DKO mice is due to spatiotemporal alterations of Shh signaling following inactivation or activation of Smo function by sterol precursors.

4.2.1.5. The Wnt signaling pathway Until recently (Chiquet *et al.*, 2008; Juriloff *et al.*, 2006; Lan *et al.*, 2006), the involvement of Wnt signaling in lip and palate normal and abnormal development remained unsung as compared to other signaling pathways. Evidence, however, suggests a role for this signaling pathway for normal growth of PS. *Wnt5a* null mice display a CPO (Yang *et al.*, 2003). However, this could be secondary to their abnormal craniofacial skeleton, therefore warranting further investigations. Some of the *Icat* mutants (discussed in Section 4.1) have CP with a minor median CL (Satoh *et al.*, 2004). Similar to midline facial clefts found in some of these mutants, CP might be caused by increased apoptosis in palate primordia as a consequence of aberrant activation of Wnt signaling (Satoh *et al.*, 2004). On the other hand, mutants lacking *R-spondin2* (*Rspo2*) encoding a secreted protein able to activate *β*-catenin signaling via the Frizzled/LRP5/6 receptor complex develop limb anomalies and craniofacial malformations, including CPO (Nam *et al.*, 2007). However, the molecular and morphological defects leading to CPO in *Rspo2* mutants are not known. *In vitro* manipulation of the mouse palate indicates that Fgfr1b-mediated inhibition of *Wnt11* expression in the PS epithelium is necessary for palatal growth (Lee *et al.*, 2008). *Pygo2* mice deficient in the function of a mammalian ortholog of the *pygopus* gene of *Drosophila*, which encodes an essential factor of the canonical Wnt signaling, display growth retardation, renal and ocular anomalies. Some embryos exhibit exencephaly and CPO (Schwab *et al.*, 2007). *Pygo2* mutants show decreased canonical Wnt signaling in the first branchial arch and other structures, including the developing kidneys. However, the molecular and morphogenetic defects leading to CPO in *Pygo2* mice need further studies.

Glycogene synthase kinase 3β (GSK3β) functions in several signaling cascades, including repression of the Wnt canonical pathway. Drug-dependent alleles are useful in studying the function of genes in a time-controlled manner. Recently, this technology has been implemented to create a rapamycin-dependent allele of *GSK3β* which produces GSK3β protein fused to a tag (FRB*) (FKBP/rapamycin binding). The tag has an amino acid domain that is thermally unstable, and thus such a tag fused to a protein renders it unstable and rapidly (<1 h) degradable. However, the instability of chimeric proteins is reversed by the binding of the drug rapamycin to the FRP* domain (Liu *et al.*, 2007; Stankunas and Crabtree, 2007). This technique has been exploited to determine the developmental temporal windows for GSK3β during palate and skeletal development (Liu *et al.*, 2007). Both mice homozygous for the tagged allele *GSK3$\beta^{FRP\star}$* and *GSK3β* null mice displayed craniofacial skeletal anomalies, CPO and sternal defects. The PS in these mutants were unable to meet at the midline. Interestingly, maternal rapamycin treatment induced stabilization of GSK3βFRP* protein and rescued the CPO when rapamycin was administered between E13.5 and E15. Maternal rapamycin treatment after E14.5 was unable to rescue the clefting. These data indicate that E13.5–E15 is the essential time window of GSK3β action during palatogenesis (Liu *et al.*, 2007).

However, the CPO in the above mutants could be a result of the craniofacial skeletal anomalies, and thus rapamycin-induced rescue of palatal clefting could be merely secondary to rescue of cranial bone defects. These two scenarios could be disentangled by checking the patterns of expression of *GSK3β* in the developing palate and determination of the molecular and cellular mechanisms leading to CPO in this mouse model, a pursuit that may be fruitful given the involvement of GSK3β in several signaling pathways implicated in palatogenesis. Wnt signaling during palatogenesis may also involve Ryk (discussed in Section 4.2.2.2). Further studies of mouse models are required to delineate the functions of Wnt proteins during lip and palate development.

4.2.1.6. A tiny SUMO with an impressive impact Posttranslational modification of proteins, following attachment of phosphate, lipid, acetyl or methyl groups, sugars or lipids, or covalent links to other small proteins such as ubiquitin greatly influence their activity, stability, and subcellular localization. Reversible protein modification by sumoylation following covalent attachment of a SUMO protein (small ubiquitin-related modifier) has emerged as an important event that controls a range of biological processes, including apoptosis, gene expression, cell proliferation, metabolism, genome integrity, organelle biogenesis, signal transduction, and ion transport (Meulmeester and Melchior, 2008). Mammalians have three SUMO proteins SUMO1, SUMO2, and SUMO3 capable of conjugating to proteins. Humans have a fourth *SUMO4* gene, however the function of its product is still unclear.

Recent findings indicate that SUMO1 plays a key role during palatogenesis in humans and mice. Human *SUMO1* haploinsufficiency engenders CL and CP as shown in an infant with a 2q breakpoint interrupting the *SUMO1* gene (Alkuraya *et al.*, 2006). In addition to RNA and protein analysis from the affected infant, further evidence for haploinsufficiency was demonstrated in *Sumo1*$^{Gt/+}$ heterozygous mice which displayed cleft palate or oblique facial cleft. SUMO1 involvement in palate development was further established by the demonstration of *Sumo1* expression in both the palatal epithelium and mesenchyme (Alkuraya *et al.*, 2006). Importantly, several direct or indirect key players in palate formation have been shown to be sumoylated. These include Satb2, Smad4, Msx1, Sox9, Eya1, p53, and p63 (Alkuraya *et al.*, 2006; Dobreva *et al.*, 2003; FitzPatrick *et al.*, 2003; Gupta and Bei, 2006; Huang *et al.*, 2004; Lin *et al.*, 2003; Rodriguez *et al.*, 1999; Taylor and Labonne, 2005; Lee *et al.*, 2006).

Thus, SUMO1's influence on key players of palate development may underlie sensitivity of this organ to *SUMO1/Sumo1* gene dosage. This was elegantly shown in compound heterozygotes for *Eyal* encoding a transcription factor and *Sumo1*, which displayed a significant increase in the occurrence of CPO (Alkuraya *et al.*, 2006). Eyal is not only a substrate for sumoylation by SUMO1 (Alkuraya *et al.*, 2006), it is also expressed in the developing palate and appears to play a role during palatogenesis as shown in mutants lacking *Eyal* (Xu *et al.*, 1999).

The involvement of sumoylation in regulating key players in palatogenesis has been further extended by the recent important demonstration of Tbx22 as a target for SUMO1 (Andreou *et al.*, 2007). Interestingly, this study showed that all CPX missense mutations studied compromise DNA-binding and result in loss of SUMO1 modification. Furthermore, Tbx22 sumoylation was shown to be necessary for its activity, which was demonstrated to be transcriptional repression of target genes (Andreou *et al.*, 2007).

In vitro cell culture studies have shown that the levels of sumoylation are altered following a range of stresses, including oxidative, osmotic, heat shock, and viral infection, and it seems that protein sumoylation is increased in brain, liver, and kidney of hibernating squirrels (Meulmeester and Melchior, 2008). Strikingly, several of the above stresses have been implicated in orofacial clefting. Altogether, these thrilling new insights point to a crucial role for sumoylation in regulating a molecular network during palate development and suggest defects in SUMO1 modification as causal factors in Mendelian and isolated clefting disorders.

As indicated above, Satb2 is also a target for sumoylation and a key player in palate development in humans (Brewer *et al.*, 1999; Fitzpatrick *et al.*, 2003) and mice (Britanova *et al.*, 2006; Dobreva *et al.*, 2006). Satb2 belongs to a family of transcription factors which bind to nuclear matrix attachment regions and are likely involved in the regulation of the tissue-specific organization of chromatin, making it accessible to long-range activity of enhancers (Britanova *et al.*, 2005; Dobreva *et al.*, 2003; FitzPatrick *et al.*, 2003). In humans *SATB2* haploinsufficiency generates CPO (Brewer *et al.*, 1999; Fitzpatrick *et al.*, 2003). This situation has been replicated in mice, where *Satb2* heterozygotes exhibit CPO and craniofacial defects similar to those that occur in humans (Britanova *et al.*, 2006). *Satb2* homozygous embryos display exacerbation of craniofacial defects, indicating *Satb2* gene dosage sensitivity in craniofacial and palate development (Britanova *et al.*, 2006). Mechanistic analyses indicated that the defects following altered Satb2 function are likely caused by increased cell death in the *Satb2* expression domains, including early palatal primordia, and altered expression of *Alx4*, *Msx1*, and *Pax9*, all key players in craniofacial and palate development (Beverdam *et al.*, 2001; Peters *et al.*, 1998; Satokata and Maas, 1994; Qu *et al.*, 1999; Zhang *et al.*, 2002). However, another study (Dobreva *et al.*, 2006) found that *Satb2* heterozygous mice were phenotypically normal, whereas homozygotes exhibited CPO owing to abnormal bulges and patterning defects in the PS, including altered expression of *Lhx8* encoding a transcription factor involved in palate development (Zhao *et al.*, 1999). Notwithstanding these differences that may be due to different targeting strategies, both studies point to a crucial role for Satb2 in palate development and link its function to that of other palatal protagonists.

P63, another target of SUMO1 (Huang *et al.*, 2004) and a member of the p53 family, causes a range of human developmental anomalies following mutations of its encoding gene, *TP63*. These include ectrodactyly–ectodermal dysplasia–clefting syndrome 3 and ankyloblepharon-ectodermal dysplasia–clefting syndrome (reviewed in van Bokhoven and Mckeon, 2002). Mutations in *TP63* are also implicated in nonsyndromic clefting (Leoyklang *et al.*, 2006). P63 is an epithelial factor regulating numerous targets involved in cell proliferation, survival, and integrity (Carroll *et al.*, 2006; Mills *et al.*, 1999; Yang *et al.*, 1999). Homozygous loss-of-function of *p63* in mice generate limb anomalies, defects in ectodermal organs and truncation of the maxilla and palate (Mills *et al.*, 1999; Yang *et al.*, 1999). However, it is not clear whether a patent CPO develops in these mutants. Thus, the function, regulation and targets of this factor during palate and lip development need to be examined. Some evidence suggests that *p63* is a target of Bmp signaling as indicated by downregulation of *p63* in the fusing nasal processes of *Bmpr1a* mutants (Liu *et al.*, 2005) similar to findings in zebrafish (Bakkers *et al.*, 2002). Deciphering the function of p63 is a complicated endeavour because of the existence of at least six different isoforms, some of which exert anatagonistic activities (van Bokhoven and Brunner, 2002; Yang *et al.*, 1998). This could be the reason behind the wide spectrum of anomalies encountered in the above syndromes.

4.2.1.7. Growth and extension of the horizontal palatal shelves Following their elevation into a horizontal position, a second phase of active growth and extension of the PS takes place and brings about contact of the opposing PS. Defects in this step generate CPO with a smaller gap between the PS as compared to clefts due to lack of elevation of the PS. Loss-of-function of *single-minded2* (*Sim2*) in mice engenders either a complete CPO or a cleft confined to its posterior-most segment (Shamblott *et al.*, 2002). The PS are able to elevate but fail to abut due to defects in their extension. Analyses revealed that the PS are hypocellular (hypoplastic) and contain abnormally high levels of hyaluronan (Shamblott *et al.*, 2004), an ECM component known to influence cell proliferation, migration, and differentiation. Similarly, failure of horizontal extension of the PS underlies the CPO in mice lacking *Tgfbr2* in the PS mesenchyme, and evidence suggests that this growth phase requires Tgfβ3 inputs (Ito *et al.*, 2003). The importance of horizontal PS extension is further well illustrated in mutants lacking the function of Pdgfc which show normal development of the PS up to E13.5; however, following a delayed elevation the PS fail to make contact owing to their abnormal small size (Ding *et al.*, 2004).

4.2.1.8. Craniofacial skeletal and tongue anomalies and the genesis of cleft palate Severe craniofacial skeletal anomalies and abnormal tongue development may indirectly affect palate development by generating crowded conditions within the oral cavity that impede PS growth and/or elevation, abnormal insertion of the lingual muscles to the hyoid bone, a hypotrophic mandible, a wide space between the PS, etc.

As indicated in Table 2.3, several mutants exhibit CPO coupled with other craniofacial malformations. It is therefore difficult to discern between clefting subsequent to intrinsic defects within the palate proper and clefting induced by anatomical and/or functional defects of other craniofacial structures. However, many of the genes that generate such malformations when targeted are expressed in the developing palate of wild-type mice, suggesting endogenous functions for these genes in this organ. These include *Alx4* (Beverdam *et al.*, 2001; Qu *et al.*, 1999), *Foxf2* (Wang *et al.*, 2003) *Gli2* (Mo *et al.*, 1997; Rice *et al.*, 2006), *Pax9* (Peters *et al.*, 1998), *Pitx1* (Lanctôt *et al.*, 1999; Szeto *et al.*, 1999), *Pitx2* (Lu *et al.*, 1999a), *Tbx1* (Jerome and Papaiannou, 2001; Zoupa *et al.*, 2006), *Pdgfra* (Soriano, 1997; Tallquist and Soriano, 2003) as well as the Pdgf signaling immediate early genes *Arid5b*,

BC055757, and *Plekha1* (Schmahl *et al.*, 2007). These Pdgf targets have been shown to modulate cell migration, implying a function for Pdgf signaling during the morphogenetic events regulating palatogenesis. These events likely affect the behavior of the *Pdgfra*-expressing PS mesenchymal cells in response to the epithelially derived Pdgfc (Ding *et al.*, 2004).

4.2.2. Molecular and cellular control of PS elevation

Delayed PS elevation is detrimental to successful palate formation. Mouse models show that this defect can arise following altered cell signaling, hindrance by the tongue, abnormal craniofacial skeletal development, or abnormal adhesion of the PS to other oral structures.

4.2.2.1. How do palatal shelves elevate? The mechanisms instigating elevation of the PS have fascinated scientists for several decades. Although PS elevation is now well accepted as a rapid event, it does not occur as a sudden rotation from a vertical to a horizontal position as suggested earlier (Lazzaro, 1940; Peter, 1924). Rather, and consistent with previous findings (Greene and Pratt, 1976; Walker and Fraser 1956), PS elevation proceeds through a progressive deformation, and according to Walker and Fraser (1956) "in a wave-like manner anteriorly, until the whole shelf lies dorsal to the tongue" (Figs. 2.5A–F). The occurrence of a progressive wave-like elevation instead of a rapid rotation is further evidenced by the presence of an intermediate state of PS movement (Figs. 2.5C and E), consistent with earlier observations (Walker and Fraser, 1956).

One of the mechanisms suggested to provide the impetus for PS elevation was an active function of the tongue coupled with its ptosis (Lazarro, 1940; Peter, 1924). However, although the tongue plays a role in hindering PS elevation when its development and/or mobility are defectuous, it seems to be rather a passive organ under normal conditions as evidenced by its position (Figs. 2.5B and C), again lending credence to previous findings (Walker and Fraser, 1956). Another argument against an active role for the tongue is the ability of PS to elevate following its removal (Greene and Pratt, 1976).

Lingual functional anomaly may be the cause of clefting of mice lacking the function of *Kir2.1*, encoding an inwardly rectifying K^+ current channel (Zaritsky *et al.*, 2008). The CPO in *muscular dysgenesis* (*mdg*) mutant mice is due to failure of elevation of the PS owing to steric hindrance by a totally paralyzed tongue (Pai, 1965). *Mdg* mice harbor a recessive mutation of the gene encoding the α1S subunit of the L-type voltage-dependent calcium channel (*Cacna1s*) leading to complete paralysis of the skeletal muscle (Chaudhari, 1992). The role of the tongue in generating CPO is also well exemplified in *Hoxa2* null mice, which have defects in the insertion of the hyoglossus muscle into the hyoid bone (Gendron-Maguire *et al.*, 1993; Rijli *et al.*, 1993). This defect is corrected in compound mutants lacking both

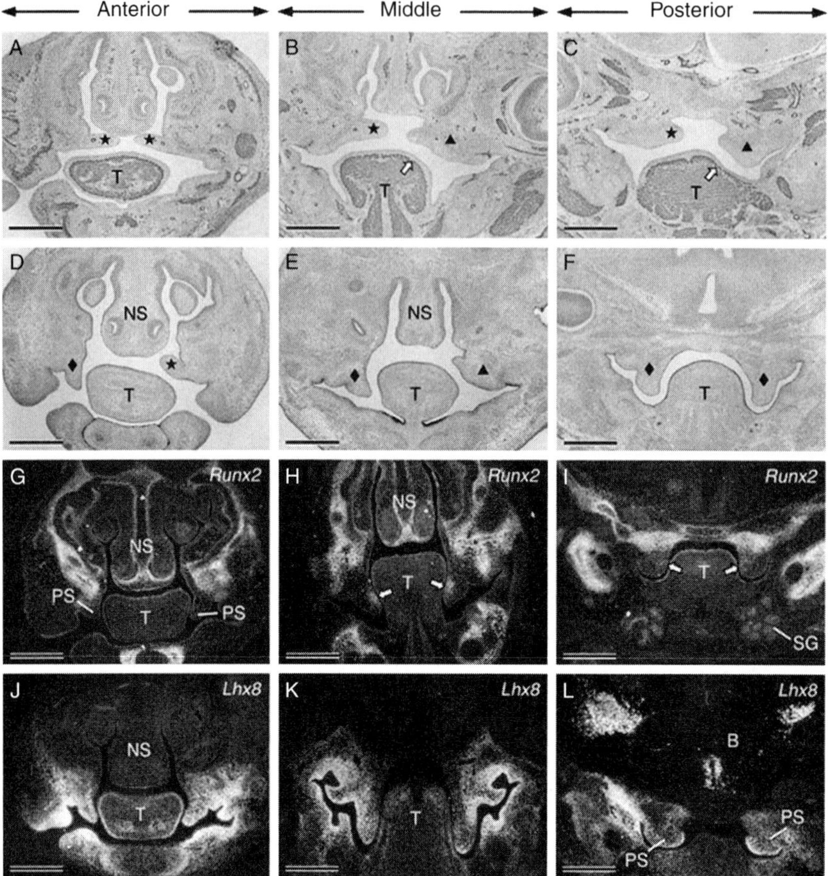

Figure 2.5 Sections from mouse embryo heads. Serial sections from the same specimen at E14.5–E15 (A–C) showing different positions of the PS along the anterior-posterior axis of the palate. The PS have elevated above the tongue in the anterior palate (asterisks in A). In the middle (B) and posterior (C) parts of the palate, one PS has assumed an elevated horizontal position (asterisks), whereas the contralateral PS displays an intermediate position (triangles). Note that posteriorly, one PS displays an early intermediate state between vertical and horizontal orientations (triangle in C). The tongue (T) is squeezed (arrows in B and C) on the side of the PS that shows an intermediate state (PS that was in the process of elevating at the time of specimen fixation). Serial sections at E14.5–E15 from the same specimen at the anterior (D), middle (E), and posterior (F) regions of the secondary palate in which the PS are either elevated (asterisk in D), in an intermediate state between a vertical and a horizontal positions (triangle in E), or still in a vertical position (diamonds in D–F), suggesting the occurrence of a "wave-like" morphogenetic movement during PS elevation which starts in the anterior palate. (G–L) Dark-field images of sections from E14 to E14.5 embryos at the anterior (G and J), middle (H and K), and posterior (I and L) regions of the palate showing expression of *Runx2* (G–I) and *Lhx8* (J–L). Note the presence of a medio-lateral gradient of expression of *Runx2* (arrows in G–I), with mesenchymal cells of the medial half of the PS displaying the highest *Runx2* expression levels. By contrast, *Lhx8* transcripts show a homogenous distribution in the Ps mesenchyme (J–L). B, brain; NS, nasal septum, SG, salivary glands. Scale bars: 500 μm. (See Color Insert.)

Hoxa1 and *Hoxa2* resulting in partial rescue of the clefting (Barrow and Capecchi, 1999).

Elevation of the PS occurs in an anterior–posterior (AP) wave (Figs. 2.5 A-F) and initiates at the level of the second ruga palatina. Subsequently, closure proceeds anteriorly and posteriorly (Ferguson, 1988). Despite this sequence of PS elevation and fusion along the AP axis, closure of the posterior palate is independent of that of the anterior segment. This is well portrayed by the occurrence of cleft of the anterior palate with the posterior palate being intact in *Shox2* mice (Yu *et al.*, 2005). Similarly, abnormal development of the soft palate does not impede that of the hard palate, as reflected in mice lacking the function of the transcription factors Sall3 (Parrish *et al.*, 2004), Sim2 (Shamblott *et al.*, 2002), and Tshz1 (Coré *et al.*, 2007). In contrast to Shox2, which is specific to the presumptive hard palate, other palatal key factors such as Bmp2, Bmp4, Msx1, Fgfr2b, Tbx22, and Meox2 (Gritli-Linde, 2007; Hilliard *et al.*, 2005; Jin and Ding, 2006a) have been shown to be differentially expressed along the AP axis of the palate and, with the exception of *Shox2* and possibly *Meox2*, these patterns are not related to the specification of the hard palate and soft palate. For instance *Tbx22* is not expressed in the anterior-most segment of the palate but is found more posteriorly in a region encompassing both the presumptive hard and soft palates. Rather, these differences may reflect regional differences related to acquired cellular properties, cell–cell and cell-ECM interactions as indicated by *in vitro* cultures of explants from the anterior and posterior palatal segments, which were found to respond differently to growth factors (Hilliard *et al.*, 2005).

The finding that some embryos display one horizontally elevated PS, while the contralateral one is still hanging vertical from the roof of the oral cavity (Figs. 2.5D–F), incited early investigators (Lazzaro, 1940; Peter, 1924) to formulate the concept of rapid "erectile" forces within the PS proper, bringing about their rotation from a vertical to a horizontal position. According to Lazzaro (1940), local increases in the ECM induce swelling of the PS, thus triggering their elevation. At present, changes in the distribution of ECM components, especially proteoglycans and glycosaminoglycans, just prior to PS lifting is accepted as an important determinant of their elevation (Ferguson, 1988; Greene and Pratt, 1976). Other factors have been suggested as playing a role in PS elevation. These are either intrinsic factors within the PS, such as the vascular network, collagen fibers, the PS epithelium, and rapid changes in cell proliferation, or extrinsic determinants such as growth and straightening of the cranial base and mandibular elongation (Ferguson, 1988; Greene and Pratt, 1976). Mandibular and/or craniofacial skeletal malformations generate CPO due to delayed PS elevation in several mouse models (Table 2.3). In these, the tongue is always found lodged between the vertical PS. However, this lingual position could either be caused by its inability to flatten as a result of crowding conditions in the oral cavity, or simply be due to the fact that the tongue slithers into any available space. Therefore, insertion of the tongue between the PS is not a proof of it being a cause of clefting.

4.2.2.2. Cell signaling Despite the existence of mouse models showing failure of or delayed PS elevation, the exact mechanisms driving PS elevation are still elusive. However, as discussed below, size alteration (hypotrophic or hypertrophic PS), changes in cellularity (hypolastic or hyperplastic PS), and patterning defects have all been shown to delay or prevent PS lifting.

The PS within a palatal segment are not homogenous but display a medio-lateral (M-L) regional specification which translates into regional differences in differentiation of the PS epithelium. The PS epithelium is thus regionalized into a squamous stratified oral epithelium on the lateral/oral side (towards the oral cavity) with formation of rugae palatinae (Fig. 2.4C), a pseudostratified, ciliated respiratory epithelium on the medial side (towards the nasal cavity), a medial edge epithelium (MEE), and a stretch of epithelium on the medial side which are fated to fuse with the contralateral MEE and vomerine epithelium, respectively, and create transient epithelia (MES and nasoplatine junction epithelium) that degenerate to allow successful fusion of the PS with each other and with the nasal septum. Within the PS mesenchyme, these medio-lateral differences are not visible in histological sections. However, they are revealed by specific molecular markers as differences in ECM distribution, cell proliferation profiles, and differential gene expression patterns (Gritli-Linde, 2007; Hilliard *et al.*, 2005 for reviews).

In addition to influencing the fate of epithelial cells, the M-L differences in the PS mesenchyme have an important role in controlling cell proliferation and morphogenetic events that are crucial for PS growth and elevation. This is well exemplified in mice lacking the function of the gene encoding the transcription factor Odd-skipped related2 (*Osr2*) (Lan *et al.*, 2004). *Osr2* mutants have a CPO owing to M-L patterning defects and delayed elevation of the PS (Lan *et al.*, 2004). *Osr2* transcripts show a preferential accumulation in the lateral mesenchyme, and Osr2 together Osr1 has been shown to regulate differential cell growth and dynamic expression patterns of *Pax9* (Lan *et al.*, 2004), another transcription factor crucial for palatogenesis in mice and implicated in human orofacial clefting (Peters *et al.*, 1998; Tables 2.1 and 2.2). Thus, dynamic molecular alterations take place in the PS just prior to their lifting. *Runx2* (discussed in Section 4.2.1.3) is expressed in a restricted domain in the medial half of the PS (Figs. 2.5G–I). The significance of this distribution is at present unclear, though it may underlie early specification of osteoblast progenitors within the PS mesenchyme that later will form the palatine and MxP of the palatal bone. However, *Runx2* is also expressed posteriorly (Fig. 2.5I) in the PS of the presumptive soft palate as well as in salivary gland and tongue epithelia and may thus have other roles than specifying osteoblasts.

Decreased growth or overgrowth of the PS impede their elevation. The CPO in mice lacking the function of Spry2 is due to failure of elevation of the PS owing to increased cell proliferation and patterning defects secondary to ectopic high expression levels of Fgf targets such as *Etv5*, *Shh*, *Barx1*, and *Msx1* (Welsh *et al.*, 2007). Again, this model reiterates the importance of a stringent control of the amplitude and site of Fgf signaling for normal palate formation. It would be interesting to determine whether CPO in this model is rescued following genetic reduction of Fgf signaling as shown for the tooth phenotype of *Spry2* null mice (Klein *et al.*, 2006).

Delayed PS elevation subsequent to altered mesenchymal proliferation patterns is also seen in *Dancer* mutants (Bush *et al.*, 2004) and *Pdgfc* null mutants (Ding *et al.*, 2004). Altogether, these findings point to the importance of differential growth patterns in elevation of the PS.

The atypical tyrosine kinases Derailed and its mammalian homolog Ryk have a Wnt binding domain similar to WIF proteins (Wnt inhibitory factors), a kinase domain devoid of catalytic activity, and a PDZ binding domain. Ryk has been shown to directly bind Wnt1 and Wnt3a via its WIF region and to be required for Wnt1-induced TCF activation. Furthermore, whereas the extracellular domain forms a ternary complex with Frizzled and Wnt1, the atypical kinase domain binds Dishevelled, through which it activates the canonical Wnt pathway (Lu *et al.*, 2004). In *Drosophila* and mice *Derailed/Ryk* are required for axon guidance, and Ryk is required for Wnt3a-induced neurite outgrowth (Lu *et al.*, 2004; Yoshikawa *et al.* 2003). *Ryk*-deficient mice have craniofacial skeletal defects and CPO (Halford *et al.*, 2000). The PS of the mutants show a patent delay in elevation, however they are able to fuse *in vitro* (Halford *et al.*, 2000). *Ryk* is expressed in PS epithelium and mesenchyme (Halford *et al.*, 2000).

Mutant mice lacking both the genes encoding Eph receptors, EphB2 (Nuk) and Ephb3 (Sek4) (*Ephb2/Ephb3*), exhibit CPO, abnormal corpus callosum, and defects in axon tracts but with no obvious craniofacial skeletal anomalies (Orioli *et al.*, 1996). Importantly, *Ephb3* has been shown to be expressed in the epithelium and mesenchyme of the vertical PS, while β-galactosidase staining portraying EphB2 distribution has been detected in the PS epithelium including the MES. Furthermore, the ephrin ligands, ephrin-B1 and ephrin-B2, have also been detected in the PS (Orioli *et al.*, 1996). Thus, all the ingredient are present for endogenous action of Eph signaling during palatogenesis. The PS of *Ephb2/Ephb3* mutants are hypotrophic, but seem to manage to elevate. Eph receptors bind to cell surface-associated ephrin ligands on adjacent cells. In addition, Eph receptors and ephrins are often expressed within the same cell, prompting the idea of possible interactions in *cis* (in the same cell) that alter signaling in *trans* (between opposing cells) (Himanen *et al.*, 2007).

Interestingly, EphB receptors and Ryk have been shown to physically interact during craniofacial development, neuroprogenitor migration, and axon guidance (Arvanitis and Davy, 2008; Halford *et al.*, 2000). In the developing palate, such interaction is further supported by the expression patterns of *Ryk*, *Ephb2*, and *Ephb3*. The defects found in *Ryk* mutants were proposed to be secondary to altered cross talk between Ryk and Eph receptors, where Ryk is required for the recruitment of AF-6 (afadin), a junctional protein, to Eph receptors leading to downstream events such as cell migration (Halford *et al.*, 2000). However, another study using human cells found no interaction between AF-6 and Ryk (Trivier and Ganessan, 2002).

Eph/ephrin signaling regulates an array of cellular activities including cell-ECM adhesion, cell–cell contacts, cell shape, migration, proliferation, survival, differentiation, and secretion (Pasquale, 2008). In epithelial cells, there exists a reciprocal regulation between E-cadherin and Eph receptors, where E-cadherin-dependent cell–cell adhesion is able to regulate the expression, cell surface localization and ephrin-dependent activation of Eph receptors. In turn, EphB signaling escorts E-cadherin to the cell membrane, a necessary step for adherens junction formation, thus promoting EphB-ephrin-B mediated cell sorting and aggregation phenomena. Prevention of Eph-ephrin-B binding affects the integrity of adherens junctions (AJs) (Pasquale, 2008 and references therein).

Although EphB-ephrin-B signaling may be important for both palate growth and fusion, this signaling pathway is discussed in this section because of the relationships with Ryk. All these observations make us wonder how Ryk, EphB-ephrin-B, and Wnt signalings are integrated during palatogenesis, and which cellular events they regulate. Things are further compounded by the recent findings showing that both EphB receptors and B-type ephrins are able to signal through the noncanonical Wnt pathway. On the other hand, the canonical Wnt signaling activates *EphB* transcription and downregulates that of *ephrin-B*. In addition, there is evidence of antagonistic interactions between Wnt/Ryk and Eph/ephrin, at least in the central nervous system (CNS) (Arvanitis and Davy, 2008; Pasquale, 2008). Nevertheless, these mouse models are worthy of further exploration, as Ryk and EphB/ephrin-B, some Wnt proteins, and E-cadherin are all expressed in the developing palate.

*4.2.2.3. **The GABA saga*** The opinions as to the involvement of GABA, the major inhibitory neurotransmitter in the CNS, as a factor regulating PS elevation has been subject to fluctuations throughout the years. Early studies incriminated neurotransmitters, including serotonin, acetylcholine, and GABA in PS elevation (Ferguson, 1988; Wee and Zimmerman, 1985). At present, it is well established that aside from functioning as a neurotransmitter in the CNS, GABA also modulates cell proliferation, survival, differentiation, and migration in both neuronal and nonneuronal cells (Andäng *et al.*, 2008; Varju *et al.*, 2001).

Previous teratological studies in rodents suggested that GABA or its agonists inhibit PS elevation, whereas GABA antagonists promote the process (Gritli-Linde, 2007 and references therein). In addition to the presence of a GABA uptake system, the presence of GABA as well as one of its biosynthetic enzymes, Gad67 (encoded by *Gad1*), has been demonstrated in the PS (Asada *et al.*, 1997; Hagiwara *et al.*, 2003; Wee and Zimmerman, 1985; Wee *et al.*, 1986). GABA's role during palatogenesis gained further credence since genetic manipulation in the mouse showed that loss-of-function of either Gad67 or the $\beta3$ subunit of $GABA_A$ receptor ($GABA_A\beta3$ encoded by *Gabrb3*) generates CPO without other craniofacial malformations (Asada *et al.*, 1997; Culiat *et al.*, 1993, 1995; Condie *et al.*, 1997; Homanics *et al.*, 1997). The PS in mutants lacking Gad67 or $GABA_A\beta3$ are elevated above the tongue (Asada *et al.*, 1997; Hagiwara *et al.*, 2003). However, a recent study showed that the PS of *Gad1* mutants fail to elevate (Iseki *et al.*, 2007). Inasmuch as GABA exerts a major function in the brain, the clefting of *Gad1* and *Gabrb3* mutants may be secondary to altered neuronal function. On the other hand, transgenic rescue of $GABA_A\beta3$ in the brain but not the palate still produced mice with CPO (Hagiwara *et al.*, 2003), implying an endogenous function for GABA signaling within the developing palate.

A twist came from a recent study showing that mice lacking the neuronal vesicular inhibitory amino acid transporter (viaat) involved in synaptic corelease of GABA and glycine have a CPO owing to tongue immobility (Vojcik *et al.*, 2006). However, this does not rule out a role for GABA signaling within the palate, since tongue anomalies are known to generate clefting, as demonstrated in *mdg* and *Hoxa2* mutants. In addition, PS from *Gad1* mice were found to elevate *in vitro* following removal of the tongue (Iseki *et al.*, 2007). Nevertheless, the latter two studies cast some doubts about a direct involvement of GABA signaling within the palate. To complicate matters, significant associations of *GAD1* (Kanno *et al.*, 2004) and *GABRB3* (Inoue *et al.*, 2008; Scapoli *et al.*, 2002) with human nonsyndromic oral clefts have been reported. These aspects call for further in depth studies and the use of molecular markers to determine the exact causes of CPO in *Gad1* and *Gabr3b* mutants.

4.2.2.4. ***Intraoral adhesions*** During palatogenesis epithelia are programmed to undergo a transient adhesion followed by degeneration at sites of contact of the PS with each other, with the primary palate and with the vomerine epithelium. Unfortunately, this fascinating event is also subject to aberrations seen as failure of fusion or fusion of PS epithelia at forbidden territories. These ectopic adhesions result in delayed or total failure of PS elevation. Cell signalings play an important function in regulating epithelial differentiation, survival, and apoptosis as revealed by several mutant mice.

Loss of Fgf10 leads to aberrant fusions of PS epithelia with those of the tongue and mandible at sites of increased apoptosis (Alappat *et al.*, 2005; Rice *et al.*, 2004). These are likely generated following a severe reduction in expression of *Jag2*, encoding a ligand for the Notch receptors, and ectopic production of Tgfβ3 in the oral epithelia (Alappat *et al.*, 2005). These mechanistic explanations are reasonable in light of the well recognized role of Tgfβ3 in palatal fusion, and the occurrence of abnormal adhesions of the PS with the tongue and mandible of *Jag2* null mutant mice which exhibit a CPO (Jiang *et al.*, 1998). Further analyses of *Jag2* mutants revealed that the Jag2-Notch signaling prevents aberrant epithelial adhesions through a stringent control of oral epithelial differentiation, as reflected by high and low expression levels of keratin 17 at sites with low and high Jag2-Notch activation, respectively (Casey *et al.*, 2006). Increased apoptosis has been documented in areas of fusion of the PS with the tongue in *Jag2* mutants. However, in contrast to *Fgf10* mutants (Alappat *et al.*, 2005), there was no ectopic expression of *Tgfβ3* or its target, matrix metalloproteinase 13 (*Mmp 13*), in the ectopic fusion areas of *Jag2* mutant oral structures (Casey *et al.*, 2006). Remarkably, PS epithelia at prospective orthotopic sites of fusion show reduced Jag2-Notch activity as compared to other oral epithelia (Casey *et al.*, 2006), prompting the speculation that aberrant increases in Jag2-Notch signaling in the MEE of PS may prevent their fusion. These observations emphasize the role of regional differentiation of oral keratinocytes and apoptosis in epithelial fusion.

Mutations of *IRF6* cause van der Woude (VWS) and popliteal pterygium (PPS) syndromes in humans (Table 2.1), and *IRF6* variants have been implicated in nonsyndromic oral clefting (Table 2.2). The recently generated mice bearing the most common mutation in PPS (R84C) (Richardson *et al.*, 2006) and a gene trap null allele of *Irf6* (Ingraham *et al.*, 2006) provide insights into the cellular and molecular mechanisms underlying these syndromes. *Irf6*$^{+/R84C}$ heterozygotes showed intraoral ectopic adhesions between epithelia of different oral structures, replicating oral syngnathia that occurs in humans inflicted with PPS (Richardson *et al.*, 2006). Some but not all the ectopic intraoral adhesion sites were also seen in embryos heterozygous (*Irf6*$^{+/gt1}$) for the gene trap allele (Ingraham *et al.*, 2006), though with reduced penetrance as compared to *Irf6*$^{+/R84C}$ heterozygotes. Inasmuch as the above *Irf6* mutant strains are on a similar genetic background, these differences may be due to allele differences. *Irf6*$^{+/R84C}$ heterozygous mutants are exempt from cleft lip, cleft palate, and syndactyly which are part of the clinical features of human PPS (Richardson *et al.*, 2006). However, this could be due to influences from stochastic factors and/or modifier genes, as suggested from observations of clinical variations within single VWS and PPS families (Kondo *et al.*, 2002). Mice homozygous for the R84C variant and the gene trap allele display CPO, skin, and skeletal defects (Ingraham *et al.*, 2006; Richardson *et al.*, 2006). The latter are likely secondary to skin anomalies, since *Irf6* is not expressed in osteoblasts and chondrocytes. This situation has also been reported in *Chuk* mice lacking the function of Ikkα, an important factor for skin differentiation, which also display CPO (Li *et al.*, 1999). However, Ikkα is not downregulated in $Irf6^{R84C/R84C}$ and $Irf6^{gtl/gtl}$ homozygotes (Ingraham *et al.*, 2006; Richardson *et al.*, 2006). The CPO in $Irf6^{R84C/R84C}$ and $Irf6^{gtl/gtl}$ homozygous embryos is caused by ectopic adhesions between the tongue and PS epithelia which prevented PS elevation. Further histological and biochemical analyses revealed that skin keratinocytes of $Irf6^{R84/R84C}$ and $Irf6^{gtl/gtl}$ embryos are unable to exit the cell cycle, terminally differentiate and to form a cornified layer. Thus, the presence of subrabasal layers devoid of a cornified overlay and bearing desmosomes is likely a contributing factor to aberrant epithelial adhesions in adjoining structures in these mutants.

Interestingly, the anomalies (including CPO) found in the above *Irf6* homozygous mutants are to a large extent similar to those described in *repeated epilation* (*Er*) mutant mice (Tassin *et al.*, 1983) which harbor a mutation of *Sfn*, encoding stratifin, a phosphoserine–threonine binding protein required for cell cycle arrest and expressed in differentiated keratinocytes (Ingraham *et al.*, 2006; Richardson *et al.*, 2006 and references therein). Importantly, compound heterozygous embryos for both $Sfn^{+/Er}$ and $Irf6^{R84C}$ display a range of defects, including CPO, that are not seen in single heterozygotes for the above alleles (Richardson *et al.*, 2006), indicating that *Er* and *Irf6* interact genetically to regulate the differentiation program of keratinocytes. Murine *Irf6* is expressed in epithelia of the palate and lip primordia, including the MES and areas of lip fusion (Knight *et al.*, 2006). However, keratinocytes of the MES do not form a cornified layer, but are fated to undergo apoptosis (discussed in Section 4.2.3). Remarkably, apoptotic cells were not detected in suprabasal layer keratinocytes of *Irf6* null mutants (Ingraham *et al.*, 2006). These observations prompt the speculation that *Irf6* is required for apoptosis of the MES and epithelia of lip primordia for successful fusion and mesenchymal confluence. This assumption is supported by the fact that in *Irf6* mutants, adhesions of different structures are not followed by degeneration of the epithelia that join them.

4.2.3. Molecular and cellular control of palate fusion

4.2.3.1. Fate of the medial epithelial seam In addition to growth and elevation of the PS, successful palatogenesis entails fusion of the bilateral PS. Anteriorly, the PS fuse with the ventral side of the nasal septum to create the nasopalatine junction and with the primary palate. Fusion of the opposing PS occurs through a sequence of events, including contact of the bilateral PS, adhesion of the opposing MEE, creating the MES which ultimately disappears to allow mesenchymal confluence. The fate of the MES has been subject to discordance, and three mechanisms underlying MES removal have been suggested: apoptosis, epithelial–mesenchymal transformation (EMT), and migration of MES cells towards the periphery of the palatal midline (for reviews see Dudas *et al.*, 2007; Gritli-Linde, 2007).

However, recent evidence from genetic marking of progenitors and progeny of MES cells ruled out the occurrence of EMT in the MES (Dudas *et al.*, 2006; Vaziri Sani *et al.*, 2005; Xu *et al.*, 2006). Following crosses of transgenic mice harboring the *ShhGFPCre* allele and/or mice expressing *Cre* under the regulation of the *Keratin 14* promoter (*K14-Cre*) with the *Rosa-loxP-stop-lacZ* reporter mice (*R26R*), Cre-mediated recombination activates the *lacZ* gene in epithelial cells that express or have expressed *Shh* or *K14* as well as their progeny. These cells stain blue due to formation of a blue precipitate following enzyme histochemistry for *β*-galactosidase, the gene product of *lacZ*. Using the above system, no blue-stained mesenchymal cells were detected in the palate of *ShhGFPCre*; *R26R* or *K14-Cre*; *R26R* mice, thus ruling out the occurrence of EMT of MES cells (Dudas *et al.*, 2006; Vaziri Sani *et al.*, 2005; Xu *et al.*, 2006). However, in a recent study using *K14-Cre*; *R26R* mice, Jin and Ding (2006b) reported the presence of blue-stained mesenchymal cells in the fusing palate and concluded that the MES undergoes EMT. Unfortunately, such conclusions are erroneous and based on artefacts as indicated by the staining of the entire palate mesenchyme, including the ECM in the specimens shown by Jin and Ding (2006b). Indeed, we have consistently documented artefactual staining of the mesenchyme adjacent to *lacZ*-expressing epithelia, including the palate, tooth, salivary glands, skin, etc. This is due to diffusion of the *β*-galactosidase reaction product in ill-prepared specimens following inadequate fixation or washing or as a result of overstaining. This aspect should be kept in mind in fate-mapping studies by this technique to avoid misinterpretations and confusions.

Compelling evidence over several decades provided morphological and molecular evidence for apoptosis of cells of the MES, indicating that apoptosis is a major mechanism eliciting degeneration of the MES as well as the disappearance of transient seams that form in different fusing areas of orofacial structures (for reviews see Dudas *et al.*, 2007; Gritli-Linde, 2007; Jiang *et al.*, 2006; Rice, 2005). Notwithstanding the existence of alternative apoptotic pathways that may act in conjunction with and parallel to the apoptotic protease activating factor1 (Apaf1) (Honarpour *et al.*, 2000; Yoshida *et al.*, 1998), the importance of apoptosis in MES degeneration is reflected in embryos lacking Apaf1, which show delayed removal of the MES (Ceconni *et al.*, 1998).

Migration of cells of the MES towards the nasal and oral surfaces of the palatal midline has also been suggested as a mechanism contributing to MES removal (Dudas *et al.*, 2007; Gritli-Linde, 2007). However, whether cells of all layers of the MES or only peridermal cells undergo this journey is still unclear. *In vitro* cell culture studies have suggested that MES cell migrate in response to Tgfβ3 and ultimately undergo apoptosis, and that cell migration is facilitated by loss of E-cadherin and expression of vimentin, fibronectin, and α-smooth muscle actin (Ahmed *et al.*, 2007). However, in this particular instance it is difficult to extrapolate the findings from *in vitro* conditions to the *in vivo* situation, as keratinocytes are notorious in undergoing activation and migratory skills following isolation and culture *in vitro*. Nevertheless, some of the effects of Tgfβ3 found in the above study, such as its requirement for apoptosis of the MES, have been well established from studies in Tgfβ3 mutants. While some studies suggested that peridermal cells (cells that cover the MEE) are shed before contact of the opposing PS, others showed that the periderm is necessary for the first contact of the PS, and that peridermal cell migration towards the periphery of the palatal midline is a necessary step for triggering apoptosis in both the periderm and MES (Dudas *et al.*, 2007; Gritli-Linde, 2007). The use of specific markers of the periderm will certainly clarify some of these aspects.

4.2.3.2. Junctional complexes and cell signaling Formation of the MES entails adhesion of the MEE of the opposing PS. Junctional complexes, including AJs, tight junctions, and desmosomes are key structures for epithelial cell–cell adhesion. AJs consist of at least two types of cell adhesion molecules (CAMs), cadherins and nectins (Gritli-Linde, 2007 and references therein). E-cadherin is expressed in epithelia of lip and palate primordia, including the MES and its remnants (Figs. 2.2G and H; Gritli-Linde, 2007 and references therein). Recently, deletion of the extracellular cadherin repeat domains required for cell–cell adhesion following mutations of *CDH1/E-cadherin* has been associated with CL/P in families with hereditary diffuse cancer (Frebourg *et al.*, 2006). Inasmuch as E-cadherins are known to work as dimers, the mutant proteins have been suggested to exert trans-dominant negative effects over the wild-type proteins (Frebourg *et al.*, 2006). Germ line targeted mutation of *E-cadherin* in mice is not compatible with embryogenesis beyond the morula stage (Riethmacher *et al.*, 1995). Evidence from conditional targeting of *E-cadherin* in the murine skin has suggested that basal layer keratinocytes are able to maintain AJs by upregulating P-cadherin in response to E-cadherin loss, whereas suprabasal layers are unable to terminally differentiate. Loss of E-cadherin in these mice leads to epidermal hyperplasia subsequent to both hyperproliferation and altered differentiation. By contrast, due to absence of a compensatory system for E-cadherin loss, some cell layers of the hair follicles disintegrate (Tinkle *et al.*, 2004). However, conditional removal of E-cadherin in epithelia does not seem to impede lip or palate development since mutant mice reach adulthood, implying that compensatory mechanisms exist in lip and palatal epithelia for AJ maintenance. Alternatively, occurrence of a palatal clefting might have been overlooked or requires breeding onto different genetic backgrounds to be expressed. As discussed in Section 4.2.2.2, Ryk and EphB/ephrin-B interact with E-cadherin and may thus be involved in PS fusion.

The immunoglobulin-like CAMs, nectins, make up a family of four members capable of bringing about cell–cell adhesion and subsequently recruiting cadherins to the nectin-based adhesion spots through afadin and catenins (Gritli-Linde, 2007 and references therein). Human mutations of the *PVRL1* gene encoding nectin1 cause CL/P-ectodermal dysplasia1 syndrome (Table 2.1) and have been implicated in nonsyndromic orofacial clefting (Table 2.2). Furthermore, missense mutations of *PVRL2* encoding nectin2 have been incriminated in isolated CL/P as well (Table 2.2). In both human and mouse embryos nectin1 and E-cadherin are coexpressed in epithelia, including the palatal MEE (Ding *et al.*, 2004; Suzuki *et al.*, 2000). However, similar to E-cadherin, the role of nectin1 in lip and palate formation is at present unclear, since most tissues of mice lacking either *Nectin1*, *Nectin2*, or *Nectin3* do not show defects in AJs and tight junctions. In addition, *Nectin1* and *Nectin2* null mutants do not exhibit orofacial clefting, perhaps as a consequence of functional redundancy between nectin family members (Gritli-Linde, 2007 and references therein).

Mutations of *NF2* encoding the tumor suppressor Merlin cause the autosomal dominant disorder neurofibromatosis type 2, and heterozygous loss-of-function of *Nf2* in mice leads to development of highly metastatic tumors (reviewed in Curto and McClatchey, 2008). Merlin coordinates membrane receptor signaling and cell–cell contact and plays a key role in controlling differentiation and contact-dependent inhibition of cell proliferation. In mammalian cells, Merlin is recruited to AJs and is regulated by cell–cell adhesion (Curto and McClatchey, 2008). Recently, the generation of *Nf2* mutant mice with a progressive loss-of-function of Merlin revealed a key role for this protein in regulating tissue fusion (McLaughlin *et al.*, 2007). Importantly, in mouse embryos, *Nf2* expression has been found to dwindle at the leading front of the eye lids and the tips of the neural fold before fusion, and to be upregulated in the respective tissue bridges that form following eye lid and neural tube closure (McLaughlin *et al.*, 2007). Furthermore, Merlin loss-of-function has been found to engender fusion defects affecting a range or organs and tissues known to rely on tissue sheet fusion for normal development. These include the eye (retinal coloboma, lens herniation, and failure of closure of the eye lids), neural tube (neural tube defects), body wall (omphalocele), heart (ventricular septal defects), and palate (cleft palate). The PS of *Nf2* mutants elevate and abut, but are unable to adhere to each other and to the vomerine epithelium. Tissues from Merlin-deficient mice revealed that Merlin is required for the assembly but not maintenance of tight junctions and AJs. From analyses of the defects seen in *Nf2* mutants it has been proposed that Merlin functions as a developmental switch to loosen and tighten cell–cell boundaries.

Without Merlin, cells at the leading fronts of adhering tissue sheets undergo abnormal detachment and apoptosis (McLaughlin *et al.*, 2007). The regulation of the function and expression of Merlin during tissue adhesion and fusion awaits further studies. Candidate signaling pathways involved include those generating neural tube defects and cleft palate when disrupted, including Bmp and Wnt signalings.

Mice harboring a transgenic insertional mutation disrupting the CASK gene and mutants with a targeted loss-of-function of CASK exhibit a CPO without other craniofacial malformations (Atasoy *et al.*, 2007; Laverty and Wilson, 1998). The PS of CASK mutants elevate but remain apart (Atasoy *et al.*, 2007), possibly due to failure of fusion. CASK (calcium/calmodulin-dependent serine protein kinase) is a member of the membrane-associated guanylate kinase (MAGUK) family, which function as scaffolds involved in organizing specialized plasma membrane domains and in coupling extracellular signals to the cytoskeleton and intracellular signal transduction cascades (Dimitratos *et al.*, 1999). MAGUKs consist of three domains, including a Src homology 3 domain (SH3), a domain homolog to guanylate kinase (GUK), and a PDZ domain (PSD-95/Dlg/ZO-1). Another MAGUK protein, discs large (Dlg), seems to function during palatal fusion as suggested from *Dlg* mutant mice (Caruana and Bernstein, 2001). As indicated by β-galacosidase staining, Dlg is present in mesenchyme and epithelia of the developing lip and palate, including the MEE and MES (Caruana and Bernstein, 2001). However, despite these interesting expression patterns and the palate phenotype following loss-of-function, the role and regulation of Dlg in palatal fusion are at present unclear and deserve further attention.

4.2.3.3. ***Cell signaling pathways*** Several embryonic epithelia undergo a transient fusion between tissue sheets that either lasts for a short period, such as the MES of the palate, or are maintained for a longer period, such as the eye lids. Others undergo definitive fusions, as exemplified by closure of the neural tube (neuroepithelium), ventral body wall, and urogenital sinus (in male embryos).

Targeted gene inactivation in mice have unveiled the requirement for several factors involved in cell signaling for adequate palatal fusion, including Tgfβ3 (Kaartinen *et al.*, 1995; Proetzel *et al.*, 1995; Yang and Kaartinen, 2007), Foxe1 (De Felice *et al.*, 1998), epidermal growth factor receptor (Egfr, Miettinen *et al.*, 1999), Pdgfc (Ding *et al.*, 2004), Eya1 (Xu *et al.*, 1999), Tgfβr type II (TgfβRII; Xu *et al.*, 2006), Alk5 (Dudas *et al.*, 2006), Shox2 (Gu *et al.*, 2008), and Slug/Snail (Murray *et al.*, 2007). Loss-of-function of Alk5 (in the epithelium), Foxe1, Pdgfc, Tgfβ3, and Shox2 generate CPO with minor or no other craniofacial anomalies, and PS explants from *Tgfβ3*, *Pdgfc*, and *Egfr* mutant embryos fail to fuse owing to persistence of the MES (Ding *et al.*, 2004; Kaartinen *et al.*, 1995; Miettinen *et al.*, 1999).

Amongst the above factors, the role of the Tgf*β* signaling pathway in palate fusion is perhaps the best understood. Formation of the MES following adhesion of the opposing MEE lies at the crux of palatal fusion. Tgf*β*3 is expressed in the MEE as well as in the MES; thus, it is present just before and during PS fusion. Previous work revealed that Tgf*β*3 signaling acts both during the early and later steps of palatal fusion. Firstly, Tgf*β*3 mediates adhesion of the opposing MEE by promoting the formation of filopodia and accumulation of chondroitin sulfate proteoglycans at their apical surface, thereafter Tgf*β*3 signaling induces apoptosis of the MES (Dudas and Kaartinen, 2005; Gritli-Linde, 2007 and references therein). Strong evidence from studies on fusion processes in different systems indicates the presence of filopodia (cell extensions) at the leading front of fusing epithelial sheets (Fristrom, 1988). Further studies demonstrated that E-cadherin is crucial for fusion, whereas filipodia are necessary for adequate alignment and guidance of tissue sheets that are fated to fuse, but not for the fusion process itself (Schöck and Perrimon, 2002). Interestingly, lack of function of Tgf*β*3 results in altered distribution of junctional proteins, including E-cadherin, and is deleterious for cell–cell adhesion (Dudas and Kaartinen, 2005; Gritli-Linde, 2007 and references therein).

Tgf*β*3 operates not only in the palatal epithelium but also in the palatal mesenchyme, where it mediates epithelial–mesenchymal interactions leading to remodeling of the ECM (Blavier *et al.*, 2001; Xu *et al.*, 2006). Gain- and loss-of-function studies *in vitro* and *in vivo* have demonstrated that Tgf*β*3-induced regression of the MES is mediated by the Tgf*β*RII and the Alk5/Smad pathway (Cui *et al.*, 2005; Dudas *et al.*, 2004a, 2006; Xu *et al.*, 2006). Loss-of-function of Alk5 and Tgf*β*RII specifically in the palatal epithelium generates a cleft palate secondary to persistence of a MES that is unable to undergo apoptosis (Dudas *et al.*, 2006; Xu *et al.*, 2006). Interestingly, ablation of the function of *Alk5* in cells that express *Tgfβ3* and their progeny generates a CPO similar to that in *Tgfβ3* mutants (Yang *et al.*, 2008). Furthermore, epithelial Tgf*β* signaling through Tgf*β*RII seems to be necessary for cessation of proliferation of the MEE, as in its absence MEE cells keep proliferating and form overgrowths that impede fusion (Xu *et al.*, 2006). Importantly, further studies from mouse models revealed that Tgf*β*3 plays a unique role within the palatal epithelium and fusing process that cannot be fulfilled by Tgf*β*1, another family member (Yang and Kaartinen, 2007). These interesting findings were gleaned from mice that express *Tgfb1* in the *Tgfβ3* locus following a knockin manipulation. The homozygous knockin mice displayed only partial rescue of the palatal clefting. Explant cultures demonstrated that reduced apoptosis and a sluggish breakdown of the basement membrane of the MES as well as general delay in epithelial fusion are responsible for the clefting (Yang and Kaartinen, 2007).

Mice harboring mutations in *Irf6* show CPO as a result of ectopic PS adhesion to other oral structures (Ingraham *et al.*, 2006; Richardson *et al.*, 2006), thus preventing accurate *in vivo* studies of its role in palatal fusion. *Irf6* is expressed in the palate epithelia, including the MEE and MES (Ding *et al.*, 2004; Knight *et al.*, 2006). Interestingly, *Irf6* expression has been found to require Tgfβ3 and TgfβRII function (Knight *et al.*, 2006; Xu *et al.*, 2006). By contrast, despite their inability to fuse, PS from *Pdgfc* mutant mice show no alterations in the expression of *Irf6*, *Tgfβ3*, and *Pvrl1*. Conversely, *Pdgfc* expression is not altered in *Tgfβ3*-deficient palates (Ding *et al.*, 2004). Similar to Pdgfc, Pdgfa is produced in the palate epithelium and both act on the PS mesenchyme through binding to PdgfR-α. However, Pdgfc acts nonredundantly in the developing palate, since the presence of Pdgfa activity is unable to compensate for loss-of-function of Pdgfc, and Pdgfa loss-of-function does not cause clefting (Ding *et al.*, 2004). Together, these important insights point to a unique function of Pdgfc in coordinating epithelial–mesenchymal interactions during palate fusion. Thus, Pdgfc and Tgfβ3 signalings seem to operate through different paths in controlling palate fusion.

The pivotal role of apoptosis in palate fusion has been further highlighted in mutant mice lacking the function of the transcription factors Snail and Slug, encoded by *Snai1* and *Snai2*, respectively (Murray *et al.*, 2007). *Snai2*-deficient neonates exhibit an incompletely penetrant CPO, which becomes fully penetrant upon removal of one *Snai1* allele on a *Snai2* mutant background (*Snai1*$^{+/-}$; *Snai2*$^{-/-}$), implying functional redundancy between the two genes. This has been further indicated by their overlapping expression domains (Murray *et al.*, 2007). The PS of *Snai2*$^{-/-}$ and *Snai1*$^{+/-}$; *Snai2*$^{-/-}$ mutants elevate and abut but fail to form a MES and fuse owing to decreased apoptosis and failure of periderm cells to migrate towards the periphery. However, expression of *Tgfβ3* and *Irf6* were not altered in these mutants, suggesting that either Snail and Slug function downstream of the Tgfβ3 pathway (Murray *et al.*, 2007), or that they act independently of Tgfβ3 signaling. Perhaps a comparison of palates from wild-type embryos in which the PS have just abutted, before MES formation, with the mutant ones would provide a more accurate appreciation of the molecular defects in *Snai2*$^{-/-}$ and *Snai1*$^{+/-}$; *Snai2*$^{-/-}$ mutants.

Evidence from *in vitro* manipulation of the palate suggests the existence of interactions between Wnt11 and Fgfr1b mediated by cellular events during palate growth and fusion. During the growth phase, Fgfr1b signaling stimulates cell proliferation partly by inhibiting *Wnt11* expression. Subsequently, upon PS contact, cell proliferation may lead to inhibition of *Fgfr1b* expression, thus allowing the gradual expression of *Wnt11* in the MES. Epithelial Wnt11 function seems to be crucial for repressing expression of *Fgfr1b* in the mesenchyme and induction of apoptosis in the MES as well as in the mesenchyme. Following MES removal, the gradual disappearance of Wnt11 in conjunction with seam disintegration leads to restoration of *Fgfr1b* expression, which is necessary for further growth and differentiation of the palatal mesenchyme (Lee *et al.*, 2008). These studies add other key player in PS fusion.

5. Caveats for Using Mouse Models for Orofacial Clefting

It is evident that mouse models for orofacial clefting have proven and will continue to prove their usefulness in unraveling the mechanisms governing lip and palate development and the etiopathogenesis of orofacial clefting. In certain instances, however, the genetically modified mice do not faithfully replicate the human phenotype, and are either exempt of a phenotype, show a more severe phenotype, or only part of the phenotype. In addition, early embryonic lethality, before development of the organ of interest, is a major hurdle. In the particular case of mouse models for CL/P and CPO, the rarity of mice that specifically exhibit CL/P is another frustrating situation. Similarly problematic is the occurrence of severe craniofacial anomalies concomitant to clefting, which complicates studies of the pathogenesis of cleft palate. Finally, different phenotypes have been produced following functional ablation of a given gene, owing to allele differences and/or differences in genetic backgrounds. However, these caveats have incited researchers to look for alternative ways to avoid them or to evaluate interacting biological pathways, influences from the genetic background, modifier genes, and the nature of mutations in the genesis of malformations.

5.1. Early embryonic lethality

Targeted mutations in mice may leads to early embryonic lethality, if the targeted gene(s) has an important nonredundant function during early embryogenesis. For example, *Alk3*-deficient mice die at E9 before lip formation (Nie *et al.*, 2006b). However, generation of mutants lacking *Alk3* in a subset of cells of lip and palate primordia allowed functional analyses of this gene in the formation of these structures (Table 2.3). Mice lacking either *Bmp5* or *Bmp7* dot not exhibit lip or palate malformations despite expression of *Bmp7* in lip and palate primordia. This could be caused by functional redundancy between Bmp7 and other family members. Compound mutant embryos for both *Bmp5* and *Bmp7* die at E10.5 and thus are not of use for evaluating the function of these genes during lip and palate development. The generation of compound conditional knockouts of the above genes may provide a way around this impasse.

5.2. Severe craniofacial defects

Conventional genetic alteration into the mouse germline may lead to severe craniofacial defects as exemplified by *Shh* null mutants, thus ruling out studies of the function of this gene during lip and palate development.

Tissue-specific gene inactivation is a good alternative as reflected by several mouse models (Table 2.3), including conditional knockouts for *Shh*. Another alternative is thc usc of drug-dependent alleles (Liu at al., 2007; Stankunas and Crabtree, 2007). Many mutant mice lacking the function of genes expressed in the developing lip and/or palate exhibit severe craniofacial skeletal anomalies in addition to orofacial clefting (Table 2.3) making the delineation of the role of a gene in lip or palate development a difficult task. This problem is perceivable in the case of CPO, as formation of the secondary palate is influenced by neighboring orofacial and head structures, including the tongue, mandible, basicranium, and maxilla. These hurdles could be overcome by careful analyses of the molecular and cellular defects within the lip and palate proper at different stages of their development.

5.3. Scarcity of mouse models for cleft lip with or without cleft palate

An impressive number of mouse mutants displaying CL, CL/P, or CPO has been produced (Table 2.3). Remarkably, an overwhelming majority of these models show CPO, whereas only a few mouse mutations cause CL or CL/P. The clefting in the latter two conditions is almost always associated with other major craniofacial anomalies, such as those resulting from CNCC defects. Other cases of CL/P in mice are those caused by alterations of genes encoding factors that regulate the activity of several key players in lip and palate development, such as *Sumo1*.

Similarly, experiments with teratogen-induced clefting have generated a tremendous number of mice with CPO (Johnston and Bronsky, 1995 and references therein), whereas teratogens have been shown to increase the incidence of CL/P, only in strains that show susceptibility to spontaneous CL/P (Johnston and Bronsky, 1995). These observations prompt the question about the reasons behind the scarcity of mutant mice harboring CL/P versus those with CPO. Notwithstanding the situation in teratogen-induced clefting in mice (Diehl and Erickson, 1997), human and mouse studies indicate that many of the genes and signaling pathways that instigate CL/P are distinct from those that cause CPO (Juriloff and Harris, 2008). Furthermore, development of the upper lip and primary palate are largely independent of secondary palate formation.

Findings from ENU mutagenesis on the progeny of male mice suggest that CL/P might be caused by point mutations, whereas recessive genes may be involved in isolated CPO (Yamada *et al.*, 2005). With the exception of inactivating mutations in *FGF8* and *FGFR1* (Riley *et al.*, 2007a,b), the functional outcome of missense mutations in genes implicated in nonsyndromic CL/P (see Section 2) is unknown. It is thus unclear whether the variants identified lead to reduced activity or loss-of-function, or are simply silent changes (Vieira, 2008). In addition, some mutations may confer new unknown functions to proteins. These issues may underlie the differences in the phenotypic expression of clefting in humans (CL/P) as a consequence of missense mutations and in mice (CPO) following homozygous loss-of-function. It may be surmised that generation of mice bearing similar mutations to those identified in isolated cleftings may enhance our understanding of the pathogenesis of these malformations. However, the involvement of some variants in candidate genes in the pathogenesis of nonsyndromic clefting is uncertain (Vieira, 2008), thus discouraging researchers from embarking onto recapitulating human variants in mice. In the case of *FGFR1*, the nonsense mutation turned out to be carried by an individual presenting features of Kallmann syndrome, and his father was born with cleft lip and palate, but without the cardinal features of Kallmann syndrome. This case suggests that some nonsyndromic CL/P instances may be part of a syndrome with variable expression (Vieira, 2008).

In addition, the "resistance" of mice to CL/cleft primary palate as compared to humans might be due to differences between human and mouse genetic backgrounds and to different environmental exposures. Another explanation could be that in mice, development of the upper lip and primary palate is a rapid process that lasts for only 2.5 days-from the appearance of the nasal placodes at E10 to completion at E12.5- as compared to the longer time of development of the secondary palate (4–4.5 days)- from the appearance of the PS at E11 until fusion around E15.5. In humans, however, lip and primary palate formation lasts for a longer time (from the 4th to the 7th week of gestation) and secondary palate development requires about 2.5 weeks. In both humans and mice fusion of the two mandibular processes that form the lower lip and mandible is a rapid event, and human and mouse cases displaying cleft of the lower lip and mandible are rarely encountered. This observation lends credence to the above suggested explanation about the rarity of CL/P in mice. Therefore, altered function of important mitogenic factors within lip and primary palate primordia would not have any severe impact on development of these primordia in mice, given that in general the length of the cell cycle in mammalian fast-dividing cells is ~24 h, and that during early organogenesis cells are not synchronized.

5.4. Genetic background, allele differences and differences in targeting strategies

Several mutant mice exhibit different phenotypes on different genetic backgrounds that are likely due to modifier loci. This high degree of phenotypic variability may constitute a major risk factor for overlooking the involvement of a gene in a given tissue, unless the genetic mutation is tested on different genetic backgrounds. For example, in a mixed CD1–129Sv background, *Gas1* null mutants develop to adulthood, whereas on a mixed 129Sv–C57BL/6 background they develop microform HPE, clefting and craniofacial skeletal defects and die perinatally (Seppala *et al.*, 2007 and references therein). 129/Sv and Balb/C *Eya1*-deficient mice develop a fully penetrant CPO. However, in the C57BL/6J background, all the *Eya1* null mutant show minor defects manifested as abnormal fusion of the PS with the nasal septum (Xu *et al.*, 1999). Similarly, *Ski* mutant mice revealed strong influences by the genetic background on the effects of the mutation. On mixed backgrounds consisting of either 129P2 and Swiss black, or 129P2 and C57BL/6J, 85% of *Ski* null mutants exhibit exencephaly, with the remaining 15% displaying midline facial clefting without exencephaly. Following extensive backcrosses and intercrosses, progressive enrichment in the C57BL/6J genome content was correlated with a progressive decrease in the occurrence of excenphaly and a proportional gain in facial clefting (Colmenares *et al.*, 2002).

Recently, mice lacking the function of Satb2 have been generated by two different laboratories. Heterozygous $Satb2^{+/-}$ have been reported to be viable and fertile (Dobreva *et al.*, 2006). However, the mutants generated by Britanova *et al.* (2006) showed craniofacial defects, including CPO, due to *Satb2* haploinsufficiency, which faithfully replicated those engendered by 2q32 deletions and translocations in humans. It is not clear whether these differences are due to differences in genetic backgrounds of the mouse lines used or to differences between the gene targeting strategies. Another example is illustrated by mice heterozygous for the *Irf6* missense mutation R84C ($Irf6^{+/R84C}$), which display intraoral adhesions (Richardson *et al.*, 2006) that are more severe and have a higher penetrance than those found in heterozygous for a null gene trap allele ($Irf6^{+/gtl}$) (Ingraham *et al.*, 2006). This occurred despite the fact that the mice were on a similar genetic background, implying that these phenotypic dissimilarities may be due to allele differences.

Influences of the mouse genetic background on the phenotype are also exemplified in *Tcof1* heterozygotes generated following breeding of male chimeras with C57BL/6J females, which exhibit not only features of TCS, but also have additional anomalies. However, TCS features were faithfully replicated following outcrosses between a $Tcof1^{+/-}$ DBA line and C57BL/6J mice (Dixon *et al.*, 2006).

5.5. Differences between mice and humans

Several causal genes of syndromic orofacial clefts (Table 2.1) and gene variants implicated in nonsyndromic clefts (Table 2.2) generate CL/P in humans. However, with rare exceptions, inactivating mutations in the mouse generate CPO (Table 2.3). Possible reasons behind these differences are discussed above (Section 5.3).

Furthermore, with a few exceptions (e.g., *Tcof1*, *Satb2*, and *Sumo1* heterozygous mice), some genes show a clefting phenotype in mice only when the two copies are inactivated, whereas in humans clefting is generated as a consequence of haploinsufficiency. Examples include *MSX1/Msx1*, *TBX1/Tbx1*, and *PAX9/Pax9* (Tables 2.1 and 2.3).

IRF6 mutations cause van de Woude (VWS) and popliteal pterygium (PPS) syndromes. *Irf6*$^{+/R84C}$ mice heterozygous for the most common mutation found in PPS mimick the human phenotype only partially, since they do not develop cleft lip, cleft palate, and syndactyly which are commonly observed traits in affected individuals (Richardson *et al.*, 2006). However, strong phenotypic variability has been reported within single VWS and PPS families, implying that other factors contribute to the wide spectrum of phenotypes (Kondo *et al.*, 2002). It would be interesting to see the phenotypic outcome of the above mutation in mice on a different genetic background.

Functional redundancy or involvement of compensatory loci might also underlie differences between human and mice. As discussed in Section 4.2.3.2., mutations in *PVRL1* encoding nectin1 are involved in both syndromic and nonsyndromic CL/P. Furthermore, *PVRL2* encoding nectin2 has been incriminated in nonsyndromic clefting cases (Tables 2.1 and 2.2). *Nectin1* and *Nectin2* null mutant mice have no clefting phenotype. This could be a result of functional redundancy between nectin family members. Perhaps generation of compound mutants of different *Nectin* family members would shed light on the function of these molecules in lip and palate development.

In humans, *FLNA* (encoding filamin A) mutations creating loss-of-function are lethal in males and cause X-linked periventricular nodular heterotopia (PVNH) in females. However, CPO does not seem to be part of the PVNH anomalies. In contrast, missense mutations believed to confer gain-of-function cause most instances of otopalatodigital syndromes type 1 (OPD1) and 2 (OPD2), Melnick–Needles syndrome (MNS), and frontometaphyseal dysplasia, and CPO occurs in OPD1, OPD2, and MNS (Robertson *et al.*, 2003). As we have learned from numerous developmental mouse models, overexpression/overactivation of a factor is not necessarily the opposite of loss-of-function, and both instances can lead to cellular dysfunction and similar organ malformation. This takes us back in time to the old adage by Theophrastus Bombastus von Hohenheim, alias Paracelsus (1493–1541): "Rien n'est poison, tout est poison: seule la dose fait le poison".

As described above (Section 4.2.1.1), it is remarkable that all the murine null mutations in genes (*Hic1*, *Mnt*, *Crk*, and *Ovca1*) implicated in MDS result in CPO among other anomalies. However, CPO has not been described in MDS patients which display other craniofacial defects. Furthermore, the craniofacial malformations in MDS patients are thought to be caused by haploinsufficiency in either one or multiple contiguous genes. By contrast, *Hic1*, *Mnt*, *Crk*, and *Ovca1 heterozygous mice* show no craniofacial malformations, but with the exception of *Crk* heterozygotes, they are prone to malignancy. Perhaps combined haploinsufficiency of *HIC1*, *MNT*, *OVCA1* (or *OVCA2*), *CRK*, and other genes in the MDS critical region is necessary for development of the full-blown craniofacial dysmorphism, including CPO. Nevertheless, the above mouse models point to an involvement of these genes in palatogenesis. Unraveling their exact mechanisms of action requires a detailed survey of the morphological and molecular defects at different stages of palate development and would benefit from studies of compound heterozygotes.

A score of teratogens such as X-radiation, hypoxia, drugs, toxins, hormones, etc., have been shown to induce CPO and other anomalies in mice, and some of these factors seem to increase the incidence of CL/P in susceptible mice (Johnston and Bronsky, 1995 and references therein). However, there is little evidence linking these to human clefting. In humans, maternal cigarette smoking and folic acid supplementation are likely important contributors that may influence the genetic risks for orofacial clefting (Vieira, 2008). Finally, it is noteworthy that in addition to clefting, the quasi-totality of mouse models for CL/P or CPO display other defects in other organs that range from minor anatomical or functional alterations to major defects. This should be kept in mind when making analogies with nonsyndromic clefting.

6. Concluding Remarks

Although we still have more clues than conclusions, a substantial amount of knowledge about the developmental biology of the lip and palate and the etiopathogenesis of their clefting has been accumulated. Mouse models in conjunction with human studies are pivotal in gathering information that will be of use in the clinical context.

Despite a palpable progress in lip and palate research, there are challenges ahead. The interplay between the environment and genes in the genesis of lip and palate clefting constitutes an important and to a large extent unknown variable. The myriad of molecules expressed during lip and palate development is humbling. Many of these factors clearly play nonredundant roles, since their dysfunction is disastrous. However, we know relatively little about their regulation and downstream targets. Identifying targets of key factors in lip and palate development may reveal novel risk factors for clefting. Furthermore, there is an urgent need to determine the functional significance of mutations of candidate genes implicated in nonsyndromic CL/P. The generation of mice carrying the same mutations in such genes might be crucial for unraveling their effects. Other challenges entail elucidating how, when, and where signaling pathways intersect and/or converge. To achieve these goals, the mouse remains the model of choice.

ACKNOWLEDGMENTS

I express my gratitude to Dr. Anders Linde for preparing the figures and reading the manuscript and to Drs. Martyn T. Cobourne and Sara Rizell for kindly providing the clinical photographs. My gratitude to members of the Oral Biochemistry laboratory for their support. I also express my gratitude and admiration to all scientists who contributed with hard work and perseverance to advance cleft lip and palate research. This work was supported by the Swedish Research Council-Medicine (grants 15181 and 20614) and the Institute of Odontology, Sahlgrenska Academy.

Articles published after May 2008 were not cited in this chapter.

REFERENCES

Åberg, T., Cavender, A., Gaikwad, J. S., Bronckers, A. L., Wang, X. P., Waltimo-Sirén, J., Thesleff, I., and D'Souza, R. N. (2004a). Phenotypic changes of *Runx2* homozygote-null mutant mice. *J. Histochem. Cytochem.* **52,** 131–139.

Åberg, T., Wang, X. P., Kim, J. H., Yamashiro, T., Bei, M., Rice, R., Ryoo, H. M., and Thesleff, I. (2004b). Runx2 mediates Fgf signaling from epithelium to mesenchyme during tooth development. *Dev. Biol.* **270,** 76–93.

Abidi, F. E., Miano, M. G., Murray, J. C., and Schwartz, C. E. (2007). A novel mutation in the *PHF8* gene is associated with X-linked mental retardayion with cleft lip/cleft palate. *Clin. Genet.* **72,** 19–22.

Abu-Issa, R., Smyth, G., Smoak, I., Yamamura, K. I., and Meyers, E. N. (2002). Fgf8 is required for pharyngeal arch and cardiovascular development in the mouse. *Development* **129,** 4613–4625.

Ahmed, S., Liu, C. C., and Nawshad, A. (2007). Mechanisms of palatal epithelial seam disintegration by transforming growth factor (TGF)β3. *Dev. Biol.* **309,** 193–207.

Alappat, S. R., Zhang, Z., Suzuki, K., Zhang, X., Liu, H., Jiang, R., Yamada, G., and Chen, Y. P. (2005). The cellular and molecular etiology of the cleft secondary palate in *Fgf10* mutant mice. *Dev. Biol.* **277,** 102–113.

Alkuraya, F., Saadi, I., Lund, J., Turbe-Doan, A., Morton, C. C., and Maas, R. L. (2006). SUMO1 haploinsufficiency leads to cleft lip and palate. *Science* **313,** 1751.

Andäng, M., Hjerling-Leffler, J., Moliner, A., Lundgren, T. K., Castelo-Branco, G., Nanou, E., Pozas, E., Bryja, V., Halliez, S., Nishimaru, H., Wilbertz, J., Arenas, E., *et al.* (2008). Histone H2AX-dependent GABA(A) receptor regulation of stem cell proliferation. *Nature* **451,** 460–464.

Andreou, A. M., Pauws, E., Jones, M. C., Singh, M. K., Bussen, M., Doudney, K., Moore, G. E., Kispert, A., Brosens, J. J., and Stanier, P. (2007). TBX22 missense mutations found in patients with X-linked cleft palate affect DNA binding, sumoylation, and transcriptional repression. *Am. J. Hum. Genet.* **82,** 700–712.

Argawal, S. K., Kennedy, P. A., Scacheri, P. C., Novotny, E. A., Hickman, A. B., Cerrato, A., Rice, T. S., Moore, J. B., Rao, S., Ji, Y., Mateo, C., Libutti, S. K., *et al.* (2005). Menin molecular interactions: insights into normal functions and tumorigenesis. *Horm. Metab. Res.* **37,** 369–374.

Arvanitis, D., and Davy, A. (2008). Eph/ephrin signaling: Networks. *Genes Dev.* **22,** 416–429.

Asada, H., Kawamura, Y., Maruyama, K., Kume, H., Ding, R. G., Kanbaea, N., Kuzume, H., Sanbo, M., Yagi, T., and Obata, K. (1997). Cleft palate and decreased brain γ-aminobutyric acid in mice lacking the 67-kDa isoform of glutamic acid decarboxylase. *Proc. Natl. Acad. Sci. USA* **94,** 6496–6499.

Atasoy, D., Schoch, S., Ho, A., Nadasy, K. A., Liu, X., Zhang, W., Mukherjee, K., Nosyreva, E. D., Fernandez-Chacon, R., Missler, M., Kavalati, E. T., and Südhof, T. C. (2007). Deletion of CASK in mice is lethal and impairs synaptic function. *Proc. Natl. Acad. Sci. USA* **104,** 2525–2530.

Avila, J. R., Jezewski, P. A., Vieira, A. R., Orioli, I. M., Castilla, E. E., Christensen, K., Daack-Hirsch, S., Romitti, P. A., and Murray, J. C. (2006). PVRL1 variants contribute to non-syndromic cleft lip and palate in multiple populations. *Am. J. Med. Genet. A* **140,** 2562–2570.

Bachler, M., and Neubüser, A. (2001). Expression of members of the Fgf family and their receptors during midfacial development. *Mech. Dev.* **100,** 313–316.

Bakkers, J., Hild, M., Kramer, C., Furutani-Seiki, M., and Hammerschmidt, M. (2002). Zebrafish deltaNp63 is a direct target of Bmp signaling and encodes a transcriptional repressor blocking neural specification in the ventral ectoderm. *Dev. Cell.* **2,** 617–627.

Barrow, J. R., and Capecchi, M. R. (1999). Compensatory defects associated with mutations in *Hoxa1* restore normal palatogenesis to *Hoxa2* mutants. *Development* **126,** 5011–5026.

Bei, M., Kratochwil, K., and Maas, R. (2000). BMP4 rescue a non-cell-autonomous function of Msx1 in tooth development. *Development* **127,** 4711–4718.

Bell, S. M., Schreiner, C. M., Waclaw, R. R., Campbell, K., Potter, S. S., and Scott, W. J. (2003). Sp8 is crucial for limb outgrowth and neuropore closure. *Proc. Natl. Acad. Sci. USA* **100,** 12195–12200.

Berk, M., Desai, S. Y., Heyman, H. C., and Comenares, C. (1997). Mice lacking the *ski* protooncogene have defects in neurulation, craniofacial, patterning, and skeletal muscle development. *Genes Dev.* **11,** 2029–2039.

Beverdam, A., Brouwer, A., Reijnen, M., Korving, J., and Meijlink, F. (2001). Severe nasal clefting and abnormal embryonic apoptosis in Alx3/Alx4 double mutant mice. *Development* **128,** 3975–3986.

Bi, W., Huang, W., Whitworth, D. J., Deng, J. M., Zhang, Z., Behringer, R. R., and de Crombrugghe, B. (2001). Haploinsufficiency of *Sox9* results in defective cartilage primordia and premature skeletal mineralization. *Proc. Natl. Acad. Sci. USA* **98,** 6698–6703.

Biddle, F. G., and Fraser, F. C. (1976). Genetics of cortisone-induced cleft palate in the mouse. Embryonic and maternal effects. *Genetics* **84,** 743–754.

Blavier, L., Lazaryev, A., Groffen, J., Heisterkamp, N., DeClerck, Y., and Kaartinen, V. (2001). TGF-β3-induced palatogenesis requires matrix metalloproteinases. *Mol. Biol. Cell* **12,** 1457–1466.

Brewer, C. M., Leek, J. P., Green, A. J., Holloway, S., Bonthron, D. T., Markham, A. F., and FitzPatrick, D. R. (1999). A locus for isolated cleft palate, located on human chromosome 2q32. *Am. J. Hum. Genet.* **65,** 387–396.

Bosman, E. A., Penn, A. C., Ambrose, J. C., Kettleborough, R., Stemple, D. L., and Steel, K. P. (2005). Multiple mutations in mouse Chd7 provide models for Charge syndrome. *Hum. Mol. Genet.* **14,** 3463–3476.

Braybrook, C., Doudney, K., Marcano, A. C. B., Arnason, A., Björnsson, A., Patton, M. A., Goodfellow, P. J., Moore, G., and Stanier, P. (2001). The T-box transcription factor gene TBX22 is mutated in X-linked cleft palate and ankyloglossia. *Genes Dev.* **29,** 179–183.

Braybrook, C., Lisgo, S., Doudney, K., Henderson, D., Marcano, A. C. B., Strachan, T., Patton, M. A., Villard, L., Moore, G. E., Stanier, P., and Lindsay, S. (2002). Craniofacial expression of human and murine TBX22 correlates with the cleft palate and ankyloglossia phenotype observed in CPX patients. *Hum. Mol. Genet.* **11,** 2793–2804.

Brewer, S., Feng, W., Huang, J., Sullivan, S., and Williams, T. (2004). Wnt1-Cre-mediated deletion of AP-2alpha causes multiple neural crest-related defects. *Dev. Biol.* **267,** 135–152.

Britanova, O., Akopov, S., Lykyanov, S., Gruss, P., and tarabykin, V. (2005). Novel transcription factor Satb2 interacts with matrix attachment region DNA elements in a tissue-specific manner and demonstrates cell-type-dependent expression in the developing mouse CNS. *Eur. J. Neurosci.* **21,** 658–668.

Britanova, O., Depew, M. J., Schwark, M., Thomas, B. L., Miletich, I., Sharpe, P. T., and Tarabykin, V. (2006). *Satb2* haploinsifficiency phenocopies 2q32-q33 deletions, whereas loss suggests a fundamental role in the coordination of jaw development. *Am. J. Hum. Genet.* **79,** 668–678.

Bush, J. O., Lan, Y., and Jiang, R. (2004). The cleft lip and palate defects in *Dancer* mutant mice result from gain of function of the Tbx10 gene. *Proc. Natl. Acad. Sci. USA* **101,** 7022–7027.

Calzolari, E., Pierini, A., Astolfi, G., Bianchi, F., Neville, A. J., and Rivieri, F. (2007). Associated anomalies in multi-malformed infants with cleft lip and palate: an epidemiologic study of nearly 6 million births in 23 EUROCAT registries. *Am. J. Med. Genet. A* **143,** 528–537.

Carter, M. G., Johns, M. A., Zeng, X., Zhou, L., Zinc, M. C., Mankowski, J. L., Donovan, D. M., and Baylin, S. B. (2000). Mice deficient in the candidate tumor suppressor gene Hic1 exhibit developmental defects of structures affected in the Miller–Dieker syndrome. *Hum. Mol. Genet.* **9,** 413–419.

Carroll, T. J., Park, J. S., Hayashi, S., Majumdar, A., and McMahon, A. P. (2005). Wnt9b plays a central role in the regulation in mesenchymal to epithelial transitions underlying organogenesis of the mammalian urogenital system. *Dev. Cell.* **9,** 283–292.

Carroll, D. K., Carroll, J. S., Leong, C.-O., Cheng, F., Brown, M., Mills, A. A., Brugge, J. S., and Ellisen, L. W. (2006). p63 regulates an adhesion programme and cell survival in epithelial cells. *Nat. Cell Biol.* **8,** 551–561.

Caruana, G., and Bernstein, A. (2001). Craniofacial dysmorphogenesis including cleft palate in mice with an insertional mutation in the *discs large* gene. *Mol. Cel. Biol.* **21,** 1475–1483.

Casey, L. M., Lan, Y., Cho, E. S., Maltby, K. M., Gridley, T., and Jiang, R. (2006). Jag2-Notch1 signaling regulates oral epithelial differentiation and palate development. *Dev. Dyn.* **235,** 1830–1844.

Caspary, T., Clearly, M. A., Perlman, E. J., Zhang, P., Elledge, S. J., and Tilghman, S. M. (1999). Oppositely imprinted genes $p57^{kip2}$ and *Igf2* interact in a mouse model for Beckwith–Wiedemann syndrome. *Genes Dev.* **13,** 3115–3124.

Castanet, M., Park, S. M., Smith, A., Bost, M., Léger, J., Lyonnet, S., Pelet, A., Czernichow, P., Chatterjee, K., and Polak, M. (2002). A novel loss-of-function mutation in *TTF-2* is associated with congenital hypothyroidism, thyroid agenesis and cleft palate. *Hum. Mol. Genet.* **11,** 2051–2059.

Cecconi, F., Alvarez-Bolado, G., Meyer, B. I., Roth, K. A., and Gruss, P. (1998). Apaf1 (CED-4 homolog) regulates programmed cell death in mammalian development. *Cell* **94,** 727–737.

Celli, J., Dujif, P., Hamel, B. C., Bamshad, M., Kramer, B., Smits, A. P., Newbury-Ecob, R., Hennekam, R. C., van Buggenhout, G., van Haeringen, A., Qoods, C. G., van Essen, A. J., *et al.* (1999). Heterozygous germline mutations in the p53 homolog p63 are the cause of EEC syndrome. *Cell* **99,** 143–153.

Chai, Y., and Maxon, R. E. (2006). Recent advances in craniofacial morphogenesis. *Dev. Dyn.* **235,** 2353–2375.

Chaudhari, N. (1992). A single nucleotide deletion in the skeletal muscle-specific calcium channel transcript of muscular dysgenesis (*mdg*) mice. *J. Biol. Chem.* **267,** 25636–25639.

Chen, C. M., and Behringer, R. R. (2004). Ovca1 regulates cell proliferation, embryonic development, and tumorigenesis. *Genes Dev.* **18,** 320–332.

Chiang, C., Litingtung, Y., Lee, E., Young, K., Corden, J., Westphal, H., and Beachy, P. A. (1996). Cyclopia and defective axial patterning in mice lacking sonic hedgehog gene function. *Nature* **383,** 407–423.

Chiquet, B. T., Blanton, S. H., Burt, A., Ma, D., Stal, S., Mulliken, J. B., and Hecht, J. T. (2008). Variation in Wnt genes is associated with nonsyndromic cleft lip with or without cleft palate. *Hum. Mol. Genet.* **17,** 2212–2218.

Christensen, K., and Mortensen, P. B. (2002). Facial clefting and psychiatric diseases: A follow-up of the Danish 1936–1987 Facial cleft cohort. *Cleft Palate Craniofac. J.* **39,** 392–396.

Christensen, K., Juel, K., Herskind, A. M., and Murray, J. C. (2004). Long term follow up of survival associated with cleft lip and palate birth. *B.M.J.* **328,** 1405.

Clifton-Bligh, R. J., Wentworth, J. M., Heinz, P., Crisp, M., Johns, R., Lazarus, J. H., Ludgate, M., and Chatterjee, V. K. (1998). Mutations of the gene encoding human TTF-2 associated with thyroid agenesis, cleft palate and choanal atresia. *Nat. Genet.* **19,** 399–401.

Cobourne, M. T. (2004). The complex genetics of cleft lip and palate. *Eur. J. Orthodont.* **26,** 7–16.

Colmenares, C., Heilstedt, H. A., Shaffer, L. G., Schwartz, S., Berk, M., Murray, J. C., and Stavnezer, E. (2002). Loss of the *SKI* proto-oncogene in individuals affected with 1p36 deletion syndrome is predicted by strain-dependent defects in $Ski^{-/-}$ mice. *Nat. Genet.* **30,** 106–109.

Colvin, J. S., White, A. C., Pratt, S. J., and Ornitz, D. M. (2001). Lung hypoplasia and neonatal death in *Fgf9*-null mice identify this gene as an essential regulator of lung mesenchyme. *Development* **128,** 2095–2106.

Condie, B. G., Bain, G. M., Gottlieb, D. I., and Capecchi, M. (1997). Cleft palate in mice with a argeted mutation in the γ-aminobutyric acid-producing enzyme glutamic acid decarboxylase 67. *Proc. Natl. Acad. Sci. USA* **94,** 11451–11455.

Cooper, M. K., Wassif, C. A., Krakowiak, P. A., Taipale, J., Gong, R., Kelley, R. I., Porter, F. D., and Beachy, P. A. (2003). A defective response to Hedgehog signaling in disorders of cholesterol biosynthesis. *Nat. Genet.* **33,** 508–513.

Corcoran, R. B., and Scott, M. P. (2006). Oxysterols stimulate Sonic hedgehog signal transduction and proliferation of medulloblastoma cells. *Proc. Natl. Acad. Sci. USA* **103,** 8408–8413.

Coré, N., Caubit, X., Metchat, A., Boned, A., Djabali, M., and Fasano, L. (2007). *Tshz1* is required for axial skeleton, soft palate and middle ear development in mice. *Dev. Biol.* **308,** 407–420.

Cox, G. A., Mahaffey, C. L., Nystuen, A., Letts, V. A., and Frankel, W. N. (2000). The mouse fidgetin gene defines a new role for AAA family proteins in mammalian development. *Nat. Genet.* **26,** 198–202.

Cui, X. M., Shiomi, N., Chen, J., Saito, T., Yamamoto, T., Ito, Y., Bringas, P., Chai, Y., and Shuler, C. F. (2005). Overexpression of Smad2 in *Tgfβ3*-null mutant mice rescues cleft palate. *Dev. Biol.* **278,** 193–202.

Culiat, C. T., Stubbs, L., Nicholls, R. D., Montgomery, C. S., Russell, L. B., Johnson, D. K., and Rinchik, E. M. (1993). Concordance between isolated cleft palate in mice and alterations within a region including the gene encoding the $\beta 3$ subunit of the type A γ-aminobutyric acid receptor. *Proc. Natl. Acad. Sci. USA* **90,** 5105–5109.

Culiat, C. T., Stubbs, L. J., Woychik, R. P., Russell, L. B., Johnson, D. K., and Rinchik, E. M. (1995). Deficiency of the $\beta 3$ subunit of the type A γ-aminobutyric acid receptor causes cleft palate in mice. *Nat. Genet.* **11,** 344–346.

Curtis, E. J., Fraser, F. C., and Warburton, D. (1961). Congenital cleft lip and palate. *Am. J. Dis. Child.* **102,** 853–857.

Curto, M., and McClatchey, A. (2008). Nf2/Merlin: A coordinator of receptor signaling and intercellular contact. *Br. J. Cancer* **98,** 256–262.

Dai, P., Shinagawa, T., Nomura, T., Harada, J., Kaul, S. C., Wadhwa, R., Khan, M. M., Akimaru, H., Sasaki, H., Colmenares, C., and Ishii, S. (2002). Ski is involved in trascriptional regulation by the repressor and full-length forms of Gli3. *Genes Dev.* **16,** 2843–2848.

De Felice, M., Ovitt, C., Biffali, E., Rodriguez-Mallon, A., Arra, C., Anastassiadis, K., Macchia, P. E., Mattei, M. G., Mariano, A., Scöler, H., Macchia, V., and Di Lauro, R. (1998). A mouse model for hereditary thyroid dysgenesis and cleft palate. *Nat. Genet.* **19,** 395–398.

De Moerlooze, L., Spencer-Dene, B., Revest, J., Hajihosseni, M., Rosewell, I., and Dickson, C. (2000). An important role for the IIIb isoform of fibroblast growth factor receptor 2 (FGFR2) in mesenchymal–epithelial signaling during mouse organogenesis. *Development* **127,** 483–492.

Derry, J. M., Gormally, E., Means, G. D., Zhao, W., Meindl, A., Kelley, R. I., Boyd, Y., and Heman, G. E. (1999). Mutations in a Δ^8–Δ^7 sterol isomerase in the tattered mouse and X-linked dominant chondrodysplasia punctata. *Nat. Genet.* **22,** 286–290.

Diehl, S. R., and Erickson, R. P. (1997). Genome scan for teratogen-induced clefting susceptibility loci in the mouse: Evidence of both allelic and locus heterogeneity distinguishing cleft lip and cleft palate. *Proc. Natl. Acad. Sci. USA* **94,** 5231–5236.

Dimitratos, S. D., Woods, D. F., Stathakis, D. G., and Bryant, P. J. (1999). Signaling pathways are focused at specialized regions of the plasma membrane by scaffolding proteins of the MAGUK family. *Bioessays* **21,** 912–921.

Ding, H., Wu, X., Boström, H., Kim, I., Wong, N., Tsoi, B., O'Rourke, M., Koh, G. H., Soriano, P., Betsholtz, C., Hart, T. C., Marazita, M. L., *et al.* (2004). A specific requirement for PDGF-C in palate formation and PDGFR-α signaling. *Nat. Genet.* **36,** 1111–1116.

Dixon, J., Jones, N. C., Sandell, L. L., Jayasinghe, S. M., Crane, J., Rey, J. P., Dixon, M. J., and Trainor, P. A. (2006). *Tcof1*/Treacle is required for neural crest cell formation and proliferation deficiencies that cause craniofacial abnormalities. *Proc. Natl. Acad. Sci. USA* **103,** 13403–13408.

Dobreva, G., Dambacher, J., and Grosschedl, R. (2003). SUMO modification of a novel MAR-binding protein, SATB2, modulates immunoglobulin μ gene expression. *Genes Dev.* **17,** 3048–3061.

Dobreva, G., Chahrour, M., Dautzenberg, M., Chirivella, L., Kanzler, B., Farinas, I., Karsenty, G., and Grosschedl, R. (2006). SATB2 is a multifunctional determinant of craniofacial patterning and osteoblast differentiation. *Cell* **125,** 971–986.

Dodé, C., Levilliers, J., Dupont, J. M., De Paepe, A., Le Du, N., Soussi-Yanicostas, N., Coimbra, R. S., Delmaghani, S., Compain-Nouaille, S., Baverel, F., Pecheux, C., Le Tessier, D., *et al.* (2003). Loss-of-function mutations in *FGFRI* cause autosomal dominant Kallmann syndrome. *Nat. Genet.* **33,** 463–465.

Dodé, C., Fouveaut, C., Mortier, G., Janssens, S., Bertherat, J., Mahoudeau, J., Kottler, M. L., Chabrolle, C., Gancel, A., François, I., Devriendt, K., Wolczynski, S., *et al.* (2007). Novel FGFR1 sequence variants in Kallmann syndrome, and genetic evidence that the FGFR1c isoform is required in olfactory bulb and palate morphogenesis. *Hum. Mutat.* **28,** 97–98.

Donner, A. L., and Williams, T. (2006). Frontal nasal prominence expression driven by Tcfap2a relies on a conserved binding site for STAT proteins. *Dev. Dyn.* **235,** 1358–1370.

Ducy, P., Starbuck, M., Priemel, M., Shen, J., Pinero, G., Geoffroy, V., Amling, M., and Karsenty, G. (1999). A Cbfa1-dependent genetic pathway controls bone formation beyond embryonic development. *Genes Dev.* **13,** 1025–1036.

Dudas, M., and Kaartinen, V. (2005). Tgf-beta superfamily and mouse craniofacial development: interplay of morphogenetic proteins and receptor signaling controls normal formation of the face. *Curr. Top. Dev. Biol.* **66,** 65–133.

Dudas, M., Nagy, A., Laping, N. J., Moustakas, A., and Kaartinen, V. (2004a). Tgf-beta3-induced palatal fusion is mediated by Alk-5/Smad pathway. *Dev. Biol.* **266,** 96–108.

Dudas, M., Sridurongrit, S., Nagy, A., Okazaki, K., and Kaartinen, V. (2004b). Craniofacial defects in mice lacking BMP type I receptor Alk2 in neural crest cells. *Mech. Dev.* **121,** 173–182.

Dudas, M., Kim, J., Li, W. Y., Nagy, A., Larsson, J., Karlsson, S., Chai, Y., and Kaartinen, V. (2006). Epithelial and ectomesenchymal role of the type I TGF-β receptor ALK5 during facial morphogenesis and palatal fusion. *Dev. Biol.* **296,** 298–314.

Dudas, M., Li, W. Y., Kim, J., Yang, A., and Kaartinen, V. (2007). Palatal fusion-Where do the midline cells go? A review on cleft palate, a major human birth defect. *Acta Histochem.* **109,** 1–14.

Dyer, M. A., and Cepko, C. L. (2000). p57 (Kip2) regulates progenitor cell prolifearation and amacrine interneuron development in the mouse retina. *Development* **127,** 3593–3605.

Edwards, S. J., Gladwin, A. J., and Dixon, M. J. (1997). The mutational spectrum in Treacher Collins syndrome reveals a predominance of mutations that create a premature-termination codon. *Am. J. Hum. Genet.* **60,** 515–524.

Engelking, L. J., Evers, B. M., Richardson, J. A., Goldstein, J. L., Brown, M. S., and Liang, G. (2006). Severe facial clefting in Insig-deficient mouse embryos caused by sterol accumulation and reversed by lovastatin. *J. Clin. Invest.* **116,** 2356–2365.

Engleka, K. A., Meilin, W., Zhang, M., Antonucci, N. B., and Epstein, J. A. (2007). Menin is required in cranial neural crest for palatogenesis and perinatal viability. *Dev. Biol.* **311,** 524–537.

Espenshade, P. J., and Hughes, A. L. (2007). Regulation of sterol synthesis in eukaryotes. *Annu. Rev. Genet.* **41,** 401–427.

Eswarakumar, V. P., Horowitz, M. C., Locklin, R., Morriss-Kay, G. M., and Lonai, P. (2004). A gain-of-function mutation of Fgfr2c demonstrates the roles of this receptor variant in osteogenesis. *Proc. Natl. Acad. Sci. USA* **101,** 12555–12560.

Felix, T. M., Hanshaw, B. C., Mueller, R., Bitoun, P., and Murray, J. C. (2006). *CHD7* gene and non-syndromic cleft lip and palate. *Am. J. Med. Genet.* **140A,** 2110–2114.

Ferguson, M. W. J. (1988). Palate development. *Development* **103,** 41–61.

Fitzky, B. U., Moebius, F. F., Asaoka, H., Waage-Baudet, H., Xu, L., Xu, G., Maeda, N., Kluckman, K., Hiller, S., Yu, H., Batta, A. K., Shefer, S., *et al.* (2001). 7-Dehydrocholesterol-dependent proteolysis of HMG-CoA reductase suppresses sterol biosynthesis in a mouse model of Smith–Lemli–Opitz/RSH syndrome. *J. Clin. Invest.* **108,** 905–915.

FitzPatrick, D. R., Carr, I. M., McLaren, L., Leek, J. P., Wighman, P., Williamson, K., Gautier, P., McGill, N., Hayward, C., Firth, H., Markham, A. F., Fantes, J. A., *et al.* (2003). Identification of *SATB2* as the cleft palate gene on 2q32–q33. *Hum. Mol. Genet.* **12,** 2491–2501.

Frank, D. U., Fotheringham, L. K., Brewer, J. A., Muglia, L. J., Tristani-Firouzi, M., Capecchi, M. R., and Moon, A. M. (2002). An Fgf8 mouse mutant phenocopies human 22q11 deletion syndrome. *Development* **129,** 4591–4603.

Fraser, F. C. (1970). The genetics of cleft lip and cleft palate. *Am. J. Hum. Genet.* **22,** 336–352.

Frebourg, T., Oliveira, C., Hochain, P., Karam, R., Manouvrier, S., Graziado, C., Vekemans, M., Hartman, A., Baert-Desurmont, S., Alexandre, C., Lejeune Dumoulin, S., Marroni, C., *et al.* (2006). Cleft lip/palate and CDH1/E-cadherin mutations in families with hereditary diffuse gastric cancer. *J. Med. Genet.* **43,** 138–142.

Fristrom, D. (1988). The cellular basis of epithelial morphogenesis. A review. *Tissue Cell* **20,** 645–690.

Furuta, Y., Piston, D. W., and Hogan, B. L. (1997). Bone morphogenetic proteins (BMP's) as regulators of dorsal forebrain development. *Development* **124,** 2203–2212.

Gendron-Maguire, M., Mallo, M., Zhang, M., and Gridley, T. (1993). *Hoxa2* mutant mice exhibit homeotic transformation of skeletal elements derived from cranial neural crest. *Cell* **75,** 1317–1331.

Georgia, S., Soliz, R., Li, M., Zhang, P., and Bhushan, A. (2006). p57 and Hes coordinate cell cycle exit with self-renewal of pancreatic progenitors. *Dev. Biol.* **298,** 22–31.

Gong, S. G., and Eulenberg, R. L. (2001). Palatal development in twirler mice. *Cleft Palate Craniof. J.* **38,** 622–628.

Gong, S. G., and Guo, C. (2003). *Bmp4* gene is expressed at the putative site of fusion in midfacial region. *Differentiation* **71,** 228–236.

Gong, S. G., White, N. J., and Sakasegawa, A. Y. (2000). The Twirler mouse, a model for the study of cleft lip and palate. *Arch. Oral Biol.* **45,** 87–94.

Greene, R. M., and Pratt, R. (1976). Developmental aspects of secondary palate formation. *J. Embryol. Exp. Morphol.* **36,** 225–245.

Gritli-Linde, A. (2007). Molecular control of secondary palate development. *Dev. Biol.* **301,** 309–326.

Gritli-Linde, A., Lewis, P., McMahon, A. P., and Linde, A. (2001). The whereabouts of a morphogen: Direct evidence for short- and graded long-range activity of Hedgehog signaling peptides. *Dev. Biol.* **236,** 364–386.

Gu, S., Wei, N., Yu, X., Jiang, Y., Fei, J., and Chen, Y. P. (2008). Mice with an anterior cleft of the palate survive neonatal lethality. *Dev. Dyn.* **237,** 1509–1516.

Gupta, V., and Bei, M. (2006). Modification of Msx1 by SUMO-1. *Biochem. Biophys. Res. Commun.* **345,** 74–77.

Hagiwara, N., Katarova, Z., Siracusa, L., and Brilliant, M. H. (2003). Nonneuronal expression of the $GABA_A\beta3$ subunit gene is required for normal palate development in mice. *Dev. Biol.* **254,** 93–101.

Halford, M. M., Armes, J., Buchert, M., Meskenaite, V., Grail, D., Hibbs, M. L., Wilks, A. F., Farlie, P. G., Newgreen, D. F., Hovens, C. M., and Stacker, S. A. (2000). Ryk-deficient mice exhibit craniofacial defects associated with perturbed Eph receptor crosstalk. *Nat. Genet.* **25,** 414–418.

Hart, A. W., Morgan, J. E., Schneider, J., West, K., McKie, L., Bhattacharya, S., Jackson, I. J., and Cross, S. H. (2006). Cardiac malformations and midline skeletal defects in mice lacking filamin A. *Hum. Mol. Genet.* **15,** 2457–2467.

Homanics, G. E., DeLorey, T. M., Firestone, L. L., Quinlan, J. J., Handforth, A., Harrison, N. L., Krasowski, M. D., Rick, C. E., Korpi, E. R., Mäkelä, R., Brilliant, M. H., Hagiwara, N., *et al.* (1997). Mice devoid of γ-aminobutyrate type A receptor $\beta3$ subunit have epilepsy, cleft palate, and hypersensitive behavior. *Proc. Natl. Acad. Sci. USA* **94,** 4143–4148.

Honarpour, N., Du, C., Richardson, J. A., Hammer, R. E., Wang, X., and Hertz, J. (2000). Adult Apaf1-deficient mice exhibit male infertility. *Dev. Biol.* **218,** 248–258.

Hilliard, S. A., Yu, L., Gu, S., Zhang, Z., and Chen, Y. P. (2005). Regional regulation of palatal growth and patterning along the anterior-posterior axis in mice. *J. Anat.* **207,** 655–667.

Himanen, J. P., Nayanendu, S., and Nikolov, D. B. (2007). Cell–cell signaling via Eph receptors and ephrins. *Curr. Opin. Cell Biol.* **19,** 534–542.

Huang, Y. P., Wu, G., Guo, Z., Osada, M., Fomenkov, T., Park, H. L., Trink, B., Sidransky, D., Fomenkov, A., and Ratovitski, E. A. (2004). Altered sumoylation of p63α contributes to the split-hand/foot malformation phenotype. *Cell Cycle* **3,** 1587–1596.

Ianakiev, P., Kilpatrick, M. W., Toudjarska, I., Basel, D., Beighton, P., and Tsiporas, P. (2000). Split-hand/split foot malformation is caused by mutations in the p63 gene on 3q27. *Am. J. Hum. Genet.* **67,** 59–66.

Ichikawa, E., Watanabe, A., Nakano, Y., Akita, S., Hirano, A., Kinoshita, A., Kondo, S., Kishino, T., Uchiyama, T., Niikawa, N., and Noshiura, K. (2006). *PAX9* and *TGFB3* are linked to susceptibility to nonsyndromic cleft lip with or without cleft palate in the Japanese: population-based and family-based candidate gene analyses. *J. Hum. Genet.* **51,** 38–46.

Incardona, J. P., Gaffield, W., Kapur, R. P., and Roelink, H. (1998). The teratogenic veratrum cyclopamine inhibits sonic hedgehog signal transduction. *Development* **125,** 3553–3562.

Ingraham, C. R., Kinoshita, A., Kondo, S., Yang, B., Sajan, S., Trout, K. J., Malik, M. I., Dunnwald, M., Goudy, S. L., Lovett, M., and Murray, J. C. (2006). Abnormal skin, limb and craniofacial morphogenesis in mice deficient for interferon regulatory factor 6 (*Irf6*). *Nat. Genet.* **38,** 1335–1340.

Inoue, H., Kayano, S., Aoki, Y., Kure, S., Yamada, A., Hata, A., Matsubara, Y., and Suzuki, Y. (2008). Association of the *GABRB3* gene with nonsyndromic oral clefts. *Cleft Palate Craniofac. J.* **45,** 261–266.

Iseki, S., Ishii-Suzuki, M., Tsunekawa, N., Yamada, Y., Eto, K., and Obata, K. (2007). Experimental induction of palate shelf elevation in glutamate decarboxylase 67-deficient mice with cleft palate due to vertically oriented palatal shelf. *Birth Defects Res. (PartA)* **79,** 688–695.

Ito, Y., Yeo, J. Y., Chytil, A., Han, J., Bringas, P., Nakajima, A., Shuler, C. F., Moses, H. L., and Chai, Y. (2003). Conditional inactivation of *Tgfbr2* in cranial neural crest causes cleft palate and calvarial defects. *Development* **130,** 5269–5280.

Jeong, J., Mao, J., Tenzen, T., Kottmann, A. H., and McMahon, A. P. (2004). Hedgehog signaling in the neural crest cells regulates the patterning and growth of facial primordia. *Genes Dev.* **18,** 937–951.

Jerome, L. A., and Papaioannou, V. E. (2001). DiGeorge syndrome phenotype in mice mutant for the T-box gene, *Tbx1*. *Nat. Genet.* **27,** 286–291.

Jezewski, P. A., Vieira, A. R., Nishimura, C., Ludwig, B., Johnson, M., O'Brien, S. E., Daack-Hirsch, S., Schultz, R. E., Weber, A., Nepomucena, B., Romitti, P. A., Christensen, K., *et al.* (2003). Complete sequencing shows a role for *MSX1* in non-syndromic cleft lip and palate. *J. Med. Genet.* **40,** 399–407.

Jiang, R., Lan, Y., Chapman, H. D., Shawber, C., Norton, C. R., Serreze, D. V., Weinmaster, G., and Gridley, T. (1998). Defects in limb, craniofacial, and thymic development in *Jagged2* mutant mice. *Genes Dev.* **12,** 1046–1057.

Jiang, R., Bush, J. O., and Lidral, A. (2006). Development of the upper lip: Morphogenetic and molecular mechanisms. *Dev. Dyn.* **235,** 1152–1166.

Jin, J. Z., and Ding, J. (2006a). Analysis of Meox-2 mutant mice reveals a novel postfusion-based cleft palate. *Dev. Dyn.* **235,** 539–546.

Jin, J. Z., and Ding, J. (2006b). Analysis of cell migration, transdifferentiation and apoptosis during mouse secondary palate fusion. *Development* **133,** 3341–3347.

Johnston, M. C., and Sulik, K. K. (1979). Some abnormal patterns of development in the craniofacial region. *Birth Defects* **15,** 23–42.

Johnston, M. C., and Bronsky, P. T. (1995). Prenatal craniofacial development: New insights on normal and abnormal mechanisms. *Crit. Rev. Oral Biol. Med.* **6,** 368–422.

Jones, N. C., Lynn, M. L., Gaudenz, K., Sakai, D., Aoto, K., Rey, J. P., Glynn, E. F., Ellington, L., Du, C., Dixon, J., and Trainor, P. A. (2008). Prevention of the neurocristopathy Treacher Collins Syndrome through inhibition of p53 function. *Nat. Med.* **14,** 125–133.

Jugessur, A., and Murray, J. C. (2005). Orofacial clefting: recent insights into a complex trait. *Curr. Opin. Genet. Dev.* **15,** 270–278.

Juriloff, D. M., and Harris, M. J. (2008). Mouse genetic models of cleft lip with or without cleft palate. *Birth Defects Res. (Part A)* **82,** 63–77.

Juriloff, D. M., Harris, M. J., McMahon, A. P., Carroll, T. J., and Lidral, A. C. (2006). Wnt9b is the mutated gene involved in multifactorial nonsyndromic cleft lip with or without cleft palate in A/wySn mice, as confirmed by a genetic complementation test. *Birth Defects Res. A Clin. Mol. Teratol.* **76,** 574–579.

Kaartinen, V., Volcken, J. W., Shuler, C., Warburton, D., Bu, D., Heisterkamp, N., and Groffen, J. (1995). Abnormal lung development and cleft palate in mice lacking TGF-β3 indicates defects of epithelial–mesenchymal interactions. *Nat. Genet.* **11,** 415–421.

Kanno, K., Suzuki, Y., Yamada, A., Aoki, Y., and Kure, S. (2004). Association between nonsyndromic cleft lip with or without cleft palate and the glutamic acid decarboxylase 67 gene in the japanese population. *Am. J. Med. Genet.* **127,** 11–16.

Klein, O. D., Minowada, G., Peterkova, R., Kangas, A., Yu, B. D., Lesot, H., Peterka, M., Jernvall, J., and Martin, G. R. (2006). Sprouty genes control diastema tooth development via bidirectional antagonism of epithelial–mesenchymal FGF signaling. *Dev. Cell* **11,** 181–190.

Knight, A. S., Schutte, B. C., Jiang, R., and Dixon, M. J. (2006). Developmental expression analysis of the mouse and chick orthologues of *IRF6*: The gene mutated in van der Woude syndrome. *Dev. Dyn.* **235,** 1441–1447.

Kondo, S., Schutte, B. C., Richardson, R. J., Bjork, B. C., Knight, A. S., Watanabe, Y., Howard, E., de Lima, R. L., Daack-Hirsh, S., Sander, A., McDonald-McGinn, D. M., Zackai, E. H., *et al.* (2002). Mutations in *IRF6* cause Van der Woude and popliteal pterygium syndromes. *Nat. Genet.* **32,** 285–289.

Krakowiak, P. A., Wassif, C. A., Kratz, L., Cozma, D., Kovárová, M., Harris, G., Grinberg, A., Yang, Y., Hunter, A. G. W., Tsokos, M., Kelley, R. I., and Porter, F. D. (2003). Lathosterolosis: an inborn error of human and murine cholesterol synthesis due to lathosterol 5-desaturase deficiency. *Hum Mol. Genet.* **12,** 1631–1641.

Kreiborg, S., and Cohen, M. M., Jr. (1992). The oral manifestations of Apert syndrome. *J. Craniofac. Genet. Dev. Biol.* **12,** 41–48.

Lan, Y., Ovitt, C. V., Cho, E. S., Maltby, K. M., Wang, Q., and Jiang, R. (2004). Odd-skipped related 2 (*Osr2*) encodes a key intrinsic regulator of secondary palate growth and morphogenesis. *Development* **131,** 3207–3216.

Lan, Y., Ryan, R. C., Zhang, Z., Bullard, S., Bush, J. O., Maltby, K. M., Lidral, A. C., and Jiang, R. (2006). Expression of Wnt9b and activation of canonical Wnt signaling during midfacial morphogenesis in mice. *Dev. Dyn.* **235,** 1448–1454.

Lanctôt, C., Moreau, A., Chamberland, M., Trembley, M. L., and Drouin, J. (1999). Hindlimb patterning and mandible development requires *Ptx1* gene. *Development* **126,** 1805–1810.

Laumonnier, F., Holbert, S., Ronce, N., Faravelli, F., Lenzner, S., Schwartz, C. E., Lespinasse, J., Van Esch, H., Lacombe, D., Goizet, C., Phan-Dinh Tuy, F., van Bokhoven, H., *et al.* (2005). Mutations in *PHF8* are associated with X linked mental retardation and cleft lip/cleft palate. *J. Med. Genet.* **42,** 780–786.

Laverty, H. G., and Wilson, J. B. (1998). Murine CASK is disrupted in a sex-linked cleft palate mouse mutant. *Genomics* **53,** 29–41.

Lazzaro, C. (1940). Sul meccanismo di chiusura del palato secondario. *Monit. Zool. Ital.* **51,** 249–273.

Lee, H., Quinn, J. C., Prasanth, K. V., Swiss, V. A., Economides, K. D., Camacho, M. M., Spector, D. L., and Abate-Shen, C. (2006). PIAS1 confers DNA-binding specificity on the Msx1 homeoprotein. *Genes Dev.* **20,** 784–794.

Lee, J. M., Kim, J. Y., Cho, K. W., Lee, M. J., Cho, S. W., Kwak, S., Cai, J., and Jung, H. S. (2008). Wnt/Fgfr1b cross talk modulates the fate of cells in palate development. *Dev. Biol.* **314,** 341–350.

Leoyklang, P., Siriwan, P., and Shotelersuk, V. (2006). A mutation of the p63 gene in non-syndromic cleft lip. *J. Med. Genet.* **43,** 28–32.

Leoyklang, P., Suphapeetiporn, K., Siriwan, P., Desudchit, T., Chaowanapanja, P., Gahl, W. A., and Shotelersuk, V. (2007). Heterozygous nonsense mutation in *SATB2* associated with cleft palate, osteoporosis, and cognitive defects. *Hum. Mutat.* **28,** 732–738.

Li, Q., Lu, Q., Hwang, J. Y., Buscher, D., Lee, K. F., Izpisua-Belmonte, J. C., and Verma, I. M. (1999). IKK1-deficient mice exhibit abnormal development of skin and skeleton. *Genes Dev.* **13,** 1322–1328.

Lidral, A. C., Romitti, P. A., Basart, A. M., Doetschman, T., Leysens, N. J., Daak-Hirsch, S., Semina, E. V., Johnson, L. R., Machida, J., Burds, A., Parnell, T. J., Rubenstein, J. L., *et al.* (1998). Association of MSX1 and TGFB3 with nonsyndromic clefting in humans. *Am. J. Hum. Genet.* **63,** 557–568.

Liu, Z., Xu, J., Colvin, J. S., and Ornitz, D. M. (2002). Coordination of chondrogenesis by fibroblast growth factor 18. *Genes Dev.* **16,** 859–869.

Lin, X., Liang, M., Liang, Y. Y., Brunicardi, F. C., Melchior, F., and Feng, X. H. (2003). Activation of transforming growth factor-beta signaling by SUMO-1 modification of tumor suppressor Smad4/DPC4. *J. Biol. Chem.* **278,** 18714–18719.

Liu, W., Sun, X., Braut, A., Mishina, Y., Behringer, R. R., Mina, M., and Martin, J. F. (2005). Distinct functions for Bmp signaling in lip and palate fusion in mice. *Development* **132,** 1453–1461.

Liu, H., Liu, W., Maltby, K. M., Lan, Y., and Jiang, R. (2006). Identification and developmental expression analysis of a novel homeobox gene closely linked to the mouse *Twirler* mutation. *Gene. Exp. Patterns* **6,** 632–636.

Liu, K. J., Arron, J. R., Stankunas, K., Crabtree, G. R., and Longaker, M. (2007). Chemical rescue of cleft palate and midline defects in conditional GSK-3β mice. *Nature* **446,** 79–82.

Loevy, H., and Fenyes, V. (1968). Spontaneous cleft palate in a family of siamese cats. *Cleft Palate J.* **5,** 57–60.

Loeys, B. L., Chen, J., Neptune, E. R., Judge, D. P., Podowski, M., Holm, T., Meyers, J., Leitch, C. C., Katsanis, N., Sharif, N., Xu, F. L., Myers, L. A., *et al.* (2005). A syndrome of altered cardiovascular, craniofacial, neurocognitive and skeletal development caused by mutations in TGFBR1 or TGFBR2. *Nat. Genet.* **37,** 275–281.

Lohnes, D., Mark, M., Mendelson, C., Dollé, P., Dietrich, A., Gorry, P., Gansmuller, A., and Chambon, P. (1994). Function of the retinoic acid receptors (RARs) during development. (I) Craniofacial and skeletal anomalies in RAR double mutants. *Development* **120,** 2723–2748.

Lu, M. F., Pressman, C., Dyer, R., Johnson, R. L., and Martin, J. F. (1999a). Function of Rieger syndrome gene in left-right asymmetry and craniofacial development. *Nature* **401,** 276–278.

Lu, W., Yamamoto, V., Ortega, B., and Baltimore, D. (2004). Mammalian Ryk is a Wnt coreceptor required for stimulation of neurite outgrowth. *Cell* **119,** 97–108.

Lupas, A. N., and Martin, J. (2002). AAA proteins. *Curr Opin. Struct. Biol.* **12,** 746–753.

Mann, M. B., Hodges, C. A., Barnes, E., Vogel, H., Hassold, T. J., and Guangbin, L. (2005). Defective sister-chromatid cohesion, aneuploidy and cancer predisposition in a mouse model of type II Rothmund–Thomson syndrome. *Hum. Mol. Genet.* **14,** 813–825.

Mansilla, M. A., Cooper, M. E., Goldstein, T., Castilla, E. E., Lopez-Camelo, J. S., Marazita, M. L., and Murray, J. C. (2006). Contributions of PTCH gene variants to isolated cleft lip and palate. *Cleft Palate Craniofac. J.* **43,** 21–29.

Marazita, M. L. (2002). Genetic etiologies of facial clefting. *In* "Understanding craniofacial anomalies. The etiopathogenesis of craniosynostoses and facial clefting" (M. P. Mooney and M. I. Siegel, eds.), pp. 147–161. Wiley-Liss, New York.

Marazita, M. L., Murray, J. C., Lidral, A. C., Arcos-Burgos, M., Cooper, M. E., Golstein, T., Maher, S. B., Daack-Hirch, S., Schultz, R., Mansilla, M. A., Field, L. L., Liu, Y.-E., *et al.* (2004). Meta-analysis of 13 genome scans reveals multiple cleft lip/palate genes with novel loci on 9q21 and 2q32–35. *Am. J. Hum. Genet.* **75,** 161–173.

Marçano, A. C. B., Doudney, K., Braybrook, C., Squires, R., Patton, M. A., Lees, M. M., Richieri-Costa, A., Lidral, A. C., Murray, J. C., Moore, G. E., and Stanier, P. (2004). TBX22 mutations are a frequent cause of cleft palate. *J. Med. Genet.* **41,** 68–74.

Martinelli, M., Di Stazio, M., Scapoli, L., Marchesini, J., Di Bari, F., Pezzetti, F., Carinci, F., Palmieri, A., Carinci, P., and Savoia, A. (2007). Cleft lip with or without cleft palate: implication of the heavy chain of non-muscle myosin IIA. *J. Med. Genet.* **44,** 387–392.

Marx, S. J. (2005). Molecular genetics of multiple endocrine neoplasia type 1 and 2. *Nat. Rev. Cancer* **5,** 367–375.

McGrath, J. A., Dujif, P. H., Doetsch, V., Irvine, A. D., de Waal, R., Vanmolkot, K. R., Wessagowit, V., Kelly, A., Atherton, D. J., Griffiths, W. A., Orlows, S. J., van Haeringen, A., *et al.* (2001). Hay–Wells syndrome is caused by heterozygous missense mutations in the SAM domain of p63. *Hum. Mol. Genet.* **10,** 221–229.

McLaughlin, M. E., Kruger, G. M., Slocum, K. L., Crowley, D., Michaud, N. A., Huang, J., Magendantz, M., and Jacks, T. (2007). The Nf2 tumor suppressor regulates cell–cell adhesion during tissue fusion. *Proc. Natl. Acad. Sci. USA* **104,** 3261–3266.

McMahon, A. P., Ingham, P. W., and Tabin, C. J. (2003). Developmental roles and clinical significance of hedgehog signaling. *Curr. Top. Dev. Biol.* **53,** 1–114.

Meulmeester, E., and Melchior, F. (2008). SUMO. *Nature* **452,** 709–711.

Miettinen, P. J., Chin, J. R., Shum, L., Slavkin, H. C., Shuler, C. F., Derynck, R., and Werb, Z. (1999). Epidermal growth factor receptor function is necessary for normal craniofacial development and palate closure. *Nat. Genet.* **22,** 69–73.

Mills, A. A., Zheng, B., Wang, X. J., Vogel, H., Roop, D. R., and Bradley, A. (1999). p63 is a p53 homologue required for limb and epidermal morphogenesis. *Nature* **398,** 708–713.

Milunsky, J. M., Maher, T. A., Zhao, G., Roberts, A. E., Stalker, H. J., Zori, R. T., Burch, M. N., Clemens, M., Mulliken, J. B., and Lin, A. E. (2008). TFAP2A mutations result in Branchio-oculo-facial syndrome. *Am. J. Hum. Genet.* **82,** 1171–1177.

Ming, J. E., Kaupas, M. E., Roessler, E., Brunner, H. G., Golabi, M., Tekin, M., Stratton, R. F., Sujansky, E., Bale, S. J., and Muenke, M. (2002). Mutations in PATCHED-1, the receptor for SONIC HEDGEHOG, are associated with holoprosencephaly. *Hum. Genet.* **110,** 297–301.

Mitchell, L. E. (1997). Transforming growth factor α locus and nonsyndromic cleft lip with or without cleft palate. A reappraisal. *Genet. Epidemiol.* **14,** 231–240.

Mo, R., Freer, A. M., Zinyk, D. L., Crackower, M. A., Michaud, J., Heng, H. H. Q., Chik, K. W., Shi, X. M., Tsui, L. C., Cheng, S. H., Joyner, A. L., and Hui, C. C. (1997). Specific and redundant function of *Gli2* and *Gli3* zinc finger genes in skeletal patterning and development. *Development* **124,** 113–123.

Moloney, D. M., Slaney, S. F., Oldridge, M., Wall, S. A., Sahlin, P., Stenman, G., and Wilkie, A. O. M. (1996). Exclusive paternal origin of new mutations in Apert syndrome. *Nat. Genet.* **13,** 48–53.

Mori-Akiyama, Y., Akiyama, H., Rowitch, D. H., and de Crombrugghe, B. (2003). Sox9 is required for determination of chondrogenic cell lineage in the cranial neural crest. *Proc. Natl. Acad. Sci. USA* **100,** 9360–9365.

Muenke, M. (1995). Holoprosencephaly as a genetic model for normal craniofacial development. *Semin. Dev. Biol.* **5,** 293–301.

Muenke, M. (2002). The pit, the cleft and the web. *Nat. Genet.* **32,** 219–220.

Muenke, M., and Beachy, P. A. (2000). Genetics of ventral forebrain development and holoprosencephaly. *Curr. Opin. Genet. Dev.* **10,** 262–9.

Mulvihill, J. J., Mulvihill, C. G., and Priester, W. A. (1980). Cleft palate in dimestic animals: Epidemiologic features. *Teratology* **21,** 109–112.

Murray, S. A., Oram, K. F., and Gridley, T. (2007). Multiple functions of snail family genes during palate development in mice. *Development* **134,** 1789–1797.

Nam, J. S., Park, E., Turcotte, T. J., Palencia, S., Zhan, X., Lee, J., Yun, K., Funk, W. D., and Yoon, J. K. (2007). Mouse *R-spondin2* is required for apical ectodermal ridge maintenance in the hind limb. *Dev. Biol.* **311,** 124–135.

Nie, X., Luukko, K., and Kettunen, P. (2006a). FGF signaling in craniofacial development and developmental disorders. *Oral Dis.* **12,** 102–111.

Nie, X., Luukko, K., and Kettunen, P. (2006b). BMP signaling in craniofacial development. *Int. J. Dev. Biol.* **50,** 511–521.

Nottoli, T., Hagopian-Donaldson, S., Zhang, J., Perkins, A., and Williams, T. (1998). *AP-2* null cells disrupt morphogenesis of the eye, face, and limbs in chimeric mice. *Proc. Natl. Acad. Sci. USA* **95,** 13714–13719.

Ohbayashi, N., Shibayama, M., Kurotaki, Y., Imanishi, M., Fujimori, T., Itoh, N., and Takada, S. (2002). FGF18 is required for normal cell proliferation and differentiation during osteogenesis and chondrogenesis. *Genes Dev.* **16,** 870–879.

Orioli, D., Hekenmeyer, M., Lemke, G., Klein, R., and Pawson, T. (1996). Sek4 and Nuk receptors cooperate in guidance of commissural axons and in palate formation. *EMBO J.* **15,** 6035–6049.

Osoegawa, K., Vessere, G. M., Utami, K. H., Mansilla, M. A., Johnson, M. K., Riley, B. M., L'Heureux, J., Pfundt, R., van der Vliet, W. A., Lidral, A. C., Schoenmakers, E. F. P. M., Borg, A., *et al.* (2008). Identification of novel candidate genes associated with cleft lip and palate using array comparative genomic hybridization. *J. Med. Genet.* **45,** 81–86.

Pai, A. C. (1965). Developmental genetics of a lethal mutation, muscular dysgenesis (MDG), in the mouse. I. Genetic analysis and gross morphology. *Dev. Biol.* **11,** 82–92.

Park, W. J., Theda, C., Maestri, N. E., Meyers, G. A., Fryburg, J. S., Difresne, C., Cohen, M. M., Jr., and Jabs, E. W. (1995). Analysis of phenotypic features and FGFR2 mutations in Apert syndrome. *Am. J. Hum. Genet.* **57,** 321–328.

Park, T. J., Boyd, K., and Curran, T. (2006). Cardiovascular and craniofacial defects in Crk-null mice. *Mol. Cell Biol.* **26,** 6272–6282.

Parrish, M., Ott, T., Lance-Jones, C., Schuetz, G., Schwaeger-Nickolenko, A., and Monaghan, A. P. (2004). Loss of the Sall3 gene leads to palate deficiency, abnormalities in cranial nerves, and perinatal lethality. *Mol. Cell Biol.* **24,** 7102–7112.

Pasquale, E. B. (2008). Eph-Ephrin bidirectional signaling in physiology and disease. *Cell* **133,** 38–52.

Peter, K. (1924). Die entwicklung des Säugetiergaumens. *Ergebn. Anat. EntwGesch.* **25,** 448–564.

Peters, H., Neubuser, A., Kratochwil, K., and Balling, R. (1998). *Pax9*-deficient mice lack pharyngeal pouch derivatives and teeth and exhibit craniofacial and limb anomalies. *Genes Dev.* **12,** 2735–2747.

Piedrahita, J. A., Oetama, B., Bennett, G. D., van Waes, J., Kamen, B. A., Richardson, J., Lacey, S. W., Anderson, R. G., and Finnell, R. H. (1999). Mice lacking the folic acid-binding protein Folbp1 are defective in early embryonic development. *Nat. Genet.* **23,** 228–232.

Porter, F. D. (2006). Cholesterol precursors and facial clefting. *J. Clin. Invest.* **116,** 2322–2325.

Proetzel, G., Pawlowski, S. A., Wiles, M. V., Yin, M., Boivin, G. P., Howles, P. N., Ding, J., Ferguson, M. W. J., and Doetschman, T. (1995). Transforming growth factor-β 3 is required for secondary palate fusion. *Nat. Genet.* **11,** 409–414.

Qu, S., Tucker, S. C., Zhao, Q., de Crombrugghe, B., and Wisdom, R. (1999). Physical and genetic interactions between Alx4 and Cart1. *Development* **126,** 359–369.

Radhakrishnan, A., Ikeda, Y., Kwon, H. J., Brown, M. S., and Golstein, J. L. (2007). Sterol-regulated transport of SREBs from endoplasmic reticulum to Golgi: Oxysterols block transport by binding to Insig. *Proc. Natl. Acad. Sci. USA* **104,** 6511–6518.

Ribeiro, L. A., Murray, J. C., and Richieri-Costa, A. (2006). PTCH mutations in four Brazilian patients with holoprosencephaly and in one with holoprosencephaly-like features and normal MRI. *Am. J. Med. Genet. A* **140,** 2584–2586.

Rice, D. P. C. (2005). Craniofacial anomalies: From development to molecular pathogenesis. *Curr. Mol. Med.* **5,** 669–722.

Rice, R., Spencer-Dene, B., Connor, E. C., Gritli-Linde, A., McMahon, A. P., Dickson, C., Thesleff, I., and Rice, D. P. C. (2004). Disruption of *Fgf10/Fgfr2b*-coordinated epithelial–mesenchymal interactions causes cleft palate. *J. Clin. Invest.* **113,** 1692–1700.

Rice, R., Connor, E., and Rice, D. P. C. (2006). Expression patterns of hedgehog signaling pathway members during palate development. *Gene Exp. Patterns* **6,** 206–212.

Richardson, R. J., Dixon, J., Malhotra, S., Hardman, M. J., Knowles, L., Boot-Handford, R. P., Shore, P., Witmarsh, A., and Dixon, M. J. (2006). Irf6 is a key determinant of the keratinocyte proliferation-differentiation switch. *Nat. Genet.* **38,** 1329–1334.

Riethmacher, D., Brinkmann, V., and Birchmeier, C. (1995). A targeted mutation in the mouse *E-cadherin* gene results in defective preimplantation development. *Proc. Natl. Acad. Sci. USA* **92,** 855–859.

Rijli, F. M., Mark, M., Lakkaraju, S., Dierich, A., Dolle, P., and Chambon, P. (1993). A homeotic transformation is generated in the rostral branchial region of the head by disruption of Hoxa-2, which acts as a selector gene. *Cell* **75,** 1333–1349.

Riley, B. M., Schultz, R. E., Cooper, M. E., Golstein-McHenry, T., Daack-Hirsch, S., Lee, K. T., Dragan, E., Vieira, A. R., Lidral, A. C., Marazita, M. L., and Murray, J. C. (2007a). A genome-wide linkage scan for cleft lip and cleft palate identifies a novel locus on 8p11–23. *Am. J. Med. Genet. A* **143A,** 846–852.

Riley, B. M., Mansilla, M. A., Ma, J., Daack-Hirsch, S., Maher, B. S., Raffensperger, L. M., Russo, E. T., Vieira, A. R., Dodé, C., Mohammadi, M., Marazita, M. L., and Murray, J. C. (2007b). Impaired FGF signaling contribute to cleft lip and palate. *Proc. Natl. Acad. Sci. USA* **104,** 4512–4517.

Robertson, S. P. (2005). Filamin A: Phenotypic diversity. *Curr. Opin. Genet. Dev.* **15,** 301–307.

Robertson, S. P., Twigg, S. R., Sutherland-Smith, A. J., Biancalana, V., Gorlin, R. J., Horn, D., Kenwrick, S. J., Kim, C. A., Morava, E., Newbury-Ecob, R., Orstavik, K. H., Quarrell, O. W., *et al.* PD-spectrum Disorders Clinical Collaborative Group. (2003). Localized mutations in the gene encoding the cytoskeletal protein filamin A cause diverse malformations in humans. *Nat. Genet.* **33,** 487–491.

Rodriguez, M., Desterro, J., Lain, S., Midgley, C., Lane, D., and Hay, R. (1999). SUMO-1 modification activates the transcriptional response of p53. *EMBO J.* **18,** 6455–6561.

Roessler, E., Du, Y. Z., Mullor, J. L., Casas, E., Allen, W. P., Gillessen-Kaesbach, G., Roeder, E. R., Ming, J. E., Ruiz i Altaba, A., and Muenke, M. (2003). Loss-of-function mutations in the human Gli2 gene are associated with pituitary anomalies and holoprosencephaly-like features. *Proc. Natl. Acad. Sci. USA* **100,** 13424–13429.

Sanlaville, D., Etchevers, H. C., Gonzales, M., Martinovic, J., Clément-Ziza, M., Delezoide, A. L., Aubry, M. C., Pelet, A., Chemouny, S., Cruaud, C., Audollent, S.,

Esculpavit, C., *et al.* (2006). Phenotypic spectrum of CHARGE in fetuses with CHD7 truncating mutations correlates with expression during human development. *J. Med. Genet.* **43,** 211–217.

Satoh, K., Kasai, M., Ishidao, T., Ohwada, S., Hasegawa, Y., Senda, T., Takada, S., Nada, S., Nakamura, T., and Akiyama, T. (2004). Anteriorization of neural fate by inhibitor of β- catenin and T-cell factor (ICAT), a negative regulator of Wnt signaling. *Proc. Natl. Acad. Sci. USA* **101,** 8017–8021.

Satokata, I., and Maas, R. (1994). Msx1 deficient mice exhibit cleft palate and abnormalities of craniofacial and tooth development. *Nat. Genet.* **6,** 348–356.

Schliekelman, P., and Slatkin, M. (2002). Multiplex relative risk and estimation of the number of loci underlying an inherited disease. *Am. J. Hum. Genet.* **71,** 1369–1385.

Scapoli, L., Martinelli, M., Pezzetti, F., Carinci, F., Bodo, M., Tognon, M., and Carinci, P. (2002). Linkage desequilibrium between *GABRB3* gene and non-syndromic familial cleft lip with or without cleft palate. *Hum. Genet.* **110,** 15–20.

Scapoli, L., Palmieri, A., Martinelli, M., Pezzetti, F., Carinci, P., Tognon, M., and Carinci, F. (2005). Strong evidence of linkage desequilibrium between polymorphisms at the *IRF6* locus and nonsyndromic cleft lip with or without cleft palate, in an italian population. *Am. J. Hum. Genet.* **76,** 180–183.

Scapoli, L., Palmieri, A., Martinelli, M., Vaccari, C., Marchesini, J., Pexxetti, F., Baciliero, U., Padula, E., Carinci, P., and Carinci, F. (2006). Study of the *PVRL1* gene in Italian nonsyndromic cleft lip patients with or without cleft palate. *Ann. Hum. Genet.* **70,** 410–413.

Schmahl, J., Raymond, C. S., and Soriano, P. (2007). PDGF signaling specificity is mediated through multiple immediate early genes. *Nat. Genet.* **39,** 52–60.

Schöck, F., and Perrimon, N. (2002). Molecular mechanisms of epithelial morphogenesis. *Annu. Rev. Cell Dev. Biol.* **18,** 463–493.

Schorle, H., Meier, P., Buchert, M., Jaenisch, R., and Mitchell, P. J. (1996). Transcription factor AP-2 essential for cranial closure and craniofacial development. *Nature* **381,** 235–238.

Schuffenhauer, S., Leifheit, H. J., Peters, H., Murken, J., and Emmerich, P. (1999). *De novo* deletion (14)(q11.2q13) including PAX9: Clinical and molecular findings. *J. Med. Genet.* **36,** 233–236.

Schüpbach, P. M. (1983). Experimental induction of an incomplete hard-palate cleft in the rat. *Oral Surg. Oral Med. Oral Pathol.* **55,** 2–9.

Schutte, B. C., and Murray, J. C. (1999). The many faces and factors of orofacial clefts. *Hum. Mol. Genet.* **8,** 1853–1859.

Schwab, K. R., Patterson, L. T., Hartman, H. A., Song, N., Lang, R. A., Lin, X., and Potter, S. S. (2007). *Pygo1* and *Pygo2* roles in Wnt signaling in mammalian kidney development. *BMC Biol.* **5,** 15.

Seppala, M., Depew, M. J., Martinelli, D. C., Fan, C. M., Sharpe, P. T., and Cobourne, M. T. (2007). *Gas1* is a modifier for holoprosencephaly and genetically interacts with sonic hedgehog. *J. Clin. Invest.* **117,** 1575–1584.

Shamblott, M. J., Bugg, E. M., Lawler, A. M., and Gearhart, J. D. (2002). Craniofacial abnormalities resulting from targeted disruption of the murine *Sim2* gene. *Dev. Dyn.* **224,** 373–380.

Shimamura, K., and Rubenstein, J. L. (1997). Inductive interactions direct early regionalization of the mouse forebrain. *Development* **124,** 2709–2718.

Shupe, J. L., Lynn, F. J., Binns, W., and Keeler, R. F. (1968). Cleft palate in cattle. *Cleft Palate J.* **5,** 346–355.

Sock, E., Rettig, S. D., Enderich, J., Bösl, M. R., Tamm, E. R., and Wegner, M. (2004). Gene targeting reveals a widespread role for the high-mobility-group transcription factor Sox11 in tissue remodeling. *Mol. Cell. Biol.* **24,** 6635–6644.

Solloway, M. J., and Robertson, E. J. (1999). Early embryonic lethality of Bmp5/Bmp7 double mutant mice suggests functional redundancy within the 60A subgroup. *Development* **126,** 1753–1768.
Song, Q., Mehler, M. F., and Kessler, J. A. (1998). Bone morphogenetic proteins induce apoptosis and growth factor dependence of cultured sympathoadrenal progenitor cells. *Dev. Biol.* **196,** 119–127.
Soriano, P. (1997). The PDGF α receptor is required for neural crest cell development and for normal patterning of the somites. *Development* **124,** 2691–2700.
Sözen, M. A., Suzuki, K., Tolarova, M. M., Bustos, T., Fernandez Iglesias, J. E., and Spitz, R. A. (2001). Mutation of *PVRL1* is associated with sporadic, nonsyndromic cleft lip/palate in northern Venezuela. *Nat. Genet.* **29,** 141–142.
Spiegelstein, O., Mitchell, L. E., Merriweather, M. Y., Wicker, N. J., Zhang, Q., Lammer, E. J., and Finnell, R. H. (2004). Embryonic development of folate binding protein-1 (Folbp1) knockout mice: Effect of the chemical form, dose, and timing of maternal folate supplementation. *Dev. Dyn.* **231,** 221–231.
Stanier, P., and Moore, G. E. (2004). Genetics of cleft lip and palate: Syndromic genes contribute to the incidence of non-syndromic clefts. *Hum. Mol. Genet.* **13,** R73–R81.
Stankunas, K., and Crabtree, G. R. (2007). Exploiting protein destruction for constructive use. *Proc. Natl. Acad. Sci. USA* **104,** 11511–11512.
Sull, J. W., Liang, K.-Y., Hetmanski, J. B., Fallin, M. D., Ingersoll, R. G., Park, J., Wu-Chou, Y.-H., Chen, P. K., Chong, S. S., Cheah, F., Yeow, V., Park, B. W., *et al.* (2008). Differential parental transmission of markers in *RUNX2* among cleft case-parent trios from four populations. *Genet. Epidemiol.* **32,** 505–512.
Sun, Y., Liu, X., Eaton, E. N., Lane, W. S., Lodish, H. F., and Weinberg, R. A. (1999). Interaction of the ski oncogene with Smad3 regulates TGF-beta signaling. *Mol. Cell* **4,** 499–509.
Suzuki, K., Hu, D., Bustos, T., Zlotogora, J., Ricchieri-Costa, A., Helms, J. A., and Spitz, R. A. (2000). Mutations of *PVRL1*, encoding a cell-cell adhesion molecule/herpesvirus receptor, in cleft lip/palate-ectodermal dysplasia. *Nat. Genet.* **25,** 427–430.
Suzuki, Y., Jezewski, P. A., Machida, J., Watanabe, Y., Shi, M., Cooper, M. E., Viet le, T, Nguyen, T, Hai, H., Natsume, N., Shimozato, K., Marazita, M. L., *et al.* (2004). In a Vietnamese population, MSX1 variants contribute to cleft lip and palate. *Genet. Med.* **6,** 117–125.
Szeto, D. P., Rodriguez-Esteban, C., Ryan, A. K., O'Connell, S. M., Liu, F., Kioussi, C., Gleiberman, A. S., Izpisua-Belmonte, J. C., and Rosenfeld, M. G. (1999). Role of the bicoid-related homeodomain factor Pitx1 in specifying hindlimb morphogenesis and pituitary development. *Genes Dev.* **13,** 484–494.
Tallquist, M. D., and Soriano, P. (2003). Cell autonomous requirement for PDGFRα in populations of cranial and cardiac neural crest cells. *Development* **130,** 507–518.
Tang, L. S., and Finnell, R. H. (2003). Neural and orofacial defects in *Folp1* knockout mice. *Birth Defects Res. A Clin. Mol. Teratol.* **67,** 209–218.
Tassin, M. T., Salzgeber, B., and Guenet, J. L. (1983). Studies on "repeated epilation" mouse mutant embryos: I. Development of facial malformation. *J. Craniofac. Genet. Dev. Biol.* **3,** 289–307.
Taylor, K. M., and Labonne, C. (2005). SoxE factors function equivalently during neural crest and inner ear development and their activity is regulated by SUMOylation. *Dev. Cell.* **9,** 593–603.
Tinkle, C. L., Lechler, T., Pasoli, A. H., and Fuchs, E. (2004). Conditional targeting of E-cadherin in skin: Insights into hyperproliferative and degenerative responses. *Proc. Natl. Acad. Sci. USA* **101,** 552–557.
Tongkobpetch, S., Siriwan, P., and Shotelersuk, V. (2006). MSX1 mutations contribute to nonsyndromic cleft lip in a Thai population. *J. Hum. Genet.* **51,** 671–676.

Toyo-Oka, K., Hirotsune, S., Gambello, M. J., Zhou, Z.-Q., Olson, L., Rosenfeld, M. G., Eisenman, R., Hurlin, P., and Wynshaw-Boris, A. (2004). Loss of the Max-interacting protein Mnt in mice results in decreased viability, defective embryonic growth and craniofacial defects: Relevance to Miller-Dieker syndrome. *Hum. Mol. Genet.* **13,** 1057–1067.

Trivier, E., and Ganessan, T. S. (2002). RYK, a catalytically inactive receptor tyrosine kinase, associates with Eph-B2 and Eph-B3 but does not interact with AF-6. *J. Biol. Chem.* **277,** 23037–23043.

Trokovic, N., Trokovic, R., Mai, P., and Partanen, J. (2003). Fgfr1 regulates patterning of the pharyngeal region. *Genes Dev.* **17,** 141–153.

Valdez, B. C., Henning, D., So, R. B., Dixon, J., and Dixon, M. J. (2004). The Treacher Collins syndrome (TCOF1) gene product is involved in ribosomal DNA gene transcription by interacting with upstream binding factor. *Proc. Natl. Acad. Sci. USA* **101,** 10709–10714.

van Bokhoven, H., and Brunner, H. G. (2002). Splitting p63. *Am. J. Hum. Genet.* **71,** 1–13.

van Bokhoven, H., and McKeon, F. (2002). Mutations in the p53 homolog p63: Allele-specific developmental syndromes. *Trends Mol. Med.* **8,** 133–139.

van Bokhoven, H., Hamel, B. C., Bamshad, M., Sangiorgi, E., Gurrieri, F., and Duijf, P. H. (2001). p63 gene mutations in EEC syndrome, limb-mammary syndrome, and isolated split hand-split foot malformation suggests a genotype–phenotype correlation. *Am. J. Hum. Genet.* **69,** 481–492.

van den Boogaard, M. J., Dorland, M., Beemer, F. A., and van Amstel, H. K. (2000). *MSX1* mutation is associated with orofacial clefting and tooth agenesis in humans. *Nat. Genet.* **24,** 342–343.

Varju, P., Katarova, Z., Madarasz, E., and Szabo, G. (2001). GABA signaling during development: New data and old questions. *Cell Tissue Res.* **305,** 239–246.

Vaziri Sani, F., Hallberg, K., Harfe, B. D., McMahon, A. P., Linde, A., and Gritli-Linde, A. (2005). Fate-mapping of the epithelial seam during palatal fusion rules out epithelial–mesenchymal transformation. *Dev. Biol.* **285,** 490–495.

Vekemans, M., and Fraser, F. C. (1979). Stage of palate closure as one indication of liability to cleft palate. *Am. J. Med. Genet.* **4,** 95–102.

Vieira, A. R. (2008). Unraveling human cleft lip and palate research. *J. Dent. Res.* **87,** 119–125.

Vieira, A. R., Avila, J. R., Daack-Hirsch, S., Dragan, E., Felix, T. M., Rahimov, F., Harrington, J., Schultz, R. R., Watanabe, Y., Johnson, M., Fang, J., O'Brien, S. E., *et al.* (2005). Medical sequencing of candidate genes for nonsyndromic cleft lip and palate. *PLOS Genet.* **1,** e64.

Vissers, L. E., van Ravenswaaij, C. M., Admiraal, R., Hurst, J. A., de Vries, B. B., Janssen, I. M., van der Vliet, W. A., Huys, E. H., de Jong, P. J., Hamel, B. C., Schoenmakers, E. F., Brunner, H. G., *et al.* (2004). Mutations in a new member of the chromodomain gene family cause CHARGE syndrome. *Nat. Genet.* **36,** 955–957.

Walker, B. E., and Fraser, F. C. (1956). Closure of the secondary palate in three strains of mice. *J. Embryol. Exp. Morphol.* **4,** 176–189.

Walker, B. E., and Fraser, F. C. (1957). The embryology of cortisone-induced cleft palate. *J. Embryol. Exp. Morphol.* **5,** 201–209.

Wang, W., Mariani, F. V., Harland, R. M., and Luo, K. (2000). Ski represses bone morphogenetic protein signaling in Xenopus and mammalian cells. *Proc. Natl. Acad. Sci. USA* **97,** 14394–14399.

Wang, T., Tamakoshi, T., Uezato, T., Shu, F., Kanzai-Kato, N., Fu, Y., Koseki, H., Nobuaki, Y., Sygiyama, T., and Miura, N. (2003). Forkhead transcription factor Foxf2 (LUN)-deficient mice exhibit abnormal development of secondary palate. *Dev. Biol.* **259,** 83–94.

Wang, Y., McMahon, A. P., and Allen, B. L. (2007). Shifting paradigms in hedgehog signaling. *Curr. Opin. Cell Biol.* **19,** 1–7.
Warrington, A., Vieira, A. R., Christensen, K., Orioli, I. M., Castilla, E. E., Romitti, P. A., and Murray, J. C. (2006). Genetic evidence for the role of loci at 19q13 in cleft lip and palate. *J. Med. Genet.* **43,** e26.
Wassif, C. A., Maslen, C., Kachilele-Linjewile, S., Lin, D., Linck, L. M., Connor, W. E., Steiner, R. D., and Porter, F. D. (1998). Mutations in the human sterol delta7-reductase gene at 11q12–13 cause Smith-Lemli-Opitz syndrome. *Am. J. Hum. Genet.* **63,** 55–62.
Wassif, C. A., Zhu, P., Kratz, L., Krakowiak, P. A., Bataille, K. P., Weight, F. F., Grinberg, A., Steiner, R. D., Nwokoro, N. A., Kelley, R. I., Stewart, R. R., and Porter, F. D. (2001). Biochemical, phenotypic and neurophysiological characterization of a genetic mouse model of RSH/Smith–Lemli–Opitz syndrome. *Hum. Mol. Genet.* **10,** 555–564.
Watanabe, A., Akita, S., Tin, N. T. D., Natsume, N., Nakano, Y., Niikawa, N., Uchiyama, T., and Yoshiura, K. I. (2006). A mutation in RYK is a genetic factor for nonsyndromic cleft lip and palate. *Cleft Palate Craniofac. J.* **43,** 310–316.
Waterston, R. H., Lindblad-Toh, K., Birney, E., Rogers, J., Abril, J. F., Agarwal, P., Agarwala, R., Ainscough, R., Alexandersson, M., An, P., Antonarakis, S. E., Attwood, J., *et al.* (2002). Initial sequencing and comparative analysis of the mouse genome. *Nature* **420,** 520–562.
Wee, E. L., and Zimmerman, E. F. (1985). GABA uptake in embryonic palate mesenchymal cells of two mouse strains. *Neurochem. Res.* **10,** 1673–1688.
Wee, E. L., Norman, E. J., and Zimmerman, E. F. (1986). Presence of γ-aminobutyric acid in embryonic palates of AJ and SWV mouse strains. *J. Craniofac. Genet. Dev. Biol.* **6,** 53–61.
Welsh, I. C., Hagge-Grenberg, A., and O'Brien, T. P. (2007). A dosage-dependent role for *Spry2* in growth and patterning during palate development. *Mech. Dev.* **124,** 746–761.
Wilkie, A. O., Slaney, S. F., Oldridge, M., Poole, M. D., Ashworth, G. J., Hockley, A. D., Hayward, R. D., David, D. J., Pulleyn, L. J., Rutland, P., Malcolm, S., Winter, R. M., *et al.* (1995). Apert syndrome results from localized mutations of FGFR2 and is allelic with Crouzon syndrome. *Nat. Genet.* **9,** 165–172.
Wojcik, S. M., Katsurabayashi, S., Guillemin, I., Friauf, E., Rosenmund, C., Brose, N., and Rhee, J. S. (2006). A shared vesicular carrier allows synaptic corelease of GABA and glycine. *Neuron* **50,** 575–587.
Woolf, C. M., Woolf, R. M., and Broadbent, T. R. (1963). A genetic study of cleft lip and palate in Utah. *Am.J. Hum. Genet.* **15,** 209–215.
Xu, P. X., Adams, J., Peters, H., Brown, M. C., Heany, S., and Maas, R. (1999). *Eya1*-deficient mice lack ears and kidney and show abnormal apoptosis of organ primordia. *Nat. Genet.* **23,** 113–117.
Xu, W., Angelism, K., Danielpour, D., Haddad, M. M., Bischof, O., Campisi, J., Stavnezer, E., and Medrano, E. E. (2000). Ski acts as a co-repressor with Smad2 and Smad3 to regulate the response to type beta transforming growth factor. *Proc. Natl. Acad. Sci. USA* **97,** 5924–5929.
Xu, X., Han, J., Yoshihiro, I., Bringas, P., Jr., Urata, M. M., and Chai, Y. (2006). Cell autonomous requirement for *Tgfbr2* in the disappearence of medial edge epithelium during palatal fusion. *Dev. Biol.* **297,** 238–248.
Yagi, H., Furutani, Y., Hamada, H., Sasaki, T., Asakawa, S., Minoshima, S., Ichida, F., Joo, K., Kimura, M., Imamura, S., Kamatani, N., Momma, K., *et al.* (2003). Role of *TBX1* in human del22q11.2 syndrome. *Lancet* **362,** 1366–1373.
Yamada, T., Fujiwara, K., Mishima, K., and Sigahara, T. (2005). Effect of ENU (ethylnitrosourea) mutagenesis in cleft lip and/or palate pathogenesis in mice. *Int. J. Oral Maxillofac.Surg.* **34,** 74–77.

Yan, Y., Frisén, J., Lee, M. H., Massagué, J., and Barbacid, M. (1997). Ablation of the CDK inhibitor *p57Kip2* results in increased apoptosis and delayed differentiation during mouse development. *Genes Dev.* **11,** 973–983.

Yang, L. T., and Kaartinen, V. (2007). *Tgfb1* expressed in the *Tgfβ3* locus partially rescues the cleft palate phenotype of *Tgfβ3* null mutants. *Dev. Biol.* **312,** 384–395.

Yang, A., Kaghad, M., Wang, Y., Gillett, E., Fleming, M. D., Doetsch, V., Andrews, N. C., Caput, D., and McKeon, F. (1998). *p63*, a *p53* homolog at 3q27–29, encodes multiple products with transactivating, death-inducing, and dominant-negative activities. *Mol. Cell* **2,** 305–316.

Yang, A., Schweitzer, R., Sun, D., Kaghad, M., Walker, N., Bronson, R. T., Tabin, C., Sharpe, A., Caput, D., Crum, C., and McKeon, F. (1999). p63 is essential for regenerative proliferation in limb, craniofacial and epithelial development. *Nature* **398,** 714–718.

Yang, Y., Topol, L., Lee, H., and Wu, J. (2003). Wnt5a and Wnt5b exhibit distinct activities in coordinating chondrocyte proliferation and differentiation. *Development* **130,** 1003–1015.

Yang, Y., Mahaffey, C. L., Bérubé, N., and Frankel, W. N. (2006). Interaction between Fidgetin and protein kinase A-anchoring protein AKAP95 is critical for palatogenesis in the mouse. *J. Biol. Chem.* **281,** 22352–22359.

Yang, L. T., Li, W. Y., and Kaartinen, V. (2008). Tissue-specific expression of Cre recombinase from the Tgf*β*3 locus. *Genesis* **46,** 112–118.

Yingling, J., Toyo-Oka, K., and Wynshaw-Boris, A. (2003). Miller-Dieker syndrome: Analysis of a human contiguous gene syndrome in the mouse. *Am. J. Hum. Genet.* **73,** 475–488.

Yoshida, H., Kong, Y. Y., Yoshida, R., Elia, A. J., Hakem, A., Penninger, J. M., and Mak, T. W. (1998). Apaf1 is required for mitochondrial pathways of apoptosis and brain develoment. *Cell* **94,** 739–750.

Yoshikawa, S., McKinnon, R. D., Kokel, M., and Thomas, J. B. (2003). Wnt-mediated axon guidance via the *Drosophila* Derailed receptor. *Nature* **422,** 583–588.

Yu, L., Gu, S., Alappat, S., Song, Y., Yan, M., Zhang, X., Zhang, G., Jiang, Y., Zhang, Z., Zhang, Y., and Chen, Y. P. (2005). *Shox2*-deficient mice exhibit a rare type of incomplete clefting of the secondary palate. *Development* **132,** 4393–4406.

Yu, L., Liu, H., Yan, M., Yang, J., Long, F., Muneoka, K., and Chen, Y. P. (2007). Shox2 is required for chondrocyte proliferation and maturation in proximal limb skeleton. *Dev. Biol.* **306,** 549–559.

Zaritsky, J. J., Eckman, D. M., Wellman, G. C., Nelson, M. T., and Schwartz, T. L. (2000). Targeted disruption of Ki2.1 and Kir2.2 genes reveals the essential role of the inwardly rectifying K + current in K + -mediated vasodilation. *Circ. Res.* **87,** 160–166.

Zhang, J., Hagopian-Donaldson, S., Serbedzija, G., Elsemore, J., Plehn-Dujowich, D., McMahon, A. P., Flavell, R. A., and Williams, T. (1996). Neural tube, skeletal and body wall defects in mice lacking transcription factor Ap-2. *Nature* **381,** 238–241.

Zhang, P., Liégeois, N. J., Wong, C., Finegold, M., Hou, H., Thompson, J. C., Silverman, A., Harper, W., DePinho, R. A., and Elledge, S. J. (1997). Altered cell differentiation and proliferation in mice lacking p57^{kip2} indicates a role in Beckwith-Wiedemann syndrome. *Nature* **387,** 151–158.

Zhang, Y., Zhang, Z., Zhao, X., Yu, X., Hu, Y., Benedicto, G., Fromm, S. H., and Chen, Y. P. (2000). A new function of BMP4: dual role for BMP4 in regulation of Sonic hedgehog expression in the mouse tooth germ. *Development* **127,** 1431–1443.

Zhang, Z., Song, Y., Zhao, X., Fermin, C., and Chen, Y. P. (2002). Rescue of cleft palate in Msx1-deficient mice by transgenic *Bmp4* reveals a network of BMP and Shh signaling in the regulation of mammalian palatogenesis. *Development* **129,** 4135–4146.

Zhang, B., Jain, S., Song, H., Fu, M., Heuckeroth, R. O., Erlich, J. M., Jay, P. Y., and Milbrandt, J. (2007). Mice lacking sister chromatid cohesion protein PDS5B exhibit developmental abnormalities reminiscent of Cornelia de Lange syndrome. *Development* **134,** 3191–3201.

Zhao, Y., Guo, Y. J., Tomac, A. C., Taylor, N. R., Grinberg, A., Lee, E. J., Huang, S. P., and Westphal, H. (1999). Isolated cleft palate in mice with targeted mutation of the LIM homeobox gene Lhx8. *Proc. Natl. Acad. Sci. USA* **96,** 15002–15006.

Zoupa, M., Seppala, M., Mitsiadis, T., and Cobourne, M. T. (2006). *Tbx1* is expressed at multiple sites of epithelial–mesenchymal interactions during early development of the facial complex. *Int. J. Dev. Biol.* **50,** 504–510.

Zucchero, T. M., Cooper, M. E., Maher, B. S., Daack-Hirch, S., Nepomuceno, B., Ribeiro, L., Caprau, D., Christensen, K., Suzuki, Y., Machida, J., Natsume, N., Yoshiura, K., *et al.* (2004). Interferon regulatory factor 6 (*IRF6*) gene variant and the risk of isolated cleft lip or palate. *N. Engl. J. Med.* **351,** 769–780.

CHAPTER THREE

Murine Models of Holoprosencephaly

Karen A. Schachter *and* Robert S. Krauss

Contents

Abstract

Holoprosencephaly (HPE), the most common developmental defect of the forebrain and midface, is caused by a failure to delineate the midline in these structures. Both genetic and environmental etiologies exist for HPE, and clinical presentation is highly variable. HPE occurs in sporadic and inherited forms, and even HPE in pedigrees is characterized by incomplete penetrance and variable expressivity. Heterozygous mutations in eight different genes have been identified in human HPE, and disruption of Sonic hedgehog expression and/or signaling in the rostroventral region of the embryo is a major common effect of these mutations. An understanding of the mechanisms whereby genetic defects and teratogenic exposures become manifest as developmental anomalies of varying severity requires experimental models that accurately reproduce the spectrum of defects seen in human HPE. The mouse has emerged as such a model, because of its ease of genetic manipulation and similarity to humans in development of the forebrain and face. HPE is generally observed in mice homozygous for mutations in orthologs of human HPE genes though, unlike humans, rarely in mice with heterozygous mutations. Moreover, reverse

Department of Developmental and Regenerative Biology, Mount Sinai School of Medicine, New York 10029

Current Topics in Developmental Biology, Volume 84
ISSN 0070-2153, DOI: 10.1016/S0070-2153(08)00603-0

genetics in the mouse has provided a wealth of new candidate human HPE genes. Construction of hypomorphic alleles, interbreeding to produce double mutants, and analysis of these mutations on different genetic backgrounds has generated multiple models of HPE and begun to provide insight into the conundrum of the HPE spectrum. Here, we review forebrain development with an emphasis on the pathways known to be defective in HPE and describe the strengths and weaknesses of various murine models of HPE.

1. Introduction

The earliest known accounts of cyclopia, a single centrally located eye, date to Greek antiquity. In Homer's *Odyssey*, the hero Odysseus lands on the island of Cyclopes, inhabited by Polyphemus and other dangerous, one-eyed giants; Odysseus and his men escape through trickery. Although cyclopia in people is not compatible with survival into adulthood, it seems quite likely that the concept of the Cyclopes may have been stimulated by descriptions of stillborn babies with severe holoprosencephaly (HPE). HPE is the most common birth defect of the forebrain in humans. It is characterized by the complete or partial failure to separate the forebrain into bilateral hemispheres. Defects in facial midline patterning, in the most severe instances including cyclopia, accompany most cases. Clinical presentation of HPE is marked by broad phenotypic heterogeneity in both familial and sporadic cases, with the range of midline defects extending from most to least severe in a continuum called the HPE spectrum. Signaling pathways and networks that pattern the midline of the forebrain and midface have been identified, and mutations in genes encoding specific components of these pathways have been identified in human HPE cases. However, clear genotype–phenotype correlations have been difficult to establish due to the heterogeneous spectrum of defects.

The observation of mutant strains of mice does not go back as far as Greek antiquity, but by the nineteenth century, many breeds of "fancy" mice were available to collectors. Because of the existence of many inbred strains, ease of genetic manipulation, and genetic kinship with people, the mouse has become the model organism of choice for the study of human disease. A large number of mouse mutants display HPE, including in some cases the phenotypic heterogeneity characteristic of human HPE. Continued development and analysis of such models is expected to illuminate the etiology and complexities of this severe and common birth defect. This chapter reviews aspects of human HPE, and the strengths and weaknesses of murine HPE models.

2. Human HPE

2.1. Types and frequency

The most recent estimates indicate that HPE has a live birth prevalence of 1.3 in 10,000, but an incidence of at least 1 in 250 conceptuses, revealing that HPE occurs frequently in early embryogenesis with most embryos eliminated by spontaneous abortion (Leoncini *et al.*, 2008; Muenke and Beachy, 2000; Yamada *et al.*, 2004).

HPE is classically categorized into three groups according to the level of severity: (1) alobar HPE, where no lateral separation of the brain occurs and a single cerebral vesicle forms; (2) semilobar HPE, in which the frontal and lateral lobes are fused while the posterior end is divided; and (3) lobar HPE, where only the most rostral portion of the telencephalon is fused, and separation occurs posteriorly and laterally (reviewed in Cohen, 2006; Dubourg *et al.*, 2007; Muenke and Beachy, 2001). In all these cases, defects in patterning of the ventral forebrain are manifest. Greater than 80% of HPE cases also display craniofacial midline anomalies, which range from cyclopia with overlying proboscis in the most severe cases of alobar HPE to solitary median maxillary central incisor in the mildest cases. Mild facial midline abnormalities may occur in the absence of overt brain malformations; in such cases they are sometimes referred to as microforms or microsigns of HPE. A distinct form of HPE, midline interhemispheric HPE (MIH HPE; sometimes called syntelencephaly), is characterized by failure to divide the hemispheres in the posterior frontal and parietal regions with normal separation of the basal forebrain, anterior frontal lobes and occipital regions (Barkovich and Quint, 1993; Simon *et al.*, 2002). Generally, MIH HPE affects more dorsal structures of the forebrain, generally without overt craniofacial and ventral forebrain pathology, and it may be considered a separate class of HPE from the "classical" alobar/semilobar/lobar categories (Fernandes and Hebert, 2008).

2.2. Etiology of human HPE—Genetics

Approximately 25% of HPE cases are syndromic, that is, present as an aspect of a broader syndrome (Dubourg *et al.*, 2007). These disorders are rare and include Smith–Lemli–Opitz (SLO), Pallister–Hall, and Rubinstein–Taybi syndromes. Nonsyndromic, isolated, forms of HPE are mainly sporadic but pedigrees exist in which HPE can be inherited in an autosomal dominant manner with partial penetrance and variable expressivity (Muenke and Beachy, 2001).

Chromosomal abnormalities have been detected in 25–50% of affected patients, and include trisomy 13, trisomy 18, and triploidy (Dubourg *et al.*, 2007). Genetic analyses of recurrent chromosomal anomalies allowed the identification of 12 loci associated with nonsyndromic HPE, and four of these loci have been definitively assigned to specific genes: *SHH, ZIC2, SIX3*, and *TGIF* (Roessler and Muenke, 1998; Wallis and Muenke, 2000). Analyses of candidate genes have revealed mutations in four additional genes: *PTCH1, GLI2, TDGF1/CRIPTO*, and *FOXH1/FAST1* (Cohen, 2006; Dubourg *et al.*, 2007; Ming and Muenke, 2002). Disruption of Sonic hedgehog (SHH) signaling is likely to be a major common theme in the development of classical HPE (see below). *PTCH1* and *GLI2* encode components of the SHH signaling pathway. TDGF1/CRIPTO and FOXH1/FAST1 are components of the Nodal signaling pathway, which regulates formation of the prechordal mesendoderm, an early developmental structure that produces SHH and is required for the specification of the ventral forebrain and formation of the midline. SIX3 regulates *SHH* expression in the ventral forebrain, and ZIC2 plays an early role in specification of the prechordal mesendoderm and a later role in dorsal patterning of the CNS. However, how loss of TGIF function results in HPE is unclear.

Mutations in these eight genes account for approximately 25% of HPE cases examined. In addition to cytogenetically detectable chromosomal abnormalities, types of mutations observed include missense and nonsense mutations, and small deletions that result in frameshifts or loss of amino acids (Dubourg *et al.*, 2007). All the mutations identified have been found in a heterozygous state; the ones that have been tested functionally correspond to partial or complete loss-of-function alleles and presumably result in haploinsufficiency (El-Jaick *et al.*, 2007; Gripp *et al.*, 2000; Maity *et al.*, 2005; Roessler *et al.*, 2003, 2008; Schell-Apacik *et al.*, 2003; Traiffort *et al.*, 2004).

2.3. Etiology of human HPE—Environmental factors

Environmental factors are also implicated in the genesis of HPE. Maternal diabetes, alcohol consumption during pregnancy, and prenatal exposure to other potential teratogens, such as retinoic acid, plant alkaloids, and pharmaceutical drugs may increase the risk of HPE (Barr *et al.*, 1983; Croen *et al.*, 2000). The incidence of HPE in infants of diabetic mothers is around 1%, but the molecular mechanism by which maternal diabetes impairs fetal forebrain development is not clear. Ethanol exposure results in a variety of malformations that include HPE and craniofacial defects, and may be caused by decreased Shh signaling (Ahlgren *et al.*, 2002; Li *et al.*, 2007).

Defective cholesterol biosynthesis and/or utilization have also been associated with HPE. Patients with SLO carry recessive mutations in 3β-hydroxysterol-Δ^7-reductase, which catalyzes the final step in cholesterol biosynthesis (Kelley *et al.*, 1996). Approximately, 5% of these patients display HPE. Furthermore, some cell lines derived from nonsyndromic HPE patients display abnormally low cholesterol biosynthesis (Haas *et al.*, 2007). In all these cases, the decrease in cholesterol levels most likely affects Shh processing and/or signaling. Shh is generated as a precursor protein that enters the secretory pathway where it is palmitoylated (Buglino and Resh, 2008; Pepinsky *et al.*, 1998). It subsequently undergoes autocatalytic processing to generate an N-terminal signaling-competent fragment; the C-terminal portion of the Shh precursor catalyzes this proteolysis, as well as the addition of a cholesterol moiety to the C-terminus of the signaling fragment (Cooper *et al.*, 2003). The dually lipid-modified Shh, termed ShhNp, corresponds to the biologically active molecule. Lipidation of Shh is required to produce a soluble multimeric protein complex, and it has important roles in facilitating signaling in target cells, as well as limiting the spatial extent of Shh effects (Mann and Beachy, 2004; van Den Heuvel, 2001). Furthermore, cholesterol is required for transduction of the Shh signal, although the precise mechanism for this requirement is not clear (Cooper *et al.*, 2003). It is suspected that ingestion of cholesterol-lowering drugs like statins during pregnancy could be a risk factor for human HPE (Edison and Muenke, 2004).

2.4. The HPE spectrum—A conundrum

Clinical expression of HPE is extremely variable. The range of midline defects that extends from alobar HPE through facial microsigns of HPE in the absence of brain defects exists as an uninterrupted phenotypic continuum (Muenke and Beachy, 2001). Remarkably, the entire spectrum of HPE phenotypes is observed in familial forms, where severely affected and clinically normal relatives carry the same mutation (Cohen, 1989; Ming and Muenke, 2002). Figure 3.1 reprints a case report (Hennekam *et al.*, 1991) of a woman of normal intelligence and no brain anomalies as determined by CT scan, that presented microsigns of HPE [solitary median maxillary central incisor, absence of nasal septal cartilage, hypotelorism (close-set eyes)]; and her offspring, which displayed alobar HPE with a single nostril and strong midface hypoplasia. The partial penetrance and variable expressivity seen in familial HPE, that is, the lack of genotype–phenotype correlation, suggests that heterozygous mutation of the known HPE genes contributes to the induction of HPE, but is alone insufficient to cause this birth defect (Ming and Muenke, 2002). Potential explanations for this phenomenon include (1) multiple hit models, such as heterozygous mutations in two HPE genes (digenic mutation) or interaction between heterozygous mutation of an HPE gene with otherwise silent modifier genes; (2) stochastic events; and (3) gene–environment interactions (Krauss, 2007).

Digenic heterozygous mutations have been found in a small number of HPE patients (two with mutations in both *SHH* and *TGIF*, and one each with mutations in *SHH* and *ZIC2* and *PTCH1* and *GLI2*) (Nanni *et al.*, 1999; Rahimov *et al.*, 2006). Additionally, multiple chromosomal rearrangements in

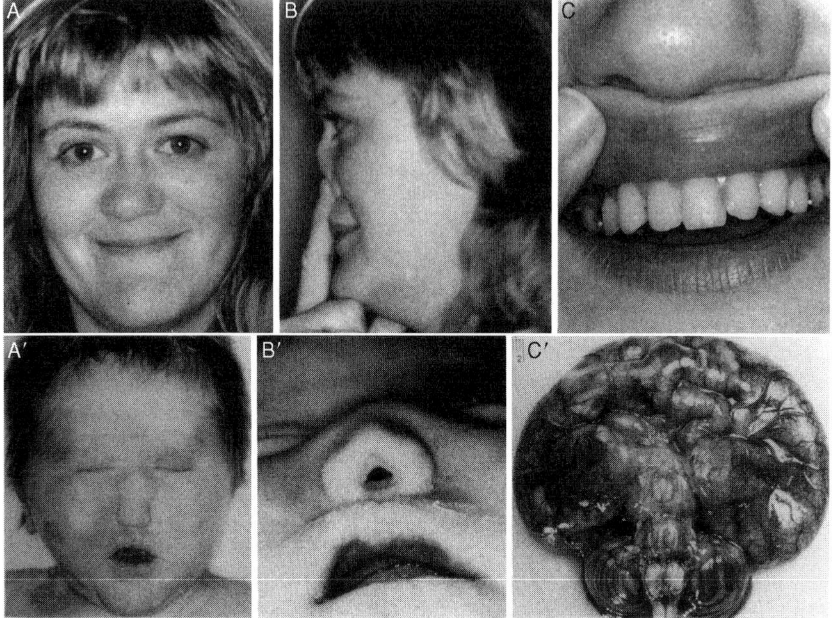

Figure 3.1 Spectrum of HPE phenotypes in a pedigree. Top panel: individual of normal intelligence with facial microforms of HPE. (A) Hypotelorism. (B) Absence of nasal septal cartilage. (C) Solitary median maxillary central incisor. Bottom panel: offspring of individual shown in top panel with a severe form of HPE. (A) Midfacial hypoplasia. (B) Single nostril. (C) Alobar HPE. Reprinted with permission of John Wiley & Sons, Inc. from Hennekam *et al.* (1991).

the same patient have also been identified in HPE cases (Bendavid *et al.*, 2007). However, other than these examples, there is little direct evidence for mechanisms to explain the HPE spectrum.

The identification of human HPE genes and potential HPE-inducing teratogens has provided enormous insight into the etiology of this common and devastating developmental anomaly. However, understanding how the proteins encoded by these genes function in signaling networks to control normal development; how these processes go awry in HPE; and analysis of potential mechanisms of the HPE spectrum requires the use of model organisms. While mouse, chick, frog, and zebrafish models have all been important in understanding development of the forebrain, the availability of targeted mutagenesis and multiple inbred genetic strains has made the mouse particularly valuable in producing models of HPE. HPE is generally observed in mice homozygous for mutations in orthologs of human HPE genes though, unlike humans, rarely in mice with heterozygous mutations in these genes. Moreover, reverse genetics in the mouse has provided a wealth of new candidate human HPE genes. Construction of hypomorphic alleles, interbreeding to produce double mutants, and analysis of these mutations on different genetic backgrounds has generated multiple models of HPE and begun to provide insight into the conundrum of the HPE spectrum. In the following sections, we review forebrain development with an emphasis on the pathways known to be defective in HPE and then describe the strengths and weaknesses of various mouse models of HPE.

3. Development of the Forebrain—A Delicate Balance

HPE is a developmental disorder. To explain how mutations in HPE genes result in disease, their role in normal forebrain development must be understood. The generation of animal models has provided great insight into the molecular events that control these processes. Forebrain development is initiated by signals derived from non-neural structures, such as the node, anterior visceral endoderm, and prechordal mesendoderm. Prior to and during gastrulation, signals emanating from these centers induce neural fates, initiate specification of anterior and posterior forebrain, and induce signaling centers within the forebrain itself (Wilson and Houart, 2004). Cells from the node migrate rostrally along the midline. Caudally, these cells form the notochord, which lies beneath the neural plate up to the midbrain-forebrain margin. Under the forebrain, the node-derived cells form the prechordal mesendoderm. Manual removal of the prechordal mesendoderm from vertebrate embryos results in severe HPE, including cyclopia, highlighting its importance as a key signaling center for specification of rostroventral structures (Li *et al.*, 1997; Mathieu *et al.*, 2002; Pera and Kessel, 1997). Formation of the prechordal mesendoderm requires signaling by Nodal, a member of the transforming growth factor beta (TGF-β) superfamily (Gritsman *et al.*, 2000). The prechordal mesendoderm in turn secretes Shh, which then specifies the ventral forebrain and is required for formation of the midline.

Nodal binds to type I and type II activin receptors (ActRIB/ALK4 and ActRIIA/B), along with the EGF-CFC cofactors Tdgf1 (Cripto) and Cfc1 (Cryptic). Upon Nodal ligand binding, the transcription factors Smad2 and Smad3 are phosphorylated by ActRIB; they subsequently dimerize with Smad4 and translocate to the nucleus where they interact with members of the FoxH family of transcription factors to drive gene expression (reviewed in Shen, 2007). Homozygous hypomorphic mutation of Nodal in the mouse leads to failure to form the prechordal mesendoderm, and consequently the expression of *Shh* in the anterior forebrain is lost (Lowe *et al.*, 2001). These data, together with experiments performed in zebrafish (Rohr *et al.*, 2001), strongly argue for a role of Nodal signaling upstream of *Shh* in forebrain development, by establishing a field of *Shh* expression in the rostral regions of the embryo.

Shh activates a complex signaling pathway (reviewed in Bertrand and Dahmane, 2006; Ingham, 2008; Ogden *et al.*, 2004). In the absence of ligand, the principle Shh receptor, the twelve transmembrane domain protein Patched1 (Ptch1), functions to inhibit the activity of Smoothened (Smo), a seven transmembrane domain protein. Binding of Shh to Ptch1 releases Smo inhibition, allowing the regulation of gene expression through the Gli transcription factors, Gli1, Gli2, and Gli3 (Huangfu and Anderson, 2006; Wang *et al.*, 2007). Gli2 performs the majority of the transcriptional activation response, with contributions from Gli1 and Gli3. However, in the absence of Shh, Gli3 is proteolytically processed into a transcriptional repressor that functions as a Shh pathway antagonist (Koebernick and Pieler, 2002; Sasaki *et al.*, 1999; Wang *et al.*, 2000).

Mice with a targeted mutation of the Shh gene display severe HPE, including cyclopia (Chiang *et al.*, 1996). They lack ventral forebrain and spinal cord structures but dorsal progenitors remain, and the expression of dorsal genes is expanded to encompass the ventral regions of highly deformed embryos (Ohkubo *et al.*, 2002; Pabst *et al.*, 2000). *Gli3* mutant mice, on the other hand, display a ventralized forebrain and exencephaly (Grove *et al.*, 1998; Tole *et al.*, 2000). Reduction of *Gli3* dosage in *Shh* null mice largely restores dorsoventral patterning of the telencephalon (Rallu *et al.*, 2002). These data are consistent with the idea that a major role of Shh during forebrain development is to block the actions of the Gli3 repressor. In addition, it suggests the existence of a mechanism whereby the ventral forebrain may be patterned in the absence of Shh signaling. The most likely factors to fill this role are fibroblast growth factors (Fgfs).

Fgfs, and in particular Fgf8, are initially expressed in the anterior neural ridge, at the rostral-most tip of the neural plate. Subsequent expression of Fgf8 in the prosencephalon is maintained in a Shh-dependent fashion (Ohkubo *et al.*, 2002). Interestingly, mice deficient for both Fgf receptors 1 and 2 lack ventral progenitor cells (Gutin *et al.*, 2006). Therefore, ventral forebrain specification requires signaling through both Shh and Fgf. In agreement with this idea, *Fgf8* hypomorphic mutant mice have smaller telencephalic vesicles, lack ventral structures and develop HPE (Storm *et al.*, 2006). Consistent with its role as a repressor of Shh signaling, *Gli3* mutant mice have an expanded *Fgf8* expression domain (Aoto *et al.*, 2002). However, *Gli3* deletion does not rescue the phenotype generated by loss of Fgf8 signaling (Gutin *et al.*, 2006), revealing that Fgf8 may function downstream of Shh/Gli3 signaling in ventral forebrain development. Consistent with this idea, Fgf8 can act in a Shh-independent manner; Fgf8 induces the expression of ventral telencephalic markers when applied to dorsal telencephalic neuroectoderm explants in the absence of Shh (Storm *et al.*, 2006).

Bone morphogenetic proteins (Bmps) regulate the specification of the dorsal midline. Bmp receptor 1a/1b double mutant mice fail to specify dorsal cell progenitors and do not separate the dorsal telencephalon into bilateral hemispheres, resulting in MIH HPE (Fernandes *et al.*, 2007). Bmp ligands also induce cell death and suppress cell proliferation, events that are required during the invagination of the midline and separation of the telencephalic vesicles (Furuta *et al.*, 1997). Interestingly, *Gli3* mutant mice lack dorsal structures like the choroid plexus and the hippocampus (Grove *et al.*, 1998; Tole *et al.*, 2000). The loss of these structures correlates with a decrease in Bmp signaling, and the aforementioned expansion in Fgf8 expression (Kuschel *et al.*, 2003). However, this process is unlikely to be dependent on Shh, since *Shh* expression is not disrupted in the *Bmpr1a/1b* mutant mice (Fernandes *et al.*, 2007). Taken together, these data suggest Bmp signaling lies downstream of Gli3 during the specification of the dorsal midline. However, while certain experimental approaches reveal hierarchical order to Shh, Fgf, and Bmp signaling that are consistent with "upstream" versus "downstream" roles, it is also clear that these pathways interact in complex ways to achieve patterning of the forebrain. Shh, Fgfs, and Bmps regulate each other's expression in a complex and incompletely understood feedback network that patterns the forebrain (Hebert, 2005; Ohkubo *et al.*, 2002). Partition of the forebrain and formation of the midline, therefore, requires a delicate balance between these signals that also involves an appropriate level of cell proliferation and apoptosis of the midline cells, while the neighboring neuroepithelial cells expand to form the bilateral telencephalic vesicles (Hebert, 2005). Ultimately, several important structures are specified from the midline, including the choroid plexus, the cortical hem, and the medial ganglionic eminences.

4. Murine Models of HPE

4.1. Genetic models of HPE in the mouse

As mentioned above, mouse models have been generated in which the eight known human HPE genes have been mutated by targeted disruption. These mice have given us clues about the genetic interactions that control forebrain development and how it goes awry in HPE. Additional mouse models have been generated for other genes that participate in the same signaling pathways, and some of these mutant lines have shed light on the complexities of human HPE. These genes are, therefore, good candidates for human HPE genes. Finally, HPE has been observed in mice carrying mutations in genes whose pathways have not yet been implicated by mutation in human HPE; these mice provide further insight into processes that pattern and partition the forebrain.

4.1.1. Nodal pathway genes

The Nodal pathway is required for formation of the prechordal mesendoderm, and mice with mutations in several genes in this pathway display HPE. Strong *Nodal* hypomorphs fail to form prechordal mesendoderm and a fraction of these mice are holoprosencephalic, while others have more severe phenotypes (Lowe *et al.*, 2001). Mice with mutations in the Nodal receptor *ActRIIA*, and the pathway-responsive transcription factors *Smad2* and *FoxH1*, also exhibit HPE due to failure in the formation of the prechordal mesendoderm (Heyer *et al.*, 1999; Hoodless *et al.*, 2001; Nomura and Li, 1998; Song *et al.*, 1999). *Tdgf1/Cripto* null mice display embryonic lethality due to failure of anterior–posterior axis positioning (Ding *et al.*, 1998a), but mice with a low level of *Tdgf1/Cripto* function display a broad spectrum of HPE phenotypes with partial penetrance (Chu *et al.*, 2005). As seen with *Nodal* mutant mice, *Shh* expression is reduced or missing in *Tdgf1/Cripto* hypomorphs, and the prechordal mesendoderm does not form properly.

Mice carrying mutations in Nodal pathway genes offer examples of roles for digenic mutation and stochastic events in the etiology of HPE. Growth and differentiation factor 1 (Gdf-1) is, like Nodal, a member of the TGF-β superfamily; it regulates *Nodal* expression in lateral mesoderm, but not in the node (Rankin *et al.*, 2000). While heterozygosity for *Nodal* does not result in HPE, removal of one copy of *Nodal* in *Gdf-1* null mice (which are also not holoprosencephalic) does, due to failure in the specification of the prechordal mesendoderm and, therefore, formation of anterior neural structures (Andersson *et al.*, 2006). Interestingly, triple heterozygosity for *Nodal, Gdf-1*, and *Alk4* (which encodes a component of the Nodal receptor) results in HPE; likewise, removing one copy of *Alk4* on a *Gdf-1* null background also produces HPE (Andersson *et al.*, 2006). These data strongly suggest that Nodal and Gdf-1 function with at least partial redundancy and signal via Alk4 in the node to initiate the signaling events that regulate ventral forebrain formation. Furthermore, these studies make it clear that heterozygosity at multiple loci can be sufficient to induce HPE.

Studies with an allelic series of *Tdgf1/Cripto* mutations suggest that stochastic events may also underlie some of the variability seen in human HPE. While fully null animals die prior to gastrulation, and heterozygotes are phenotypically normal, compound heterozygotes that carry one null and one hypomorphic *Tdgf1/Cripto* allele display highly variable phenotypes that range from early embryonic lethality similar to true nulls, to severe HPE with cyclopia, to less severe HPE-related phenotypes, to viability (Chu *et al.*, 2005). These results are probably best interpreted to indicate that a key developmental event in patterning of the rostroventral midline is dosage-sensitive and may occur normally or fail in stochastic fashion when the Nodal pathway is partially compromised.

4.1.2. Shh pathway genes

In addition to the induction of ventral neural fates, Shh maintains the proliferation and survival of the progenitor cells (Fuccillo *et al.*, 2006). Interestingly, midline structures, like the choroid plexus and the hippocampus, fail to form in *Shh*$^{-/-}$ mice (Chiang *et al.*, 1996). These structures arise from the invagination of cells at the dorsal midline during the partitioning of the brain into left and right hemispheres, suggesting that Shh can modulate cell fate at a distance, likely through a combination of direct and indirect effects.

Mutation of *SHH* is the most common known genetic cause of human HPE (Muenke and Beachy, 2001). Mutations in the *GLI2* gene have also been described in HPE patients, affecting mainly the formation of the pituitary gland (Roessler *et al.*, 2003). Targeted disruption of *Gli2* is neonatal lethal in the mouse, and results in defects in early brain and spinal cord development, the most striking defect being absence of the floor plate of the neural tube (Ding *et al.*, 1998b; Matise *et al.*, 1998; Mo *et al.*, 1997). *Gli1* mutant mice do not have any apparent phenotypic defect, and formation of the forebrain midline is unaltered in *Gli1;Gli2* double mutant mice, which display a much milder phenotype than *Shh*$^{-/-}$ mice in all tissues. These data underscore the importance of Shh-mediated removal of Gli3 repressor function in forebrain patterning (Park *et al.*, 2000).

Several mutations in *PTCH1* have been observed in HPE patients (Ming *et al.*, 2002; Ribeiro *et al.*, 2006). In the absence of ligand, Ptch1 keeps the Shh signaling cascade off. Loss of Ptch1 activity results in constitutive pathway activation, and *Ptch1*$^{-/-}$ mice display ventralization of the neural tube and a dorsal expansion in the expression of *Shh* (Goodrich *et al.*, 1997). Therefore, in contrast to other known HPE genes, mutations in *PTCH1* that result in HPE are not expected to be loss-of-function mutations; rather, mutations that enhance its activity or render it insensitive to ligand would be predicted to result in HPE. Seven different missense mutations have been found in *PTCH1* in human HPE, and the expectation is that they alter the ability of Ptch1 to bind Shh or to become inactivated upon Shh binding. At present, these ideas have not been tested in experimental models.

Dispatched1 (Disp1) is a 12-transmembrane domain protein required for transport of lipid-modified Shh from Shh-producing cells (Ma *et al.*, 2002). Analysis of three independent *Disp1* mouse mutant lines revealed that homozygous mutant mice die at E9.5 with HPE, including cyclopia (Caspary *et al.*, 2002; Kawakami *et al.*, 2002; Ma *et al.*, 2002). Loss of Disp1 also prevents specification of ventral cell types in the neural tube. Mice lacking Smo display very similar phenotypes to those seen in *Disp1* null mice (Caspary *et al.*, 2002; Kawakami *et al.*, 2002; Ma *et al.*, 2002). While Disp1 is required in Shh-producing cells, Smo is required cell-autonomously for transduction of the Shh signal; however, mutations in *SMO* or *DISP1* have not yet been reported in human HPE.

A recently identified secreted protein of unknown function that participates in Shh signaling is Tectonic (Tect). *Tect* mutant mice generated by insertion of a gene trap in the *Tect* locus die between E13.5 and E16.5, displaying HPE (Reiter and Skarnes, 2006). *Tect* mutant mice do not form a neural tube floor plate and lack expression of ventral neural markers. Moreover, expression of the Shh target genes *Gli1* and *Ptch1* are reduced. Additional studies indicate that Tect modulates Shh signaling downstream of Smo. Interestingly, *Shh;Tect* double mutant mice look like *Tect* mutant mice and regain a level of *Ptch1* expression, which is lost in *Shh*$^{-/-}$ mice. Overall, the data suggest that Tect can positively and negatively modulate Shh signaling in a manner dependent on the concentration of Shh (Reiter and Skarnes, 2006).

Cdo, Boc, and Gas1 are each Shh-binding proteins that may function as Hedgehog coreceptors with Ptch1 (Kang *et al.*, 2007). Mice with mutations in these genes provide evidence for digenic mutation and silent modifier genes in modulation of phenotypic variation across the HPE spectrum.

Cdo and Boc are members of the Ig superfamily and act as putative coreceptors for Shh to positively regulate Shh signaling (Tenzen *et al.*, 2006; Yao *et al.*, 2006; Zhang *et al.*, 2006). Homozygous *Cdo* mutant mice on a 129/Sv background display facial microforms of HPE with partial penetrance and without overt brain defects (Cole and Krauss, 2003; Zhang *et al.*, 2006). In contrast, on a C57BL/6 background, *Cdo* mutant mice show semilobar HPE with a single nostril and absent or rudimentary olfactory bulbs (Zhang *et al.*, 2006) (Fig. 3.2). Shh-dependent gene expression is reduced but not lost in the ventral forebrains of *Cdo*$^{-/-}$ mice, which likely explains the lack of a cyclopic or fully alobar phenotype. Thus, *Cdo* mutants on a 129/Sv background closely resemble the woman shown in Fig. 3.1, while *Cdo* mutants on a C57BL/6 background are similar to her more severely affected offspring. These results implicate strain-dependent modifier genes in the variability of the phenotype and, by inference, strongly suggest that otherwise silent modifier genes in the human population may underlie aspects of the human HPE spectrum (El-Jaick *et al.*, 2007; Zhang *et al.*, 2006). Furthermore, removal of one copy of *Shh* from *Cdo*$^{-/-}$ mice of a relatively resistant hybrid background strongly enhances the phenotype (Tenzen *et al.*, 2006). However, unlike *Shh*$^{-/-}$ mice, *Cdo*$^{-/-}$ mice do not present limb defects. The strain-dependent spectrum of HPE phenotypes, in combination with the spatial restriction of this phenotype, makes *Cdo*$^{-/-}$ mice an accurate model of human HPE and suggests *CDO* is a strong candidate HPE gene (Krauss, 2007).

Boc and *Cdo* show generally similar expression patterns; one exception is the prechordal mesendoderm, where *Cdo* but not *Boc* is expressed. Consistent with this observation, targeted disruption of *Boc* in the mouse does not result in HPE (Okada *et al.*, 2006). Instead, *Boc*$^{-/-}$ mice show defects in

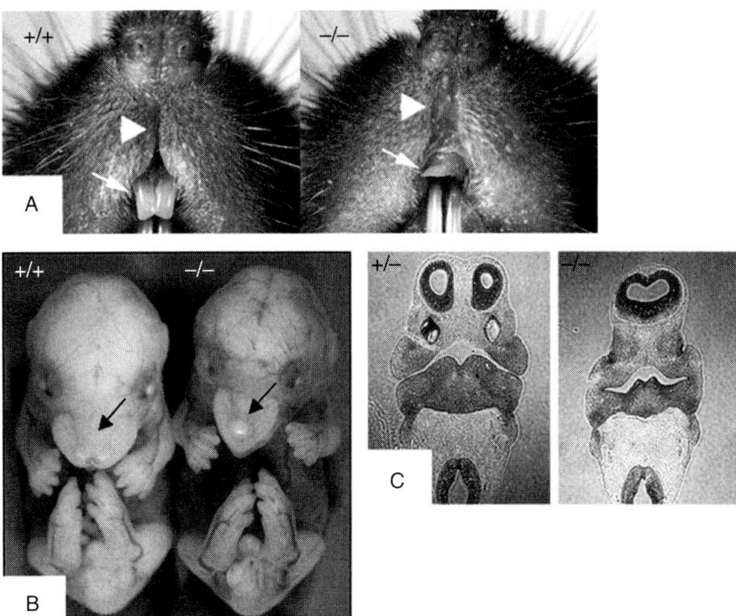

Figure 3.2 *Cdo* mutant mice display variable HPE phenotypes dependent on genetic background. (A) Frontal views of the midfacial region of 4-week-old adult mice of the indicated *Cdo* genotypes on the 129/Sv background. The $Cdo^{-/-}$ animal in the right panel shows microforms of HPE, including a single, central maxillary incisor (arrows) and a dysgenic philtrum (arrowheads). $Cdo^{-/-}$ mice of the 129/Sv background show such craniofacial midline phenotypes with ~50% penetrance. (B) Whole embryos of the indicated *Cdo* genotypes on the C57BL/6 background at embryonic day (E) 15.5. The arrows indicate that, while the $Cdo^{+/+}$ embryo has two nostrils and a normal midface, the $Cdo^{-/-}$ embryo displays a single nostril and hypoplastic midface. (C) Sections of E11.5 embryos of the indicated *Cdo* genotypes on the C57BL/6 background. Note the paired lateral ventricles in the $Cdo^{+/-}$ (control) embryo and the single ventricle in the $Cdo^{-/-}$ embryo. $Cdo^{-/-}$ mice of the C57BL/6 background show the more severe HPE phenotypes shown in B and C with ~80% penetrance. Reprinted with permission from Cole and Krauss (2003) and Zhang *et al.* (2006).

Shh-mediated guidance of commissural axons in the developing spinal cord, where *Cdo* is not expressed.

Gas1 is a GPI-anchored protein that has been recently characterized as a positive regulator of Shh and displays a similar expression pattern to *Boc* and *Cdo* during embryogenesis (Allen *et al.*, 2007; Lee and Fan, 2001; Mulieri *et al.*, 2000, 2002; Tenzen *et al.*, 2006). *Gas1* null mice have developmental defects consistent with a reduction in Shh signaling, including incomplete specification of the floor plate, and defects in digit patterning and craniofacial development (Allen *et al.*, 2007; Martinelli and Fan, 2007). Importantly, $Gas1^{-/-}$ animals display microform HPE, and removal of one copy of *Shh* from *Gas1*$^{-/-}$ mice worsens the craniofacial defects seen with loss of Gas1 alone (Allen *et al.*, 2007; Martinelli and Fan, 2007; Seppala *et al.*, 2007). Interestingly, *Gas1;Cdo* double mutant mice show modulation of Shh signaling in a dosage-dependent manner in the neural tube, with severe HPE in double null animals (Allen *et al.*, 2007). Mice carrying mutations in Cdo and Gas1, therefore, provide genetic models of how Shh pathway genes and strain-dependent modifier genes may interact to produce HPE phenotypes of varying severity, similar to the human situation.

4.1.3. Six3

Six3 is a homeodomain transcription factor. *Six3* null mice lack head structures anterior to the brain, including the eyes and nose (Lagutin *et al.*, 2003). A likely explanation is that, in *Six3*$^{-/-}$ embryos, *Wnt1* expression is expanded rostrally into the entire anterior region of the brain; Wnt signaling must be inhibited for development of the rostral telencephalon or the prospective forebrain will acquire a caudal diencephalic identity (Yamaguchi, 2001). These data indicate that lack of Six3 leads to partial caudalization of the head, but how this phenotype relates to the role of Six3 in HPE is not clear.

A very recent study links *Six3* haploinsufficiency in mice to defective *Shh* expression in the ventral midline of the rostral diencephalon (Geng *et al.*, 2008). *Six3*$^{+/-}$ mice display semilobar HPE with strain-dependent penetrance, the C57BL/6 background being more sensitive than the 129/Sv background. This is intriguing as *Six3* is the only HPE gene to show clear haploinsufficiency in both humans and mice. Furthermore, double heterozygosity for *Six3* and *Shh* raised the HPE penetrance to 100% on each background. *Shh* expression is directly activated by Six3 in the ventral forebrain (but not the prechordal mesendoderm) and was reduced in *Six3*$^{+/-}$ embryos; conversely, *Six3* expression is at least partially dependent on Shh in this structure, suggesting the existence of a positive feedback regulatory loop between these factors in the ventral forebrain that is disrupted in HPE.

4.1.4. Zic2

Zic2 is a transcription factor with a zinc-finger DNA-binding motif very similar to that of the Gli proteins. *Zic2* is expressed in both ventral and dorsal midline regions, and unlike the genes in the Nodal and Shh pathways, is implicated in both classical and MIH HPE, without facial involvement (Brown *et al.*, 1998). A partial loss-of-function *Zic2* mutation has been generated in the mouse, where *Zic2* levels are reduced by ~80% (Nagai *et al.*, 2000). The most frequent abnormality observed is spina bifida, which is also observed in some HPE patients with a *ZIC2* mutation. Rostral regions of the neural tissue were also affected, giving rise to an undivided cerebral cortex with missing or hypoplastic and contracted dorsal forebrain structures and no obvious abnormalities of the face; this resembles MIH HPE (Nagai *et al.*, 2000).

How loss of Zic2 results in classical HPE is not clear. *Zic2* expression is lost in *Shh*$^{-/-}$ mouse embryos, and addition of HhAntag, a Shh pathway chemical antagonist (Romer *et al.*, 2004) to *in vitro* embryo cultures suppressed *Zic2* expression (Hayhurst *et al.*, 2008). Despite its role in development of dorsal brain structures and MIH HPE (Brown *et al.*, 1998; Nagai *et al.*, 2000), *Zic2* expression is maintained in *Bmpr1a;Bmpr1b* double mutant mice, which also display MIH HPE (Fernandes *et al.*, 2007). In contrast, beads coated with Fgf8 induced strong *Zic2* expression in tissue explants from the E9.5 mouse lateral telencephalon, while Shh- or Bmp4-coated beads failed to do so (Hayhurst *et al.*, 2008). These results suggest that Fgf8 induces *Zic2* expression *in vivo*, and the loss of *Zic2* expression in *Shh*$^{-/-}$ and HhAntag-treated embryos corresponds to a secondary effect of Shh on Fgf8 levels. Furthermore, they reveal a potential mechanism whereby Shh can indirectly affect dorsal patterning.

A very recent study with a severe, perhaps null, *Zic2* allele in the mouse revealed a different phenotype than that observed with the hypomorphic allele described above (Warr *et al.*, 2008). Mice homozygous for this mutation displayed a transient defect in node function that resulted in arrested development of prechordal mesendoderm, in turn producing (indirectly) the expected defect in Shh signaling in the forebrain. Interestingly, even on an inbred background these animals showed phenotypic variability, with more than half being cyclopic but a large fraction having only hypotelorism. As with *Tdgf1/Cripto* hypomorphs, this may arise from a potentially stochastic developmental process required for midline formation. Taken together, it seems likely that Zic2 plays an early role in development of the prechordal mesendoderm that becomes apparent only in the near-complete absence of function, and a later role in formation of the roof plate and dorsal midline that is disrupted with partial loss of function (Warr *et al.*, 2008).

4.1.5. Tgif

TGIF is the most perplexing of the established HPE genes. TGIF is an atypical homeodomain-containing protein that functions as a transcriptional corepressor of Smad2 and Smad3 to inhibit responses to TGF-β family ligands (Wotton *et al.*, 1999a,b). Loss-of-function mutations should therefore result in increased Nodal signaling (and hence, increased Shh signaling), which is the opposite of what would be predicted for an HPE-inducing mutation. It is possible that, as appears to be the case with Bmp signaling (see below), too much or too little Nodal signaling might result in HPE. TGIF also physically interacts with the retinoic acid receptor RXRα, functioning to repress transcription in the absence of retinoic acid (RA) (Bartholin *et al.*, 2006). Mice exposed to teratogenic amounts of RA display anterior neural tube defects and HPE in a strain-dependent manner (Sulik *et al.*, 1995). Hence, it is conceivable that *Tgif* mutations could sensitize embryos to HPE when in contact with teratogenic amounts of RA. *TGIF* HPE mutants behave as hypomorphic or full loss-of-function alleles in cell-based reporter assays for transcriptional corepression of both Smad3 and RXRα (El-Jaick *et al.*, 2007). Thus, it is not clear how mutation of TGIF results in HPE.

Construction of animal models, particularly mice carrying mutations in *Tgif*, might be expected to clarify this situation. However, four independently generated null alleles of *Tgif* revealed no HPE phenotype in mice, either in the homozygous or heterozygous state; in fact these are viable (Bartholin *et al.*, 2006; El-Jaick *et al.*, 2007; Jin *et al.*, 2006; Mar and Hoodless, 2006; Shen and Walsh, 2005). Each of the alleles was studied on a hybrid genetic background, but two were also analyzed on a congenic C57BL/6 background. Two human HPE cases with double heterozygosity for *TGIF* and *SHH* have been reported (see above), suggesting that perhaps loss of Tgif contributes to HPE only in a background with compromised Shh signaling. However, $Tgif^{+/-};Shh^{+/-}$ and $Tgif^{-/-};Shh^{+/-}$ mice also do not display HPE on either mixed or C57BL/6 backgrounds (El-Jaick *et al.*, 2007; Shen and Walsh, 2005). Furthermore, while $Tgif^{+/-}$ and $Tgif^{-/-}$ embryos were more sensitive to RA-induced birth teratogenesis, HPE was not among the observed defects (Bartholin *et al.*, 2006). Thus, although multiple genetic hits, modifier genes and/or gene–environment interactions remain good possible explanations for the differences between humans and mice in their respective responses to *Tgif* mutation, the mouse has not yet provided experimental support for any of these hypotheses. It should be noted that mice carrying a fifth *Tgif* allele, one predicted to produce a truncated protein product, displayed hypoplastic heads, exencephaly and, rarely, HPE on a C57BL/6 but not a mixed genetic background (Kuang *et al.*, 2006). It is possible that specific *Tgif* alleles may interact with modifier genes to produce developmental defects; how this might relate to null alleles in human HPE cases is not clear.

4.1.6. Bmp inhibitors

Additional examples of genetic interactions in murine HPE are seen with the Bmp pathway regulators Chordin (Chrd), Noggin (Nog), and Twisted gastrulation (Twsg1); these are secreted antagonists of Bmps that are expressed in the node (reviewed in Gazzerro and Canalis, 2006). Chrd and Nog have similar biochemical activities and expression domains, found at the prechordal mesendoderm and anterior neural ridge; thus, they could well compensate for each other during development. $Chrd^{-/-};Nog^{-/-}$ double mutant mice displayed embryonic lethality but the few embryos found at late stages of development showed HPE, including cyclopia (Bachiller *et al.*, 2000). Similarly, while most $Chrd^{-/-};Nog^{+/-}$ animals were viable, a fraction died prenataly, some of them displaying severe HPE while others had more dramatic rostral truncations (Anderson *et al.*, 2002). These interactions, not seen with double heterozygotes, strongly suggest redundant or overlapping function between Chrd and Nog (Anderson *et al.*, 2002; Bachiller *et al.*, 2000, 2003; Brunet *et al.*, 1998; McMahon *et al.*, 1998).

Twsg1$^{-/-}$ mice also have marked forebrain defects that include rostral truncations and severe HPE (Petryk *et al.*, 2004). In the case of *Twsg1*$^{-/-}$ mice, HPE phenotypes are strain-dependent, with 44% penetrance on a C57BL/6 background and almost no phenotype on other backgrounds. In these models, HPE phenotypes correlate with loss of *Shh* expression in the prechordal mesendoderm and Shh-dependent gene expression in the surrounding tissue. Furthermore, there was a reduction in *Fgf8* expression in the anterior neural ridge and increased Bmp signaling in the rostral tissue, which is consistent with the strong rostral-most malformations. Although no mutations in Bmp inhibitor genes have been reported in human HPE, taken together, results with mice carrying mutations in Bmp inhibitors suggest that these factors interact genetically with each other and strain-specific modifier genes in a complex manner in patterning of rostral structures, including the midline. Remarkably, removal of one copy of *Bmp4* from *Twsg1*$^{-/-}$ mice on a hybrid genetic background, where loss of Twsg1 does not give any appreciable phenotype, resulted in HPE with anophthalmia, suggesting that loss of Twsg1-mediated Bmp function in these mice is responsible for the HPE phenotype. Twsg1 may therefore regulate Bmp signaling in both positive and negative ways (Zakin and De Robertis, 2004). As gain of Bmp function can also cause HPE and cyclopia (Anderson *et al.*, 2002; Bachiller *et al.*, 2000; Golden *et al.*, 1999), these results and additional reports (Ohkubo *et al.*, 2002) highlight the notion that Bmp signaling plays multiple and complex roles in forebrain patterning.

4.1.7. Megalin

Megalin (also known as LRP2) is an endocytic receptor of the low-density lipoprotein (LDL) receptor-related protein family. It is expressed in the neuroepithelium of the embryo and binds various signaling ligands, functioning in turnover of both Shh and Bmp4 (Spoelgen *et al.*, 2005). Megalin mutant mice displayed HPE characterized by incomplete separation of the forebrain hemispheres and extension of the lateral ventricles (Spoelgen *et al.*, 2005; Willnow *et al.*, 1996). Interestingly, by E10.5, *Megalin*$^{-/-}$ embryos have lost Shh expression specifically in the anterior entopeduncular (AEP) area of the ventral forebrain, without affecting expression in notochord, floor plate, limb buds, and diencephalon. *Fgf8* expression was reduced in the rostral ventral telencephalon but extended dorsally beyond the commissural mesendoderm, while Bmp4 expression was ventrally expanded. These alterations could be readily detected at E9.5, before any changes in *Shh* expression (Spoelgen *et al.*, 2005). These data, together with the ability of cells expressing Megalin to take up Bmp4 by endocytosis (Spoelgen *et al.*, 2005), suggest that the HPE associated with loss of Megalin is caused by hyperactive Bmp4 signaling and a consequent inhibition of ventral forebrain patterning.

4.1.8. Cdc42

In addition to the identification of candidate genes that function within the pathways known to be involved in HPE, the generation of conditional mouse mutants may allow new insights into the process of forebrain partitioning and midline formation by factors that function more globally and whose complete loss from the embryo results in phenotypes too severe to visualize a specific role in these processes. One such factor appears to be Cdc42, a member of the Rho family of small GTPases.

It has been proposed that the generation of bilateral telencephalic vesicles requires the formation of a radial anisotropic neuroepithelial wall, which works against the hydrostatic pressure inside the telencephalic vesicle, allowing its expansion (Rakic, 1995). These events require correct positioning of neural progenitor cells to the apical side of the neuroepithelium, a process that involves cell and tissue polarity, which are modulated by Rho GTPases. Conditional deletion of *Cdc42* in the developing forebrain resulted in HPE, due to a disruption in the localization of adherens junctions-associated proteins and therefore a change in the architecture of the neuroepithelium (Chen *et al.*, 2006). As a consequence, bifurcation and generation of bilateral telencephalic vesicles did not occur, even though Shh target gene expression was not grossly affected.

mFat1 encodes a giant protocadherin, and mice with mutations in *mFat1* also display HPE, albeit with very low penetrance, in the absence of obvious alterations in Shh signaling (Ciani *et al.*, 2003). Because cadherins mediate cell–cell adhesion, it is possible that these animals, like the conditional *Cdc42* mutants, have disruptions in neuroepithelial architecture. Although mutations in *CDC42* (or human *FAT1*) have not been reported in human HPE, these models raise the notion that both accurate signaling in forebrain specification and regulation of tissue mechanics are crucial for normal development of a properly partitioned forebrain.

4.2. Teratogen-induced models of HPE in the mouse

The fact that mutations have been identified in only ~25% of human HPE cases suggests that not all HPE genes have been discovered; that mutations outside of coding regions (which have been analyzed much less frequently) may be prevalent; and/or that environmental factors play a major role in human HPE. The mouse is an important model organism for the study of teratogenesis. Although reports of teratogen-induced HPE in humans and mice have been more limited than those on genetics, sufficient data exist to posit that the mouse will be valuable in understanding environmental contributions to HPE. In this regard, both *in vivo* and *ex vivo* studies have been performed.

4.2.1. *In vivo* studies

Ethanol consumption during pregnancy can lead to fetal alcohol syndrome (FAS) and fetal alcohol spectrum disorders (FASD) (Itthagarun *et al.*, 2007). Although the incidence is not very high, a number of cases have been reported of children with HPE born to mothers who drank heavily during pregnancy (Bonnemann and Meinecke, 1990; Jellinger *et al.*, 1981; Ronen and Andrews, 1991). Additionally, various anomalies within the HPE spectrum have been identified in infants with FAS and FASD (Coulter *et al.*, 1993; Majewski, 1981; Peiffer *et al.*, 1979).

Ethanol is among the few chemicals known to induce HPE both in humans and mice, with an incidence in mice of around 20% (Higashiyama *et al.*, 2007). Mouse embryos exposed to ethanol for a brief but precise period of time during early development display mild HPE defects, principally midline craniofacial malformations. Ethanol-induced HPE has a very narrow temporal window of susceptibility, with administration required at day 7.0 of pregnancy, followed by a second dose four hours later (Sulik and Johnston, 1982, 1983; Sulik *et al.*, 1981). The phenotypes obtained are similar to many seen in FAS and included microcephaly, small eyes, some midfacial abnormalities that resemble microform HPE and occasional failure to partition the forebrain. Similar experiments performed with methanol indicated that it also might induce HPE when administered to pregnant mice during gastrulation stages (Rogers *et al.*, 2004).

As with mutation-induced HPE, the genetic background of the mice used for these studies strongly affects the ability of ethanol to induce HPE-like phenotypes. Although most experiments have been performed with C57BL/6 mice, analysis of other strains demonstrated that among DBA/2J, CD-1, C57BL/6J, and Swiss-Webster mice, DBA/2J mice were the most susceptible to ethanol-induced HPE, whereas CD-1 mice were the least affected (Giknis *et al.*, 1980). Likewise, methanol administration to CD-1 pregnant mice did not result in HPE, whereas C57BL/6J mice were affected (Rogers *et al.*, 2004).

In spite of all these studies, the molecular and pathological mechanisms of ethanol-induced HPE have not yet been clarified. Early studies suggested that ethanol treatment could result in abnormal mesodermal development (Sulik and Johnston, 1982). Increased cell death in the prosencephalon has also been observed (Kotch and Sulik, 1992a,b). Histological examination of ethanol-treated, holoprosencephalic embryos at E10.5 revealed loss of midline fibers in the ventral hippocampal and anterior commissures, suggesting that these mice would likely have lacked a corpus callosum, a defect often associated with HPE (Higashiyama *et al.*, 2007). Importantly, at

E11.5, *Shh* mRNA levels were lower than normal. Interestingly, the decrease in *Shh* expression is mainly detected in the anterior telencephalon as opposed to the hypothalamus, which was unaffected (Aoto *et al.*, 2008). When embryos were analyzed 3–6 h after ethanol exposure, *Shh* expression was reduced only in the anterior portion of the prechordal mesendoderm, and this correlated with increased apoptosis in that same area (Aoto *et al.*, 2008). Taken together, these data suggest that ethanol-induced HPE might be due to decreased *Shh* expression at early stages of development in a key area for the induction of midline structures.

A few case reports of potential RA-induced HPE have been reported in which RA derivatives had been prescribed to treat skin disorders, such as cystic acne and severe keratinization (Cohen and Shiota, 2002). Administration of RA to pregnant C57BL/6J mice 7 days postfertilization, produced embryos at E16 with severe craniofacial defects of the aprosencephaly spectrum, and some cases of HPE were also detected (Sulik *et al.*, 1995). While much attention has been given in recent years to the role of RA as a morphogen, the mechanisms by which RA acts as an HPE-inducing teratogen in mice remain unclear.

Finally, in an example of gene–environment interaction, dramatic reduction of cholesterol levels in mice can also produce HPE. Treatment of cholesterol-deficient mice (that carry a mutation in *Apob*, which encodes apolipoprotein B) with the 3β-hydroxysterol-Δ^7-reductase inhibitor, BM15.766 recapitulates features of SLO, including mild through severe HPE phenotypes (Lanoue *et al.*, 1997).

4.2.2. *Ex vivo* studies—Embryo culture

The corn lily *Veratrum californicum* contains alkaloids like cyclopamine and jervine that have been known for many years to cause HPE in offspring of pregnant sheep that ate this plant (James, 1999; Keeler, 1975). These compounds inhibit Shh signaling by binding to Smo and blocking its signaling function (Chen *et al.*, 2002; Cooper *et al.*, 1998). The entire HPE spectrum can be produced in chicks by exposing embryos to cyclopamine at different stages of gestation (Cordero *et al.*, 2004). While studies on the teratogenicity of cyclopamine or jervine in mice have been published since the 1970s no standard regimen for administration has been developed. Lipinski *et al.* (2008) compared multiple routes of administration of cyclopamine and found that intraperitoneal injection and oral gavage administration of cyclopamine to mice had adverse toxic effects to pregnant dams and a disadvantageous pharmacokinetic profile. The use of micro-osmotic pump infusion was superior, but produced only cleft lip and palate in ~30% of exposed embryos. Thus, despite the strong suspicion that environmental agents may be important in human HPE, no standardized protocols currently exist in which a teratogen can easily, efficiently, and reproducibly induce HPE with high penetrance in the mouse.

The whole embryo culture system is emerging as a more amenable way to solve some of these problems, and this system has long been used for teratological studies (reviewed in New, 1990). With this system, it has been possible to induce a mild form of HPE in mouse embryos, by culturing them with cyclopamine between developmental stages E8.5 and E10.5, before the two cerebral hemispheres are divided (Nagase *et al.*, 2005). The craniofacial phenotype produced consisted mainly of reduced angles and distance between the nasal placodes, accompanied by a reduction in the expression levels of the Shh target genes *Ptch1* and *Gli1* in the midbrain and forebrain. Expression of *Shh* in the precordal mesendoderm starts before E7.5, and treatment of cultured embryos with cyclopamine at E7.8 (0–1 somites) had a very strong effect on overall embryo morphology, making it impossible to assess specific forebrain and craniofacial defects; conditions might need further adjustments to phenocopy more severe forms of HPE *in vitro*. It is also possible that the conditions were such that inhibition of Shh only partially affected craniofacial development in this model. In fact, differences in time of initiation of cyclopamine administration as short as half a day had a large impact on the phenotype, which varied from defects in conformational rotation and underdevelopment of the brain with treatment at E8.0, to normal turning of the embryo and mild HPE with treatment at E8.5 (Nagase *et al.*, 2006).

In a similar approach, the small molecule inhibitor of Shh signaling, HhAntag, has been used in embryo cultures to study the formation and maintenance of the signaling centers involved in the partitioning of the brain. Like cyclopamine, HhAntag binds directly to Smo to block Shh signaling (Frank-Kamenetsky *et al.*, 2002; Romer *et al.*, 2004; Williams *et al.*, 2003). Treatment of embryos at E9.5 with HhAntag for two days resulted in smaller embryos, with smaller limb buds and altered tail shapes (Hayhurst *et al.*, 2008). However, HhAntag did not reproducibly block bifurcation of the forebrain, even when used at high concentrations, but instead generated a variety of intermediate phenotypes. These results might be explained by variations in diffusion of the drug into the brains of the embryos, or by subtle differences in the embryonic stage of the embryos. Nonetheless, in the most severely affected embryos, *Ptch1* levels were undetectable in the ventral telencephalon after two days of treatment, and expression of *Fgf8* in the anterior midline was lost (Hayhurst *et al.*, 2008). Surprisingly, in contrast to the ventral expansion of *Bmp4* expression seen in *Shh*$^{-/-}$ mice, expression of *Bmp4* at the dorsal midline was also reduced upon HhAntag treatment. Since the expression of the Fgf8-regulated gene *Foxg1* was detectable after 2 days of treatment, and given that FoxG1 negatively regulates *Bmp4* expression (Dou *et al.*, 1999, 2000), it is conceivable that expression of *Fgf8* prior to or during the first day of HhAntag treatment was sufficient to induce *Foxg1* gene expression, resulting in repression of Bmp4 expression. Undoubtedly, the *in vitro* systems provide significant information regarding the timing of the signaling events that control brain development. However, some questions still remain: for example, treatment of the embryos with HhAntag for one day was insufficient to block Shh signaling and patterning activities; all the treated embryos looked like the control treatment group. And yet, technically, the uptake of the drug is very rapid. Additionally, when E9.5 embryos were treated with HhAntag not all embryos were affected, and there was variability in the severity of the phenotype among embryos that were affected. It is unclear whether these are technical problems of the *in vitro* system, or if the variability is inherent to the embryos themselves. If the latter, this system may mimic the variability of HPE itself, but will make it harder to exploit the advantages of an *in vitro* system.

5. Conclusions and Perspectives

HPE is a common and devastating birth defect; it is also complex in etiology. While heterozygous mutations in several genes have been identified in human HPE cohorts, genotype–phenotype correlations in human HPE are poor, even in pedigrees. An understanding of the mechanisms, whereby these genetic defects become manifest as developmental defects therefore requires experimental models that accurately reproduce the spectrum of anomalies seen in human HPE. The mouse has emerged as such a model, because of its ease of genetic manipulation and its similarity to humans in development of the forebrain and face. Studies with various mutant mouse lines have linked most of the HPE genes together as regulators of either Shh expression or signaling (Fig. 3.3). The Nodal signaling pathway and Zic2 are required for normal development of the prechordal mesendoderm, a critical source of Shh. Shh produced by the prechordal mesendoderm induces its own expression in the ventral forebrain, likely in a Six3-dependent manner, to establish ventral identity and maintain the delicate balance of a network of Shh–Fgf–Bmp signaling centers that produce full dorsoventral pattern and appropriate partitioning of the forebrain.

Despite these insights and the existence of mouse mutants that display variable phenotypes within or between inbred backgrounds, a clear understanding of the issues of penetrance and expressivity that are so characteristic of human HPE remains elusive. Furthermore, no good models for teratogen-induced HPE currently exist. While this is in part due to the lack of unequivocal epidemiological evidence for specific environmental agents in human HPE, the likelihood of an environmental component to human HPE makes development of such models a priority. It seems likely that as the field progresses, the mouse will continue to shed light on mechanisms of HPE and also serve as a resource for hypothesis generation in probing the complexities of human HPE.

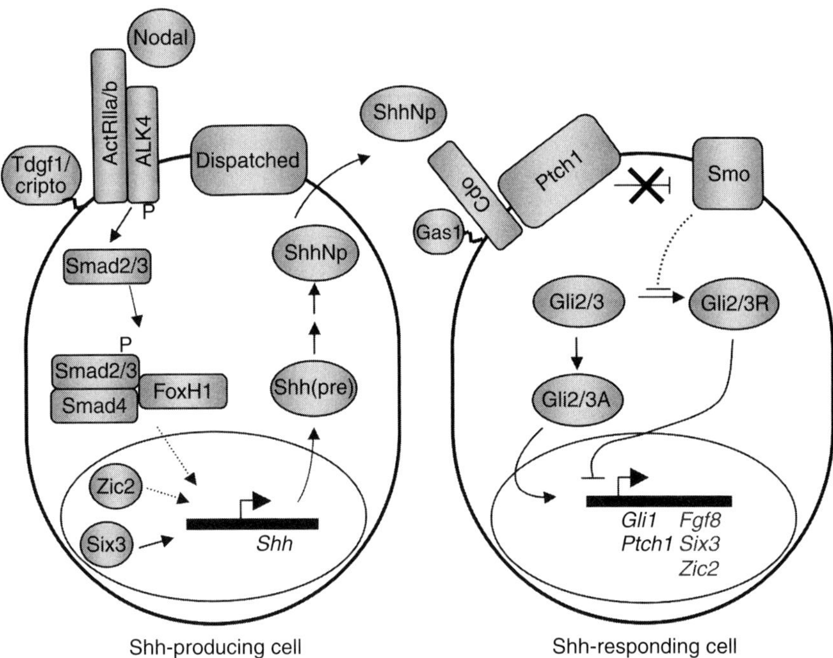

Figure 3.3 HPE genes encode direct or indirect regulators of Shh expression or signaling. The left hand cell represents a Shh-producing cell. In cells of the presumptive prechordal mesendoderm, Nodal binds to a receptor complex consisting of ALK4-ActRIIA/B and the cofactor Tdgf1/Cripto. Ligand binding results in receptor phosphorylation and subsequential phosphorylation of the transcription factors Smad2 and Smad3. Smad2/3 heterodimerize with Smad4 and translocate to the nucleus. Smad transcriptional complexes also interact with FoxH1 to regulate gene expression. Nodal signaling is required for development of prechordal mesendoderm, but it is not known whether *Shh* gene expression in this structure is a direct or indirect target of Smad complexes; the possibility of an indirect mechanism is indicated by the dashed arrows. Similarly, the defects observed in *Zic2*$^{-/-}$ mice suggest that Zic2 might regulate *Shh* gene expression, although the mechanisms may be indirect. In cells of the ventral forebrain, expression of *Shh* is a direct transcriptional target of Six3. Note that the regulators of Shh expression pictured in the left hand cell do not necessarily function in the same cell type. Shh is generated as a precursor, Shh(pre), which is then processed into a signaling-competent molecule (ShhNp) through proteolysis and lipidation, and is released from the cell in a Dispatched-dependent manner. The right hand cell represents a Shh-responding cell. Prior to Shh exposure, the Shh receptor Ptch1 inhibits Smo activity, and the Gli transcription factors are processed into their repressor forms (Gli2/3R). Upon binding of Shh to Ptch1 and other membrane proteins such as Cdo and Gas1, Ptch1-mediated repression of Smo is relieved, processing of the Gli proteins is inhibited and they function as transcriptional activators (Gli2/3A). Major direct target genes of the pathway are *Ptch1* and *Gli1*, and Shh signaling also regulates the expression of *Fgf8, Six3*, and *Zic2*, although these targets may be indirect. Note that Shh-dependent expression of these genes does not necessarily occur in the same cells. Mutations in *TDGF1/CRIPTO, FOXH1, ZIC2, SIX3, SHH, PTCH1*, and *GLI2* have all been observed in human HPE, and mice carrying mutations in these and additional components shown in the figure also display HPE.

ACKNOWLEDGMENTS

Work in the authors' laboratory on this topic is funded by the National Institutes of Health, the March of Dimes and the T. J. Martell Foundation.

REFERENCES

Ahlgren, S. C., Thakur, V., and Bronner-Fraser, M. (2002). Sonic hedgehog rescues cranial neural crest from cell death induced by ethanol exposure. *Proc. Natl Acad. Sci. USA* **99,** 10476–10481.

Allen, B. L., Tenzen, T., and McMahon, A. P. (2007). The Hedgehog-binding proteins Gas1 and Cdo cooperate to positively regulate Shh signaling during mouse development. *Genes Dev.* **21,** 1244–1257.

Anderson, R. M., Lawrence, A. R., Stottmann, R. W., Bachiller, D., and Klingensmith, J. (2002). Chordin and noggin promote organizing centers of forebrain development in the mouse. *Development* **129,** 4975–4987.

Andersson, O., Reissmann, E., Jornvall, H., and Ibanez, C. F. (2006). Synergistic interaction between Gdf1 and Nodal during anterior axis development. *Dev. Biol.* **293,** 370–381.

Aoto, K., Nishimura, T., Eto, K., and Motoyama, J. (2002). Mouse GLI3 regulates Fgf8 expression and apoptosis in the developing neural tube, face, and limb bud. *Dev. Biol.* **251,** 320–332.

Aoto, K., Shikata, Y., Higashiyama, D., Shiota, K., and Motoyama, J. (2008). Fetal ethanol exposure activates protein kinase A and impairs Shh expression in prechordal mesendoderm cells in the pathogenesis of holoprosencephaly. *Birth Defects Res. A Clin. Mol. Teratol.* **82,** 224–231.

Bachiller, D., Klingensmith, J., Kemp, C., Belo, J. A., Anderson, R. M., May, S. R., McMahon, J. A., McMahon, A. P., Harland, R. M., Rossant, J., and De Robertis, E. M. (2000). The organizer factors Chordin and Noggin are required for mouse forebrain development. *Nature* **403,** 658–661.

Bachiller, D., Klingensmith, J., Shneyder, N., Tran, U., Anderson, R., Rossant, J., and De Robertis, E. M. (2003). The role of chordin/Bmp signals in mammalian pharyngeal development and DiGeorge syndrome. *Development* **130,** 3567–3578.

Barkovich, A. J., and Quint, D. J. (1993). Middle interhemispheric fusion: An unusual variant of holoprosencephaly. *Am. J. Neuroradiol.* **14,** 431–440.

Barr, M., Jr., Hanson, J. W., Currey, K., Sharp, S., Toriello, H., Schmickel, R. D., and Wilson, G. N. (1983). Holoprosencephaly in infants of diabetic mothers. *J. Pediatr.* **102,** 565–568.

Bartholin, L., Powers, S. E., Melhuish, T. A., Lasse, S., Weinstein, M., and Wotton, D. (2006). TGIF inhibits retinoid signaling. *Mol. Cell. Biol.* **26,** 990–1001.

Bendavid, C., Dubourg, C., Pasquier, L., Gicquel, I., Le Gallou, S., Mottier, S., Durou, M. R., Henry, C., Odent, S., and David, V. (2007). MLPA screening reveals novel subtelomeric rearrangements in holoprosencephaly. *Hum. Mutat.* **28,** 1189–1197.

Bertrand, N., and Dahmane, N. (2006). Sonic hedgehog signaling in forebrain development and its interactions with pathways that modify its effects. *Trends Cell Biol.* **16,** 597–605.

Bonnemann, C., and Meinecke, P. (1990). Holoprosencephaly as a possible embryonic alcohol effect: Another observation. *Am. J. Med. Genet.* **37,** 431–432.

Brown, S. A., Warburton, D., Brown, L. Y., Yu, C. Y., Roeder, E. R., Stengel-Rutkowski, S., Hennekam, R. C., and Muenke, M. (1998). Holoprosencephaly due to mutations in ZIC2, a homologue of Drosophila odd-paired. *Nat. Genet.* **20,** 180–183.

Brunet, L. J., McMahon, J. A., McMahon, A. P., and Harland, R. M. (1998). Noggin, cartilage morphogenesis, and joint formation in the mammalian skeleton. *Science* **280,** 1455–1457.

Buglino, J. A., and Resh, M. D. (2008). Hhat is a palmitoylacyl transferase with specificity for N-palmitoylation of sonic hedgehog. *J Biol Chem.* **283,** 22076–22088.

Caspary, T., Garcia-Garcia, M. J., Huangfu, D., Eggenschwiler, J. T., Wyler, M. R., Rakeman, A. S., Alcorn, H. L., and Anderson, K. V. (2002). Mouse dispatched homolog1 is required for long-range, but not juxtacrine, Hh signaling. *Curr. Biol.* **12,** 1628–1632.

Chen, J. K., Taipale, J., Cooper, M. K., and Beachy, P. A. (2002). Inhibition of Hedgehog signaling by direct binding of cyclopamine to Smoothened. *Genes Dev.* **16,** 2743–2748.

Chen, L., Liao, G., Yang, L., Campbell, K., Nakafuku, M., Kuan, C. Y., and Zheng, Y. (2006). Cdc42 deficiency causes Sonic hedgehog-independent holoprosencephaly. *Proc. Natl Acad. Sci. USA* **103,** 16520–16525.

Chiang, C., Litingtung, Y., Lee, E., Young, K. E., Corden, J. L., Westphal, H., and Beachy, P. A. (1996). Cyclopia and defective axial patterning in mice lacking Sonic hedgehog gene function. *Nature* **383,** 407–413.

Chu, J., Ding, J., Jeays-Ward, K., Price, S. M., Placzek, M., and Shen, M. M. (2005). Non-cell-autonomous role for Cripto in axial midline formation during vertebrate embryogenesis. *Development* **132,** 5539–5551.

Ciani, L., Patel, A., Allen, N. D., and ffrench-Constant, C. (2003). Mice lacking the giant protocadherin mFAT1 exhibit renal slit junction abnormalities and a partially penetrant cyclopia and anophthalmia phenotype. *Mol. Cell. Biol.* **23,** 3575–3582.

Cohen, M. M., Jr. (1989). Perspectives on holoprosencephaly. Part I. Epidemiology, genetics, and syndromology. *Teratology* **40,** 211–235.

Cohen, M. M., Jr. (2006). Holoprosencephaly: Clinical, anatomic, and molecular dimensions. *Birth Defects Res. A Clin. Mol. Teratol.* **76,** 658–673.

Cohen, M. M., Jr., and Shiota, K. (2002). Teratogenesis of holoprosencephaly. *Am. J. Med. Genet.* **109,** 1–15.

Cole, F., and Krauss, R. S. (2003). Microform holoprosencephaly in mice that lack the Ig superfamily member Cdon. *Curr. Biol.* **13,** 411–415.

Cooper, M. K., Porter, J. A., Young, K. E., and Beachy, P. A. (1998). Teratogen-mediated inhibition of target tissue response to Shh signaling. *Science* **280,** 1603–1607.

Cooper, M. K., Wassif, C. A., Krakowiak, P. A., Taipale, J., Gong, R., Kelley, R. I., Porter, F. D., and Beachy, P. A. (2003). A defective response to Hedgehog signaling in disorders of cholesterol biosynthesis. *Nat. Genet.* **33,** 508–513.

Cordero, D., Marcucio, R., Hu, D., Gaffield, W., Tapadia, M., and Helms, J. A. (2004). Temporal perturbations in sonic hedgehog signaling elicit the spectrum of holoprosencephaly phenotypes. *J. Clin. Invest.* **114,** 485–494.

Coulter, C. L., Leech, R. W., Schaefer, G. B., Scheithauer, B. W., and Brumback, R. A. (1993). Midline cerebral dysgenesis, dysfunction of the hypothalamic–pituitary axis, and fetal alcohol effects. *Arch. Neurol.* **50,** 771–775.

Croen, L. A., Shaw, G. M., and Lammer, E. J. (2000). Risk factors for cytogenetically normal holoprosencephaly in California: A population-based case-control study. *Am. J. Med. Genet.* **90,** 320–325.

Ding, J., Yang, L., Yan, Y. T., Chen, A., Desai, N., Wynshaw-Boris, A., and Shen, M. M. (1998a). Cripto is required for correct orientation of the anterior–posterior axis in the mouse embryo. *Nature* **395,** 702–707.

Ding, Q., Motoyama, J., Gasca, S., Mo, R., Sasaki, H., Rossant, J., and Hui, C. C. (1998b). Diminished Sonic hedgehog signaling and lack of floor plate differentiation in Gli2 mutant mice. *Development* **125,** 2533–2543.

Dou, C. L., Li, S., and Lai, E. (1999). Dual role of brain factor-1 in regulating growth and patterning of the cerebral hemispheres. *Cereb. Cortex* **9,** 543–550.

Dou, C., Lee, J., Liu, B., Liu, F., Massague, J., Xuan, S., and Lai, E. (2000). BF-1 interferes with transforming growth factor beta signaling by associating with Smad partners. *Mol. Cell. Biol.* **20,** 6201–6211.

Dubourg, C., Bendavid, C., Pasquier, L., Henry, C., Odent, S., and David, V. (2007). Holoprosencephaly. *Orphanet. J. Rare Dis.* **2,** 8.

Edison, R. J., and Muenke, M. (2004). Mechanistic and epidemiologic considerations in the evaluation of adverse birth outcomes following gestational exposure to statins. *Am. J. Med. Genet. A* **131,** 287–298.

El-Jaick, K. B., Powers, S. E., Bartholin, L., Myers, K. R., Hahn, J., Orioli, I. M., Ouspenskaia, M., Lacbawan, F., Roessler, E., Wotton, D., and Muenke, M. (2007). Functional analysis of mutations in TGIF associated with holoprosencephaly. *Mol. Genet. Metab.* **90,** 97–111.

Fernandes, M., and Hebert, J. M. (2008). The ups and downs of holoprosencephaly: Dorsal versus ventral patterning forces. *Clin. Genet.* **73,** 413–423.

Fernandes, M., Gutin, G., Alcorn, H., McConnell, S. K., and Hebert, J. M. (2007). Mutations in the BMP pathway in mice support the existence of two molecular classes of holoprosencephaly. *Development* **134,** 3789–3794.

Frank-Kamenetsky, M., Zhang, X. M., Bottega, S., Guicherit, O., Wichterle, H., Dudek, H., Bumcrot, D., Wang, F. Y., Jones, S., Shulok, J., Rubin, L. L., and Porter, J. A. (2002). Small-molecule modulators of Hedgehog signaling: Identification and characterization of Smoothened agonists and antagonists. *J. Biol.* **1,** 10.

Fuccillo, M., Joyner, A. L., and Fishell, G. (2006). Morphogen to mitogen: The multiple roles of hedgehog signalling in vertebrate neural development. *Nat. Rev. Neurosci.* **7,** 772–783.

Furuta, Y., Piston, D. W., and Hogan, B. L. (1997). Bone morphogenetic proteins (BMPs) as regulators of dorsal forebrain development. *Development* **124,** 2203–2212.

Gazzerro, E., and Canalis, E. (2006). Bone morphogenetic proteins and their antagonists. *Rev. Endocr. Metab. Disord.* **7,** 51–65.

Geng, X., Speirs, C., Lagutin, O., Inbal, A., Liu, W., Solnica-Krezel, L., Jeong, Y., Epstein, D. J., and Oliver, G. (2008). Haploinsufficiency of Six3 fails to activate Sonic hedgehog expression in the ventral forebrain and causes holoprosencephaly. *Dev. Cell* **15,** 236–247.

Giknis, M., Damjanov, I., and Rubin, E. (1980). The differential transplacental effects of ethanol in four mouse strains. *Neurobehav. Toxicol.* **2,** 235–237.

Golden, J. A., Bracilovic, A., McFadden, K. A., Beesley, J. S., Rubenstein, J. L., and Grinspan, J. B. (1999). Ectopic bone morphogenetic proteins 5 and 4 in the chicken forebrain lead to cyclopia and holoprosencephaly. *Proc. Natl Acad. Sci. USA* **96,** 2439–2444.

Goodrich, L. V., Milenkovic, L., Higgins, K. M., and Scott, M. P. (1997). Altered neural cell fates and medulloblastoma in mouse patched mutants. *Science* **277,** 1109–1113.

Gripp, K. W., Wotton, D., Edwards, M. C., Roessler, E., Ades, L., Meinecke, P., Richieri-Costa, A., Zackai, E. H., Massague, J., Muenke, M., and Elledge, S. J. (2000). Mutations in TGIF cause holoprosencephaly and link NODAL signalling to human neural axis determination. *Nat. Genet.* **25,** 205–208.

Gritsman, K., Talbot, W. S., and Schier, A. F. (2000). Nodal signaling patterns the organizer. *Development* **127,** 921–932.

Grove, E. A., Tole, S., Limon, J., Yip, L., and Ragsdale, C. W. (1998). The hem of the embryonic cerebral cortex is defined by the expression of multiple Wnt genes and is compromised in Gli3-deficient mice. *Development* **125,** 2315–2325.

Gutin, G., Fernandes, M., Palazzolo, L., Paek, H., Yu, K., Ornitz, D. M., McConnell, S. K., and Hebert, J. M. (2006). FGF signalling generates ventral telencephalic cells independently of SHH. *Development* **133,** 2937–2946.

Haas, D., Morgenthaler, J., Lacbawan, F., Long, B., Runz, H., Garbade, S. F., Zschocke, J., Kelley, R. I., Okun, J. G., Hoffmann, G. F., and Muenke, M. (2007). Abnormal sterol metabolism in holoprosencephaly: Studies in cultured lymphoblasts. *J. Med. Genet.* **44,** 298–305.

Hayhurst, M., Gore, B. B., Tessier-Lavigne, M., and McConnell, S. K. (2008). Ongoing sonic hedgehog signaling is required for dorsal midline formation in the developing forebrain. *Dev. Neurobiol.* **68,** 83–100.

Hebert, J. M. (2005). Unraveling the molecular pathways that regulate early telencephalon development. *Curr. Top. Dev. Biol.* **69,** 17–37.

Hennekam, R. C., Van Noort, G., de la Fuente, F. A., and Norbruis, O. F. (1991). Agenesis of the nasal septal cartilage: Another sign in autosomal dominant holoprosencephaly. *Am. J. Med. Genet.* **39,** 121–122.

Heyer, J., Escalante-Alcalde, D., Lia, M., Boettinger, E., Edelmann, W., Stewart, C. L., and Kucherlapati, R. (1999). Postgastrulation Smad2-deficient embryos show defects in embryo turning and anterior morphogenesis. *Proc. Natl Acad. Sci. USA* **96,** 12595–12600.

Higashiyama, D., Saitsu, H., Komada, M., Takigawa, T., Ishibashi, M., and Shiota, K. (2007). Sequential developmental changes in holoprosencephalic mouse embryos exposed to ethanol during the gastrulation period. *Birth Defects Res. A Clin. Mol. Teratol.* **79,** 513–523.

Hoodless, P. A., Pye, M., Chazaud, C., Labbe, E., Attisano, L., Rossant, J., and Wrana, J. L. (2001). FoxH1 (Fast) functions to specify the anterior primitive streak in the mouse. *Genes Dev.* **15,** 1257–1271.

Huangfu, D., and Anderson, K. V. (2006). Signaling from Smo to Ci/Gli: Conservation and divergence of Hedgehog pathways from Drosophila to vertebrates. *Development* **133,** 3–14.

Ingham, P. W. (2008). Hedgehog signalling. *Curr. Biol.* **18,** R238–R241.

Itthagarun, A., Nair, R. G., Epstein, J. B., and King, N. M. (2007). Fetal alcohol syndrome: Case report and review of the literature. *Oral Surg. Oral Med. Oral Pathol. Oral Radiol. Endod.* **103,** e20–e25.

James, L. F. (1999). Teratological research at the USDA-ARS poisonous plant research laboratory. *J. Nat. Toxins* **8,** 63–80.

Jellinger, K., Gross, H., Kaltenback, E., and Grisold, W. (1981). Holoprosencephaly and agenesis of the corpus callosum: Frequency of associated malformations. *Acta Neuropathol.* **55,** 1–10.

Jin, J. Z., Gu, S., McKinney, P., and Ding, J. (2006). Expression and functional analysis of Tgif during mouse midline development. *Dev. Dyn.* **235,** 547–553.

Kang, J. S., Zhang, W., and Krauss, R. S. (2007). Hedgehog signaling: Cooking with Gas1. *Sci. STKE* **2007,** pe50.

Kawakami, T., Kawcak, T., Li, Y. J., Zhang, W., Hu, Y., and Chuang, P. T. (2002). Mouse dispatched mutants fail to distribute hedgehog proteins and are defective in hedgehog signaling. *Development* **129,** 5753–5765.

Keeler, R. F. (1975). Teratogenic effects of cyclopamine and jervine in rats, mice and hamsters. *Proc. Soc. Exp. Biol. Med.* **149,** 302–306.

Kelley, R. L., Roessler, E., Hennekam, R. C., Feldman, G. L., Kosaki, K., Jones, M. C., Palumbos, J. C., and Muenke, M. (1996). Holoprosencephaly in RSH/Smith–Lemli–Opitz syndrome: Does abnormal cholesterol metabolism affect the function of Sonic Hedgehog? *Am. J. Med. Genet.* **66,** 478–484.

Koebernick, K., and Pieler, T. (2002). Gli-type zinc finger proteins as bipotential transducers of Hedgehog signaling. *Differentiation* **70,** 69–76.

Kotch, L. E., and Sulik, K. K. (1992a). Experimental fetal alcohol syndrome: Proposed pathogenic basis for a variety of associated facial and brain anomalies. *Am. J. Med. Genet.* **44,** 168–176.

Kotch, L. E., and Sulik, K. K. (1992b). Patterns of ethanol-induced cell death in the developing nervous system of mice; neural fold states through the time of anterior neural tube closure. *Int. J. Dev. Neurosci.* **10,** 273–279.

Krauss, R. S. (2007). Holoprosencephaly: New models, new insights. *Expert Rev. Mol. Med.* **9,** 1–17.

Kuang, C., Xiao, Y., Yang, L., Chen, Q., Wang, Z., Conway, S. J., and Chen, Y. (2006). Intragenic deletion of Tgif causes defectsin brain development. *Hum. Mol. Genet.* **15,** 3508–3519.

Kuschel, S., Ruther, U., and Theil, T. (2003). A disrupted balance between Bmp/Wnt and Fgf signaling underlies the ventralization of the Gli3 mutant telencephalon. *Dev. Biol.* **260,** 484–495.

Lagutin, O. V., Zhu, C. C., Kobayashi, D., Topczewski, J., Shimamura, K., Puelles, L., Russell, H. R., McKinnon, P. J., Solnica-Krezel, L., and Oliver, G. (2003). Six3 repression of Wnt signaling in the anterior neuroectoderm is essential for vertebrate forebrain development. *Genes Dev.* **17,** 368–379.

Lanoue, L., Dehart, D. B., Hinsdale, M. E., Maeda, N., Tint, G. S., and Sulik, K. K. (1997). Limb, genital, CNS, and facial malformations result from gene/environment-induced cholesterol deficiency: Further evidence for a link to sonic hedgehog. *Am. J. Med. Genet.* **73,** 24–31.

Lee, C. S., and Fan, C. M. (2001). Embryonic expression patterns of the mouse and chick Gas1 genes. *Mech. Dev.* **101,** 293–297.

Leoncini, E., Baranello, G., Orioli, I., Annerén, G., Bakker, M., Bianch, I. F., Bower, C., Canfield, M., Castilla, E., Cocchi, G., Correa, A., De Vigan, C., *et al.* (2008). Frequency of holoprosencephaly in the International Clearinghouse Birth Defects Surveillance Systems: Searching for population variations. *Birth Defects Res. A Clin. Mol. Teratol.* **82,** 585–591.

Li, H., Tierney, C., Wen, L., Wu, J. Y., and Rao, Y. (1997). A single morphogenetic field gives rise to two retina primordia under the influence of the prechordal plate. *Development* **124,** 603–615.

Li, Y. X., Yang, H. T., Zdanowicz, M., Sicklick, J. K., Qi, Y., Camp, T. J., and Diehl, A. M. (2007). Fetal alcohol exposure impairs Hedgehog cholesterol modification and signaling. *Lab. Invest.* **87,** 231–240.

Lipinski, R. J., Hutson, P. R., Hannam, P. W., Nydza, R. J., Washington, I. M., Moore, R. W., Girdaukas, G. G., Peterson, R. E., and Bushman, W. (2008). Dose- and route-dependent teratogenicity, toxicity, and pharmacokinetic profiles of the hedgehog signaling antagonist cyclopamine in the mouse. *Toxicol. Sci.* **104,** 189–197.

Lowe, L. A., Yamada, S., and Kuehn, M. R. (2001). Genetic dissection of nodal function in patterning the mouse embryo. *Development* **128,** 1831–1843.

Ma, Y., Erkner, A., Gong, R., Yao, S., Taipale, J., Basler, K., and Beachy, P. A. (2002). Hedgehog-mediated patterning of the mammalian embryo requires transporter-like function of dispatched. *Cell* **111,** 63–75.

Maity, T., Fuse, N., and Beachy, P. A. (2005). Molecular mechanisms of Sonic hedgehog mutant effects in holoprosencephaly. *Proc. Natl Acad. Sci. USA* **102,** 17026–17031.

Majewski, F. (1981). Alcohol embryopathy: Some facts and speculations about pathogenesis. *Neurobehav. Toxicol. Teratol.* **3,** 129–144.

Mann, R. K., and Beachy, P. A. (2004). Novel lipid modifications of secreted protein signals. *Annu. Rev. Biochem.* **73,** 891–923.

Mar, L., and Hoodless, P. A. (2006). Embryonic fibroblasts from mice lacking Tgif were defective in cell cycling. *Mol. Cell. Biol.* **26,** 4302–4310.
Martinelli, D. C., and Fan, C. M. (2007). Gas1 extends the range of Hedgehog action by facilitating its signaling. *Genes Dev.* **21,** 1231–1243.
Mathieu, J., Barth, A., Rosa, F. M., Wilson, S. W., and Peyrieras, N. (2002). Distinct and cooperative roles for Nodal and Hedgehog signals during hypothalamic development. *Development* **129,** 3055–3065.
Matise, M. P., Epstein, D. J., Park, H. L., Platt, K. A., and Joyner, A. L. (1998). Gli2 is required for induction of floor plate and adjacent cells, but not most ventral neurons in the mouse central nervous system. *Development* **125,** 2759–2770.
McMahon, J. A., Takada, S., Zimmerman, L. B., Fan, C. M., Harland, R. M., and McMahon, A. P. (1998). Noggin-mediated antagonism of BMP signaling is required for growth and patterning of the neural tube and somite. *Genes Dev.* **12,** 1438–1452.
Ming, J. E., and Muenke, M. (2002). Multiple hits during early embryonic development: Digenic diseases and holoprosencephaly. *Am. J. Hum. Genet.* **71,** 1017–1032.
Ming, J. E., Kaupas, M. E., Roessler, E., Brunner, H. G., Golabi, M., Tekin, M., Stratton, R. F., Sujansky, E., Bale, S. J., and Muenke, M. (2002). Mutations in PATCHED-1, the receptor for SONIC HEDGEHOG, are associated with holoprosencephaly. *Hum. Genet.* **110,** 297–301.
Mo, R., Freer, A. M., Zinyk, D. L., Crackower, M. A., Michaud, J., Heng, H. H., Chik, K. W., Shi, X. M., Tsui, L. C., Cheng, S. H., Joyner, A. L., and Hui, C. (1997). Specific and redundant functions of Gli2 and Gli3 zinc finger genes in skeletal patterning and development. *Development* **124,** 113–123.
Muenke, M., and Beachy, P. A. (2000). Genetics of ventral forebrain development and holoprosencephaly. *Curr. Opin. Genet. Dev.* **10,** 262–269.
Muenke, M., and Beachy, P. A. (2001). "Holoprosencephaly." McGraw-Hill, New York.
Mulieri, P. J., Okada, A., Sassoon, D. A., McConnell, S. K., and Krauss, R. S. (2000). Developmental expression pattern of the cdo gene. *Dev. Dyn.* **219,** 40–49.
Mulieri, P. J., Kang, J. S., Sassoon, D. A., and Krauss, R. S. (2002). Expression of the boc gene during murine embryogenesis. *Dev. Dyn.* **223,** 379–388.
Nagai, T., Aruga, J., Minowa, O., Sugimoto, T., Ohno, Y., Noda, T., and Mikoshiba, K. (2000). Zic2 regulates the kinetics of neurulation. *Proc. Natl Acad. Sci. USA* **97,** 1618–1623.
Nagase, T., Nagase, M., Osumi, N., Fukuda, S., Nakamura, S., Ohsaki, K., Harii, K., Asato, H., and Yoshimura, K. (2005). Craniofacial anomalies of the cultured mouse embryo induced by inhibition of sonic hedgehog signaling: An animal model of holoprosencephaly. *J. Craniofac. Surg.* **16,** 80–88.
Nagase, T., Nagase, M., Yoshimura, K., Machida, M., and Yamagishi, M. (2006). Defects in aortic fusion and craniofacial vasculature in the holoprosencephalic mouse embryo under inhibition of sonic hedgehog signaling. *J. Craniofac. Surg.* **17,** 736–744.
Nanni, L., Ming, J. E., Bocian, M., Steinhaus, K., Bianchi, D. W., Die-Smulders, C., Giannotti, A., Imaizumi, K., Jones, K. L., Campo, M. D., Martin, R. A., Meinecke, P., *et al.* (1999). The mutational spectrum of the sonic hedgehog gene in holoprosencephaly: SHH mutations cause a significant proportion of autosomal dominant holoprosencephaly. *Hum. Mol. Genet.* **8,** 2479–2488.
New, D. A. (1990). Whole embryo culture, teratogenesis, and the estimation of teratologic risk. *Teratology* **42,** 635–642.
Nomura, M., and Li, E. (1998). Smad2 role in mesoderm formation, left-right patterning and craniofacial development. *Nature* **393,** 786–790.
Ogden, S. K., Ascano, M., Jr., Stegman, M. A., and Robbins, D. J. (2004). Regulation of Hedgehog signaling: A complex story. *Biochem. Pharmacol.* **67,** 805–814.

Ohkubo, Y., Chiang, C., and Rubenstein, J. L. (2002). Coordinate regulation and synergistic actions of BMP4, SHH and FGF8 in the rostral prosencephalon regulate morphogenesis of the telencephalic and optic vesicles. *Neuroscience* **111,** 1–17.

Okada, A., Charron, F., Morin, S., Shin, D. S., Wong, K., Fabre, P. J., Tessier-Lavigne, M., and McConnell, S. K. (2006). Boc is a receptor for sonic hedgehog in the guidance of commissural axons. *Nature* **444,** 369–373.

Pabst, O., Herbrand, H., Takuma, N., and Arnold, H. H. (2000). NKX2 gene expression in neuroectoderm but not in mesendodermally derived structures depends on sonic hedgehog in mouse embryos. *Dev. Genes. Evol.* **210,** 47–50.

Park, H. L., Bai, C., Platt, K. A., Matise, M. P., Beeghly, A., Hui, C. C., Nakashima, M., and Joyner, A. L. (2000). Mouse Gli1 mutants are viable but have defects in SHH signaling in combination with a Gli2 mutation. *Development* **127,** 1593–1605.

Peiffer, J., Majewski, F., Fischbach, H., Bierich, J. R., and Volk, B. (1979). Alcohol embryo- and fetopathy. Neuropathology of 3 children and 3 fetuses. *J. Neurol. Sci.* **41,** 125–137.

Pepinsky, R. B., Zeng, C., Wen, D., Rayhorn, P., Baker, D. P., Williams, K. P., Bixler, S. A., Ambrose, C. M., Garber, E. A., Miatkowski, K., Taylor, F. R., Wang, E. A., *et al.* (1998). Identification of a palmitic acid-modified form of human Sonic hedgehog. *J. Biol. Chem.* **273,** 14037–14045.

Pera, E. M., and Kessel, M. (1997). Patterning of the chick forebrain anlage by the prechordal plate. *Development* **124,** 4153–4162.

Petryk, A., Anderson, R. M., Jarcho, M. P., Leaf, I., Carlson, C. S., Klingensmith, J., Shawlot, W., and O'Connor, M. B. (2004). The mammalian twisted gastrulation gene functions in foregut and craniofacial development. *Dev. Biol.* **267,** 374–386.

Rahimov, F., Ribeiro, L. A., de Miranda, E., Richieri-Costa, A., and Murray, J. C. (2006). GLI2 mutations in four Brazilian patients: How wide is the phenotypic spectrum? *Am. J. Med. Genet. A* **140,** 2571–2576.

Rakic, P. (1995). A small step for the cell, a giant leap for mankind: A hypothesis of neocortical expansion during evolution. *Trends Neurosci.* **18,** 383–388.

Rallu, M., Machold, R., Gaiano, N., Corbin, J. G., McMahon, A. P., and Fishell, G. (2002). Dorsoventral patterning is established in the telencephalon of mutants lacking both Gli3 and Hedgehog signaling. *Development* **129,** 4963–4974.

Rankin, C. T., Bunton, T., Lawler, A. M., and Lee, S. J. (2000). Regulation of left–right patterning in mice by growth/differentiation factor-1. *Nat. Genet.* **24,** 262–265.

Reiter, J. F., and Skarnes, W. C. (2006). Tectonic, a novel regulator of the Hedgehog pathway required for both activation and inhibition. *Genes Dev.* **20,** 22–27.

Ribeiro, L. A., Murray, J. C., and Richieri-Costa, A. (2006). PTCH mutations in four Brazilian patients with holoprosencephaly and in one with holoprosencephaly-like features and normal MRI. *Am. J. Med. Genet. A* **140,** 2584–2586.

Roessler, E., and Muenke, M. (1998). Holoprosencephaly: A paradigm for the complex genetics of brain development. *J. Inherit. Metab. Dis.* **21,** 481–497.

Roessler, E., Du, Y. Z., Mullor, J. L., Casas, E., Allen, W. P., Gillessen-Kaesbach, G., Roeder, E. R., Ming, J. E., Ruiz i Altaba, A., and Muenke, M. (2003). Loss-of-function mutations in the human GLI2 gene are associated with pituitary anomalies and holoprosencephaly-like features. *Proc. Natl Acad. Sci. USA* **100,** 13424–13429.

Roessler, E., Ouspenskaia, M. V., Karkera, J. D., Velez, J. I., Kantipong, A., Lacbawan, F., Bowers, P., Belmont, J. W., Towbin, J. A., Goldmuntz, E., Feldman, B., and Muenke, M. (2008). Reduced NODAL signaling strength via mutation of several pathway members including FOXH1 is linked to human heart defects and holoprosencephaly. *Am. J. Hum. Genet.* **83,** 18–29.

Rogers, J. M., Brannen, K. C., Barbee, B. D., Zucker, R. M., and Degitz, S. J. (2004). Methanol exposure during gastrulation causes holoprosencephaly, facial dysgenesis, and cervical vertebral malformations in C57BL/6J mice. *Birth Defects Res. B Dev. Reprod. Toxicol.* **71,** 80–88.

Rohr, K. B., Barth, K. A., Varga, Z. M., and Wilson, S. W. (2001). The nodal pathway acts upstream of hedgehog signaling to specify ventral telencephalic identity. *Neuron* **29,** 341–351.

Romer, J. T., Kimura, H., Magdaleno, S., Sasai, K., Fuller, C., Baines, H., Connelly, M., Stewart, C. F., Gould, S., Rubin, L. L., and Curran, T. (2004). Suppression of the Shh pathway using a small molecule inhibitor eliminates medulloblastoma in Ptc1(+/−)p53 (−/−) mice. *Cancer Cell* **6,** 229–240.

Ronen, G. M., and Andrews, W. L. (1991). Holoprosencephaly as a possible embryonic alcohol effect. *Am. J. Med. Genet.* **40,** 151–154.

Sasaki, H., Nishizaki, Y., Hui, C., Nakafuku, M., and Kondoh, H. (1999). Regulation of Gli2 and Gli3 activities by an amino-terminal repression domain: Implication of Gli2 and Gli3 as primary mediators of Shh signaling. *Development* **126,** 3915–3924.

Schell-Apacik, C., Rivero, M., Knepper, J. L., Roessler, E., Muenke, M., and Ming, J. E. (2003). SONIC HEDGEHOG mutations causing human holoprosencephaly impair neural patterning activity. *Hum. Genet.* **113,** 170–177.

Seppala, M., Depew, M. J., Martinelli, D. C., Fan, C. M., Sharpe, P. T., and Cobourne, M. T. (2007). Gas1 is a modifier for holoprosencephaly and genetically interacts with sonic hedgehog. *J. Clin. Invest.* **117,** 1575–1584.

Shen, M. M. (2007). Nodal signaling: Developmental roles and regulation. *Development* **134,** 1023–1034.

Shen, J., and Walsh, C. A. (2005). Targeted disruption of Tgif, the mouse ortholog of a human holoprosencephaly gene, does not result in holoprosencephaly in mice. *Mol. Cell. Biol.* **25,** 3639–3647.

Simon, E. M., Hevner, R. F., Pinter, J. D., Clegg, N. J., Delgado, M., Kinsman, S. L., Hahn, J. S., and Barkovich, A. J. (2002). The middle interhemispheric variant of holoprosencephaly. *Am. J. Neuroradiol.* **23,** 151–156.

Song, J., Oh, S. P., Schrewe, H., Nomura, M., Lei, H., Okano, M., Gridley, T., and Li, E. (1999). The type II activin receptors are essential for egg cylinder growth, gastrulation, and rostral head development in mice. *Dev. Biol.* **213,** 157–169.

Spoelgen, R., Hammes, A., Anzenberger, U., Zechner, D., Andersen, O. M., Jerchow, B., and Willnow, T. E. (2005). LRP2/megalin is required for patterning of the ventral telencephalon. *Development* **132,** 405–414.

Storm, E. E., Garel, S., Borello, U., Hebert, J. M., Martinez, S., McConnell, S. K., Martin, G. R., and Rubenstein, J. L. (2006). Dose-dependent functions of Fgf8 in regulating telencephalic patterning centers. *Development* **133,** 1831–1844.

Sulik, K. K., and Johnston, M. C. (1982). Embryonic origin of holoprosencephaly: Interrelationship of the developing brain and face. *Scan Electron. Microsc.* **1,** 309–322.

Sulik, K. K., and Johnston, M. C. (1983). Sequence of developmental alterations following acute ethanol exposure in mice: Craniofacial features of the fetal alcohol syndrome. *Am. J. Anat.* **166,** 257–269.

Sulik, K. K., Johnston, M. C., and Webb, M. A. (1981). Fetal alcohol syndrome: Embryogenesis in a mouse model. *Science* **214,** 936–938.

Sulik, K. K., Dehart, D. B., Rogers, J. M., and Chernoff, N. (1995). Teratogenicity of low doses of all-trans retinoic acid in presomite mouse embryos. *Teratology* **51,** 398–403.

Tenzen, T., Allen, B. L., Cole, F., Kang, J. S., Krauss, R. S., and McMahon, A. P. (2006). The cell surface membrane proteins Cdo and Boc are components and targets of the Hedgehog signaling pathway and feedback network in mice. *Dev. Cell* **10,** 647–656.

Tole, S., Ragsdale, C. W., and Grove, E. A. (2000). Dorsoventral patterning of the telencephalon is disrupted in the mouse mutant extra-toes(J). *Dev. Biol.* **217,** 254–265.

Traiffort, E., Dubourg, C., Faure, H., Rognan, D., Odent, S., Durou, M. R., David, V., and Ruat, M. (2004). Functional characterization of sonic hedgehog mutations associated with holoprosencephaly. *J. Biol. Chem.* **279,** 42889–42897.

van Den Heuvel, M. (2001). Fat hedgehogs, slower or richer? *Sci. STKE* **2001,** PE31.

Wallis, D., and Muenke, M. (2000). Mutations in holoprosencephaly. *Hum. Mutat.* **16,** 99–108.

Wang, B., Fallon, J. F., and Beachy, P. A. (2000). Hedgehog-regulated processing of Gli3 produces an anterior/posterior repressor gradient in the developing vertebrate limb. *Cell* **100,** 423–434.

Wang, Y., McMahon, A. P., and Allen, B. L. (2007). Shifting paradigms in Hedgehog signaling. *Curr. Opin. Cell Biol.* **19,** 159–165.

Warr, N., Powles-Glover, N., Chappell, A., Robson, J., Norris, D., and Arkell, R. M. (2008). Zic2-associated holoprosencephaly is caused by a transient defect in the organiser region during gastrulation. *Hum. Mol. Genet.* **17,** 2986–2996.

Williams, J. A., Guicherit, O. M., Zaharian, B. I., Xu, Y., Chai, L., Wichterle, H., Kon, C., Gatchalian, C., Porter, J. A., Rubin, L. L., and Wang, F. Y. (2003). Identification of a small molecule inhibitor of the hedgehog signaling pathway: Effects on basal cell carcinoma-like lesions. *Proc. Natl Acad. Sci. USA* **100,** 4616–4621.

Willnow, T. E., Hilpert, J., Armstrong, S. A., Rohlmann, A., Hammer, R. E., Burns, D. K., and Herz, J. (1996). Defective forebrain development in mice lacking gp330/megalin. *Proc. Natl Acad. Sci. USA* **93,** 8460–8464.

Wilson, S. W., and Houart, C. (2004). Early steps in the development of the forebrain. *Dev. Cell* **6,** 167–181.

Wotton, D., Lo, R. S., Lee, S., and Massague, J. (1999a). A Smad transcriptional corepressor. *Cell* **97,** 29–39.

Wotton, D., Lo, R. S., Swaby, L. A., and Massague, J. (1999b). Multiple modes of repression by the Smad transcriptional corepressor TGIF. *J. Biol. Chem.* **274,** 37105–37110.

Yamada, S., Uwabe, C., Fujii, S., and Shiota, K. (2004). Phenotypic variability in human embryonic holoprosencephaly in the Kyoto Collection. *Birth Defects Res. A Clin. Mol. Teratol.* **70,** 495–508.

Yamaguchi, T. P. (2001). Heads or tails: Wnts and anterior–posterior patterning. *Curr. Biol.* **11,** R713–R724.

Yao, S., Lum, L., and Beachy, P. (2006). The ihog cell-surface proteins bind Hedgehog and mediate pathway activation. *Cell* **125,** 343–357.

Zakin, L., and De Robertis, E. M. (2004). Inactivation of mouse Twisted gastrulation reveals its role in promoting Bmp4 activity during forebrain development. *Development* **131,** 413–424.

Zhang, W., Kang, J. S., Cole, F., Yi, M. J., and Krauss, R. S. (2006). Cdo functions at multiple points in the Sonic Hedgehog pathway, and Cdo-deficient mice accurately model human holoprosencephaly. *Dev. Cell* **10,** 657–665.

CHAPTER FOUR

Mouse Models of Congenital Cardiovascular Disease

Anne Moon

Contents

School of Medicine, University of Utah, 15 North 2030 East Room 4160B, EIHG University of Utah, Salt Lake City, UT 84112

Current Topics in Developmental Biology, Volume 84
ISSN 0070-2153, DOI: 10.1016/S0070-2153(08)00604-2

Abstract

Congenital heart defects occur in nearly 1% of human live births and many are lethal if not surgically repaired. In addition, the genetic contribution to congenital or acquired cardiovascular diseases that are silent at birth, but progress to cause significant disease in later life is being increasingly appreciated. Heart development and structure are highly conserved between mouse and human. The discoveries that are being made in this model system are highly relevant to understanding the pathogenesis of human heart defects whether they occus in isolation, or in the context of a syndrome. Many of the genes required for cardiovascular development were discovered fortuitously when early lethality or structural defects were observed in mouse mutants generated for other purposes, and relevant genes continue to be defined in this manner. Candidate genes for this process are being identified by their roles other species, or by their expression in pertinent tissues in mice. In this review, I will briefly summarize heart development as currently understood in the mouse, and then discuss how complementary studies in mouse and human have identified genes and pathways that are critical for normal cardiovascular development, and for maintaining the structure and function of this organ system throughout life.

1. Introduction

Birth defects affecting the heart, whether presenting in isolation or as part of a syndrome, are the most common class of human congenital malformation. This is not surprising when one considers the complex morphogenetic events required to remodel what is initially a primitive contractile tube into the mature four-chambered mammalian heart. This pump supports serial flow of deoxygenated blood from the systemic venous system to the pulmonary capillaries via the right heart, and then returns the oxygenated blood to the left side of the heart for distribution to the systemic arterial circulation. Structural defects that perturb the correct relationships between the four chambers, and their connections to the systemic and pulmonary vasculature are often fatal if not surgically corrected in the neonatal period. The purpose of this review is to describe some of the mouse models that have been employed to dissect the cellular and molecular bases of congenital cardiovascular defects. These models represent the entire spectrum of defects seen in humans, from severe structural anomalies that present early in life to relatively subtle structural/functional defects that are even more common, and impact both quality and length of life in adulthood.

The benefits of the mouse as model system are myriad. Murine cardiovascular structure, physiology, and development are highly conserved with human. Embryos, tissues, and primary cells derived therefrom, are accessible at all stages of development for detailed analyses of molecular, cellular, physiological, and structural phenotypes. The range and subtlety of genetic manipulations that can be achieved in the mouse are unparalleled. The ability to study the consequences of a particular mutation in a uniform genetic background can reveal genotype–phenotype correlations and the impact of modifier loci. Each approach to generating gain- or loss-of-function models in the mouse has relative strengths and weaknesses. Models based on targeted null alleles are frequently not very informative, because while heterozygotes may have normal cardiovascular development, complete loss of gene function in homozygotes may severely disrupt embryogenesis; the resulting early embryonic lethality precludes their use as a model of congenital malformation. Conditional mutagenesis approaches permit more precise control of gene function in a tissue- and time-specific manner allowing gene function to be interrogated for distinct aspects of heart development.

After a brief survey of the major events in cardiac development in Section 2, this review is simply divided to emphasize how studies employing mouse models have complemented human genetic approaches to discovering key regulators of cardiac development and the molecular and cellular bases of normal and abnormal heart development. As our understanding of the genetic and molecular bases of cardiac development improves, so does our ability to define modifier genes that influence risk/susceptibility to environmental factors, to test and counsel affected families, and to identify potential therapies for a broad range of congenital and acquired cardiovascular diseases.

2. An Overview of Cardiac Development in the Mouse

For more detailed information about cardiac development and additional images, the reader is referred to recent comprehensive reviews (Kirby, 2006; Rosenthal and Harvey, 1999) and http://www.med.unc.edu/embryo_images.

2.1. Formation of the primitive heart tube, elongation, and looping

Mesodermal cells destined to contribute to the heart arise in the anterior primitive streak at approximately embryonic day 7.5 in the mouse (E7.5, Fig. 4.1A and A′). They migrate to the anterior ventral aspect of the embryo to form a bilaterally symmetric heart field called the cardiac crescent. The cardiac crescent can be considered to contain two ''fields'' or groups of cells distinguishable by their relative location in the crescent, the time at which they accrue to the heart, and their ultimate location within the heart. Cells located in a primary (or first) heart field are relatively ventral in the crescent, express the first myocardial markers, and form the primary heart tube (Fig. 4.1B and B′). The primary heart tube displays morphologic, physiologic, and molecular polarity along both its rostral-caudal and dorsoventral

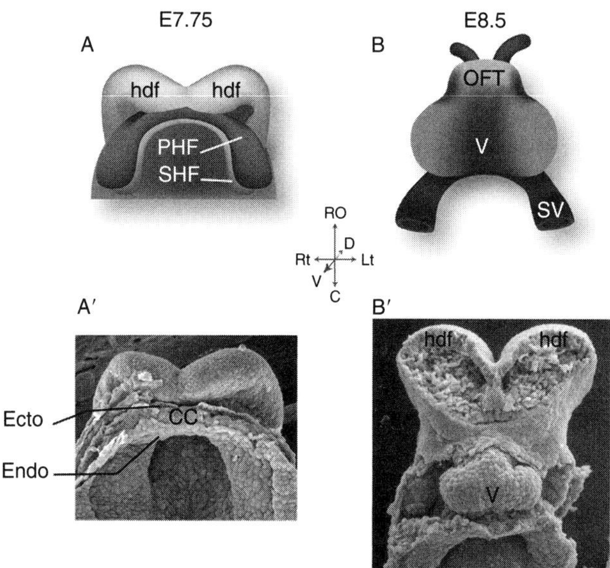

Figure 4.1 (A and A′) Schematic (A) and scanning electron micrograph (A′) showing the rostral portion of a mouse embryo in a ventral view at approximately 7.75 days gestation (E7.75), prior to formation of the heart tube. The head is at top (hdf, headfold), the primary and second heart fields (PHF, SHF) are shown in green and blue, respectively and constitute the mesoderm of the cardiac crescent (CC) between the ectodermal (Ecto) and endodermal (Endo) layers. Note the relatively more dorsal location of the SHF. (B and B′) By E8.5, the primary heart tube consists of a primitive ventricle (V) and atrium (not visible as it is dorsal to the ventricle) in series with the sinus venosus (sv, inflow, posterior pole). Cells that will contribute to the right ventricle (RV) and outflow tract (OFT) are beginning to accrue to the rostral (anterior pole) end of the primary heart tube. Axis diagram: RO, rostral; C, caudal; Rt, right; Lt, left; V, ventral; D, dorsal. (See Color Insert.)

axes. The heart tube is lined by an inner layer of endothelial cells called the endocardium. Myocardial cells secrete a specialized extracellular matrix (ECM) called cardiac jelly, which is interposed between the endocardium and myocardium. Although there are few contractile proteins and almost no organized contractile apparatus at this stage, the heart tube begins peristaltic contractions that propel blood from the inflow (venous, posterior) to outflow (arterial, anterior) poles of the heart tube (Kirby, 2006). Because the inability to form myocardium and a functional heart tube is incompatible with survival until birth, and thus not a cause of congenital heart defects in human patients, regulators of cardiac progenitor specification, differentiation, and formation of the primitive heart tube will not be discussed with respect to these early functions. However, as many of these factors also function during later chamber morphogenesis and into adulthood, the relevant mouse models will be presented in this context.

Myocardial cells accrue from precursors located in a second heart field (SHF) to both the inflow (venous) and outflow (arterial) poles of the heart tube (Figs. 4.1B and B′ and 4.2A and A′). SHF cells initially reside dorsomedial (relative to those in the primary heart field) in the cardiac crescent and are subsequently located in the pharyngeal and splanchnic mesoderm from where they differentiate and migrate into the heart after formation of the primary heart tube (Buckingham *et al.*, 2005). Some atrial progenitors reside in the caudal SHF (Galli *et al.*, 2008). An anterior (rostral) population of the SHF is crucial for formation of the right ventricle (RV) and outflow tract (OFT) at the arterial pole of the heart (Mjaatvedt *et al.*, 2001; Waldo *et al.*, 2001; Kelly *et al.*, 2001). The OFT has two segments: a proximal (near the ventricle) segment called the conus, and a distal segment called the truncus arteriosus. Thus, "conotruncus" is often used interchangeably with OFT when referring to defects at the arterial pole of the heart.

Concurrent with elongation of the heart tube by myocardium accruing at the poles, the tube loops to the right in the first visible evidence of left–right (LR) asymmetry in the embryo (Fig. 4.2A and A′). During looping the venous pole moves to nearly the same rostro-caudal level as the arterial pole (Fig. 4.2A and B); this is critical to establish correct relationships between the atria and ventricles, and their connections to the vasculature.

Signaling events that determine the direction of looping occur long before the morphogenetic event itself, at the embryonic node. The cellular processes that drive the mechanics of looping are not known, although both asymmetric cell proliferation and migration from the sides of the crescent likely play a role. Looping creates the outer and inner curvatures of the heart, which roughly correlate to the ventral and dorsal aspects of the primary heart tube. Outer curvature myocardium contributes to the chambers. The inner curvature contributes to atrioventricular (AV) canal and OFT, and retains the electrophysiologic properties of primary myocardium; these regions secrete a specialized complement of ECM and signaling

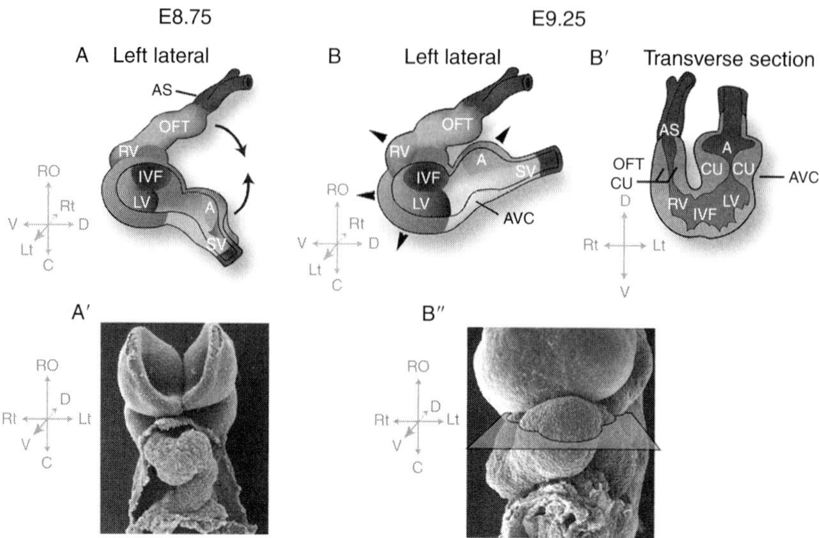

Figure 4.2 (A) Schematic left lateral view of E8.75 heart tube with left outer curvature removed. (A′) EM of similar stage embryo. The heart is beginning to loop as indicated by the black arrows indicating movement of the venous pole (SV) rostrally toward the outflow concurrent with addition of myocardium to both poles. Blue represents working myocardium that is beginning to expand. Gray regions are myocardium that maintains a nonworking myocardial phenotype. Red regions are vascular connections at the poles and include the sinus venosus (SV) and aortic sac (AS). OFT, outflow tract; RV, right ventricle; LV, left ventricle; A, atrium; IVF, interventricular foramen. (A′) EM of early looping stage embryo in ventral view; note rightward "S" curve of the heart tube. (B) E9.25, looping has brought the inflow to approximately the same rostrocaudal level as the outflow. Ventricular ballooning and atrial growth (symbolized by black arrowheads) have increased the size of these chambers. More OFT myocardium has been added. AVC, atrioventricular canal. (A) and (B) are adapted from Christoffels *et al.* (2004a,b). (B′) Schematic of transverse section (plane denoted in B″); the endocardial cushions (cu) are beginning to form in the OFT and AVC. (B″) EM of approximately E9.25 embryo in ventral view. The schematic overlay indicates the approximate plane of section in (B′). (See Color Insert.)

molecules to form internal acellular swellings called the AV and OFT cushions (Fig. 4.2B′) (Christoffels *et al.*, 2000).

Heart development is closely intertwined with that of the pharynx and the central vasculature; correct deployment of cells from the SHF in the pharynx into the heart requires signaling between the different pharyngeal cell populations. The pharyngeal arches are transient bulges of tissue in the neck of vertebrate embryos consisting of mesenchyme derived from neural crest (NC) cells and lateral plate mesoderm; these bulges are encased in ectoderm and endoderm (Fig. 4.3A). The heart is connected to the bilateral dorsal aortae by pharyngeal arch arteries (PAAs). Though initially symmetric (Fig. 4.3A and E), the most rostral vessels in the first and second arches largely

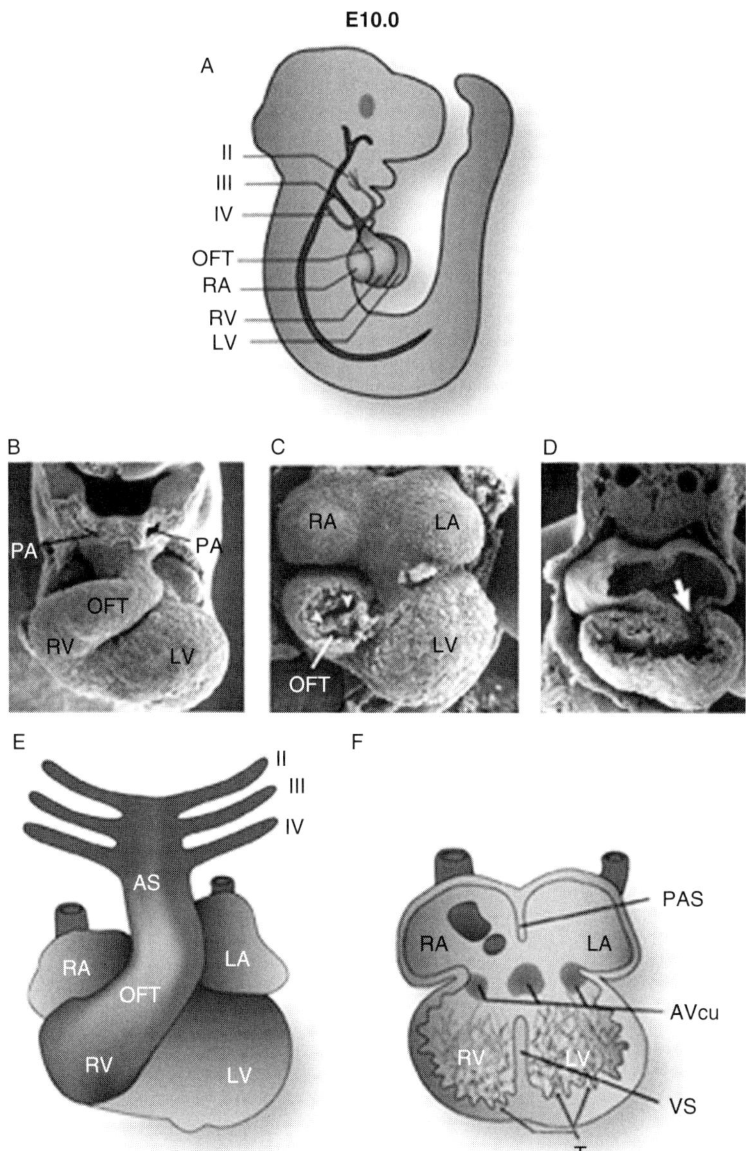

Figure 4.3 (A) Right lateral view of E10.0 embryo. The pharyngeal arches and the pharyngeal arch arteries that run through them to connect with the dorsal aorta are numbered (II–IV). The heart is in blue and the segments are labeled as in previous figures. (B) EM of similar stage embryo as in A. The view is ventral but slightly oblique; the pharyngeal arch and artery arising from the aortic sac are labeled PA. (C) Same view and stage embryo, distal OFT removed. The OFT cushions are marked with yellow arrowheads. The external walls of the right and left atria (RA, LA) and AVC are visible. (D) Same view and stage but more of the rostral heart has been removed so the AV

regress, while the caudal PAAs are extensively remodeled into adult vascular structures (Fig. 4.4B and B′). The right fourth PAA gives rise to the origin of the right subclavian artery. The left fourth PAA becomes the arch of the aorta between the left common carotid artery and left subclavian artery; this is the portion of the aorta that is absent in interruption of the aortic arch type B (IAAB). Although IAAB is compatible with prenatal survival, it is rapidly fatal after birth because there is no path for blood to reach the systemic circulation. Right aortic arch and anomalous origin of the subclavian arteries also result from defects in formation or remodeling of the fourth PAA. Since many of the genes that regulate heart development also regulate formation and/or remodeling of the pharyngeal arch arterial system, congenital vascular malformations are frequently observed in parallel with cardiac defects.

2.2. Generation of left–right asymmetry

There are numerous LR asymmetries in the vertebrate body plan. Internal organs are asymmetrically arranged (liver on the right, spleen on the left), the left and right sides of organs that span the midline are not mirror images of one another (the right lung has more lobes than the left, rotation of the gut during development generates a stereotypic arrangement of the stomach and bowel) and there are functional differences between the left and right sides of the brain (Fig. 4.5). Generating asymmetries in the cardiovascular system is crucial to establishing the separate pulmonary and systemic circulations.

The direction and process of looping influence the relative positions of the atria and ventricles, and their concordant alignment and connection to the correct circulations. It also affects identity and alignment of the AV cushions and chamber septation, and OFT alignment, rotation, and septation. Thus, failure to establish or propagate LR patterning has profound consequences for the heart and vasculature (Ramsdell, 2005). *Situs solitus* refers to the normal arrangement of organs along the LR axis. In *Situs inversus* there is mirror-image inversion of the body plan so that heart looping and subsequent LR structural heart development are concordant with the LR arrangement of the rest of the organs, and the normal

canal (white arrow) is visible, as are the atrial and ventricular chambers. (E) Schematic showing the symmetrically paired pharyngeal arch arteries (II–IV) arising from the aortic sac at this stage. Both sides are extensively remodeled and there is regression of the right-sided structures. The left fourth pharyngeal arch artery contributes to the arch of the aorta. (F) Schematic showing details similar to plane in (D). The venous inflow into the right atrium is visible in the dorsal wall. The pulmonary pit as the site of the future entrance of the pulmonary veins is shown in the left atrial wall in blue. There are 4 AV cushion (brown, Avcu), but one is out of the plane of the section. The primary atrial septum is forming (PAS) as is the interventricular septum (VS). Trabeculae (T) are forming in both ventricular chambers. (See Color Insert.)

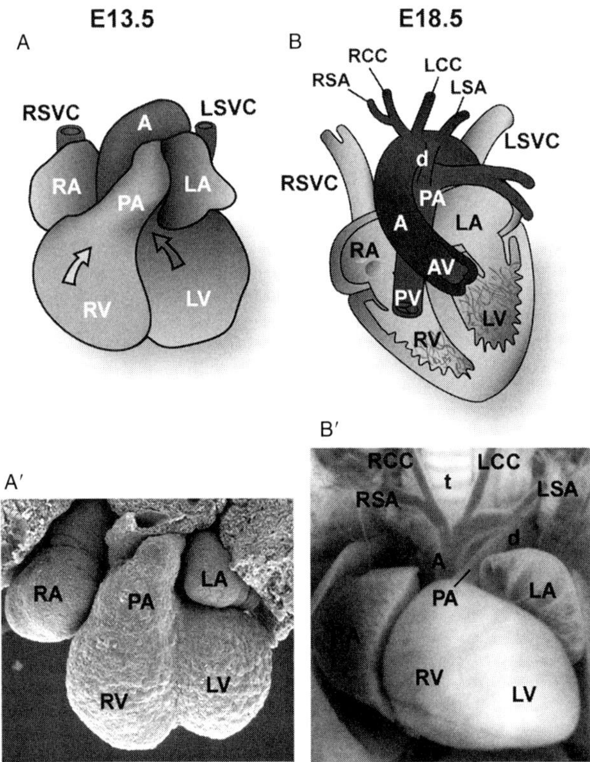

Figure 4.4 (A and A′) Schematic and EM of E13.5 heart in ventral view. Note that the OFT has septated and rotated such that the pulmonary artery (PA) arises from the RV and the aorta from the LV. (B nad B′) Schematic (with ventral half of heart removed) and whole mount photograph of E18.5 heart. The pharyngeal arch arterial system has asymmetrically remodeled to form the left arch of the aorta, right subclavian (RSA), right common carotid (RCC), left common carotid (LCC), left subclavian (LSA) arteries, and the ductus arterious (d). The ductus is a communication vital for the fetal circulation that allows relatively well-oxygenated blood returning to the RV to bypass the nonfunctioning, high resistance pulmonary circuit, and enter the aorta for distribution to the periphery. This conduit normally closes in the first few postnatal days. Egress of blood from the RV is via the pulmonic valve (PV) and from the LV, the aortic valve (AV). The AV valves are not labeled but are shown as flaps in gray between the atria and the ventricles. In mice the left superior vena cava (LSVC) persists and empties into the right atrium (not visible). t, trachea. (See Color Insert.)

connections between the pulmonary/systemic circuits and the heart are preserved. *Situs inversus* is usually asymptomatic and occurs in 1:7500 humans (Pinto and Faria, 1985). When LR heart and organ morphogenesis and positioning along the LR axis are discordant (*situs ambiguous* or heterotaxy), organ isomerism and serious cardiac structural defects occur. These include abnormalities in alignment of the OFT and great vessels with

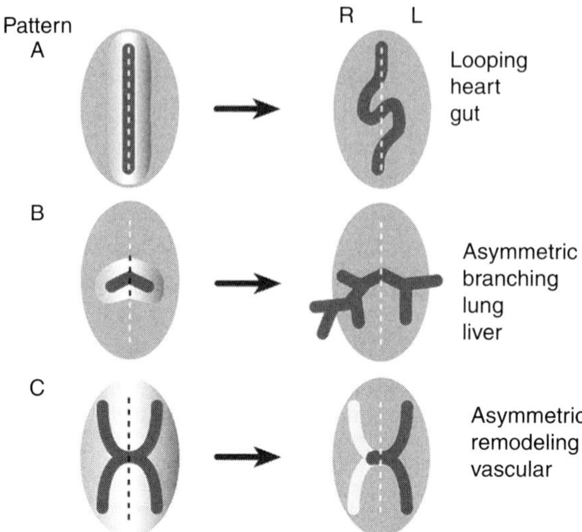

Figure 4.5 Patterns of organ left–right asymmetry. (A) A linear or tubular structure may bend asymmetrically, as in heart and gut looping to the right. (B) An initially symmetric budding tube can be converted by asymmetric branching morphogenesis as in the lung and liver. (C) An initially symmetric array can be remodeled by differential regression as occurs in the pharyngeal arch arterial system. (See Color Insert.)

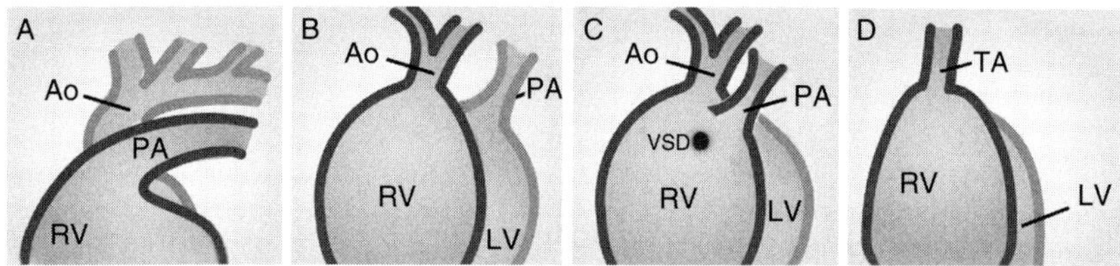

Figure 4.6 (A) Schematic showing the positions of the great arteries (aorta, Ao and pulmonary artery, PA) relative to the ventricles and to one another in the normal situation. When alignment/rotation of the OFT does not occur properly, defects such as complete transposition of the great arteries (B) or Double outlet RV with VSD (C) occur. When the OFT does not septate, persistent truncus arteriosus occurs (D)o. The case shown is the most severe type, in which the OFT is unseptated along its entire extent and there is only a singe great vessel (TA). (See Color Insert.)

respect to the ventricles (Double outlet RV, Transposition of the great arteries see Fig. 4.6), Double inlet left or RV due to abnormal remodeling and alignment of the AV canal, and vascular anomalies from abnormal remodeling of the pharyngeal arterial system or aberrant routing of venous return to the heart.

In the mouse the symmetry-breaking event has been thus far traced to the node at ~E7.5, which is later than events at early gastrulation stages (prior to node formation) identified in other vertebrate systems. Although the primary heart tube initially appears symmetric, long-standing evidence indicates that there are asymmetric contributions to the left and right sides of the heart from the cardiac mesodermal precursors as they migrate. Molecular data also indicate that heart precursors in the crescent have specific LR identity; thus symmetry-breaking in the mouse embryo may occur earlier than E7.5. Mouse models and observations in human patients with ciliary defects support a model of "nodal flow" whereby unidirectional rotation of nodal cilia creates an asymmetric gradient of morphogen(s) and/or triggers a response on a second group of "mechanosensor cilia" that alters intracellular calcium levels asymmetrically. These events activate axial signaling and transcription factor cascades (Shiratori and Hamada, 2006). The Tgfβ family-member Nodal is central in this signaling cascade, although the direct initiators of *nodal* expression in perinodal cells have not yet been identified. LR information is propagated to the lateral plate mesoderm first, by left-sided expression of *Nodal* itself, and soon after by a host of interacting factors that repress and potentiate propagation of LR signals to maintain left-sided *nodal* expression in the lateral plate mesoderm and repress it on the right (Shiratori and Hamada, 2006). Factors in the midline also prevent propagation of "left" signals to the right. Members of the LR cascade and its effectors will be discussed in light of their roles in human heart disease and mouse mutants that have exposed novel factors.

2.3. Chamber formation

Early models of the origin of atrial and ventricular chamber myocardium proposed that all the progenitors were present and arrayed segmentally from inflow to outflow along the primary heart tube (Srivastava and Olson, 2000). Lineage tracing data now indicate that only LV and a subset of atrial precursors are present in the primary heart tube, whereas myocardium of the RV, interventricular septum, OFT, and much of the atria accrues later through the poles (Meilhac *et al.*, 2004).

Neither the timing nor the regulatory details of chamber specification have been well defined. As the inner and outer curvatures are morphologically delineated by looping, the future chamber or "working" myocardial cells of the LV and atria begin to express distinct genes reflecting their chamber identities. Ventricular cells proliferate and chamber expansion occurs from the outer curvature in a morphologic process described as ballooning (Fig. 4.2B) (Christoffels *et al.*, 2000). The common atrium must expand along both the outer and inner curvatures to form the left and right atria, respectively. Molecular triggers of ballooning and concurrent changes in gene expression are also still poorly defined (Biben and Harvey, 1997). Inner curvature and OFT maintain a primary myocardial phenotype (Fig. 4.2). Chamber molecular programs are induced in ballooning regions by one set of factors, and repressed by others in primary myocardium to retain its molecular and electrophysiologic characteristics (Christoffels *et al.*, 2004a).

After ballooning, ventricular chamber development occurs concomitantly with an increasing hemodynamic load imposed by the circulation (Risebro and Riley, 2006). Differentiation of an inner trabeculated myocardial layer with distinct properties is coupled with that of an outer compact layer. Cells in the compact outer layer are arrayed concentrically, while trabecular myocardium projects into the cardiac jelly on the luminal surface subadjacent to the endocardium (Fig. 4.3D and F). Trabeculae provide a large surface area in contact with blood to support metabolic demands of the increasing ventricular mass prior to formation of the coronary vascular system, and also provide the contractile force to meet the increasing hemodynamic stress placed on the heart by the rapidly developing peripheral vasculature (Risebro and Riley, 2006). Although RV myocardium is not present in the primary heart tube, a portion of it has been added from the SHF at the time of looping and the onset of LV ballooning; thus LV and RV growth and trabeculation are largely concurrent.

The compact layer is initially 2–3 cells thick, but expands by proliferation which, combined with endocardial-derived signals, is crucial for elaboration of the trabeculae (Risebro and Riley, 2006). The epicardium (the outermost epithelial layer that ultimately invests the myocardium, see below) provides critical signals for growth and morphology of the compact layer. Blood flows through the chambers and mechanical stress on the walls provide additional inputs that influence ventricular wall growth and structure. Later, the compact myocardium thickens both by proliferation and by reincorporating trabecular myocardium in a process called compaction. At this point, the compact myocardium takes over the majority of the hemodynamic work. Compact layer maturation and remodeling continues postnatally as resistance to flow increases in the systemic circulation. Failure of trabecular remodeling causes "ventricular noncompaction" and "thin-walled myocardium" which present as cardiomyopathy in humans, manifest as heart failure and/or arrhythmias that cause sudden death (Breckenridge *et al.*, 2007; Chung *et al.*, 2003).

Severe defects in ventricular formation and myocardial remodeling in some mouse models (discussed below) result in fetal death at E13–15. These processes depend on sufficient development of the coronary circulation to nourish the growing myocardium. Disruption of coronary vascular development, or mismatch with myocardial oxygen demand disrupts survival, remodeling, and contractile function of myocardial cells. Defects in outgrowth of the ventricular myocardium cause chamber hypoplasia. Hypoplastic left heart syndrome is the most severe manifestation of chamber hypoplasia seen in live-born humans; it includes mitral stenosis, LV hypoplasia, aortic valve stenosis, and hypoplastic or atretic aortic arch.

Expansion of the atrial component of the primary heart tube begins at ~E9 and forms the trabeculated atrial appendages. Multiple sources of atrial chamber myocardium are incorporated at different developmental stages (Franco *et al.*, 2000; Kirby, 2006); contributions come from the AV canal and caudal SHF (Galli *et al.*, 2008). The common atrium receives venous inflow via the sinus venosus which drains the bilateral common cardinal veins. The dorsal right atrium incorporates the sinus venosus, thus directing the entire systemic venous return to the right side of the heart after atrial septation. Myocardium is also derived from the developing caval veins adjacent to the venous pole. As the pulmonary vein begins to form, myocardium that ensheaths the first segment adjacent to the atrium is incorporated and may contribute to the atrial septum (Snarr *et al.*, 2007). As the pulmonary circulation becomes functional, pulmonary venous blood is routed to the left atrium. Alterations in patterning that disrupt venoatrial connections cause total or partial anomalous pulmonary venous return to either the right atrium or other central veins. This defect can be lethal if not surgically corrected. It is often associated with diffuse abnormalities in the pulmonary venous bed that progress and cause lethality even after correction of the venoatrial connection.

2.4. Endocardial cushion development, outflow tract septation, and valve development

All portions of the heart must be septated to establish serial flow of deoxygenated blood from the systemic venous circulation to the pulmonary vascular bed for oxygenation, and then to the left side of the heart for distribution to the systemic circulation. Failure to septate any part of the heart permits mixing of oxygenated and deoxygenated blood. The AV (mitral, tricuspid) and OFT valves (pulmonary and aortic, Fig. 4.4B) can be considered as dynamic septae as they periodically separate the atrial and ventricular chambers, and the ventricles from their vascular outlets, respectively. This coordinates sequential filling (during chamber relaxation) and emptying (during chamber contraction), and prevents regurgitation. Abnormal valve function decreases heart efficiency and causes progressive chamber dysfunction. In severe cases such as "critical" aortic valve stenosis, patients present shortly after birth with inadequate tissue perfusion; organ damage and death occur rapidly in the absence of intervention. Formation and remodeling of the endocardial cushions that contribute to the valves and to septating the OFT and chambers are thus areas of intense interest.

The endocardial cushions are initially acellular swellings composed of cardiac jelly secreted by the primary myocardium in the AV canal (AV cushions, Figs. 4.2B′ and 4.3F) and OFT (Fig. 4.3C). These cushions prevent retrograde blood flow during the peristaltic contraction of the heart tube (Fig. 4.2B′). At ~E9.5 in the mouse, a subset of endocardial cells in the AV canal and proximal OFT are activated by signals from overlying myocardium to proliferate and undergo an epithelial–mesenchymal transformation (EMT) to invade and migrate into the cushion jelly where they activate a mesenchymal molecular program (Person *et al.*, 2005). This early phase of valve and septal development has been extensively studied *in vitro* and in many animal models; several crucial components of the jelly and myocardial signals that regulate and this process have been identified. Little is known about what specifies the subset of endothelial cells that are competent to respond to EMT-inducing myocardial signals.

The OFT endocardial cushions septate the conal portion of the OFT into the RV and LV outflows (the pulmonary infundibulum and aortic vestibule), and the truncus arteriosus (TA) into the pulmonary and aortic valve region (Fig. 4.4 A and A′). The proximal OFT cushions are initially invaded by endothelial cells via EMT while an extracardiac population of cells, the cardiac NC, migrate ventrally through the caudal pharyngeal arches and into the distal OFT cushions. Enlargement and fusion of the OFT cushions occurs concurrently with spiraling and complex rotation and alignment of the OFT myocardium relative to the ventricles (Fig. 4.4 A and A′). Differential programmed cell death and resorption of RV versus LV outlet myocardium are required for correct alignment of the aorta/aortic valve with the LV to send oxygenated blood to the systemic arterial circulation, and the RV to the pulmonary artery so deoxygenated blood reaches the lungs.

Disrupted alignment/rotation of the septated OFT relative to the ventricles causes defects such as transposition of great arteries (TGA, Fig. 4.6B) and double outlet right ventricle (DORV, Fig. 4.6C). The position of the outflow valves and vessels relative to the ventricles can be transposed in multiple ways, some of which are physiologically "correct" and well tolerated. However, if the aorta is completely transposed over the RV and the pulmonary artery over the left (called D-TGA, Fig. 4.6B), instead of serial blood flow from the venous to arterial circulations via the lungs, there are two parallel circuits. Deoxygenated blood recirculates systemically, and oxygenated blood recirculates from the lungs to the left side of the heart and back to the lungs via the aorta. With the transition from fetal to postnatal circulation, channels that permitted mixing between these parallel circuits close; this is rapidly fatal. In DORV, both great vessels arise from the right side of the heart and egress of blood from the LV is via an associated VSD; the location of the VSD is variable and determines the resulting physiology.

Defects in OFT septation range in severity from small outflow VSDs, to complete failure of OFT septation along its proximo-distal extent, called persistent truncus arteriosus (PTA, Fig. 4.6D). In this case the truncus and conus are not divided and there is a single outflow valve. This defect is commonly seen in mouse models with abnormal NC function (Hutson and Kirby, 2003, 2007; Kirby, 2006; Snider *et al.*, 2007).

Formation of the endocardial cushions and EMT in the AV canal and OFT are crucial initial steps in formation of the AV and OFT valves. The

AV valves are derived entirely from endocardial cushion tissue, but lineage studies indicate that the aortic and pulmonary semilunar valves also contain NC- and SHF-derived cells. The molecular and cellular events whereby the cushions are remodeled into the mature, thin valves is poorly understood (Armstrong and Bischoff, 2004; Schroeder *et al.*, 2003). However, it is known that differentiation of the mesenchymal cells in the AV valves begins when the cushions start to fuse to separate the AV canal into left and right channels. The transition from a mesenchymal to fibroblastic phenotype is accompanied by dramatic changes in the composition and organization of the ECM into a collagen-rich fibrous tissue. The thin leaflets contain a single endothelial layer with a core of matrix composed of collagen, elastin, and multiple glycosaminoglycans. The formation of the apparatus required to tether the valves must be coordinated with valve maturation and myocardial remodeling (Lincoln *et al.*, 2004). Many of the transcription factors, signaling molecules, and ECM structural proteins expressed during valve remodeling are shared with developing cartilage and bone in the limbs, and mouse mutants are being employed to determine how these factors interact and function during valve development (Lincoln *et al.*, 2006b).

Although valve structure is largely determined prenatally, there is continued growth and remodeling after birth. Abnormal valve development may not be symptomatic until later in life when calcification, fibrosis, or other degenerative changes cause stenosis, prolapse, or regurgitation. For example, bicuspid aortic valve is estimated to occur in 2% of the population at large and in 10% of individuals in families in which it is hereditary. Subsequent calcification and dysfunction are a common cause of adult heart disease (Garg, 2006).

2.5. Chamber septation

The primary atrial septum (septum primum, PAS Fig. 4.3F) grows from the dorsal aspect of the common atrial wall toward the AV cushions and fuses with them along its leading edge to close the opening between the left and right atrium (ostium primum). However, a portion of the primary atrial septum simultaneously disintegrates to form the ostium secundum to allow continuous right to left shunting of oxygenated blood. Relatively late in fetal development a second incomplete septum (septum secundum) grows circumferentially from the atrial wall to the right of the primary septum; the opening in this septum is called the foramen ovale. Prior to birth, relatively well-oxygenated venous blood crosses from the right atrium via the foramen ovale and ostium secundum (bypassing the nonfunctioning lungs) into the left atrium for delivery to the systemic circulation. The two septae form a valve that functionally closes when left atrial pressure exceeds right atrial pressure after birth, and anatomic fusion occurs a short time later. If the AV cushions or septum primum are hypoplastic, the resulting primum atrial septal defect (ASD) can be quite large causing significant shunting of blood from left to right, chamber dilatation, heart failure, and arrhythmias.

When caused by defective AV cushion morphogenesis (as opposed to abnormal formation of the primary atrial septum), primum ASDs are associated with AV valve abnormalities and inflow VSDs. ASDs can result from hypoplasia of the septum secundum. The most dorsal type of ASD results from deficient incorporation of sinus-venosus derived myocardium. The foramen ovale can also fail to close (patent foramen ovale); this is quite common in humans. Although patent foramen and secundum ASDs may not be hemodynamically significant, in recent years they have been surgically closed due to concern for potential long term complications such as arrhythmias. They may confer an increased risk of infection of the endocardial lining of the heart (endocarditis) and stroke, and have been postulated to contribute to migraine pathogenesis in some individuals (Tepper *et al.*, 2007).

Separation of the ventricles occurs in stages and the structural components of the interventricular septum (VS, Fig. 4.3F) derive from different sources (Kirby, 2006). The location of the VS is first apparent after looping as an external sulcus; within the lumen, the channel between the future left and RVs (the interventricular foramen, Fig. 4.2B′) becomes narrower as presumptive chamber myocardium balloons and trabeculates. The muscular portion of the VS grows from myocardium between the chambers toward the developing AV and OFT cushions. The interventricular foramen closes at ~E14.5 in the mouse when the membranous portion of the VS formed from endocardial cushion tissue fuses with the rim of the muscular septum. Most of the VSDs seen in humans and in mouse models affect the membranous septum; their dorsoventral location reflects whether the AV or proximal OFT cushion contribution to the membrane is deficient. Large inlet VSDs are seen as a part of the defect called complete AV canal in which atrial septation and AV valve development are also affected due to profound hypoplasia of the AV cushions.

2.6. Development of the epicardium and coronary vasculature

The epicardial layer of the heart envelops the external myocardial surface and in addition to providing signals that support the further growth and function of the developing myocardium, some cells undergo EMT and give rise to the smooth muscle cells in the coronary vasculature and to a subendocardial fibroblast population (Smith and Bader, 2007). Endothelial cells derived from the proepicardium were initially also thought to be EMT derived, but more recent data suggests that they may be mesenchymal from the outset (Cai *et al.*, 2008b; Manner *et al.*, 2001; Mikawa and Gourdie, 1996). Recent work also indicates that proepicardial cells give rise to a population of cardiomyocytes (Cai *et al.*, 2008b; Zhou *et al.*, 2008). It was thought that cell clusters detach from the proepicardium and float through the pericardial space and then attach to the ventricular surface. However, more recent work suggests that the primary mechanism for proepicardial cells to reach the myocardium is from villous extensions that directly contact the myocardium (Rodgers *et al.*, 2008).

Although several genes expressed in the proepicardium and required for its function have been identified in mice, the signals that induce proepicardial cells to detach from their source, adhere to and spread across the adjacent myocardium are unknown. Defective signaling or elaboration of matrix by the epicardium impairs myocardial growth and trabeculation, while failure of epicardial cells to undergo EMT prevents formation of the coronary vasculature. The signals that drive the differentiation of epicardially derived mesenchymal cells to adopt its various fates are being dissected in the mouse.

2.7. The contractile apparatus and excitation–contraction coupling

The fundamental contractile unit of myocardial cells is the sarcomere, which is composed of both thick (predominantly composed of myosin) and thin filaments (containing F-actin, tropomyosin, and multiple troponins). Titin is a huge filamentous protein ("third filament") that interacts with many proteins in the thick and thin filaments and plays a central role in coordinating sarcomere assembly (Gregorio *et al.*, 1999). Sarcomeres are arrayed in parallel and are attached at their ends by Z-discs, although at the early stages of heart development, the few myofibrils present are small and poorly organized and the contractile force is weak, reflecting both the number of sarcomeres and their composition. Myosin contains two heavy and two light chains; these subunits are encoded by multiple developmentally regulated isoforms that are differentially expressed in the atria and ventricles. Different isoforms influence the rate of muscle contraction and the sensitivity of the contractile apparatus to changing cytosolic calcium levels. Myosin heavy chains (MYH) are encoded by multigene families and in addition to tissue-specific and developmental regulation, expression of these isoforms is regulated in response to hemodynamic stresses on the myocardium (Vikstrom *et al.*, 1997).

The globular head of the myosin molecule contains the ATPase domain which allows contraction to be regulated by ATP hydrolysis and calcium. Calcium activation of the troponin–tropomyosin complex brings actin into contact with the myosin head, and sliding of thick and thin filaments causes contraction. Many human cardiomyopathies are caused by mutations in sarcomeric proteins, or by failure to correctly assemble or turn-over the contractile apparatus; these defects are being modeled in the mouse (Chung *et al.*, 2003).

Action potentials are changes in the electrical potential across the myocardial plasma membrane and include both depolarization and repolarization phases that reflect the coordinated function of several ionic currents. The characteristic action potentials displayed by different cell types in the heart at different stages of development reflect tissue-specific complements of sodium, calcium, potassium, and chloride channels. Depolarization of the cell membrane must then be coupled to myocyte contraction in a process called excitation–contraction coupling; this coupling is mediated by calcium. Calcium channels are present when the cardiomyocytes first begin to beat prior to heart tube formation, this beating is dependent on extracellular calcium for contraction as the nascent cardiomyocytes do not have a sarcoplasmic reticulum (SR, an intracellular calcium storage organelle) or T-tubule system to distribute calcium. Different types of calcium channels are present in the plasma and sarcoplasmic reticular membranes in different cell types and at different developmental stages. These different channels affect action potential duration and morphology in pacemaker, conduction system, and working myocardium. They also determine calcium distribution/reuptake to regulate duration/force of contraction, and transcription via calmodulin regulated kinases, for example.

2.8. Development of the conduction system

Regular, sequential conduction of action potentials occurs in the embryonic heart prior to the presence of an anatomically distinct conduction system (Moorman *et al.*, 1998). At the tubular stage the inflow has dominant pacemaker activity, indicating that electrophysiologic polarity due to graded ion channel expression is already present. Most cells in the mature heart retain some degree of automaticity; the baseline membrane potential, and rate of spontaneous depolarization is determined by the type of ion channels present at the cell membrane. This rate is normally fastest in the sinoatrial (SA) node in the right atrium, so it is the physiologic pacemaker. Action potentials are propagated through the atrial myocardium to the AV node. The AV node centralizes the signal from the atria and delays transmission to the ventricles to coordinate atrial diastole and ventricular systole. Electrical activity leaving the AV node is transmitted to ventricular cardiomyocytes sequentially via special muscle fibers contained in the Bundle of His, left and right bundle branches and finally, the ventricular Purkinje system. Gap junctions connect the Purkinje fibers with cardiomyocytes and the pattern of Purkinje arborization permits simultaneous contraction of the ventricles sequentially from the apex to base, which optimizes ventricular emptying and valve function. Chamber myocardium conducts impulses rapidly so that the myocardium functions as a syncytium with coordinated contractions. Primary myocardium gives rise to the slow-conducting, nonchamber myocardium that forms the SA and AV nodes of the conduction system (Gourdie *et al.*, 1995). The ventricular conduction system derives from a specialized ring of myocardium that encircles the interventricular foramen; the His bundle forms from myocardium in the dorsal portion of the ring while the cells at the apex of the developing IVS will form the bundle branches. Purkinje fibers are thought to derive from trabecular myocardium (Christoffels *et al.*, 2004a; Moorman *et al.*, 1998).

The view of heart development I have presented is organized in a temporally linear and structurally and functionally segmental manner. In some aspects, this representation is accurate because some constellations of heart defects arise from malformation of a particular embryonic structure or dysfunction of a single cell population. On the other hand, as mouse mutants are daily making clearer, developmental, and molecular programs are "reused" at multiple stages and in different tissue types during cardiovascular development, so defects are increasingly being viewed in light of their molecular origin. Furthermore, since ongoing morphogenesis of one substructure changes the hemodynamic and signaling inputs received by another, the events are much more plastic than presented here. In the end, as in most developmental processes, one is amazed that the cardiovascular system develops normally the majority of the time.

3. From Human to Mouse: Using Mouse Models to Understand How Mutations Identified in Humans Cause Congenital Cardiovascular Disease

This section discusses how identification of genes that cause familial or syndrome-associated cardiovascular defects in humans has led to the creation of informative mouse models. Sequencing of the human genome combined with improved mapping methodologies has greatly facilitated identification of mutations and allelic variants associated with congenital cardiovascular disease, and of candidate genes within chromosomal deletions or duplications. Subsequent efforts to model these defects using gene ablation and more subtle alterations of gene function in mice has provided a means to obtain a mechanistic understanding of how the gene products participate in normal and abnormal development of the cardiovascular system, and influence its function throughout life. In some cases, the pathogenesis as revealed in the mouse has led to translational studies and disease-modifying treatment strategies in humans.

3.1. DiGeorge syndrome and other 22q11 deletion disorders

The complex constellation of craniofacial, cardiovascular, and glandular phenotypes that characterize 22q11 deletion syndrome indicate that all pharyngeal arch-derived structures may be affected. Indeed, the first description of DiGeorge syndrome focused on abnormalities of the thymus, as many affected infants succumbed to their immune deficiency. Other patients present in the neonatal period with life-threatening congenital heart defects or hypocalcemia (due to parathyroid aplasia or dysfunction). In one study, del22q11 contributed to 1:68 cases of major congenital heart defects (Botto *et al.*, 2003). OFT malformations are more common in these patients than in the general CHD population, and the observation of PTA in association with interrupted aortic arch is virtually pathognomonic. Fifty percent of cases of interrupted aortic arch type B and 20% of PTA are caused by del22q11. Other aortic arch defects are also seen. Central nervous system malformations, psychiatric, and hearing defects are being increasingly appreciated, in part, because surgical correction of otherwise lethal heart defects and preventive measures to decrease mortality from infections and feeding complications have improved survival to allow these aspects of the phenotype to become manifest.

The deletion 22q11 syndromes include DiGeorge (DiGeorge, 1965), velocardiofacial (Shprintzen *et al.*, 1978) and conotruncal anomaly face syndromes (Burn *et al.*, 1993; Kelly *et al.*, 1993). Haploinsufficiency for chromosome 22q11 (del22q11) is the most common microdeletion thus far detected in humans and occurs in approximately 1:4000 live births (estimated 700 per year in the US (Botto *et al.*, 2003)). Del22q11, (most typically a three megabase deletion) accounts for the vast majority of patients with DiGeorge syndrome and greater than 60% of those with VCFS. These syndromes have also been linked to chromosome 10p13 deletions (Gottlieb *et al.*, 1998; Herve *et al.*, 1984), but candidate genes in this region have not yet been identified. The commonality of these syndromes is not only reflected by a shared genetic mechanism, but by the observation that parents with mild VCFS phenotypes have children with malformations that span the entire spectrum of 22q11 deletion phenotypes, including DiGeorge syndrome. With the advent of molecular diagnostics for del22q11, it has become apparent that there is substantial phenotypic variability among individuals with identical deletions (Lindsay, 2001).

Molecular defects resulting from deletion of genes in del22q11 are hypothesized to affect NC function adversely during cardiac and pharyngeal arch development (Kirby and Waldo, 1990; Robinson, 1975; van Mierop and Kutsche, 1986). Mechanical ablation of premigratory "cardiac" NC from the posterior hindbrain of chick embryos produces many features of del22q11, including interrupted aortic arch type B and truncus arteriosus (Bockman *et al.*, 1989; Kirby *et al.*, 1988; Kirby *et al.*, 1990). Numerous mouse models in which the function of NC specific genes is disrupted cause del22q11-like phenotypes. These observations led to the long-standing hypothesis that del22q11 represents a primary neurocristopathy.

An enormous effort has been directed toward generating informative mouse models of haploinsufficiency of the three megabase typically deleted region (TDR, deleted in 90% of patients) of chromosome 22. Approaches have ranged from deleting large segments of mouse chromosome 16 that contains orthologous genes to those in the TDR, to disrupting the function of a single orthologue (Galili *et al.*, 1997; Guris *et al.*, 2001; Jerome and Papaioannou, 2001; Kimber *et al.*, 1999; Lindsay, 2001; Lindsay *et al.*, 1999; Merscher *et al.*, 2001; Puech *et al.*, 2000; Saint-Jore *et al.*, 1998). Unlike the severe haploinsufficiency syndrome seen in humans, mice heterozygous for these alleles are either phenotypically normal or have infrequent, minor cardiovascular defects (Lindsay *et al.*, 1999).

Different dosage sensitivity between mouse and human is a recurring theme in many mouse models of CHD. Typically, the human phenotype is detected in heterozygotes, while the null mouse mutant is reportedly "phenotypically normal, healthy, and fertile"; in the case of obvious differences in gross or histologic anatomy, we can have confidence that the difference is bona fide. However, in other situations it may be an artifact resulting from current technical limitations on our ability to detect subtle functional defects.

Several groups identified *TBX1* as a critical player in the del22q11 phenotype (Jerome and Papaioannou, 2001; Lindsay *et al.*, 2001; Merscher *et al.*, 2001). *TBX1* encodes a transcription factor in the T-box family. In the developing mouse, *Tbx1* is broadly expressed in tissues that are abnormal in del22q11 patients, but notably, not in NC cells. *Tbx1*$^{-/-}$ mutants have the most severe features of del22q11: 100% of *Tbx1* null homozygotes have PTA. Tissue-specific conditional loss-of-function experiments have revealed a cell autonomous requirement for *Tbx1* in SHF mesoderm for normal OFT development (Zhang *et al.*, 2006), suggesting that NC dysfunction (with regard to the OFT defects) seen in human del22q11 and the various mouse models of del22q11 is secondary to defects in other tissues.

Based on the mouse studies, *TBX1* was resequenced in humans with DiGeorge or velocardiofacial syndrome, but without a 22q11 deletion; point mutations were found in a few individuals (Yagi *et al.*, 2003). Gain-of-function mutations also cause del22q11 phenotypes in mice and humans (Zweier *et al.*, 2007), (Vitelli and Baldini, personal communication) which suggests that interactions and competition between Tbx1 and other factors (including Tbx proteins) are sensitive to gene dosage (Liao *et al.*, 2004; Zhang and Baldini, 2008).

Since *TBX1* was identified as a critical gene, vigorous efforts are being made to understand the target genes it regulates in different domains of the developing embryo, and how it influences morphogenesis of the heart and other structures. The sensitivity of phenotypes to genetic background effects in mice, variability of affected structures, and the presence and severity of cardiovascular disease in affected humans, suggest that multiple loci on chromosome 22 or elsewhere contribute to the pathogenesis (Emanuel *et al.*, 2001; Lindsay, 2001; Scambler, 2000). Dysfunction of genes that operate in common developmental or genetic pathways with Tbx1 or other del22q11 genes may independently generate or modify the "del22q11 phenotype." Mutation of another gene in the del22q11 syntenic region in mice adversely affects cardiac and pharyngeal arch development, and NC function; *Crkl* (homologous to *CRKL* in the human TDR) homozygotes have del22q1-like malformations (Guris *et al.*, 2001). *CRKL* encodes a widely expressed adapter protein that functions downstream of several tyrosine kinase receptors. Although *CRKL* is not always within the human 22q11 deletions, it is likely that deletions influence expression of both deleted and adjacent genes. *Tbx1* and *Crkl* genetically interact in mice (Guris *et al.*, 2006). These findings indicate that del22q11 is a contiguous gene syndrome and suggest that disrupted CRKL-mediated signal transduction downstream of tyrosine kinase receptors, such as fibroblast growth factor (FGF) receptors, plays a role in pathogenesis.

Although the human *FGF8* locus (encoding Fibroblast Growth Factor 8) is not on chromosome 22, *Fgf8* deficiency in mice phenocopies del22q11 (Frank *et al.*, 2002). This finding, and the report of del22q11 phenotypes in *Crkl* mutant mice, led us and others to hypothesize that disrupted Fgf8 signaling due to CRKL deficiency could contribute to del22q11 phenotypes in humans, and that *FGF8* may be a modifier of del22q11 (Frank *et al.*, 2002; Vitelli *et al.*, 2002). Indeed *Fgf8* and *Crkl* genetically interact in mice, and some cellular responses to Fgf8 are Crkl dependent (Moon *et al.*, 2006). Evidence indicates that Fgf8 and Tbx1 operate in common developmental, and possibly molecular, pathways during development (Abu-Issa *et al.*, 2002; Ilagan *et al.*, 2006; Macatee *et al.*, 2003; Park *et al.*, 2006b). As with *Crkl*, *Fgf8* genetically interacts with *Tbx1* to increase the penetrance of aortic arch defects (Vitelli *et al.*, 2002), possibly through effects on *Gbx2* (Byrd and Meyers, 2005). *Fgf8* and *Tbx1* are both expressed in the pharyngeal epithelia and SHF mesoderm, and *Fgf receptors 1* and *2* are also expressed in the SHF and pharyngeal tissues (Moon, unpublished). *Fgfr1* expression is decreased in the pharynx of *Tbx1* null mutants (Park *et al.*, 2006b). Thus, *Fgf8* interacts with at least two genes that are deleted in human del22q11 that regulate cardiovascular, pharyngeal, and craniofacial morphogenesis. By determining the factors that regulate and interact with *Tbx1* as well as delineating the pathways downstream of Tbx1 in mice, we will discover novel candidate modifiers of the del22q11 phenotypes and potential therapeutic targets. For example, alterations in retinoic acid signaling influence the severity of defects in compound *Crkl/Tbx1* mouse mutants (Guris *et al.*, 2006).

3.2. Holt–Oram syndrome

Holt–Oram syndrome (HOS) was the first of the "heart–hand syndromes" described in which ASDs and anterior limb defects are coupled (Holt and Oram, 1960). Additional cardiovascular phenotypes have since been described in humans, most commonly conduction defects and VSDs. However, alignment defects and abnormalities in the left sided circulation such as mitral valve prolapse, aortic stenosis, and even hypoplastic left heart syndrome have also been described (Mori and Bruneau, 2004). HOS is a rare dominant syndrome (prevalence <1:100,000 in the US) with variable penetrance and severity. Nonsense, insertion, and deletion mutations presumed to cause haploinsufficiency for *TBX5* are detected in affected humans (Li *et al.*, 1997c).

In the mouse, *Tbx5* is expressed in the regions of the developing heart, and at the relevant times, to explain the structural cardiac defects seen in human patients (Bruneau *et al.*, 1999). Evaluation of different *Tbx5* mutant mice shows that multiple aspects of cardiac development are sensitive to *Tbx5* dosage (Moskowitz *et al.*, 2007). Heterozygotes for a null allele of *Tbx5* display ASDs, VSDS, LV malformations, and conduction system defects, and frequently die in the perinatal period. (Bruneau *et al.*, 2001; Hiroi *et al.*, 2001). These phenotypes are quite sensitive to genetic background. Mechanistic information from *Tbx5* mutants has fed back to provide insight to a different human syndrome, Okihiro syndrome. Like HOS, this syndrome affects the heart and limbs but is caused by mutations in the transcription factor gene, *SALL4*. Enough phenotypic similarity exists between HOS and Okihiro syndromes that misdiagnoses were made until the advent of gene-based diagnosis. Heterozygous null *Sall4* mutant mice have background-dependent limb and interventricular septal defects similar to those seen in humans; homozygotes suffer early embryonic death. *Sall4;Tbx5* compound mutants have more severe heart malformations including septal and valve defects. At a molecular level, Tbx5 and Sall4 interact positively and competitively on regulatory elements to control expression of genes that influence heart and limb development (Koshiba-Takeuchi *et al.*, 2006).

Tbx5 interacts with transcriptional cofactors (Nkx2.5, Gata4) to regulate expression of other transcription factors, signaling molecules, and ion channels in the developing heart. Interactions between TBX5, NKX2.5, and GATA4 are required for septal formation in humans. A *GATA4* missense mutation that disrupts DNA binding and the interaction between GATA4 and TBX5 was reported in a family inheriting nonsyndromic ASDs and other defects. In the same study, two *TBX5* missense mutations that cause HOS in humans were also found to disrupt the interactions between TBX5, NKX2.5 and GATA4. (Garg *et al.*, 2003).

3.3. Marfan syndrome

Marfan syndrome (MS) impacts the structure and function of multiple organs including skeleton, eyes, heart, lungs, and blood vessels. Affected individuals are tall, thin, and have joint laxity; scoliosis is common as are myopia and lens abnormalities. The major morbidities and sudden death seen in these patients are due to progressive cardiovascular disease, particularly aortic aneurysm and dissection. Mitral valve regurgitation, prolapse and calcification, tricuspid valve defects, coronary artery aneurysms, and arrhythmias have been reported and cause significant morbidity in patients as any age (Judge and Dietz, 2008; Stuart and Williams, 2007).

MS is an autosomal dominant disorder with high penetrance and variable severity. Approximately 1:5000 people worldwide are affected. Mutations in the *FBN1* gene cause for this syndrome (Dietz *et al.*, 1991). Fibrillin-1 is a major structural protein required for normal elastin deposition and organization in the extracellular space. The phenotype of MS was initially attributed to abnormal structure of microfibrils but mouse models have been crucial to revealing that the degenerative changes in cardiovascular and other connective tissues relate to both altered microfibrillar structure and dysregulated (excess) TGFβ signaling as a consequence of mutations in *Fbn1* (Isogai *et al.*, 2003; Judge and Dietz, 2005; Neptune *et al.*, 2003; Ng *et al.*, 2004; Pereira *et al.*, 1997, 1999). The TGFβ aspect of MS pathogenesis was confirmed with the discovery that mutations in *TGF*β*R 1* and *2* cause a syndrome called Loeys–Dietz that has significant phenotypic similarity to MS (Dietz *et al.*, 2005; Loeys *et al.*, 2005). Although it was originally postulated that *FBN1* missense mutations generated proteins with dominant-negative effects, expression of a wild-type *Fbn1* transgene in a gene-targeted missense mouse model rescued the histologic and clinical aortic pathology of the mutants. Conversely, overexpression of a mutant human Fibrillin-1 protein via a YAC transgene in otherwise wild-type mice did not cause any manifestations of MS. This supports a pathogenetic model of haploinsufficiency for wild type Fibrillin-1 protein (Judge *et al.*, 2004).

Mouse models have been employed to test the efficacy of pharmacologic interventions in preventing progression or decreasing severity of cardiovascular complications associated with MS. For example, three different strains of gene-targeted mice harboring different *Fbn1* point mutations display features of MS with variable severity and have increased TGFβ activity in the lungs and heart. Notably, administration of anti-TGFβ neutralizing antibodies rescued these phenotypes. Angiotensin II can directly activate R-Smads in the absence of TGFβ ligand and the angiotensin II type 1 receptor blocker Losartan decreases expression of TGFβ ligands and receptors, decreases latent TGFβ activation (Naito *et al.*, 2004), and prevents aortic aneurysms in mutant mice (Habashi *et al.*, 2006). An NIH-sponsored clinical trial of this drug is now underway in affected humans. MS is thus an extremely good example of the power of mouse models in translational research, and how their use can impact the diagnosis and treatment of human cardiovascular disease. Mouse models have also been employed to study the effects of mutations in the fibrillar collagen genes seen in humans with a related disorder called Ehlers–Danlos syndrome (Wenstrup *et al.*, 2006).

3.4. Williams syndrome and supravalvar aortic stenosis

Supravalvar aortic stenosis (SVAS) is a vasculopathy that causes narrowing or obstruction of large arteries. Any artery can be affected; when the aorta is involved, heart failure and myocardial infarction may result. Williams syndrome (WS) is complex and influences cognition, personality, craniofacial morphology, and connective tissue function. SVAS is inherited in an autosomal dominant fashion and can be seen in isolation or as a feature of WS. SVAS is estimated to occur in 1:25,000 live births. Point mutations in the elastin gene were found in isolated cases of SVAS, and in patients with WS (Li *et al.*, 1997a).

Loss-of-function of one *Eln* allele in mice causes abnormal elastic lamellar structure in the aortic walls and this pathologic finding was subsequently found in two surgical specimens from patients with SVAS (Li *et al.*, 1998b). The phenotypes of homozygous *Eln* null mouse mutants have provided insight into vascular homeostasis and revealed the mechanism of vascular obstruction in SVAS. *Eln* null mutants survive to birth but develop aortic pathology shortly before birth. As in humans with SVAS, a pathologic accumulation of cells was observed in the subendothelial space and eventually obstructed the lumen. The accumulated cells were abnormal smooth muscle cells, and although endothelial damage and inflammation had been invoked to explain human pathology, these mechanisms did not contribute to the murine phenotype (Li *et al.*, 1998a). Elastin was found to signal directly to vascular smooth muscle cells, and exogenous elastin can reduce proliferative responses in an injured arterial model (Karnik *et al.*, 2003).

3.5. Noonan syndrome

Noonan syndrome includes craniofacial abnormalities, webbed neck, scoliosis and joint laxity, and heart defects of varying severity. Pulmonary stenosis is the most common cardiac defect, but atrial and ventricular sepal defects, aortic coarctation, and mitral valve defects are also seen. Pulmonary stenosis is also frequently seen in LEOPARD syndrome, but the most common cardiovascular defect seen in this syndrome is left ventricular hypertrophy. Noonan syndrome is inherited in an autosomal dominant manner and affects at least 1:2500 people . Approximately half of the cases of this genetically heterogeneous syndrome can be attributed to mutations in *PTPN11*, which encodes the protein tyrosine phosphatase, SHP2 (Tartaglia *et al.*, 2001). Mutations in this gene also cause the related disorder, LEOPARD syndrome.

SHP2 modulates intracellular signal transduction downstream of several receptor tyrosine kinases (EGF, FGF, HGF) by interacting with adaptor proteins associated with the receptor. There is evidence that it functions upstream of RAS; it appears to both positively and negatively modulate MAPK signaling, depending on the receptor involved. It is noteworthy that SHP2 can inactivate members of the Sprouty family which are known to be negative feedback regulators of the RAS/MAPK cascade, particularly downstream of FGF receptors (Hanafusa *et al.*, 2004). Multiple lines of evidence indicate that the mutations detected in humans with Noonan syndrome are gain-of-function and increase basal phosphatase activity whereas those associated with LEOPARD decrease enzymatic activity

(Edouard *et al.*, 2007). Recent studies have also revealed somatic mutations in *PTPN11* in patients with specific forms of leukemia; interestingly, although these mutations affect the same residues as in Noonan syndrome, the substitutions are less conservative and the mutant proteins tend to generate higher constitutive enzymatic activity.

Homozygous null *Ptpn11* mouse mutants have embryonic lethal gastrulation defects (Arrandale *et al.*, 1996). The importance of *Shp2* in semilunar valve development was confirmed in compound mutant mice bearing two hypomorphic alleles of the EGF receptor and a single null *Ptpn11* allele; these animals develop aortic stenosis and thickened pulmonic valves (Chen *et al.*, 2000). Expression of a mutant Shp2 protein resulting from a point mutation in *PTPN11* seen in humans in gene targeted mice recapitulate the features of Noonan syndrome including mitral valve and septal defects due to aberrant endocardial cushion morphogenesis, and OFT alignment defects (Araki *et al.*, 2004). Cardiac defects resulting from expression of mutant Shp2 in cardiomyocytes in mice are not seen in *Erk1/2* null mutant mice (Nakamura *et al.*, 2007).

Mutations in numerous other genes encoding proteins within the Ras/MAPK signal transduction cascade have now been detected in patients with Noonan, LEOPARD and related syndromes (Aoki *et al.*, 2008). Thus, studies in humans, *in vitro,* and in mutant mice support the conclusion that dysregulated RAS–MAPK signaling is the molecular basis of these syndromes. The mouse models that have been generated will be important tools for further dissection of the different signaling and cellular processes in pathologic development of cardiovascular and other tissues.

3.6. Alagille syndrome

Alagille syndrome is a multisystem disorder although the liver is usually most severely involved (paucity of bile ducts may cause liver failure requiring transplantation). The kidneys, eyes, spine, and heart may also be affected. The most common heart defects are pulmonary valve and peripheral pulmonary artery stenosis. Alagille syndrome is an autosomal dominant disorder with variable penetrance and severity affecting 1:50,000 individuals. Mutations in *JAG1* are the cause in many families (Li *et al.*, 1997b; Oda *et al.*, 1997). This gene encodes the JAGGED1 ligand for the NOTCH receptor family. Greater than 100 mutations at this locus have now been identified in patients with Alagille syndrome and the predominant mechanism is believed to be haploinsufficiency, although some missense mutations may encode proteins with dominant-negative effects.

Mutations in other NOTCH pathway members have been associated with nonsyndromic and syndromic OFT defects in humans (Krantz *et al.*, 1999; McDaniell *et al.*, 2006). The importance of Notch function in other types of sporadic and heritable congenital heart defects in human cardiovascular disease has been confirmed by the identification of *NOTCH1* mutations in humans with autosomal dominant defects such as critical aortic valve stenosis, bicuspid aortic valve, and aortic valve calcification in adulthood (Garg *et al.*, 2005; Mohamed *et al.*, 2006). *NOTCH3* mutations have been found in patients with CADASIL, a vascular disease that results in strokes (Joutel *et al.*, 1996).

There are four Notch receptors and five ligands (in subclasses called jagged and delta-like) in mammals; receptors and ligands are transmembrane proteins and their interaction between adjacent cells results in cleavage of the receptor and release of the Notch intracellular domain (NICD) into the cytoplasm. Nuclear translocation of NICD permits formation of an active transcriptional complex with the DNA binding protein RBPJ and other coactivators. In the absence of NICD, RBPJ interacts with corepressors to inhibit transcription of Notch target genes. Notch transcriptional targets include Hairy enhancer of split (Hes) and Hes-related (Hrt) transcriptional factors. Downstream effects of Notch signaling are context dependent, and influence cell fate, proliferation and survival during cardiomyocyte differentiation, AV canal and OFT morphogenesis, valve development, and ventricular trabecular formation. The findings in humans have spurred intense investigation and numerous mouse mutants have been generated to understand how Notch signaling functions in cardiovascular development (High and Epstein, 2008). Mutants have been generated in many Notch pathway genes most of which have heart and/or vascular abnormalities (High and Epstein, 2008).

The first mouse model of Alagille syndrome was a targeted null *Jagged1* allele (Xue *et al.*, 1999). Although heterozygotes are viable, homozygotes die by E11.5 with multiple vascular anomalies. Compound null heterozygous *Jag1/Notch2* mutant mice phenocopy the liver and heart defects seen in human Alagille syndrome (McCright *et al.*, 2002). To address the mechanism underlying the most common OFT defect seen in Alagille syndrome, pulmonary artery stenosis, a conditional dominant-negative *Maml* allele was used to block signaling downstream of all four Notch receptors in different OFT tissues. *Maml* (mastermind-like) encodes one member of the Notch transcriptional complex. Conditional activation of a dominant-negative Maml specifically in NC caused pulmonary stenosis at high penetrance (High *et al.*, 2007). Malalignment OFT defects, VSDs, and vascular defects resulting from abnormal remodeling of the pharyngeal arch arterial system were also observed. The data indicate that Notch signals are required for postmigratory function of NC-derived cells, such as their ability to differentiate into vascular smooth muscle cells and invest the developing aorta and pulmonary arteries. This observation helps to explain the diverse vascular pathologies seen in patients with Alagille syndrome.

Mouse mutants have also revealed a role for Notch-mediated signaling during AV canal morphogenesis and multiple steps in valve remodeling. *Bmp2* and *4* and their receptors are expressed in the developing AV canal and OFT myocardium and are required for endocardial cushion morphogenesis (Liu *et al.*, 2004; Ma *et al.*, 2005; Park *et al.*, 2006a; Song *et al.*, 2007a). These Bmps stimulate expression of *Tbx2* which in turn, is thought to repress chamber myocardial programs in the primary myocardium of the AV canal (Christoffels *et al.*, 2004b). In contrast, expression of *Notch* and its target genes *Hrt1* and *Hrt2* is restricted to the working myocardium. Because *Bmp2* and *Tbx2* are ectopically expressed in *Hrt1/2* null mutants and conversely, downregulated when *Hrt1/2* are expressed throughout the heart (Kokubo *et al.*, 2007), Notch signaling via Hrt 1/2 is hypothesized to limit expression of *Bmp2* and *Tbx2* to primary myocardium and define the AV canal (High and Epstein, 2008). A similar regulatory process may occur in the OFT myocardium with *Bmp4* and *Tbx3* (Mommersteeg *et al.*, 2007b; Moon, unpublished). Notch signaling also stimulates expression of *Snai1* and *slug* in the cushion endocardium; these factors downregulate expression of the intercellular adhesion molecule VE-cadherin to facilitate invasion of endocardial cells during EMT. Tgf*β* signaling from the myocardium is also crucial for cushion morphogenesis, and is disrupted in Notch pathway mutants. Thus, alterations in Notch signaling in the heart disrupts EMT at multiple points (altered myocardial Bmp/Tgf*β* activity, and disrupted *Snai* function in the endocardium downstream of both Notch and BMP/Tgf*β*) resulting in acellular cushions in Notch pathway mutants (Timmerman *et al.*, 2004). Notch pathway mutants have defects in differentiation/proliferation of trabecular myocardium. The proliferative defect is associated with decreased myocardial *Bmp10* activity due to loss of Notch-mediated signaling in the endocardium. The differentiation defect is attributed to loss of secreted neuregulin1 from the endocardium (Grego-Bessa *et al.*, 2007). This phenotype provides an example of the interplay between endocardial function and myocardial development.

3.7. Down syndrome

Down Syndrome (DS) can affect morphology and function of virtually every organ system. Characteristic features of DS are typical facial dysmorphology, mental retardation, and a 50% incidence of cardiovascular defects, ranging from cleft mitral valve to complete atrioventricular canal (CAVC). Abnormal SHF and NC function are manifest as DORV, tetralogy of Fallot, aberrant subclavian artery and occasionally, coarctation of the aorta. DS is the most common known genetic cause of congenital cardiovascular disease in humans and is caused by trisomy of chromosome 21 (Ts21) or portions thereof. Nearly 220,000 infants are diagnosed with Down syndrome each year worldwide. However, the relative incidence of cardiovascular and other defects varies with the ethnic/geographic population of DS patients studied, suggesting a role for genetic modifiers. This has been demonstrated for *GATA1* and *CRELD1* with regard to hematologic and cardiac phenotypes, respectively (Maslen *et al.*, 2006; Vyas and Crispino, 2007).

Numerous hypotheses have been proposed to explain how Ts21 results in Down syndrome phenotypes. The simplest theories propose that increased gene dosage at critical loci, or of groups of genes within a "Down syndrome critical region" (DSCR), results in increased transcript/ protein product production, which disrupts specific developmental pathways and causes the degenerative processes that occur postnatally and in adulthood. Trisomic genes may function in the relevant pathways themselves, or regulate expression of disomic genes on other chromosomes, increasing the likelihood of complex genetic and molecular interactions.

The phenotypes observed in patients with partial trisomy 21 have suggested candidate genes that may contribute to congenital heart defects in DS patients. For example, variants at the *COL6A1* locus influence the risk of heart defects in DS patients (Davies *et al.*, 1995). *DSCAM* (DS cell adhesion molecule) was identified within a candidate region in partial trisomy patients with heart defects and more recently, a patient with heart disease and a small duplication containing *DSCAM*, but not *COL6A1,* was described (Kosaki *et al.*, 2005).

Modeling DS in mice to identify how dosage imbalance of specific gene (s) contributes to the cardiac malformations is challenging given that at least 230 genes are contained on the long arm of chromosome 21, and their murine orthologues are located on multiple mouse chromosomes: 144 genes on chromosome 16, 23 on chromosome 17, and 58 on chromosome 10. Multiple approaches have been employed (reviewed in (Moore and Roper, 2007)). These include generating transgenic mice that express a single gene from the orthologous region on mouse chromosome 16, generating large regions of redundant genetic material with transgenic YACS, BACs, or PACs, creating mice with complete (Ts16) or partial trisomy 16 (the best studied being Ts65Dn), and generating human chromosome 21 trans-chromosomal mice (Tc1) (Reeves, 2006). Intercrossing the various mutants to increase or decrease dosage of specific regions has also been employed. No single gene or region has been shown to recapitulate the complete spectrum of cardiac defects seen in DS.

100% of Ts16 mutants have cardiovascular defects and a high incidence of CAVC (Miyabara *et al.*, 1982). However, additional defects that are rare in DS patients such as PTA are also seen in these mice and may result from dosage imbalance of orthologues of the region of human chromosome 22 associated with del22q11 syndromes (also present on mouse chromosome 16). The Ts65Dn segmental trisomy contains orthologues for nearly half of the genes on the long arm of human chromosome 21 and these mice have many DS-like phenotypes including patent foramen ovale, VSDs, valvular defects, and right or interrupted aortic arch, but not CAVC (Williams *et al.*, 2008). In the Tc1 model, most of human chromosome 21 is present, although the trisomy is variable in different tissue types; these mice also have cardiac defects, mostly VSDs. *COL6A1* and DSCAM1 variants influence the risk of heart defects in humans but mouse mutants have not yet been published. A spontaneous deletion mouse mutant in *Dscam* was identified, but did not have heart defects (Fuerst *et al.*, 2008); however excess Dscam could contribute to DS phenotypes.

Two promising candidates that have been investigated in mouse models are *DSCR1* and *DYRK1a*, both of which may affect valve and septal morphogenesis by altering calcineurin/NFAT signaling. The NFAT (nuclear factor of activated T cells) family of transcription factors has diverse roles in organogenesis. NFATs translocate to the nucleus after dephosphorylation by the calcium-dependent phosphatase, calcineurin. NFATc nuclear translocation in endocardial endothelial cells is required for EMT and cushion development. *Nfatc* null mouse mutants die from circulatory failure with absent pulmonary and aortic valves, immature AV valves, and septal defects (de la Pompa *et al.*, 1998; Ranger *et al.*, 1998). *DSCR1* encodes a protein called "regulator of calcineurin 1" (Rcan1, also called MCIP1) which is an inhibitor of calcineurin signaling and is expressed in the developing heart, including endocardial cells in the region of cushion formation (Rothermel *et al.*, 2000). *Rcan1* expression is regulated by Nfat in mice. However, *Rcan1* null mouse mutants have abnormal myocyte responses to stress, but not structural heart disease (Sanna *et al.*, 2006). *DYRK1A* (dual-specificity tyrosine phosphorylation regulated kinase 1a) primes GSK3 targets, including NFATs, for phosphorylation/inactivation. Transgenic overexpression of *Dyrk1A* and *Rcan1* in mice results in exclusion of Nfatc1 from the nuclei of endocardial cells, and synergistically impairs Nfat-dependent transcription to disrupt an autoregulatory signaling loop (Arron *et al.*, 2006). In the Ts16 mouse, *Rcan1* expression is elevated and Nfat transcriptional activity is decreased. However, restoring disomic gene dosage of *Rcan1* in the Ts16 mouse does not rescue their cardiac malformations (Lange *et al.*, 2005).

Clearly, isolating the genes that contribute to DS cardiovascular phenotypes will require even more complex genetic approaches in mouse, in combination with additional studies in humans with smaller duplications to identify new critical genes. It will be necessary to analyze existing models in the same genetic backgrounds, and in combination with one another, to better delineate the critical regions required for the development of different tissues (Moore and Roper, 2007). New mice need to be engineered to address the roles of trisomic regions not represented in current models.

3.8. Syndromes and isolated defects related to altered left–right asymmetry

LR patterning is required to establish serial circulation by generating correct connections between the heart and the vasculature, and correct flow of blood through the heart, which are referred to as venoatrial, AV, and ventriculoarterial concordance. In addition to severe phenotypes with defective concordance (anomalous pulmonary venous connection, TGA, single ventricle), patients with situs abnormalities also have more subtle defects, such as abnormal valves, septae, and ventricular structure (see Section 4). This can be explained by two nonexclusive mechanisms. The molecules and signaling pathways that establish and propagate LR asymmetry (e.g., Shh, Cited2, Pitx2c) are also expressed and function later in the developing pharynx and cardiovascular system. It is difficult to test whether heart defects seen in heterotaxy patients are a manifestation of perturbation of the upstream LR cascade, as it requires perturbing early embryonic function and later restoration.

While heart defects attributed (currently) to abnormal laterality only account for 3% of all congenital heart defects in humans, the complex nature of the malformations and the association with other serious birth defects results in a high rate of mortality. Nearly 10% of infants with heterotaxy have a family member with congenital heart disease; both X-linked and autosomal familial cases have been reported but the locus involved has not yet been identified in the majority of cases.

Seminal observations that human patients with Kartagener syndrome (*situs inversus*, bronchiectasis, and sinusitis) have immotile cilia and ultrastructural defects in ciliary dynein arms first suggested the crucial role of ciliary function in LR patterning (Afzelius, 1976). Genes encoding ciliary components are now the subject of candidate screens in humans with primary ciliary diskinesia (with or without *situs inversus*) and recent mutational analyses demonstrated that 38% of such patients have mutations in the dynein genes *DNAI1* and *DNAH5* (Zariwala *et al.*, 2007).

Randomized LR asymmetry in the *iv* mouse was described in 1959 (Hummel and Chapman, 1959) and in 1997 found to be caused by a mutation in an axonemal dynein gene called LR dynein (*Lrd*) expressed in the mouse node (Supp *et al.*, 1997, 1999). Several elegant studies in mice have since shown that *Lrd* function is required for rotation of the cilia and that asymmetric fluid flow across the node is required to establish the LR axis (Lowe *et al.*, 1996; Nonaka *et al.*, 2002; Okada *et al.*, 1999). The list of ciliary motor or intraflagellar proteins (Kif3-a, Kif3-b, polaris, wimpl, inversin) required in mice for establishing normal expression of LR pathway genes and correct visceral situs is rapidly increasing (Huangfu and Anderson, 2005; Marszalek *et al.*, 1999; Murcia *et al.*, 2000; Nonaka *et al.*, 1998; Watanabe *et al.*, 2003).

Linkage analysis and positional cloning identified mutations in the X-linked gene *ZIC3* as causative in a large family with heritable *situs* abnormalities. Mutations were also detected in sporadic cases of *situs ambiguous* (Gebbia *et al.*, 1997). Patients with mutations in *ZIC3* display the full array of complex congenital defects associated with alterations in LR asymmetry. Males are affected with situs ambiguous, complex heart defects, pulmonary patterning defects, and intra-abdominal organ development. Some females have also been reported to have *situs inversus*. Central nervous system, neural tube and sacral defects are also seen.

ZIC3 encodes a transcription factor with significant homology to the GLI family of transcription factors. Zic3 likely functions as an effector in the Shh pathway by interacting with Gli proteins to modulate transcription in target cells; Zic3 can also bind targets independently (Herman and El-Hodiri, 2002; Koyabu *et al.*, 2001; Mizugishi *et al.*, 2001). *Zic3* null mouse mutants have defects in cardiac looping and a high rate of embryonic lethality (Purandare *et al.*, 2002). Expression of *nodal* and *Pitx2* in the lateral plate mesoderm is randomized, but since the initial expression of *nodal* at the node is intact, it appears that Zic3 participates in propagation of patterning from the node to the lateral plate mesoderm. However, Zic3 can regulate the expression of a Nodal enhancer *in vivo* and *Zic3* and *Nodal* genetically interact (Ware *et al.*, 2006). *Zic3* null heterozygotes phenocopy the tail defects seen in the spontaneous mouse mutant, *bent tail* which is a deletion of the *Zic3* gene (Carrel *et al.*, 2000; Klootwijk *et al.*, 2000). Tail kinking and neural tube defects are seen in heterozygotes, and although midline CNS abnormalities consistent with altered LR development are present, cardiac defects were not reported. It is unclear if the different cardiac phenotypes seen in the targeted versus spontaneous mutants (both of which are thought to be null alleles) are a manifestation of genetic background.

Rieger syndrome (RS) in humans is an autosomal dominant syndrome of craniofacial, eye and abdominal wall abnormalities, and relatively rare cardiac and limb defects. Its incidence is estimated to be 1:200,000. RS is predominantly due to mutations in *PITX2*, which encodes a bicoid-like homeobox transcription factor (Semina *et al.*, 1996). Other loci have since been linked to RS (*Foxc1* and *Pax6*).

Shortly after the discovery of *PITX2* mutations in humans, observations in multiple organisms revealed that Pitx2 is a critical transcriptional effector of LR patterning directly regulated by Nodal signaling (Logan *et al.*, 1998; Piedra *et al.*, 1998; Yoshioka *et al.*, 1998). Expression of *Pitx2* in the left lateral plate mesoderm, and in left organ primordia, is highly conserved across species. Asymmetric Shh signaling as the result of nodal flow has been postulated to function near the top of the LR cascade: *Shh* null mouse mutants have randomized heart looping, pulmonary left isomerism and bilateral expression of *Pitx2* and other LR determinants in the lateral plate mesoderm (Meyers and Martin, 1999). *Pitx2* expression is randomized in the mouse *inversus viscerum* mutant (50% of homozygous mice have *situs solitus* and 50%, *situs inversus*) whereas these genes are aberrantly expressed on the right in the *situs inversus* mouse mutant *inv/inv* (100% *situs inversus*) (Campione *et al.*, 1999; Lowe *et al.*, 1996; Piedra *et al.*, 1998; Ryan *et al.*, 1998; Yokoyama *et al.*, 1993).

Pitx2 $^{+/-}$ mice pheoncopy defects seen in RS. Although null homozygotes have normal heart looping, they have right pulmonary and atrial isomerism, and complex heart defects involving virtually every substructure in the heart (Kitamura *et al.*, 1999; Lin *et al.*, 1999; Lu *et al.*, 1999). *Pitx2c* is expressed in the left SHF, pharyngeal mesoderm, left atrium and the AVC;

deletion of this isoform in mice causes DORV, great vessel defects, abnormal pulmonary and systemic venous drainage, septal defects and defective endocardial cushion and valve development (Liu *et al.*, 2002). Further studies in mice demonstrate dosage-dependent as well as stage- and tissue-specific roles for the *Pitx2c* isoform in conferring axial, regional, and organ LR identity. For example, the anatomic left atrium of *Pitx2c* mutants has a node-like structure that is indistinguishable from the right SA node; therefore, Pitx2c confers atrial "leftness" at least in part by repressing a right-sided atrial program (Galli *et al.*, 2008). *Pitx2c* is also expressed and required for development of the myocardium ensleeving the pulmonary veins adjacent to the left atrium (Mommersteeg *et al.*, 2007a). Aberrant automaticity in this tissue is a source of atrial arrhythmias in humans (atrial fibrillation and flutter) and recent studies demonstrate that variants in genomic sequence adjacent to *PITX2* associate with significantly increased risk for atrial arrhythmias in humans (Gudbjartsson *et al.*, 2007; Martin, 2007).

Other determinants of LR morphogenesis elucidated in the mouse have informed the search for causative loci in humans. For example, cryptic (*Cfc1*), activin receptor 2b (*Acvr2b),* and *Lefty1* are positioned in the TGFβ signaling cascade downstream of nodal. Ablation of the function of these genes in mice disrupts propagation and maintenance of left-sided *nodal* and *Pitx2* expression in the lateral plate mesoderm (Gaio *et al.*, 1999; Meno *et al.*, 1998; Oh and Li, 1997). Mutations in these genes have now been linked to laterality defects in humans (human genes: *EGF-CFC*, *ACVR2B*, and *LEFTYA*) (Ware, 2006). Additional "mining" of the nodal pathway in humans revealed that while mutations in any single gene examined (*NODAL*, *GDF1*, *CFC1*, *TDGF1*, *FOXH1*, *SMAD2*) are rare causes of laterality defects, as a group, they constitute a genetic etiology of heart defects (particularly OFT abnormalities) as common as *del22q11* (Roessler *et al.*, 2008).

3.9. Cardiomyopathies: Hypertrophic, dilated and noncompaction

Hypertrophic cardiomyopathy (HCM) often presents as sudden death during exertion in a formerly asymptomatic individual. At the cellular level, the major features of HCM are myocyte and myofibrillar disarray, hypertrophy, and myocardial fibrosis. Fifty percentage of HCMs are inherited in an autosomal dominant fashion. Increased awareness and echocardiographic screening have improved detection and shown that HCM is more common than previously thought. The development of genetic screens for HCM will have real impact on human health. Identification of new candidates in mutant mice will contribute to the comprehensiveness of these screening tools, in addition to providing a means to understand the effects of different mutations on cardiomyocyte structure and function.

Although a cardiac failure associated with ventricular dilatation is a common end-stage result of HCM, chronic hypertension or ischemic heart disease, dilated cardiomyopathy (DCM) is a distinct clinical and histopathologic entity that can occur in an idiopathic/sporadic or inherited manner and may present in childhood. Infectious, inflammatory and toxic processes can also cause DCM. Typically, there is patchy apoptosis and necrosis of myocytes that vastly exceeds the limited proliferative capacity of the mature myocardium (Kajstura *et al.*, 1998) Fibrosis, abnormal myocyte structure, and myofibrillar disarray result in ventricular dilatation and poor contractility. Congestive heart failure and arrhythmias are common causes of death in these patients. The identification of causative genes in the heritable forms in humans, and mechanistic studies in mouse models are providing fundamental insights into pathogenesis that will benefit individuals affected with both heritable and acquired disease.

Genetic causes of HCM are now estimated to account for at least 50% of sporadic childhood onset cases (Morita *et al.*, 2008). Modifier loci appear to play a major role in the ultimate phenotype and may also contribute to the severity and potential for recovery from acquired cardiomyopathies. Hundreds of mutations affecting sarcomeric protein genes, intermediate and nuclear filament components, z-disc, and sarcolemmal proteins have been linked to both HCM and DCM in childhood- and adult-onset forms. In most cases, incorporation of the mutant proteins into the sarcomere disrupts its structure and contractile function. The enormous literature and many mouse models that have been employed in this area prevent an inclusive review here, so I will focus instead on a few that provide general mechanistic insight.

The first mouse models of HCM that were generated remain the best characterized and are targeted missense mutations in the α-MyHC gene to model different β-MyHC mutations observed in affected humans (Maass and Leinwand, 2000). While these models exhibit cardinal features of HCM, there were multiple, unexpected differences in phenotype between male and female mice. This observation is not yet understood and is interesting in a broader context: cardiac failure in general is less common and occurs later in life in women, and the penetrance of heritable HCM is also lower in females. The rates of apoptotic and necrotic cell death in the failing hearts of women are significantly lower than men (Guerra *et al.*, 1999). Further study of these mouse models may shed light on the bases of these gender differences, which could well result in different screening and treatment strategies in men and women.

Numerous other models have been generated to examine the effects of disrupting specific functional domains in MyHC genes and mutations affecting other structural proteins listed above. The effects of these mutations on the histologic, excitation–contraction coupling, calcium sensitivity, contractile and electrophysiologic properties of the myocardium, as well as life span, cardiac function and morbidity in these models demonstrate the strengths of the mouse for exhaustive *in vivo* and *ex vivo* studies in well-defined genetic backgrounds. These investigations have provided a more detailed understanding of the complex interplay between altered sarcomere structure, progressive hypertrophy, cardiac dysfunction and fibrosis, and abnormal molecular myocyte "signatures" (Hoshijima and Chien, 2002). Understanding the pathways that are dysregulated in genetic models of cardiomyopathy will identify candidate loci that influence the severity and rate of progression of acquired cardiomyopathies, such as those from chronic hypertension and ischemic heart disease, which are much more common.

By dissecting the signal transduction pathways that are pivotal for different aspects of the hypertrophic response, new therapeutic targets will be identified (Hunter and Chien, 1999). For example, defects in calcium cycling have been shown to contribute to both hypertrophic and dilated cardiomyopathies. Overexpression of the calcium/calmodulin dependent protein phosphatase calcineurin in transgenic mice generates an HCM (Molkentin *et al.*, 1998). Furthermore, the hypertrophic response in several mouse models can be prevented by treating the animals with calcineurin inhibitors (Sussman *et al.*, 1998). Transgenic overexpression of the calcineurin inhibitory protein Rcan1 has similar beneficial effects (Rothermel *et al.*, 2001). A dominant DCM in humans was shown to be due to a mutation in phospholamban that prevents its phosphorylation. Unphosphorylated phospholamban binds and decreases the sensitivity of SERCA to calcium. Overexpressing this mutant phospholamban in mice reproduces the cardiomyopathy and revealed that the mutant protein had a dominant negative effect to block phosphorylation of wild type phospholamban and constitutively depress SERCA2 function (Schmitt *et al.*, 2003). The LIM-only protein MLP is associated with actin based cytoskeletal structures in striated muscle cells; MLP deficient mice (*Csrp3*$^{-/-}$) develop DCM and heart failure postnatally (Arber *et al.*, 1997). Ablating phospholamban activity in *MLP* null mice by genetic complementation with the *phospholamban* null allele prevented the echocardiographic, histologic, gene expression, and fibrotic features of DCM (Minamisawa *et al.*, 1999).

Ventricular noncompaction is a cardiomyopathy that represents a defect in myocardial maturation/remodeling; large trabeculae project into the ventricular lumen and the intertrabecular recesses extend deep into the ventricular wall; the compact layer may be abnormally thin. Improvements in diagnosis have increased the frequency of this diagnosis from infancy through adulthood. It usually presents with heart failure and arrhythmias, and may be accompanied by neuromuscular and craniofacial abnormalities, as well as other cardiac defects, particularly VSDs and LV outflow obstruction. Few disease-causing mutations have been identified in humans (some of which have also been reported in patients with DCM) and there are several mouse mutants with this phenotype. Mutations in the *TAZ* gene have been described in humans with familial noncompaction (Moric-Janiszewska and Markiewicz-Loskot, 2008). Tafazzin is a poorly understood protein thought to catalyze transacylation of a mitochondrial phospholipid called cardiolipin (Schlame and Ren, 2006). Barth syndrome is a rare X-linked cardiomyopathy associated with cyclic neutropenia, skeletal myopathy, mitochondrial dysfunction and an inborn aciduria. It has also been mapped to *TAZ*. Although models have been generated in fish and fly that support the hypothesis that the primary molecular defect relates to abnormal cardiolipin processing and mitochondrial function, no mouse model has yet been reported.

LMNA (lamin A/C) encodes intermediate filament proteins that form the inner layer of the nuclear membrane. *LMNA* mutations have been associated with a variety of myopathies and neuromuscular diseases: DCM with conduction system disease, Emery–Dreifuss muscular dystrophy, limb girdle muscular dystrophy, Charcot–Marie–Tooth syndrome, Dunnigan familial partial lipodystrophy, mandibuloacral dysplasia, and progeria (Moric-Janiszewska and Markiewicz-Loskot, 2008). A point mutation in *Lmna* in mice designed to model Emery–Dreifuss phenocopies aspects of these human diseases. These mice have small hearts with increased fibroblasts but DCM was not described (Mounkes *et al.*, 2003).

Alpha-dystobrevin is a part of the dystrophin-containing glycoprotein complex (DGC) that has structural and signaling functions in muscle. Mutations in α-dystrobrevin were found in patients with congenital heart defects including ventricular noncompaction (Moric-Janiszewska and Markiewicz-Loskot, 2008). Ablation of α-dystrobrevin in gene-targeted mice caused a cardiomyopathy with degenerating cardiomyocytes and fibrosis but not ventricular noncompaction (Grady *et al.*, 1999).

The PDZ and LIM domain-containing cytoskeletal protein called Cypher or Ldb3 in mouse (ZASP in human) is localized to the z-disc. *Cypher* null mice have a severe congenital myopathy that prevents analysis of cardiac function beyond the first few days of life however, their postnatal phenotype included biventricular hypertrophy, dilated ventricles, and abnormal trabeculae (Zhou *et al.*, 2001). There were no detectable Z-discs in the mutant cardiac muscle, likely reflecting progressive degeneration throughout the embryonic and fetal period. Further studies indicated that Cypher interacts with alpha-actinin to maintain structural integrity of the z-disc during contraction. These observations in mutant mice led to the discovery of mutations in this gene in humans with dilated and noncompaction cardiomyopathies (Sheikh *et al.*, 2007)

These are but a few examples of the use of mouse models to understand the etiology and identify potential targets for intervention in human heritable and acquired cardiomyopathies.

3.10. Arrhythmias and conduction system defects

As noted above, arrhythmias are a common complication of cardiomyopathies and can in part be attributed to fibrosis, myocyte drop out, and alterations in response to autonomic input. Mechanistic understanding of heritable primary arrhythmias, such as Long QT syndrome based on modeling of human mutations in the mouse is simultaneously improving our understanding of how alterations in ion channel function influence susceptibility and progression of arrhythmias (Salama and London, 2007).

Long QT syndrome is a genetically heterogeneous inherited arrhythmia characterized by prolongation of the repolarization phase of the myocardium (reflected by the QT interval on an electrocardiogram). It can present with syncope and sudden death due to arrhythmias stimulated by stress, exercise, or even sleep in some patients. Mutations in genes encoding potassium channels that regulate different phases of the repolarization current account for most of the LQT syndromes described to date (*KCNQ1*, *KCNE1*, *KCNH2*, *KCNE2*). Mutations in channels affecting resting potential (*KCNJ2*), action potential duration (SCNA5, sodium), atrial repolarization (*KCNA5*), and calcium triggering (*CACNA1C*, *CACNB2B*) are linked to other inherited arrhythmias such as short QT and Brugada syndromes. Mutations in the ryanodine receptor 2 gene (*RYR2*) disrupts SR calcium release and causes arrythmogenic RV dysplasia and polymorphic ventricular tachycardias (Nilles and London, 2007).

Oculodentaldigital dysplasia (ODDD) is a phenotypically variable, autosomal dominant syndrome characterized by hearing, limb, dental, and craniofacial defects; cardiac abnormalities are less common and include ventricular tachyarrhythmias, AV block, and ASDs (Loddenkemper *et al.*, 2002). Twenty-seven different mutations in the *GJA1* gene encoding connexin 43 have been found in affected families (mutations have also been detected in patients with heterotaxy). Connexins are gap junction proteins that are assembled into connexons and permit intercellular passage of ions and small molecules including second messengers and metabolites. Mutations identified in humans can be grouped based on their molecular effects on protein function such as conductance, trafficking, and ability of the mutant protein to form heteromeric connexons (Dobrowolski *et al.*, 2008; Paznekas *et al.*, 2003). The importance of Connexin 43 for other aspects heart development were known as targeted null mouse mutants die postnatally with OFT obstruction (Huang *et al.*, 1998; Reaume *et al.*, 1995). *G6ja1* is also expressed in the proepicardium and defects in epicardial coupling, cell migration, and coronary artery patterning were also detected in these null mutants (Li *et al.*, 2002).

Targeted mouse models of the different point/missense mutations seen in ODDD have been generated. There is also an ethyl-*N*-nitrosourea induced point mutant mouse ($Gja1^{Jrt}$). This mutation affects a highly conserved residue, although it has not been observed in ODDD patients. Both types of mutants provide excellent phenocopies of the spectrum of ODDD defects. With regard to the heart, $Gja1^{Jrt/+}$ mice have decreased myocardial gap junctions, globally depressed cardiac function, patent foramen ovale, prolonged QRS duration, and premature ventricular contractions (a harbinger of ventricular tachycardias) (Flenniken *et al.*, 2005). There was also evidence of DCM, which is interesting as alterations in connexin 43 (and connexin 45) expression and connexon function are seen in DCM and heart failure in mice and in humans, although the causal relationship remains unclear (Severs *et al.*, 2006). Ubiquitous expression of the *GJA1* G138R human mutation in heterozygous mice reproduced the phenotypes, and the interfamilial variability seen in ODDD. Conditional activation of this mutation in the heart caused postnatal death; arrhythmias were observed in isolated perfused hearts and *in vivo*. The frequency and severity of arrhythmias was dramatically increased under hypoxic conditions and when the animals were exposed to inhalational anesthetics, correlating with the previously published observation that hemiconnexons can open in stressed and hypoxic conditions (Evans *et al.*, 2006). Cardiomyocytes derived from mutant embryos displayed increased ATP-release and this too was exacerbated by hypoxia. The authors hypothesize that the mechanism of enhanced automaticity and profound tendency to arrhythmia in these mutants as compared to other mouse models may be due to increased conductance across the plasma membrane which increases the resting potential.

Polymorphisms and mutations in *GJA5* (Connexin 40) have also been associated with human arrhythmias; interest in this locus first arose from evaluation of a large kindred in which autosomal dominant conduction abnormalities and DCM were mapped to an interval containing connexin 40 (Kass *et al.*, 1994). Polymorphisms were shown to increase risk of atrial and ventricular tachycardias in humans (Firouzi *et al.*, 2004; Hauer *et al.*, 2006). A mouse model of *Gja5* deficiency has altered myocyte intercellular coupling, slower AV and interventricular conduction, although patterning of the conduction system appears normal. Since *Gja5* is not expressed in adult myocardium, the authors hypothesize that delayed propagation via abnormal Purkinje fibers causes ventricular delay (Simon *et al.*, 1998). Of note, *Gja5* null mutants have OFT, septal, and endocardial cushion defects (Gu *et al.*, 2003; Kirchhoff *et al.*, 2000). Their limb defects, while not identical to *Tbx5* heterozygotes, affect the some of the same structures (Pizard *et al.*, 2005). Tbx5 has now been shown to regulate *Gja5* expression in heart and limb, and mutations in *TBX5* cause HOS (see above). This pathway may be the mechanism underlying some HOS phenotypes. However, patterning of the AV conduction system is normal in *Gja5* mutants, so the cause of structural defects in the conduction system observed in *Tbx5* mutants cannot be attributed to a decrement in *Gja5* expression but functional defects may be (Moskowitz *et al.*, 2004).

Mouse models of the mutations that cause conduction system dysfunction in humans do not always phenocopy the arrhythmias. For example, minK (encoded by *KCNE1* in humans) is a beta subunit of the potassium channel that controls a portion of the repolarization current in human hearts. Mutations in this gene cause autosomal dominant and recessive arrhythmia syndromes, including LQT. There are both gene targeted and spontaneous mutant alleles of *minK*, and neither model displays LQT or ventricular arrhythmias. However, the ability to simultaneously ablate the function of *minK* and label the relevant cell types in which it is expressed using a lacZ reporter knock-in revealed the basis for the different phenotypes in mouse and human: *KCNE1* is expressed widely in the human heart, but only in the conduction system cells of mouse (Kupershmidt *et al.*, 1999).

4. From Mouse to Human: Using Mouse Models to Discover Novel Factors and Pathways That Regulate Cardiovascular Development

Although nonhypothesis-driven experiments are routinely decried in the scientific community, the identity of many of the genes required for cardiovascular development (as in other organ systems) was discovered fortuitously in knockout mice generated for other reasons. Relevant genes continue to be defined in this manner although increasingly, candidate genes are identified by their roles in other species, or by their expression in pertinent tissues in mice, and are then subject to germline or conditional ablation studies.

As will become clear in this section, pathways built on inferred relationships from shared anatomic phenotypes or gene expression profiles are standard in the literature. This approach is inherently limited and conclusions derived therefrom are potentially flawed by the resolution with which we can, or choose to examine a particular phenotype. For example, anatomic malformations of the OFT may be grouped into a limited number of categories (absent, abnormal size, defective septation, or alignment); these are relatively superficial endpoints of enormously complex morphogenetic events. Yet it is common to group genes or infer functional relationships because two mutants both display PTA. As a means for hypothesis generation, this is a reasonable first start; unfortunately, relationships derived in this manner often become dogma before they are rigorously tested and the resulting edifice is a "house of cards". However, as technical barriers are overcome, our ability to reliably construct transcriptional and signaling architectures that reflect the truc situation *in vivo* will dramatically improve.

There are literally hundreds of mouse mutants with structural cardiovascular defects; rather than assemble a compendium of lists, this discussion will be limited to a small subset that illustrate some crucial transcription factor networks, signaling cascades and ECM proteins involved in structural defects of the chambers, OFT and valves. I have tried to choose examples for which the pathway data are robust and may have been reproduced in multiple labs; however, in many cases, the position of genes in the pathway presented is still based on indirect evidence. Pertinent reviews are suggested in addition to primary literature.

4.1. Models of defective chamber specification, formation, and remodeling

Studies in mice are identifying members of a complex regulatory network that regulates chamber formation and identity, and stimulates region-specific myogenic programs (Bruneau, 2008; Peterkin *et al.*, 2005; Risebro and Riley, 2006; Stennard and Harvey, 2005). This network includes Nkx2.5, GATA, Tbx, serum response factor (SRF) and others. Requirements for the continued function of this network in the adult myocardium are also being discovered in mice (and in humans, see above).

Nkx2.5 encodes a homeobox transcription factor with a central role in the transcriptional network that regulates cardiomyocyte differentiation and heart morphogenesis; it is expressed in precardiac mesoderm prior to activation of myogenic programs and continues to be expressed in the pharynx and diffusely throughout the myocardium. Nkx2.5 negatively regulates the expression of some myocardial progenitor genes in the cardiac crescent, the primary heart tube and the SHF; these genes continue to be expressed in "differentiated" myocardium of the primary heart tube in *Nkx2.5* null mutants. A negative autoregulatory loop from Nkx2.5 via Bmp2 limits the number of cells specified to the cardiac lineage and is required for proliferation of RV and OFT progenitors in the SHF (Prall *et al.*, 2007). These findings may explain the phenotypes of *Nkx2.5*$^{-/-}$ mutants; although they form a primitive heart tube, they die by E10.5 with an untrabeculated left ventricle, and a hypoplastic RV and OFT (Tanaka *et al.*, 1999). Additional studies in Nkx2.5 null heterozygotes, hypomorphs, and conditional mutants revealed later roles for Nkx2.5 in OFT remodeling, ventricular compaction/remodeling, septal formation, and for ongoing survival, identity and function of conduction system cells in adult mice (Biben *et al.*, 2000; Jay *et al.*, 2004; Prall *et al.*, 2007). Ablation of Nkx2.5 only in the ventricle after E10.5 does not cause structural defects, but the mice develop heart block and massive trabecular muscle overgrowth (Pashmforoush *et al.*, 2004). At birth, mutant mice display a hypoplastic AV node and then dropout of nodal and conduction system cells. Aberrant expression of the proliferative factor *Bmp10* in myocardium of adult mutants accounts for the hypertrophic ventricular walls and trabeculae observed in Nkx2.5 tissue-specific mutants.

The importance of *Nkx2.5* in multiple steps of early heart development and later cardiac function in mice and other model organisms made it an obvious candidate for investigations in humans with sporadic and heritable heart disease (Prall *et al.*, 2002). Mutations were found in patients with ASDs, conduction abnormalities, and less frequently, defects in OFT alignment (Benson *et al.*, 1999; McElhinney *et al.*, 2003; Schott *et al.*, 1998). Mutations in *NKX2.5* are estimated to contribute to 2% of structural heart defects in humans. At present, few downstream targets of Nkx2.5 and its partners are known. Mutations in one target, *MYH6* (encodes α-myosin heavy chain in humans) have been reported in patients with ASDs (Ching *et al.*, 2005). Abnormalities in ventricular remodeling in *Nkx2.5* mutant mice suggest *NKX2.5* as a candidate gene for cardiomyopathy in humans, but thus far, this has not been demonstrated.

Gata4 is expressed in cardiac progenitors and subsequently in cardiomyocytes, endocardium, proepicardium, and epicardium. Although cardiomyocyte specification and differentiation are unaffected in *Gata4* null mutants, Gata4 is required for formation of a single heart tube in mice (Kuo *et al.*, 1997; Molkentin *et al.*, 1997). *Gata4* mutants with rescued endodermal activity (using tetraploid complementation) have aberrantly looped hearts, a thin-walled myocardium lacking trabeculae, and no proepicardium (absent coronary vasculature); the myocardial defect was initially attributed to loss of proepicardial signaling (Watt *et al.*, 2004). However, conditional *Gata4* ablation in *Nkx2.5*-expressing cardiomyocytes (which preserves proepicardial *Gata4* function) causes a similar phenotype and associated RV hypoplasia. Later myocyte-restricted ablation caused OFT alignment defects (Zeisberg *et al.*, 2005). Differential sensitivity of individual aspects of heart development to *Gata4* gene dosage was elucidated with *Gata4* hypomorphic mice (produce <50% of the normal quantity of protein); mutants died at 14 days of gestation with AV canal, DORV, and hypoplastic ventricular myocardium; the coronary vasculature was normal (Pu *et al.*, 2004).

Srf is another key transcription factor that functions in the Nkx2.5/Gata4 transcriptional network (Sepulveda *et al.*, 2002) during both early and late ventricular development. Because null mouse mutants die shortly after gastrulation and do not make mesoderm (Arsenian *et al.*, 1998), conditional ablation has been used to determine its contribution to heart development at different stages. *Srf* null ES cells fail to express myogenic markers or form beating cardiomyocytes in culture in spite of normal expression of Gata4 and Nkx2.5, indicating that Srf is required for postspecification events in this lineage (Niu *et al.*, 2005). Ablating *Srf* function postcardiomyocyte differentiation specifically in the heart disrupted chamber maturation and remodeling resulting in midgestational lethality. The ventricular myocardium was thin with few trabeculae and the expression of *Nkx2.5*, *Gata4*, and other ventricular muscle markers were decreased (Niu *et al.*, 2005; Parlakian *et al.*, 2004). Although heritable human heart disease has not yet been linked to mutations in *SRF*, this is one of the fetal genes that are reactivated in human heart failure and has also been posited to regulate the CFTR gene (mutated in cystic fibrosis) (Rene *et al.*, 2005).

Tbx20 has both transcriptional activator and repressor functions. It interacts with Nkx2.5 and Gata4 to regulate target gene expression (Stennard *et al.*, 2003). It is initially expressed strongly throughout the crescent, diffusely in the myocardium of the primitive heart tube and continues to be expressed broadly in the heart, although at lower levels in the OFT and atrioventricular canal. Several groups ablated *Tbx20* in mice and found that chamber morphogenesis is profoundly impaired; the heart is arrested at the primary tube stage, working myocardial marker gene expression is lost, and *Tbx2* is ectopically expressed throughout the myocardium (Cai *et al.*, 2005; Singh *et al.*, 2005; Stennard *et al.*, 2005). *Bmp2* is normally expressed in the AVC and OFT and likely required for the normal pattern of *Tbx2* expression in these tissues (Ma *et al.*, 2005). Consistent with this hypothesis, the regionalized *Bmp2* expression in the AVC is also absent in *Tbx20* null mutants. Since Tbx2 can directly repress chamber program genes, findings are consistent with normal Tbx20 functioning to repress expression of *Tbx2* in outer curvature cells that form chamber myocardium. Based on these findings in mice, mutations in *TBX20* were sought among patients with familial heart defects. Two missense mutations have been identified in patients with chamber hypoplasia, septation, and valve defects (Kirk *et al.*, 2007). Some patients have DCM, which is interesting because *Tbx20* heterozygotes have decreased cardiac contractility (Stennard *et al.*, 2005).

Hand1 and Hand2 are basic helix-loop-helix transcription factors in the ventricular morphogenesis network downstream of Nkx2.5. *Hand1* is expressed in the outer curvature of the primitive ventricles and AV canal (Christoffels *et al.*, 2000) and later is enriched in the LV and distal OFT; its expression is dependent on Nkx2.5. *Hand1*$^{-/-}$ mutants form a primitive heart tube but do not loop or develop a ventricular segment (Firulli *et al.*, 1998). *Hand1* null ES cells cannot contribute to the LV of chimeric mice (Riley *et al.*, 2000). Cardiomyocyte-restricted ablation of *Hand1* results in biventricular hypoplasia, VSDs and surprisingly, hyperplastic AV valves (McFadden *et al.*, 2005). Very recently, 24/31 patients with defects in the spectrum of hypoplastic left heart syndrome were reported to have a frameshift mutation in *HAND1* that disrupts its regulatory properties (Reamon-Buettner *et al.*,2008). *Hand2* is initially expressed broadly in the primitive heart tube and lateral plate mesoderm and then its expression in the heart becomes largely restricted to the RV. Ablation of *Hand2* results in RV hypoplasia, decreased LV trabeculation, and vascular malformations (the latter may be due to abnormal NC function as this gene is also expressed in NC). *Hand2* expression in the RV is dependent on an enhancer with functional GATA binding sites (McFadden *et al.*, 2000) which may, at least in part, explain the RV hypoplasia observed in *Gata4* conditional mutants (above). No mutations in *HAND2* have yet been associated with human congenital heart disease.

Another set of transcription factors downstream of Nkx2.5 are important for formation of the RV and OFT myocardium from the mesoderm in the SHF. Islet1 (*Isl1*) is a LIM homeodomain protein first expressed (relevant to heart development) at the crescent stage with continuing expression in SHF cells in the pharynx until they are added to the poles of the heart. Although the *Isl1* lineage was initially reported to primarily contribute myocardium to the RV and OFT (Cai *et al.*, 2003), it is now clear that this lineage contributes extensively to all four chambers, including many of the cells in the primary heart tube and a majority of myocardial cells in the mature heart (Park *et al.*, 2006b). Nonetheless, *Isl1* null mutants are able to form a primary heart tube with atrial and ventricular components; the ventricle expresses *Hand1*, but the tube does not loop and no additional myocardium accrues. No markers of nascent OFT or RV myocardium can be detected and *Isl1*$^{-/-}$ cells do not migrate into the heart. Isl1 is required for proliferation and survival of SHF cells and to prevent premature differentiation of myocardial progenitors (Cai *et al.*, 2003). This observation has potential therapeutic implications for human heart disease in that an Isl1+ population of undifferentiated precursors is maintained in the heart into adulthood and can be recruited to differentiate into functional cardiomyocytes *in vitro* (Laugwitz *et al.*, 2005). The *Isl1* lineage also contributes to epicardial, coronary vascular, and pacemaker cells. Thus, discovering both the regulators of *Isl1*, and its transcriptional targets will provide new insight into congenital heart defects that affect the RV and OFT, as well as pathways that may be harnessed to allow regeneration of cardiomyocytes to treat human congenital and acquired heart disease.

Mef2c (Myocyte enhancer factor 2c) is a likely target of Islet1 regulation. *Mef2c* is expressed in the cardiac crescent shortly after *Nkx2.5* and *Gata4* and continues to be expressed in cells in the SHF in the pharynx. *Mef2c* null mutants die at E9.5 with cardiac defects that are similar to *Isl1* mutants in that the RV and OFT fail to form and there is no looping. However, the phenotype is more severe because the primary heart tube is also hypoplastic and the endocardium is abnormal (Bi *et al.*, 1999; Lin *et al.*, 1997). These findings suggest that Mef2c has roles during formation of the primary heart tube and development of SHF-derived structures. Conditional ablation of *Mef2c* in differentiated myocardium from E8.5 does not disrupt cardiac morphogenesis or survival, and no studies have been performed that dissociate the requirements for Mef2c in the progenitors of the primary heart tube versus those in the SHF. An intronic *Mef2c* enhancer is a target of Isl1 and Gata factors and regulates expression in a subset of SHF progenitors that give rise to the RV and OFT (Dodou *et al.*, 2004). *Mef2c* is also regulated by Foxh1, a forkhead-domain transcription factor that interacts with R-Smads to effect TGFβ signaling and is required for RV/OFT formation (von Both *et al.*, 2004). Mef2c directly activates expression of the histone methyltransferase Smyd1 (previously called Bop) in the precursors of the RV and OFT; *Smyd1* null mutants have RV hypoplasia although the defects are not as severe as seen in *Mef2c* mutants (Gottlieb *et al.*, 2002). In turn, *Hand1* expression appears to be regulated by Smyd1.

Isl1 and *Mef2c* are downregulated in the SHF of *Fgf8* mutants that either fail to form the RV and OFT, or have hypoplastic RVs and abnormal OFT remodeling (Park *et al.*, 2006b). A direct link between the Pea3 subfamily of ets transcription factors effect Fgf8 signal transduction and *Isl1* or *Mef2c* expression has not yet been established but, an enhancer that regulates SHF *Mef2c* expression contains conserved Pea3 binding sites (Moon, unpublished).

Novel factors are being identified in mice that are required for ventricular expansion and remodeling. Differential proliferation between the trabecular (lower) and compact myocardial layers (higher) during early ventricular expansion is needed to permit thickening of the compact layer prior to taking over the hemodynamic workload. Numerous factors that regulate cell cycle and chromatin dynamics have been identified. In the developing heart, zones of low proliferation have low levels of the proto-oncogene product, Nmyc1, and of cyclind2. Ablation of *Nmyc1* function results in midgestational lethality and organ hypoplasia; heart development arrests at E9.5 with minimal ventricular ballooning, a thin compact layer and few trabeculae (Charron *et al.*, 1992). Tbx proteins regulate cell cycle progression and proliferation in many cells types during development and in renewing tissues. Tbx2 and 3 regulate expression of cyclin-dependent kinases, interact with Myc and Ras (Carlson *et al.*, 2002), and can inhibit senescence (Brummelkamp *et al.*, 2002). The OFT and AVC (regions of *Tbx2/3* expression) are regions of low proliferation. Ectopic expression of *Tbx2* throughout the heart of *Tbx20* mutants is associated with decreased *Nmyc1* and *cyclind2* expression. Using chromatin immunoprecipitation (ChIP) on embryonic mouse heart DNA, Tbx2 was shown to bind and repress an intronic *Nmyc1* enhancer (Cai *et al.*, 2005).

Bmp signaling regulates regional *Tbx* gene expression in the heart. The R-Smads are transcriptional effectors of Bmp signaling; Smad4 mediates and balances Bmp/Tgfβ signaling by interacting with R-Smads in each pathway. Ablation of *Smad4* function in nascent cardiomyocytes disrupts proliferation and survival during chamber expansion in association with decreased expression of *Nmyc1*, cyclins, and Id2 (Song *et al.*, 2007b). An *Nmyc* regulatory element is bound and regulated by Smad4. Thus, Bmp signaling regulates myocardial proliferation (and *Nmyc*) directly via Smad4, and indirectly via effects on Tbx factors. Bmp10 is a proliferative factor required for chamber expansion and trabecular development; this was first discovered in FK506 binding protein 12 (FKBP12) mutant mice. FKBP12 is an enzyme that interacts with the intracellular domains of type I Tgfβ receptors and with intracellular calcium channels. FKBP12 mutants have thin compact layers, markedly enlarged and complex trabeculae in both ventricles, and muscular VSDs; most die from cardiac dysfunction prior to birth. RNA differential display on embryonic hearts from *FKBP12 −/−* mutants revealed marked overexpression of *Bmp10*. *Bmp10* is normally expressed only transiently in the developing trabecular myocardium (Shou *et al.*, 1998). Ablation of *Bmp10* results in defective trabeculation, thin compact layer and overproduction of the negative cell cycle regulator, $p57^{kip2}$; this finding confirms the role of Bmp10 in the pathogenesis of the *FKBP12 −/−* phenotype (Chen *et al.*, 2004). These studies provide mechanistic insight into HCM in transplant patients treated with FK506 (Tacrolimus) for immunosuppression (Atkison *et al.*, 1995).

Signals from the epicardium to myocardium are critical for ventricular expansion, trabecular development, and chamber wall remodeling (Risebro and Riley, 2006). Wilms tumor homolog (*Wt1*) encodes a transcription factor with both tumor suppressor and oncogenic activities; mutations in this gene are associated with numerous cancers and genitourinary syndromes in humans, but congenital heart disease has not been reported. *Wt1* expression in epicardium is required for normal expansion of the compact layer. This requirement is due in part to its activation of α4 integrin production. α4 integrin is an intercellular and ECM-cell adhesion molecule required in mice for proepicardial cells to adhere to the myocardium; loss of α4 integrin disrupts both coronary vascular and ventricular wall development (Yang *et al.*, 1994). α4 integrin is thought to interact with VCAM-1 present on myocardial cells and VCAM mutants have a similar thin myocardial phenotype (Kwee *et al.*, 1995). Ablation of erythropoietin (Epo, produced in the epicardium) or its receptor (expressed in endocardium, cushion mesenchyme, and epicardium) causes thin compact layer and heart failure by E12.5 due to failure of adequate myocardial proliferation (Wu *et al.*, 1999). How epicardial Epo signaling is transduced into a myocardial proliferative stimulus is unknown. Similarly, retinoic acid signaling via RXRα within the epicardium is required for compact myocardial growth (Merki *et al.*, 2005). Fgf9 was explored as a candidate signal to the myocardium, as it is produced in the epicardium at the time of compact layer proliferation and trabecular elaboration. Expression of *Fgf9* in the epicardium (and endocardium) is retinoic acid responsive and loss of *Fgf9* function causes a thin compact zone and DCM (Lavine *et al.*, 2005). Podoplanin (*Pdpn*) is a transmembrane sialoprotein that has been primarily studied for its role in renal epithelial function, tumor metastasis and lymphatic endothelial function. It is expressed in the proepicardium and *Pdpn* null mutant mice have abnormal epicardial adhesion, spreading and EMT. Fewer WT1+ epicardial cells are present and these cells overproduce E-cadherin; downregulation of E-cadherin is necessary for both epicardial and endothelial EMT. By E12.5, surviving mutants have thin myocardium, few trabeculae, and abnormal AV cushion remodeling (Mahtab *et al.*, 2008).

Epicardial-derived cells and signals are crucial to the formation of the coronary circulation which in turn, is required for normal ventricular growth and remodeling. The capacity to invest the heart in an epicardial layer is separable from these functions, as evident from the phenotype of *FOG-2* (now called *Zfpm2*) mutant mice. *Zfpm2* encodes a "Friend-of-GATA" cofactor. In *Zfpm2* mutants, ventricular myocardium is poorly developed and coronary vessels are absent but the epicardium itself appears normal (Tevosian *et al.*, 2000). Myocardial defects were initially attributed to abnormal epicardial EMT however, recent experiments indicate that a subset of cardiomyocytes derive from *Tbx18-* and *Wt1*-expressing epicardial lineage(s) (Cai *et al.*, 2008a; Zhou *et al.*, 2008) and so the pathogenesis of this phenotype may need to be reconsidered in this light.

Endocardial signaling to the myocardium also plays a role in ventricular growth and remodeling as discussed above with regard to Notch signaling upstream of ephrinB2 and neuregulin in human heart disease (see Section 3.6). The angiogenic Flk1/Vegf and Tie2/Angiopoietin-1 receptor/ligand pairs also mediate myocardial-endocardial signaling required for normal ventricular growth (Risebro and Riley, 2006).

Formation of the muscular interventricular septum is coupled to ventricular growth and remodeling, thus it is not surprising that VSDs are frequently seen in association with thin myocardium or ventricular noncompaction in these mouse models.

4.2. Mouse mutants with defective OFT remodeling

As discussed above, in some models early disruption of SHF function results in failure to form the RV and OFT resulting in early embryonic death. The number of mutants that have defects in OFT remodeling (elongation, alignment, septation, and valve structure or positioning) that phenocopy commonly seen OFT defects in humans is much larger. The many genes involved reflect the fact that remodeling occurs over a relatively long window of developmental time and requires complex signaling and migratory interactions among many different cell types. In some cases, the primary defect occurs in OFT myocardial precursors, either while they still reside in the SHF, or later after they have differentiated and accrued into the myocardial wall of the primitive OFT. This can result in abnormal secretion of signaling or ECM proteins by the SHF-derived myocardial cells which secondarily affects endothelial EMT or NC invasion of the OFT. Other mutations have cell autonomous effects in NC or in OFT endothelium. Mutations in some genes that regulate laterality also cause OFT defects. Reviews focused on OFT and NC are available (High and Epstein, 2008; Hutson and Kirby, 2003, 2007; Person *et al.*, 2005; Snider *et al.*, 2007; Xu and Baldini, 2007).

4.2.1. Primary defects in OFT progenitors in the SHF

Foxc1 and *Foxc2* encode transcription factors that are not only expressed in the cardiac crescent and SHF, but also expressed in NC. Thus it may not be surprising that all aspects of OFT morphogenesis are sensitive to the dosage of *Foxc1* and *Foxc2* in mice. The cardiovascular phenotype of compound heterozygotes is similar to that caused by ablating the function of either gene alone (aortic arch interruptions, VSDs, and valve dysplasias) (Iida *et al.*, 1997; Winnier *et al.*, 1999). *Foxc1*$^{-/+}$; *Foxc2*$^{-/-}$ mutants have short OFTs with hypocellular cushions and small RVs, while *Foxc1*$^{-/-}$; *Foxc2*$^{-/-}$ have severe defects (similar to *Isl1* mutants) with no morphologically detectable RV or OFT (Seo and Kume, 2006). Double null mutants have fewer Isl1+ cells in the SHF and pharyngeal endoderm at E8.5 and the cells that are present express abnormally low levels of *Tbx1* and *Fgf8*. As discussed in Section 3.1, complete loss of *Tbx1* function causes failed OFT septation (PTA) and OFT malalignment in 100% of mouse mutants. Conditional mutagenesis experiments have shown that both *Tbx1* and *Fgf8* are required for normal proliferation of Isl1+ SHF mesodermal cells prior to their accrual to the OFT (Ilagan *et al.*, 2006; Park *et al.*, 2006b; Xu *et al.*, 2004).

Early loss of *Fgf8* function in cardiac precursors when they still reside in the primitive streak prevents OFT and RV formation (as seen in *Isl1*$^{-/-}$ mutants), whereas later inactivation in the SHF perturbs OFT alignment (causing DORV and TGA). Simultaneous ablation of *Fgf8* in the SHF and pharyngeal endoderm causes 100% PTA and OFT malalignment (Park *et al.*, 2006). In both *Tbx1* and *Fgf8* mutants, there is abnormal NC survival and invasion into the distal OFT. Since *Tbx1* is not expressed in NC, this finding is by definition non-cell autonomous to the crest. In the case of Fgf8, we conditionally ablated *Fgf receptor* genes to determine which of cell population(s) are direct targets of Fgf signaling required for OFT morphogenesis and surprisingly, the NC are not. Rather, an autocrine Fgf signaling loop in the SHF is required for its OFT derivatives to establish the myocardial identity necessary for secretion of signaling molecules (*Bmp4*, Tgfβ pathway members) and ECM required for NC survival and migration into the OFT and endothelial EMT (Park *et al.*, 2008; Zhang *et al.*, 2008).

Loss of *Bmp4* expression in the OFT of *Fgf8* conditional mutants in itself would be predicted to significantly disrupt OFT remodeling because Bmp4 is required for endothelial EMT and cushion mesenchymal cell proliferation; ablation of *Bmp4* in myocardium causes a proximal OFT septation defect called AP window. NC and endocardium are direct targets of OFT myocardial-derived Bmp andTgfβ signals (Mercado-Pimentel and Runyan, 2007). Ablation of *Alk2* (activin A receptor I), *Alk5* (Tgfβ receptor I), and *Alk3* (Bmp receptor 1a) or the Bmp/Tgf*b* effector Smad4 in NC cells disrupts OFT alignment and septation at high penetrance (Jia *et al.*, 2007; Kaartinen *et al.*, 2004; Stottmann *et al.*, 2004; Wang *et al.*, 2006).

Hypomorphic *Bmpr2* mutants have proximal OFT defects and no semilunar valves (Delot *et al.*, 2003). The endothelial receptor that mediates EMT in the OFT has not been identified; Bmpr1a and Bmpr1b may have redundant functions.

Activation and regulation of Tgf*β* ligand bioavailability by latent Tgf*β* binding protein 1 (*Ltbp1*) is absolutely required for OFT septation (Todorovic *et al.*, 2007). *Tgfβ2* is expressed in OFT (and AV canal) myocardium and null mutants most frequently have DORV associated with decreased cushion jelly (Bartram *et al.*, 2001). Since this phenotype is less severe than that of the *Ltbp1* mutants, the redundant ligand(s) required remain to be elucidated. Tgf*β*2 has recently been placed in a regulatory/signaling cascade initiated by canonical Wnt signaling in the SHF that regulates OFT formation and remodeling. *β*catenin directly regulates *Isl1* expression in the SHF (Lin *et al.*, 2007) and loss of *β*catenin in the SHF results in a marked decrease in *Cyclind2* expression (Ai *et al.*, 2007). Multiple groups have shown that canonical Wnt signaling is required for normal expansion of SHF progenitors to support OFT and RV formation (Ai *et al.*, 2007; Cohen *et al.*, 2007; Kwon *et al.*, 2007; Lin *et al.*, 2007). Pitx2 is also a downstream target of canonical Wnt signaling (Kioussi *et al.*, 2002). *β*catenin and Pitx2 coregulate expression of the noncanonical/planar cell polarity pathway Wnt ligand, *Wnt11* (Lin *et al.*, 2007). The importance of the noncanonical/PCP pathway in OFT remodeling had been previously suggested by the observation of DORV in the spontaneous mouse mutant *Looptail* which is a mutation in *Vangl2* (Henderson *et al.*, 2006), the murine orthologue of Drosophila *Van Gogh* (not to be confused with the zebrafish *van gogh*, which is a mutation in *Tbx1*). *Wnt11* null mutants were found to have OFT remodeling defects (DORV, TGA) associated with decreased expression of Tgf*β*2 and a Tgf*β* target gene, *Hspg2* (encodes perlecan, these null mutants have TGA, (Costell *et al.*, 2002)). A pathway whereby Wnt11 signaling activates JNK to phosphorylate and activate the ATF2 histone acetyltransferase ATF2/CREB, which in turn regulates the Tgf*β*2 promoter was proposed based on *in vitro* studies in cultured cells and embryonic ChIP, which demonstrate recruitment of ATF2 to the Tgf*β*2 promoter in response to Wnt11. Furthermore, mutation of a conserved series of ATF2 sites in the Tgf*β*2 promoter abolishes expression of a reporter in the SHF/OFT of transgenic mice (Lin *et al.*, 2007).

4.2.2. Cell autonomous defects in the NC that disrupt OFT remodeling

Splotch spontaneous mouse mutants (five different alleles) have mutations of *Pax3* and have variable cardiovascular and other phenotypes. PTA is 100% penetrant in $sp^{2H};sp^{2H}$ homozygotes (Conway *et al.*, 1997; Koushik *et al.*, 2002; Snider *et al.*, 2007). *Pax3* encodes a paired-box-containing transcription factor expressed in premigratory and migrating NC, and in some postmigratory tissues, but not OFT cushion mesenchyme (Epstein *et al.*, 2000). There is controversy surrounding the basis of decreased numbers of cardiac NC en route to the heart and their delayed migration in sp^{2H}; sp^{2H} mutants (Conway *et al.*, 1997; Epstein *et al.*, 2000). Survival and proliferation of migratory NC is not abnormal (Conway *et al.*, 2000). Transplantation studies between wild type and sp^{2H};sp^{2H} mutants suggest that abnormal migratory behavior is only present when both the NC and the environment are from the mutant background (Chan *et al.*, 2004). The main cell intrinsic defect appears to be expansion of premigratory NC (Conway *et al.*, 2000). Indeed, expression of the canonical Wnt ligand *Wnt1*, a known proliferative factor for premigatory crest, is abnormal in sp^{2H};sp^{2H} mutants (Conway *et al.*, 2000). However, altered ECM production in the pharynx and OFT may explain inhibition of wild type NC migration in the sp^{2H};sp^{2H} mutant environment (Henderson *et al.*, 1997). Because sp^{2H};sp^{2H} mutants do not have cranial NC defects, these findings indicate that the subset of NC destined for the heart is already specified in the neural tube, and raise the as yet unrealized possibility that these cells can be isolated for experimental and therapeutic purposes.

Another interesting spontaneous mouse mutant is *Patch* (*Ph*), a deletion of the coding region of the Pdgf receptor α subunit (Stephenson *et al.*, 1991). Most of these mutants have PTA (Morrison-Graham *et al.*, 1992) and conditional ablation has confirmed that this is due to loss of receptor function in the NC (Soriano, 1997; Tallquist and Soriano, 2003). Simultaneous conditional ablation of both PdgfR α and β increases the penetrance of PTA to 100% (Richarte *et al.*, 2007). Although Pdgf signaling has diverse cell-type specific effects, it is interesting to note that in an early description of the Patch phenotype, the authors noted abnormal "matrix granules" in the ECM (Morrison-Graham *et al.*, 1992). Subsequent investigation revealed that expression and activity of the metalloprotease MMP-2 and its activator MT–MMP are dramatically decreased in OFT tissue from *Ph* homozygotes and that explanted mesenchymal cells have decreased migratory capacity (Robbins *et al.*, 1999). An enormous literature exists on the effects of Pdgf signaling (in some cases in partnership with Tgfβ, which regulates *mmp2* expression in some cell types) to modulate ECM production and remodeling and promote migration of diverse cell types including VSMCs and endothelial cells. The source of MMP2 and MT–MMP in the OFT, while likely to be NC, has not been determined, nor have potential effects of altered ECM remodeling on endothelial-derived mesenchymal cells in this model. It is known that type IV collagen remodeling by MMP-2 and MT–MMP along endothelial basement membrane is required for initiation of endothelial EMT and postinvasion migration in other species (Song *et al.*, 2000).

Plexins are receptors for the semaphorin family of transmembrane and secreted proteins best studied with regard to their axonal guidance properties. More recently, the roles of Sema/Plexin pathways in migration and morphology of other cell types during development have been appreciated (Tran *et al.*, 2007). There are numerous co-receptors in vertebrates (neuropilins, ErbB2, VEGFR2, and others), which increases the complexity and tissue-specificity of this signaling system. Cardiac NC and NC-derived cells in the distal OFT express *PlexinA2* while OFT myocardium expresses the secreted Sema3c ligand. *Sema3c* null mutants have a 75% incidence of PTA and greater than 90% have aortic arch interruptions; markedly less PlexinA2 staining is detected in the distal OFT of *Sema3c* mutants. Neuropilin1 is a coreceptor for PlexinA2 (and for VEGF); *Nrp1* mutants also have a high incidence of PTA (Kawasaki *et al.*, 1999). Ablation of *Gata6* in NC or in vascular smooth muscle cells (VSMCs, most of which are NC-derived) causes PTA and aortic arch interruption. However, smooth muscle differentiation of the NC is normal in these mutants. Notably, *Sema3c* is expressed in OFT VSMCs and is downregulated in *Gata6* conditional mutants while myocardial expression was preserved (Lepore *et al.*, 2006). This is consistent with observations that a dominant-negative Gata6 decreases *Sema3c* expression in VSMCs (Lepore *et al.*, 2005) and regulates a portion of the *Sema3c* promoter containing conserved Gata6 binding sites (Lepore *et al.*, 2006). It is interesting to note that another Gata6 target is endothelin-1; this ligand is expressed in the epithelia of the pharyngeal arches and its receptor in NC; null mutants of either type have aortic arch and OFT defects (Kurihara *et al.*, 1995; Yanagisawa *et al.*, 1998).

4.2.3. Mutations in genes that regulate laterality resulting in OFT defects

A link between genes that affect LR axis determination (but are also expressed later during heart development; *Foxc1*, *Shh*, *Pitx2*, *Zic3*, *Fgf8*, see above and Section 3.8) and OFT development is becoming increasingly apparent (Schneider and Brueckner, 2000). It has been suggested that OFT malformations may be the sole manifestation of disrupted laterality in some patients (Bamforth *et al.*, 2004). For example, Cited2 is a hypoxia-inducible transcriptional cofactor; it has both activator and repressor activities. It competitively inhibits the interaction between HIF1α and Ep300 and thereby inhibits induction of hypoxia-responsive genes. It also coactivates Tfap2 and numerous other factors by linking them to Ep300 and Crebbp. Tfap2α is expressed in NC and null mouse mutants have DORV and PTA (Brewer *et al.*, 2002). Mutations in both *EP300* and *CREBBP2* have been detected in patients with Rubinstein–Taybi syndrome, which frequently includes OFT and other cardiac defects (Stevens and Bhakta, 1995). Char syndrome has been linked to mutations in *TFAP2*. Defects in *Cited2* null mice are embryonic lethal; DORV, PTA, VSDs are found. These defects were initially attributed to abnormal Tfap2 activity and possibly to disruption of hypoxia-mediated transcription (Bamforth *et al.*, 2001; Yin *et al.*,

2002). Local hypoxic signals are known to be important for OFT remodeling as well as regulation of VEGF signaling (Sugishita *et al.*, 2004). Additional investigations of *Cited2* mutants on an inbred background revealed evidence of laterality defects and altered LR gene expression. Tfap2 binding sites were identified in the *Pitx2* promoter and ChIP revealed binding of Tfap2 and Cited2 to this region.

In some genetic backgrounds, *Shh* null mutants survive to birth and have a single OFT vessel. This appears to be complete pulmonary atresia rather than PTA, although the outflow valve characteristics have not been reported. The size of the OFT cushions is unequal and the spiraling process that aligns the OFT correctly with the ventricles appears to be disrupted (Washington Smoak *et al.*, 2005). Foxc1 and Foxc2 have been shown to directly regulate *Tbx1* downstream of Shh signaling in mice (Yamagishi *et al.*, 2003). In this case, conditional ablation of the obligate hedgehog receptor Smoothened (*Smo*) in either NC or SHF/OFT myocardium reproduce the phenotype (Goddeeris *et al.*, 2007). Whether local laterality alterations and/or abnormal *Tbx1* function contribute to the OFT phenotype in *Shh* pathway mutants remains to be determined.

4.3. Valve and septation mutants

Defects in septal formation and valvulogenesis are the most common type of congenital heart defect seen in humans (Hoffman and Kaplan, 2002). As discussed in Section 1, formation of the endocardial cushions is the first step in the morphogenesis of these structures. More than 100 genes have thus far been shown to play a role in cushion formation and subsequent endocardial EMT; extensive reviews are available (Norris *et al.*, 2008; Person *et al.*, 2005; Schroeder *et al.*, 2003).

As noted above, the cushion forming regions (AV canal and OFT) retain a primary myocardial phenotype which is dependent on regionalized expression of Tbx proteins in nonchamber versus chamber myocardium. Null and conditional mouse models have clearly established the importance of Bmp signaling in maintaining the regional identity of the AV canal, as well as for later signaling between the myocardium and endocardium in endocardial cushion development/EMT in both the AV canal and OFT. Ablation of *Bmp2* specifically in the AV myocardium profoundly decreases expression of *Tbx2* resulting in ectopic expression of chamber myocardial genes (Nppa and others) in this region. In these mutants, and in those with loss of Bmp receptor 1a (*Alk3*) function restricted to the endocardium, the cushions are severely hypoplastic with little cardiac jelly, and few mesenchymal cells by E9.5. Expression of numerous genes also known to play a role in endocardial EMT is abnormal (*Nfatc1*, *twist1*, *Msx1*, *Snai1*, *Notch*), and VE-cadherin persisted in the endothelium, consistent with the failure of these cells to undergo EMT. Expression of the Bmp/Tgfβ signaling antagonist *Smad6* was decreased (Ma *et al.*, 2005). Bmp5 and 7 also function during cushion morphogenesis (Solloway and Robertson, 1999). Interruption of Bmp (and possibly Tgf*β*) signaling in *Alk2* endothelial-specific mutants results in AV cushion and septal defects associated with similar alterations in gene expression as seen in *Bmp2/Bmpr1a* mutants (Wang *et al.*, 2005). Loss of *Tgfβ2* permits normal formation of the endocardial cushions but later morphogenesis is abnormal causing septal defects, malalignment of the AV canal (double inlet left ventricle), and defective valve remodeling (Sanford *et al.*, 1997). In the converse, ablation of *Smad6* function causes both AV and outflow valve hyperplasia (Galvin *et al.*, 2000).

VEGF and NFAT signaling are highly dynamic in the developing heart and both pathways participate in multiple stages of valve development. Different levels of VEGF production in subsets endothelial cells in the OFT and AV canal may distinguish which endothelial cells that are competent to undergo EMT (Armstrong and Bischoff, 2004; Lambrechts and Carmeliet, 2004). Endocardial EMT is disrupted in VEGF over- and under-expression mutants, and endothelial proliferation/differentiation is sensitive to the ratio of different VEGF isoforms: loss of one copy of the *VEGF*^{164} causes septal, valve, and OFT defects in mice (Stalmans *et al.*, 2003). NFATc/calcineurin signaling (see Section 3.7) is required to repress myocardial *VEGF* expression during early cushion morphogenesis/EMT. Valve defects reported in *NFATc2/3/4* mutants are associated with early ectopic VEGF production (Chang *et al.*, 2004). NFAT activity is required again during valve elongation.

Proteoglycans, collagens, and other fibrillar ECM components are present in cardiac jelly, although few restricted to the cushion-forming region have been described. Bmp and Tgf*β* signaling regulate, and are modulated by, ECM composition. For example, the proteoglycans versican, decorin, and biglycan influence the fibrous structure of collagen in the ECM and its physical interaction with cells, as well as bioavailability and cellular responses to Tgf*β*/Bmp (Macri *et al.*, 2007). The importance of cardiac jelly composition in the cushion regions is evident by a loss of function of the enzyme Has2, which synthesizes hyaluronan. This glycosaminoglycan has multiple ECM functions: structural, as a facilitator of intercellular interactions, and as a modulator of signaling factor availability/diffusion (Schroeder *et al.*, 2003). *Has2* null mice die early in development with heart failure and no visible cardiac jelly. A random transgene insertion in mice in the *Vcan* gene, which encodes versican (previously Cspg2, hdf) profoundly disrupts cushion formation in the AV canal and OFT (Mjaatvedt *et al.*, 1998). Versican is a proteoglycan that interacts with hyaluronan and other ECM components and with cell surface receptors to influence migration and other cell behaviors. Its ability to interact with other components of the ECM is regulated by ADAMTS (metalloproteases). The cleaved and intact forms of versican are localized differently at different stages of cushion morphogenesis (Kern *et al.*, 2006).

Direct (as opposed to receptor mediated) intercellular signaling also plays a role in cushion development. The gap junction protein Connexin 45 (*Cx45*, *Gjc1*) is the first to be expressed throughout the developing heart; its endocardial expression is limited to those cells overlying the AV canal and OFT cushions. Modulation of intracellular calcium levels that regulate calcineurin activity (and by extension, nuclear NFATc1) in the endocardium is achieved in part via connexon function. Gjc1$^{-/-}$ mutants die by E10.5 with heart failure, dyscoordinated focal contractions, and failure to propagate contraction from the atrium to the ventricles. Few mesenchymal cells are present in the cushion jelly at E9.5, although the quantity of jelly appears normal. NFATc1 immunostaining revealed that the antigen was detectable primarily in the endothelial cytoplasm in mutants, but was highly concentrated in the nucleus of controls (Kumai *et al.*, 2000).

So far, there are fewer models in which valve remodeling is disrupted. Because many of the molecules involved in early valve development may also function during remodeling, this situation will be remedied as more tools for conditional mutagenesis are developed that can be temporally regulated so that expression of these genes can be maintained during early the early cushion development, and inactivated later. Many of the genes that are expressed and thought to play a role in valve remodeling are expressed in developing cartilage and bone.

Periostin is an "osteoblast-specific" adhesion molecule expressed in the endocardial cushions post-EMT and in maturing valves (Kruzynska-Frejtag *et al.*, 2001). It regulates collagen fibrillogenesis and is produced in response to Bmp/Tgf*β*. Null mutants have septal defects, hypoplastic AV and OFT valves, aberrant expression of myocardial markers within the cushion mesenchyme, and histologic defects in valve laminar structure and ECM organization. Periostin may promote the differentiation of mesenchymal cells into fibroblasts (in part by repressing myocardial programs), which then secrete a family of components specifically required for valve remodeling post EMT (Norris *et al.*, 2008; Snider *et al.*, 2008).

Few studies have identified critical ECM proteins and their functions in remodeling valves. As discussed in Section 3.3, fibrillar ECM components play structural and signaling roles in vascular homeostasis. These functions are also important for valve remodeling. Loss of *fibulin-4* (now *Efemp2*, epidermal growth factor-containing fibulin-like ECM protein 2) function in mice causes aortic valve hyperplplasia in association with dysregulated Tgf*β* and Bmp signaling. These signaling alterations were sensitive to the gene dosage of *fibulin-4*. Altering the balance of different collagens in the ECM that interact with complex glycoproteins also disrupts valve remodeling (Hanada *et al.*, 2007). *Collagen Va1*$^{+/-}$; *Collagen XIa1*$^{-/-}$ mutant mice have overproduction of Collagen I and III in their hyperplastic valve leaflets. It is not known whether mechanical signals are the stimulus for the collagen type dysregulation or whether the altered ECM structure disrupts Tgfβ bioavailability or signal modulation (Lincoln *et al.*, 2006a).

Sox9 is expressed in developing cartilage and at multiple stages during valve formation and remodeling; many of its transcriptional targets are ECM proteins. Mutations in *SOX9* in humans cause autosomal sex reversal and campomelic dysplasia, cardiac septal and OFT defects have been reported (Foster *et al.*, 1994; Houston *et al.*, 1983; Wagner *et al.*, 1994). Sox9 is produced in post-EMT mesenchymal cells; germline loss of function in mice causes cushion hypoplasia and decreased endothelial transformation and subsequent proliferation of mesenchymal cells. Increased expression of NFATc1 is detected in the cushion endocardium and ectopically in post-EMT mesenchymal cells while *ErbB3* expression was aberrantly absent in the mesenchyme (Akiyama *et al.*, 2004). The Erb family of receptors is crucial for subsequent cushion remodeling into the thin fibrous mature valve (see below). Conditional ablation to examine the roles of Sox9 in later valve morphogenesis was recently reported (Lincoln *et al.*, 2007). Loss of function in endothelial cells at the time of their differentiation causes similar cushion defects to those seen in germline mutants. Inactivation in valve mesenchymal cells from E15.5 causes thickening of the valve leaflets due to increased cell number. The quantity of ECM is decreased and there is loss of cartilage link protein (CLP) and type II collagen immunoreactivity in the matrix; these are known transcriptional targets of Sox9. Ablation of a single copy of Sox9 in valve mesenchyme caused abnormal valve calcification and thickening in adult mice; *SOX9* levels are elevated in human myxomatous and calcifying valves.

Members of the EGF ligand family and ErbB tyrosine kinase receptors play a role in valve formation and remodeling. *Neuregulin* (ligand) and *ErbB3* (receptor) are expressed in cushion endothelium and mesenchymal cells, respectively. *Neuregulin*$^{-/-}$ mutants die at E10.5 with profoundly disrupted trabecular formation and myocyte differentiation-organization in the compact layer, and hypoplastic cushions (Meyer and Birchmeier, 1995). *ErbB3* mutants have more cardiac jelly production and minimal ventricular defects. Their cushions are thin and few cells have undergone EMT; they survive to E13.5 but the valves do not develop (Erickson *et al.*, 1997). ErbB2-4 mutants all have abnormal trabecular formation although of varying severities due to functional redundancy between some of the receptors with regard to ability to bind ligand. Hyaluronan has been shown to stimulate ErbB2/ErbB3-mediated transformation and invasion of endothelial cells in AV canal explants and direct stimulation of ErbB2/ErbB3 signaling rescues EMT defects in explants from *Has2*$^{-/-}$ mutants (Camenisch *et al.*, 2002). The EGF-family ligand heparin-binding EGF (HB–EGF) activates ErbB1 and ErbB4. Less than half of the expected *Hbegf* null mutants survive to weaning and all newborns have hyperplastic/stenotic AV and outflow valves; similar valve defects were seen in Egf receptor and Adam17 (Tace, a disintegrin metalloproteinase that processes and activates the EGF-family ligand TNFα) null mutants. These remodeling defects appear to be a result of persistent, elevated BMP signaling in valve mesenchyme during the fetal period, evidenced by decreased levels of the inhibitory Smad6 and elevated immunoreactivity for phospho-Smad1/5/8 in valve mesenchymal cells (Jackson *et al.*, 2003).

The importance of Phospholipase Cε signal transduction downstream of Ras and Rap GTPases (activated by numerous receptor tyrosine kinases, including Egf, Pdgf, Fgf receptors) during valve remodeling is revealed in PLCε deficient mice (Tadano *et al.*, 2005). These mice survive and develop massive dilated cardiomegaly due to semilunar valve regurgitation and stenosis. The aortic valve and pulmonary valves never remodel and progressively thicken by excessive cell proliferation beginning at E16.5 and continuing postnatally. In some cases, the aortic valve fuses. The AV valves are not affected. The authors noted similarities to the *Hb-Egf* and *Egf receptor* mutant phenotypes, and as seen in these mutants, there is evidence of increased Bmp signaling in PLCε mutants. However, it is not clear whether other signaling pathways that converge on Ras play a role in the phenotype as well. This is additional evidence that ongoing signaling is required to maintain valve homeostasis and structure throughout life.

5. Conclusion

Our concept of what constitutes a congenital cardiovascular defect is evolving as the group of genes required for normal cardiovascular development and function burgeons, and the diverse nature of the relevant proteins and processes is exposed. Although many of the morphogenetic defects in cardiovascular development in genetically altered mice are detectable *in utero* or at birth, more are being described that are initially silent (or undetected) but progress during postnatal life, mirroring what is being discovered in humans.

At present little is known about how rapidly changing embryonic and fetal hemodynamics and physiology modulate the regulatory programs that control cardiac morphogenesis and remodeling in normal and pathological situations. Similarly, while progress is being made toward repopulating the failing heart by recruiting resident stem cells or introducing stem cells via the circulation, significant barriers remain to assuring that such cells integrate into the existing myocardium in a beneficial manner rather than forming additional scar or an arrhythmogenic focus. However, the technologies to examine the interactions between hemodynamics, intercellular signaling and changes in gene expression, and characterize the electrophysiology of stem-cell derived myocardium *in vivo* are now becoming available in the mouse. Genetic models that reproducibly create specific cardiovascular defects and heart failure are imperative for these studies. Such investigations will reveal the complex feedback loops that integrate cardiac form with function during development, and will allow us to develop therapies that harness the beneficial features (and circumvent the maladaptive effects) of reactivating these loops to treat cardiac dysfunction related to structural, ischemic or infiltrative heart diseases.

As of 2007, nearly 10,000 mouse genes had been disrupted by gene targeting in mice or gene trapping in ES cells (http://grants.nih.gov/grants/guide/rfa-files/RFA-RR-06-005.html). Enormous resources are being invested by international consortia and by the NIH (Knockout Mouse Project, KOMP) (Collins *et al.*, 2007) based on the fundamental concept that a systematic, comprehensive library of null and conditional mutant ES cells and mice will allow us to dissect the functions of each gene in normal development and homeostasis and determine how mutations (or allelic variations) cause human malformations and disease, influence risk of disease, or predict therapeutic outcomes. Identifying and mutating relevant genes will no longer be a rate-limiting step. Advancing technologies in gene discovery, high resolution evaluation of immediate and long-term phenotypes, *in vivo* biochemistry, ascertainment of direct, tissue-specific transcriptional factor complexes and targets are rapidly being developed in the mouse. These approaches will improve our ability to discover direct regulatory and functional hierarchies and determine how a cell integrates, modulates and responds to the complex morphogen gradients, intercellular signals, and mechanical inputs it receives in its unique temporospatial environment to make a "correct" developmental decision. Answering these challenging but fundamental questions will allow us to realize the enormous potential of these libraries, and better apply our better understanding of normal and pathological heart development to improving human health.

ACKNOWLEDGMENTS

I am very grateful to Drs. Margaret Kirby, Bruce Gelb, and Kirk Thomas for critical reading and inciteful suggestions on the manuscript. Dr. Kathleen Sulik provided the beautiful electron micrographs of developing mouse hearts used in the figures and the reader is referred to http://www.med.unc.edu/embryo_images for a dynamic view of heart development in the mouse. My thanks to Diana Lim for her artistic rendering of early heart morphogenesis (Figs. 4.1–4.5), and to Dr. Yukio Saijoh who created Fig. 4.6. In spite of the length of this tome, it is a woefully incomplete synthesis of the literature on mouse mutants that cause congenital heart defects or heart dysfunction and how these models have led to new mechanistic insights. While I have cited the primary literature in many instances, it is impossible to recognize the efforts that have been expended by so many investigators in these fields. I hope that the focused reviews I have recommended are helpful for those seeking more comprehensive details and lists of mutants.

REFERENCES

Abu-Issa, R., Smyth, G., Smoak, I., Yamamura, K., and Meyers, E. N. (2002). Fgf8 is required for pharyngeal arch and cardiovascular development in the mouse. *Development* **129,** 4613–4625.

Afzelius, B. A. (1976). A human syndrome caused by immotile cilia. *Science* **193,** 317–319.

Ai, D., Fu, X., Wang, J., Lu, M. F., Chen, L., Baldini, A., Klein, W. H., and Martin, J. F. (2007). Canonical Wnt signaling functions in second heart field to promote right ventricular growth. *Proc. Natl. Acad Sci. USA* **104,** 9319–9324.

Akiyama, H., Chaboissier, M. C., Behringer, R. R., Rowitch, D. H., Schedl, A., Epstein, J. A., and de Crombrugghe, B. (2004). Essential role of Sox9 in the pathway that controls formation of cardiac valves and septa. *Proc. Natl. Acad Sci. USA* **101,** 6502–6507.

Aoki, Y., Niihori, T., Narumi, Y., Kure, S., and Matsubara, Y. (2008). The RAS/MAPK syndromes: Novel roles of the RAS pathway in human genetic disorders. *Hum. Mutat.*

Araki, T., Mohi, M. G., Ismat, F. A., Bronson, R. T., Williams, I. R., Kutok, J. L., Yang, W., Pao, L. I., Gilliland, D. G., Epstein, J. A., and Neel, B. G. (2004). Mouse model of Noonan syndrome reveals cell type- and gene dosage-dependent effects of Ptpn11 mutation. *Nat. Med.* **10,** 849–857.

Arber, S., Hunter, J. J., Ross, J., Jr., Hongo, M., Sansig, G., Borg, J., Perriard, J. C., Chien, K. R., and Caroni, P. (1997). MLP-deficient mice exhibit a disruption of cardiac cytoarchitectural organization, dilated cardiomyopathy, and heart failure. *Cell* **88,** 393–403.

Armstrong, E. J., and Bischoff, J. (2004). Heart valve development: Endothelial cell signaling and differentiation. *Circ. Res.* **95,** 459–470.

Arrandale, J. M., Gore-Willse, A., Rocks, S., Ren, J. M., Zhu, J., Davis, A., Livingston, J. N., and Rabin, D. U. (1996). Insulin signaling in mice expressing reduced levels of Syp. *J. Biol. Chem.* **271,** 21353–21358.

Arron, J. R., Winslow, M. M., Polleri, A., Chang, C. P., Wu, H., Gao, X., Neilson, J. R., Chen, L., Heit, J. J., Kim, S. K., Yamasaki, N., Miyakawa, T., *et al.* (2006). NFAT dysregulation by increased dosage of DSCR1 and DYRK1A on chromosome 21. *Nature* **441,** 595–600.

Arsenian, S., Weinhold, B., Oelgeschlager, M., Ruther, U., and Nordheim, A. (1998). Serum response factor is essential for mesoderm formation during mouse embryogenesis. *EMBO J.* **17,** 6289–6299.

Atkison, P., Joubert, G., Barron, A., Grant, D., Paradis, K., Seidman, E., Wall, W., Rosenberg, H., Howard, J., Williams, S., *et al.* (1995). Hypertrophic cardiomyopathy associated with tacrolimus in paediatric transplant patients. *Lancet* **345,** 894–896.

Bamforth, S. D., Braganca, J., Eloranta, J. J., Murdoch, J. N., Marques, F. I., Kranc, K. R., Farza, H., Henderson, D. J., Hurst, H. C., and Bhattacharya, S. (2001). Cardiac malformations, adrenal agenesis, neural crest defects and exencephaly in mice lacking Cited2, a new Tfap2 co-activator. *Nat. Genet.* **29,** 469–474.

Bamforth, S. D., Braganca, J., Farthing, C. R., Schneider, J. E., Broadbent, C., Michell, A. C., Clarke, K., Neubauer, S., Norris, D., Brown, N. A., Anderson, R. H., and Bhattacharya, S. (2004). Cited2 controls left-right patterning and heart development through a Nodal-Pitx2c pathway. *Nat. Genet.* **36,** 1189–1196.

Bartram, U., Molin, D. G., Wisse, L. J., Mohamad, A., Sanford, L. P., Doetschman, T., Speer, C. P., Poelmann, R. E., and Gittenberger-de Groot, A. C. (2001). Double-outlet right ventricle and overriding tricuspid valve reflect disturbances of looping, myocardialization, endocardial cushion differentiation, and apoptosis in TGF-beta(2)-knockout mice. *Circulation* **103,** 2745–2752.

Benson, D. W., Silberbach, G. M., Kavanaugh-McHugh, A., Cottrill, C., Zhang, Y., Riggs, S., Smalls, O., Johnson, M. C., Watson, M. S., Seidman, J. G., Seidman, C. E., Plowden, J., *et al.* (1999). Mutations in the cardiac transcription factor NKX2.5 affect diverse cardiac developmental pathways. *J. Clin. Invest.* **104,** 1567–1573.

Bi, W., Drake, C. J., and Schwarz, J. J. (1999). The transcription factor MEF2C-null mouse exhibits complex vascular malformations and reduced cardiac expression of angiopoietin 1 and VEGF. *Dev. Biol.* **211,** 255–267.

Biben, C., and Harvey, R. P. (1997). Homeodomain factor Nkx2-5 controls left/right asymmetric expression of bHLH gene eHand during murine heart development. *Genes Dev.* **11,** 1357–1369.

Biben, C., Weber, R., Kesteven, S., Stanley, E., McDonald, L., Elliott, D. A., Barnett, L., Koentgen, F., Robb, L., Feneley, M., and Harvey, R. P. (2000). Cardiac septal and valvular dysmorphogenesis in mice heterozygous for mutations in the homeobox gene Nkx2-5. *Circ. Res.* **87,** 888–895.

Bockman, D. E., Redmond, M. E., and Kirby, M. L. (1989). Alteration of early vascular development after ablation of cranial neural crest. *Anat. Rec.* **225,** 209–217.

Botto, L. D., May, K., Fernhoff, P. M., Correa, A., Coleman, K., Rasmussen, S. A., Merritt, R. K., O'Leary, L. A., Wong, L. Y., Elixson, E. M., Mahle, W. T., and Campbell, R. M. (2003). A population-based study of the 22q11.2 deletion: Phenotype, incidence, and contribution to major birth defects in the population. *Pediatrics* **112,** 101–107.

Breckenridge, R. A., Anderson, R. H., and Elliott, P. M. (2007). Isolated left ventricular non-compaction: The case for abnormal myocardial development. *Cardiol. Young* **17,** 124–129.

Brewer, S., Jiang, X., Donaldson, S., Williams, T., and Sucov, H. M. (2002). Requirement for AP-2alpha in cardiac outflow tract morphogenesis. *Mech. Dev.* **110,** 139–149.

Brummelkamp, T. R., Kortlever, R. M., Lingbeek, M., Trettel, F., MacDonald, M. E., van Lohuizen, M., and Bernards, R. (2002). TBX-3, the gene mutated in Ulnar-Mammary Syndrome, is a negative regulator of p19ARF and inhibits senescence. *J. Biol. Chem.* **277,** 6567–6572.

Bruneau, B. G. (2008). The developmental genetics of congenital heart disease. *Nature* **451,** 943–948.

Bruneau, B. G., Logan, M., Davis, N., Levi, T., Tabin, C. J., Seidman, J. G., and Seidman, C. E. (1999). Chamber-specific cardiac expression of Tbx5 and heart defects in Holt-Oram syndrome. *Dev. Biol.* **211,** 100–108.

Bruneau, B. G., Nemer, G., Schmitt, J. P., Charron, F., Robitaille, L., Caron, S., Conner, D. A., Gessler, M., Nemer, M., Seidman, C. E., and Seidman, J. G. (2001). A murine model of Holt-Oram syndrome defines roles of the T-box transcription factor Tbx5 in cardiogenesis and disease. *Cell* **106,** 709–721.

Buckingham, M., Meilhac, S., and Zaffran, S. (2005). Building the mammalian heart from two sources of myocardial cells. *Nat. Rev. Genet.* **6,** 826–835.

Burn, J., Takao, A., Wilson, D., Cross, I., Momma, K., Wadey, R., Scambler, P., and Goodship, J. (1993). Conotruncal anomaly face syndrome is associated with a deletion within chromosome 22q11. *J. Med. Genet.* **30,** 822–224.

Byrd, N. A., and Meyers, E. N. (2005). Loss of Gbx2 results in neural crest cell patterning and pharyngeal arch artery defects in the mouse embryo. *Dev. Biol.* **284,** 233–245.

Cai, C. L., Liang, X., Shi, Y., Chu, P. H., Pfaff, S. L., Chen, J., and Evans, S. (2003). Isl1 identifies a cardiac progenitor population that proliferates prior to differentiation and contributes a majority of cells to the heart. *Dev. Cell* **5,** 877–889.

Cai, C. L., Martin, J. C., Sun, Y., Cui, L., Wang, L., Ouyang, K., Yang, L., Bu, L., Liang, X., Zhang, X., Stallcup, W. B., Denton, C. P., *et al.* (2008). A myocardial lineage derives from Tbx18 epicardial cells. *Nature* **454,** 104–108.

Cai, C. L., Zhou, W., Yang, L., Bu, L., Qyang, Y., Zhang, X., Li, X., Rosenfeld, M. G., Chen, J., and Evans, S. (2005). T-box genes coordinate regional rates of proliferation and regional specification during cardiogenesis. *Development* **132,** 2475–2487.

Camenisch, T. D., Schroeder, J. A., Bradley, J., Klewer, S. E., and McDonald, J. A. (2002). Heart-valve mesenchyme formation is dependent on hyaluronan-augmented activation of ErbB2-ErbB3 receptors. *Nat. Med.* **8,** 850–855.

Campione, M., Steinbeisser, H., Schweickert, A., Deissler, K., van Bebber, F., Lowe, L. A., Nowotschin, S., Viebahn, C., Haffter, P., Kuehn, M. R., and Blum, M. (1999). The homeobox gene Pitx2: Mediator of asymmetric left-right signaling in vertebrate heart and gut looping. *Development* **126,** 1225–1234.

Carlson, H., Ota, S., Song, Y., Chen, Y., and Hurlin, P. J. (2002). Tbx3 impinges on the p53 pathway to suppress apoptosis, facilitate cell transformation and block myogenic differentiation. *Oncogene* **21,** 3827–3835.

Carrel, T., Purandare, S. M., Harrison, W., Elder, F., Fox, T., Casey, B., and Herman, G. E. (2000). The X-linked mouse mutation Bent tail is associated with a deletion of the Zic3 locus. *Hum. Mo.l Genet.* **9,** 1937–1942.

Chan, W. Y., Cheung, C. S., Yung, K. M., and Copp, A. J. (2004). Cardiac neural crest of the mouse embryo: Axial level of origin, migratory pathway and cell autonomy of the splotch (Sp2H) mutant effect. *Development* **131,** 3367–3379.

Chang, C. P., Neilson, J. R., Bayle, J. H., Gestwicki, J. E., Kuo, A., Stankunas, K., Graef, I. A., and Crabtree, G. R. (2004). A field of myocardial-endocardial NFAT signaling underlies heart valve morphogenesis. *Cell* **118,** 649–663.

Charron, J., Malynn, B. A., Fisher, P., Stewart, V., Jeannotte, L., Goff, S. P., Robertson, E. J., and Alt, F. W. (1992). Embryonic lethality in mice homozygous for a targeted disruption of the N-myc gene. *Genes Dev.* **6,** 2248–2257.

Chen, B., Bronson, R. T., Klaman, L. D., Hampton, T. G., Wang, J. F., Green, P. J., Magnuson, T., Douglas, P. S., Morgan, J. P., and Neel, B. G. (2000). Mice mutant for Egfr and Shp2 have defective cardiac semilunar valvulogenesis. *Nat. Genet.* **24,** 296–299.

Chen, H., Shi, S., Acosta, L., Li, W., Lu, J., Bao, S., Chen, Z., Yang, Z., Schneider, M. D., Chien, K. R., Conway, S. J., Yoder, M. C., *et al.* (2004). BMP10 is essential for maintaining cardiac growth during murine cardiogenesis. *Development* **131,** 2219–2231.

Ching, Y. H., Ghosh, T. K., Cross, S. J., Packham, E. A., Honeyman, L., Loughna, S., Robinson, T. E., Dearlove, A. M., Ribas, G., Bonser, A. J., Thomas, N. R., Scotter, A. J., *et al.* (2005). Mutation in myosin heavy chain 6 causes atrial septal defect. *Nat. Genet.* **37,** 423–428.

Christoffels, V. M., Burch, J. B., and Moorman, A. F. (2004a). Architectural plan for the heart: Early patterning and delineation of the chambers and the nodes. *Trends Cardiovasc. Med.* **14,** 301–307.

Christoffels, V. M., Hoogaars, W. M., Tessari, A., Clout, D. E., Moorman, A. F., and Campione, M. (2004b). T-box transcription factor Tbx2 represses differentiation and formation of the cardiac chambers. *Dev. Dyn.* **229,** 763–770.

Christoffels, V. M., Habets, P. E., Franco, D., Campione, M., de Jong, F., Lamers, W. H., Bao, Z. Z., Palmer, S., Biben, C., Harvey, R. P., and Moorman, A. F. (2000). Chamber formation and morphogenesis in the developing mammalian heart. *Dev. Biol.* **223,** 266–278.

Chung, M. W., Tsoutsman, T., and Semsarian, C. (2003). Hypertrophic cardiomyopathy: From gene defect to clinical disease. *Cell. Res.* **13,** 9–20.

Cohen, E. D., Wang, Z., Lepore, J. J., Lu, M. M., Taketo, M. M., Epstein, D. J., and Morrisey, E. E. (2007). Wnt/beta-catenin signaling promotes expansion of Isl-1-positive cardiac progenitor cells through regulation of FGF signaling. *J. Clin. Invest.* **117,** 1794–1804.

Collins, F. S., Rossant, J., and Wurst, W. (2007). A mouse for all reasons. *Cell* **128,** 9–13.

Conway, S. J., Bundy, J., Chen, J., Dickman, E., Rogers, R., and Will, B. M. (2000). Decreased neural crest stem cell expansion is responsible for the conotruncal heart defects within the splotch (Sp(2H))/Pax3 mouse mutant. *Cardiovasc. Res.* **47,** 314–328.

Conway, S. J., Henderson, D. J., and Copp, A. J. (1997a). Pax3 is required for cardiac neural crest migration in the mouse: Evidence from the splotch (Sp2H) mutant. *Development* **124,** 505–514.

Costell, M., Carmona, R., Gustafsson, E., Gonzalez-Iriarte, M., Fassler, R., and Munoz-Chapuli, R. (2002). Hyperplastic conotruncal endocardial cushions and transposition of great arteries in perlecan-null mice. *Circ. Res.* **91,** 158–164.

Davies, G. E., Howard, C. M., Farrer, M. J., Coleman, M. M., Bennett, L. B., Cullen, L. M., Wyse, R. K., Burn, J., Williamson, R., and Kessling, A. M. (1995). Genetic variation in the COL6A1 region is associated with congenital heart defects in trisomy 21 (Down's syndrome). *Ann. Hum. Genet.* **59,** 253–269.

de la Pompa, J. L., Timmerman, L. A., Takimoto, H., Yoshida, H., Elia, A. J., Samper, E., Potter, J., Wakeham, A., Marengere, L., Langille, B. L., Crabtree, G. R., and Mak, T. W. (1998). Role of the NF-ATc transcription factor in morphogenesis of cardiac valves and septum. *Nature* **392,** 182–186.

Delot, E. C., Bahamonde, M. E., Zhao, M., and Lyons, K. M. (2003). BMP signaling is required for septation of the outflow tract of the mammalian heart. *Development* **130,** 209–220.

Dietz, H. C., Cutting, G. R., Pyeritz, R. E., Maslen, C. L., Sakai, L. Y., Corson, G. M., Puffenberger, E. G., Hamosh, A., Nanthakumar, E. J., Curristin, S. M., *et al.* (1991). Marfan syndrome caused by a recurrent de novo missense mutation in the fibrillin gene. *Nature* **352,** 337–339.

Dietz, H. C., Loeys, B., Carta, L., and Ramirez, F. (2005). Recent progress towards a molecular understanding of Marfan syndrome. *Am. J. Med. Genet. C Semin. Med. Genet.* **139C,** 4–9.

DiGeorge, A. M. (1965). Discussion on a new concept of the cellular basis of immunology. *Journal of Pediatrics* **67,** 907.

Dobrowolski, R., Sasse, P., Schrickel, J. W., Watkins, M., Kim, J. S., Rackauskas, M., Troatz, C., Ghanem, A., Tiemann, K., Degen, J., Bukauskas, F. F., Civitelli, R., *et al.* (2008). The conditional connexin43G138R mouse mutant represents a new model of hereditary oculodentodigital dysplasia in humans. *Hum. Mol. Genet.* **17,** 539–554.

Dodou, E., Verzi, M. P., Anderson, J. P., Xu, S. M., and Black, B. L. (2004). Mef2c is a direct transcriptional target of ISL1 and GATA factors in the anterior heart field during mouse embryonic development. *Development* **131,** 3931–3942.

Edouard, T., Montagner, A., Dance, M., Conte, F., Yart, A., Parfait, B., Tauber, M., Salles, J. P., and Raynal, P. (2007). How do Shp2 mutations that oppositely influence its biochemical activity result in syndromes with overlapping symptoms? *Cell Mol. Life Sci.* **64,** 1585–1590.

Emanuel, B. S., McDonald-McGinn, D., Saitta, S. C., and Zackai, E. H. (2001). The 22q11.2 deletion syndrome. *Adv. Pediatr.* **48,** 39–73.

Epstein, J. A., Li, J., Lang, D., Chen, F., Brown, C. B., Jin, F., Lu, M. M., Thomas, M., Liu, E., Wessels, A., and Lo, C. W. (2000). Migration of cardiac neural crest cells in Splotch embryos. *Development* **127,** 1869–1878.

Erickson, S. L., O'Shea, K. S., Ghaboosi, N., Loverro, L., Frantz, G., Bauer, M., Lu, L. H., and Moore, M. W. (1997). ErbB3 is required for normal cerebellar and cardiac development: a comparison with ErbB2-and heregulin-deficient mice. *Development* **124,** 4999–5011.

Evans, W. H., De Vuyst, E., and Leybaert, L. (2006). The gap junction cellular internet: Connexin hemichannels enter the signalling limelight. *Biochem. J.* **397,** 1–14.

Firouzi, M., Ramanna, H., Kok, B., Jongsma, H. J., Koeleman, B. P., Doevendans, P. A., Groenewegen, W. A., and Hauer, R. N. (2004). Association of human connexin40 gene polymorphisms with atrial vulnerability as a risk factor for idiopathic atrial fibrillation. *Circ. Res.* **95,** e29–e33.

Firulli, A. B., McFadden, D. G., Lin, Q., Srivastava, D., and Olson, E. N. (1998). Heart and extra-embryonic mesodermal defects in mouse embryos lacking the bHLH transcription factor Hand1. *Nat. Genet.* **18,** 266–270.

Flenniken, A. M., Osborne, L. R., Anderson, N., Ciliberti, N., Fleming, C., Gittens, J. E., Gong, X. Q., Kelsey, L. B., Lounsbury, C., Moreno, L., Nieman, B. J., Peterson, K., *et al.* (2005). A Gja1 missense mutation in a mouse model of oculodentodigital dysplasia. *Development* **132,** 4375–4386.

Foster, J. W., Dominguez-Steglich, M. A., Guioli, S., Kowk, G., Weller, P. A., Stevanovic, M., Weissenbach, J., Mansour, S., Young, I. D., Goodfellow, P. N., *et al.* (1994). Campomelic dysplasia and autosomal sex reversal caused by mutations in an SRY-related gene. *Nature* **372,** 525–530.

Franco, D., Campione, M., Kelly, R., Zammit, P. S., Buckingham, M., Lamers, W. H., and Moorman, A. F. (2000). Multiple transcriptional domains, with distinct left and right components, in the atrial chambers of the developing heart. *Circ. Res.* **87,** 984–991.

Frank, D. U., Fotheringham, L. K., Brewer, J. A., Muglia, L. J., Tristani-Firouzi, M., Capecchi, M. R., and Moon, A. M. (2002). An Fgf8 mouse mutant phenocopies human 22q11 deletion syndrome. *Development* **129,** 4591–4603.

Fuerst, P. G., Koizumi, A., Masland, R. H., and Burgess, R. W. (2008). Neurite arborization and mosaic spacing in the mouse retina require DSCAM. *Nature* **451,** 470–474.

Gaio, U., Schweickert, A., Fischer, A., Garratt, A. N., Muller, T., Ozcelik, C., Lankes, W., Strehle, M., Britsch, S., Blum, M., and Birchmeier, C. (1999). A role of the cryptic gene in the correct establishment of the left-right axis. *Curr. Biol.* **9,** 1339–1342.

Galili, N., Baldwin, H. S., Lund, J., Reeves, R., Gong, W., Wang, Z., Roe, B. A., Emanuel, B. S., Nayak, S., Mickanin, C., Budarf, M. I., and Buck, C. A. (1997). A region of mouse chromosome 16 is syntenic to the DiGeorge, velocardiofacial syndrome minimal critical region. *Genome Res.* **7,** 399.

Galli, D., Dominguez, J. N., Zaffran, S., Munk, A., Brown, N. A., and Buckingham, M. E. (2008). Atrial myocardium derives from the posterior region of the second heart field, which acquires left-right identity as Pitx2c is expressed. *Development* **135,** 1157–1167.

Galvin, K. M., Donovan, M. J., Lynch, C. A., Meyer, R. I., Paul, R. J., Lorenz, J. N., Fairchild-Huntress, V., Dixon, K. L., Dunmore, J. H., Gimbrone, M. A., Jr., Falb, D., and Huszar, D. (2000). A role for smad6 in development and homeostasis of the cardiovascular system. *Nat. Genet.* **24,** 171–174.

Garg, V. (2006). Molecular genetics of aortic valve disease. *Curr Opin Cardiol* **21,** 180–184.

Garg, V., Kathiriya, I. S., Barnes, R., Schluterman, M. K., King, I. N., Butler, C. A., Rothrock, C. R., Eapen, R. S., Hirayama-Yamada, K., Joo, K., Matsuoka, R., Cohen, J. C., *et al.* (2003). GATA4 mutations cause human congenital heart defects and reveal an interaction with TBX5. *Nature* **424,** 443–447.

Garg, V., Muth, A. N., Ransom, J. F., Schluterman, M. K., Barnes, R., King, I. N., Grossfeld, P. D., and Srivastava, D. (2005). Mutations in NOTCH1 cause aortic valve disease. *Nature* **437,** 270–274.

Gebbia, M., Ferrero, G. B., Pilia, G., Bassi, M. T., Aylsworth, A., Penman-Splitt, M., Bird, L. M., Bamforth, J. S., Burn, J., Schlessinger, D., Nelson, D. L., and Casey, B. (1997). X-linked situs abnormalities result from mutations in ZIC3. *Nat. Genet.* **17,** 305–308.

Goddeeris, M. M., Schwartz, R., Klingensmith, J., and Meyers, E. N. (2007). Independent requirements for Hedgehog signaling by both the anterior heart field and neural crest cells for outflow tract development. *Development* **134,** 1593–1604.

Gottlieb, P. D., Pierce, S. A., Sims, R. J., Yamagishi, H., Weihe, E. K., Harriss, J. V., Maika, S. D., Kuziel, W. A., King, H. L., Olson, E. N., Nakagawa, O., and Srivastava, D. (2002). Bop encodes a muscle-restricted protein containing MYND and SET domains and is essential for cardiac differentiation and morphogenesis. *Nat. Genet.* **31,** 25–32.

Gottlieb, S., Driscoll, D. A., Punnett, H. H., Sellinger, B., Emanuel, B. S., and Budarf, M. L. (1998). Characterization of 10p deletions suggests two nonoverlapping regions contribute to the DiGeorge syndrome phenotype. *Am. J. Hum. Genet.* **62,** 495–498.

Gourdie, R. G., Mima, T., Thompson, R. P., and Mikawa, T. (1995). Terminal diversification of the myocyte lineage generates Purkinje fibers of the cardiac conduction system. *Development* **121,** 1423–1431.

Grady, R. M., Grange, R. W., Lau, K. S., Maimone, M. M., Nichol, M. C., Stull, J. T., and Sanes, J. R. (1999). Role for alpha-dystrobrevin in the pathogenesis of dystrophin-dependent muscular dystrophies. *Nat. Cell. Biol.* **1,** 215–220.

Grego-Bessa, J., Luna-Zurita, L., del Monte, G., Bolos, V., Melgar, P., Arandilla, A., Garratt, A. N., Zang, H., Mukouyama, Y. S., Chen, H., Shou, W., Ballestar, E., *et al.* (2007). Notch signaling is essential for ventricular chamber development. *Dev. Cell* **12,** 415–429.

Gregorio, C. C., Granzier, H., Sorimachi, H., and Labeit, S. (1999). Muscle assembly: A titanic achievement? *Curr. Opin. Cell Biol.* **11,** 18–25.

Gu, H., Smith, F. C., Taffet, S. M., and Delmar, M. (2003). High incidence of cardiac malformations in connexin40-deficient mice. *Circ. Res.* **93,** 201–206.

Gudbjartsson, D. F., Arnar, D. O., Helgadottir, A., Gretarsdottir, S., Holm, H., Sigurdsson, A., Jonasdottir, A., Baker, A., Thorleifsson, G., Kristjansson, K., Palsson, A., Blondal, T., *et al.* (2007). Variants conferring risk of atrial fibrillation on chromosome 4q25. *Nature* **448,** 353–357.

Guerra, S., Leri, A., Wang, X., Finato, N., Di Loreto, C., Beltrami, C. A., Kajstura, J., and Anversa, P. (1999). Myocyte death in the failing human heart is gender dependent. *Circ. Res.* **85,** 856–866.

Guris, D. L., Duester, G., Papaioannou, V. E., and Imamoto, A. (2006). Dose-dependent interaction of Tbx1 and Crkl and locally aberrant RA signaling in a model of del22q11 syndrome. *Dev. Cell* **10,** 81–92.

Guris, D. L., Fantes, J., Tara, D., Druker, B. J., and Imamoto, A. (2001). Mice lacking the homologue of the human 22q11.2 gene CRKL phenocopy neurocristopathies of DiGeorge syndrome. *Nat. Genet.* **27,** 293–298.

Habashi, J. P., Judge, D. P., Holm, T. M., Cohn, R. D., Loeys, B. L., Cooper, T. K., Myers, L., Klein, E. C., Liu, G., Calvi, C., Podowski, M., Neptune, E. R., *et al.* (2006). Losartan, an AT1 antagonist, prevents aortic aneurysm in a mouse model of Marfan syndrome. *Science* **312,** 117–121.

Hanada, K., Vermeij, M., Garinis, G. A., de Waard, M. C., Kunen, M. G., Myers, L., Maas, A., Duncker, D. J., Meijers, C., Dietz, H. C., Kanaar, R., and Essers, J. (2007). Perturbations of vascular homeostasis and aortic valve abnormalities in fibulin-4 deficient mice. *Circ. Res.* **100,** 738–746.

Hanafusa, H., Torii, S., Yasunaga, T., Matsumoto, K., and Nishida, E. (2004). Shp2, an SH2-containing protein-tyrosine phosphatase, positively regulates receptor tyrosine kinase signaling by dephosphorylating and inactivating the inhibitor Sprouty. *J. Biol. Chem.* **279,** 22992–22995.

Hauer, R. N., Groenewegen, W. A., Firouzi, M., Ramanna, H., and Jongsma, H. J. (2006). Cx40 polymorphism in human atrial fibrillation. *Adv. Cardiol.* **42,** 284–291.

Henderson, D. J., Phillips, H. M., and Chaudhry, B. (2006). Vang-like 2 and noncanonical Wnt signaling in outflow tract development. *Trends Cardiovasc. Med.* **16,** 38–45.

Henderson, D. J., Ybot-Gonzalez, P., and Copp, A. J. (1997). Over-expression of the chondroitin sulphate proteoglycan versican is associated with defective neural crest migration in the Pax3 mutant mouse (splotch). *Mech. Dev.* **69,** 39–51.

Herman, G. E., and El-Hodiri, H. M. (2002). The role of ZIC3 in vertebrate development. *Cytogenet Genome. Res.* **99,** 229–235.

Herve, J., Warnet, J. F., Jeaneau-Bellego, E., Portnoi, M. F., Taillemitte, J. L., and Herve, F. (1984). [Partial monosomy of the short arm of chromosome 10, associated with Rieger's syndrome and a Di George type partial immunodeficiency]. *Ann. Pediatr. (Paris)* **31,** 77–80.

High, F. A., and Epstein, J. A. (2008). The multifaceted role of Notch in cardiac development and disease. *Nat. Rev. Genet.* **9,** 49–61.

High, F. A., Zhang, M., Proweller, A., Tu, L., Parmacek, M. S., Pear, W. S., and Epstein, J. A. (2007). An essential role for Notch in neural crest during cardiovascular development and smooth muscle differentiation. *J. Clin. Invest.* **117,** 353–363.

Hiroi, Y., Kudoh, S., Monzen, K., Ikeda, Y., Yazaki, Y., Nagai, R., and Komuro, I. (2001). Tbx5 associates with Nkx2-5 and synergistically promotes cardiomyocyte differentiation. *Nat. Genet.* **28,** 276–280.

Hoffman, J. I., and Kaplan, S. (2002). The incidence of congenital heart disease. *J. Am. Coll. Cardiol.* **39,** 1890–1900.

Holt, M., and Oram, S. (1960). Familial heart disease with skeletal malformations. *Br. Heart J.* **22,** 236–242.

Hoshijima, M., and Chien, K. R. (2002). Mixed signals in heart failure: Cancer rules. *J. Clin. Invest.* **109,** 849–855.

Houston, C. S., Opitz, J. M., Spranger, J. W., Macpherson, R. I., Reed, M. H., Gilbert, E. F., Herrmann, J., and Schinzel, A. (1983). The campomelic syndrome: Review, report of 17 cases, and follow-up on the currently 17-year-old boy first reported by Maroteaux et al in 1971. *Am. J. Med. Genet.* **15,** 3–28.

Huang, G. Y., Cooper, E. S., Waldo, K., Kirby, M. L., Gilula, N. B., and Lo, C. W. (1998). Gap junction-mediated cell-cell communication modulates mouse neural crest migration. *J. Cell. Biol.* **143,** 1725–1734.

Huangfu, D., and Anderson, K. V. (2005). Cilia and Hedgehog responsiveness in the mouse. *Proc. Natl. Acad Sci. USA* **102,** 11325–11330.

Hummel, K., and Chapman, D. (1959). Visceral inversion and associated anomalies in the mouse. *Journal Heredity* **50,** 9–13.

Hunter, J. J., and Chien, K. R. (1999). Signaling pathways for cardiac hypertrophy and failure. *N. Engl. J. Med.* **341,** 1276–1283.

Hutson, M. R., and Kirby, M. L. (2003). Neural crest and cardiovascular development: A 20-year perspective. *Birth Defects Res C Embryo Today* **69,** 2–13.

Hutson, M. R., and Kirby, M. L. (2007). Model systems for the study of heart development and disease. *Cardiac neural crest and conotruncal malformations. Semin. Cell Dev. Biol.* **18,** 101–110.

Iida, K., Koseki, H., Kakinuma, H., Kato, N., Mizutani-Koseki, Y., Ohuchi, H., Yoshioka, H., Noji, S., Kawamura, K., Kataoka, Y., Ueno, F., Taniguchi, M., *et al.* (1997). Essential roles of the winged helix transcription factor MFH-1 in aortic arch patterning and skeletogenesis. *Development* **124,** 4627–4638.

Ilagan, R., Abu-Issa, R., Brown, D., Yang, Y. P., Jiao, K., Schwartz, R. J., Klingensmith, J., and Meyers, E. N. (2006). Fgf8 is required for anterior heart field development. *Development* **133,** 2435–2445.

Jackson, L. F., Qiu, T. H., Sunnarborg, S. W., Chang, A., Zhang, C., Patterson, C., and Lee, D. C. (2003). Defective valvulogenesis in HB-EGF and TACE-null mice is associated with aberrant BMP signaling. *EMBO J.* **22,** 2704–2716.

Jay, P. Y., Harris, B. S., Maguire, C. T., Buerger, A., Wakimoto, H., Tanaka, M., Kupershmidt, S., Roden, D. M., Schultheiss, T. M., O'Brien, T. X., Gourdie, R. G., Berul, C. I., *et al.* (2004). Nkx2-5 mutation causes anatomic hypoplasia of the cardiac conduction system. *J. Clin. Invest.* **113,** 1130–1137.

Jerome, L. A., and Papaioannou, V. E. (2001). DiGeorge syndrome phenotype in mice mutant for the T-box gene, Tbx1. *Nat. Genet.* **27,** 286–291.

Jia, Q., McDill, B. W., Li, S. Z., Deng, C., Chang, C. P., and Chen, F. (2007). Smad signaling in the neural crest regulates cardiac outflow tract remodeling through cell autonomous and non-cell autonomous effects. *Dev. Biol.* **311,** 172–184.

Joutel, A., Corpechot, C., Ducros, A., Vahedi, K., Chabriat, H., Mouton, P., Alamowitch, S., Domenga, V., Cecillion, M., Marechal, E., Maciazek, J., Vayssiere, C., *et al.* (1996). Notch3 mutations in CADASIL, a hereditary adult-onset condition causing stroke and dementia. *Nature* **383,** 707–710.

Judge, D. P., Biery, N. J., Keene, D. R., Geubtner, J., Myers, L., Huso, D. L., Sakai, L. Y., and Dietz, H. C. (2004). Evidence for a critical contribution of haploinsufficiency in the complex pathogenesis of Marfan syndrome. *J. Clin. Invest.* **114,** 172–181.

Judge, D. P., and Dietz, H. C. (2005). Marfan's syndrome. *Lancet* **366,** 1965–1976.

Judge, D. P., and Dietz, H. C. (2008). Therapy of Marfan syndrome. *Annu. Rev. Med.* **59,** 43–59.

Kaartinen, V., Dudas, M., Nagy, A., Sridurongrit, S., Lu, M. M., and Epstein, J. A. (2004). Cardiac outflow tract defects in mice lacking ALK2 in neural crest cells. *Development* **131,** 3481–3490.

Kajstura, J., Leri, A., Finato, N., Di Loreto, C., Beltrami, C. A., and Anversa, P. (1998). Myocyte proliferation in end-stage cardiac failure in humans. *Proc. Natl. Acad Sci. USA* **95,** 8801–8805.

Karnik, S. K., Brooke, B. S., Bayes-Genis, A., Sorensen, L., Wythe, J. D., Schwartz, R. S., Keating, M. T., and Li, D. Y. (2003). A critical role for elastin signaling in vascular morphogenesis and disease. *Development* **130,** 411–423.

Kass, S., MacRae, C., Graber, H. L., Sparks, E. A., McNamara, D., Boudoulas, H., Basson, C. T., Baker, P. B., 3rd, Cody, R. J., Fishman, M. C., *et al.* (1994). A gene defect that causes conduction system disease and dilated cardiomyopathy maps to chromosome 1p1-1q1. *Nat. Genet.* **7,** 546–551.

Kawasaki, T., Kitsukawa, T., Bekku, Y., Matsuda, Y., Sanbo, M., Yagi, T., and Fujisawa, H. (1999). A requirement for neuropilin-1 in embryonic vessel formation. *Development* **126,** 4895–4902.

Kelly, D., Goldberg, R., Wilson, D., Lindsay, E., Carey, A., Goodship, J., Burn, J., Cross, I., Shprintzen, R. J., and Scambler, P. J. (1993). Confirmation that the velo-cardio-facial syndrome is associated with haplo-insufficiency of genes at chromosome 22q11. *Am. J. Med. Genet.* **45,** 308–312.

Kelly, R. G., Brown, N. A., and Buckingham, M. E. (2001). The arterial pole of the mouse heart forms from Fgf10-expressing cells in pharyngeal mesoderm. *Dev. Cell.* **1,** 435–440.

Kern, C. B., Twal, W. O., Mjaatvedt, C. H., Fairey, S. E., Toole, B. P., Iruela-Arispe, M. L., and Argraves, W. S. (2006). Proteolytic cleavage of versican during cardiac cushion morphogenesis. *Dev. Dyn.* **235,** 2238–2247.

Kimber, W. L., Hsieh, P., Hirotsune, S., Yuva-Paylor, L., Sutherland, H. F., Chen, A., Ruiz-Lozano, P., Hoogstraten-Miller, S. L., Chien, K. R., Paylor, R., Scambler, P. J., and Wynshaw-Boris, A. (1999). Deletion of 150 kb in the minimal DiGeorge/velocardiofacial syndrome critical region in mouse. *Hum. Mol. Genet.* **8,** 2229–2237.

Kioussi, C., Briata, P., Baek, S. H., Rose, D. W., Hamblet, N. S., Herman, T., Ohgi, K. A., Lin, C., Gleiberman, A., Wang, J., Brault, V., Ruiz-Lozano, P., *et al.* (2002). Identification of a Wnt/Dvl/beta-Catenin –> Pitx2 pathway mediating cell-type-specific proliferation during development. *Cell* **111,** 673–685.

Kirby, M. L. (1988). Role of extracardiac factors in heart development. *Experientia* **44,** 944–951.

Kirby, M. L. (2006). Cardiac Development. Oxford, UK, Oxford University Press.

Kirby, M. L., and Waldo, K. L. (1990). Role of neural crest in congenital heart disease. *Circulation* **82,** 332–340.

Kirchhoff, S., Kim, J. S., Hagendorff, A., Thonnissen, E., Kruger, O., Lamers, W. H., and Willecke, K. (2000). Abnormal cardiac conduction and morphogenesis in connexin40 and connexin43 double-deficient mice. *Circ. Res.* **87,** 399–405.

Kirk, E. P., Sunde, M., Costa, M. W., Rankin, S. A., Wolstein, O., Castro, M. L., Butler, T. L., Hyun, C., Guo, G., Otway, R., Mackay, J. P., Waddell, L. B., *et al.* (2007). Mutations in cardiac T-box factor gene TBX20 are associated with diverse cardiac pathologies, including defects of septation and valvulogenesis and cardiomyopathy. *Am. J. Hum. Genet.* **81,** 280–291.

Kitamura, K., Miura, H., Miyagawa-Tomita, S., Yanazawa, M., Katoh-Fukui, Y., Suzuki, R., Ohuchi, H., Suehiro, A., Motegi, Y., Nakahara, Y., Kondo, S., and Yokoyama, M. (1999). Mouse Pitx2 deficiency leads to anomalies of the ventral body wall, heart, extra- and periocular mesoderm and right pulmonary isomerism. *Development* **126,** 5749–5758.

Klootwijk, R., Franke, B., van der Zee, C. E., de Boer, R. T., Wilms, W., Hol, F. A., and Mariman, E. C. (2000). A deletion encompassing Zic3 in bent tail, a mouse model for X-linked neural tube defects. *Hum. Mol. Genet.* **9,** 1615–1622.

Kokubo, H., Tomita-Miyagawa, S., Hamada, Y., and Saga, Y. (2007). Hesr1 and Hesr2 regulate atrioventricular boundary formation in the developing heart through the repression of Tbx2. *Development* **134,** 747–755.

Kosaki, R., Kosaki, K., Matsushima, K., Mitsui, N., Matsumoto, N., and Ohashi, H. (2005). Refining chromosomal region critical for Down syndrome-related heart defects with a case of cryptic 21q22.2 duplication. *Congenit. Anom. (Kyoto)* **45,** 62–64.

Koshiba-Takeuchi, K., Takeuchi, J. K., Arruda, E. P., Kathiriya, I. S., Mo, R., Hui, C. C., Srivastava, D., and Bruneau, B. G. (2006). Cooperative and antagonistic interactions between Sall4 and Tbx5 pattern the mouse limb and heart. *Nat. Genet.* **38,** 175–183.

Koushik, S. V., Chen, H., Wang, J., and Conway, S. J. (2002). Generation of a conditional loxP allele of the Pax3 transcription factor that enables selective deletion of the homeodomain. *Genesis* **32,** 114–117.

Koyabu, Y., Nakata, K., Mizugishi, K., Aruga, J., and Mikoshiba, K. (2001). Physical and functional interactions between Zic and Gli proteins. *J. Biol. Chem.* **276,** 6889–6892.

Krantz, I. D., Smith, R., Colliton, R. P., Tinkel, H., Zackai, E. H., Piccoli, D. A., Goldmuntz, E., and Spinner, N. B. (1999). Jagged1 mutations in patients ascertained with isolated congenital heart defects. *Am. J. Med. Genet.* **84,** 56–60.

Kruzynska-Frejtag, A., Machnicki, M., Rogers, R., Markwald, R. R., and Conway, S. J. (2001). Periostin (an osteoblast-specific factor) is expressed within the embryonic mouse heart during valve formation. *Mech. Dev.* **103,** 183–188.

Kumai, M., Nishii, K., Nakamura, K., Takeda, N., Suzuki, M., and Shibata, Y. (2000). Loss of connexin45 causes a cushion defect in early cardiogenesis. *Development* **127,** 3501–3512.

Kuo, C. T., Morrisey, E. E., Anandappa, R., Sigrist, K., Lu, M. M., Parmacek, M. S., Soudais, C., and Leiden, J. M. (1997). GATA4 transcription factor is required for ventral morphogenesis and heart tube formation. *Genes Dev.* **11,** 1048–1060.

Kupershmidt, S., Yang, T., Anderson, M. E., Wessels, A., Niswender, K. D., Magnuson, M. A., and Roden, D. M. (1999). Replacement by homologous recombination of the minK gene with lacZ reveals restriction of minK expression to the mouse cardiac conduction system. *Circ. Res.* **84,** 146–152.

Kurihara, Y., Kurihara, H., Oda, H., Maemura, K., Nagai, R., Ishikawa, T., and Yazaki, Y. (1995). Aortic arch malformations and ventricular septal defect in mice deficient in endothelin-1. *J. Clin. Invest.* **96,** 293–300.

Kwee, L., Baldwin, H. S., Shen, H. M., Stewart, C. L., Buck, C., Buck, C. A., and Labow, M. A. (1995). Defective development of the embryonic and extraembryonic circulatory systems in vascular cell adhesion molecule (VCAM-1) deficient mice. *Development* **121,** 489–503.

Kwon, C., Arnold, J., Hsiao, E. C., Taketo, M. M., Conklin, B. R., and Srivastava, D. (2007). Canonical Wnt signaling is a positive regulator of mammalian cardiac progenitors. *Proc. Natl. Acad. Sci. USA* **104,** 10894–10899.

Lambrechts, D., and Carmeliet, P. (2004). Sculpting heart valves with NFATc and VEGF. *Cell* **118,** 532–534.

Lange, A. W., Rothermel, B. A., and Yutzey, K. E. (2005). Restoration of DSCR1 to disomy in the trisomy 16 mouse model of Down syndrome does not correct cardiac or craniofacial development anomalies. *Dev. Dyn.* **233,** 954–963.

Laugwitz, K. L., Moretti, A., Lam, J., Gruber, P., Chen, Y., Woodard, S., Lin, L. Z., Cai, C. L., Lu, M. M., Reth, M., Platoshyn, O., Yuan, J. X., *et al.* (2005). Postnatal isl1+ cardioblasts enter fully differentiated cardiomyocyte lineages. *Nature* **433,** 647–653.

Lavine, K. J., Yu, K., White, A. C., Zhang, X., Smith, C., Partanen, J., and Ornitz, D. M. (2005). Endocardial and epicardial derived FGF signals regulate myocardial proliferation and differentiation *in vivo*. *Dev. Cell* **8,** 85–95.

Lepore, J. J., Cappola, T. P., Mericko, P. A., Morrisey, E. E., and Parmacek, M. S. (2005). GATA-6 regulates genes promoting synthetic functions in vascular smooth muscle cells. *Arterioscler. Thromb. Vasc. Biol.* **25,** 309–314.

Lepore, J. J., Mericko, P. A., Cheng, L., Lu, M. M., Morrisey, E. E., and Parmacek, M. S. (2006). GATA-6 regulates semaphorin 3C and is required in cardiac neural crest for cardiovascular morphogenesis. *J. Clin. Invest.* **116,** 929–939.

Li, D. Y., Brooke, B., Davis, E. C., Mecham, R. P., Sorensen, L. K., Boak, B. B., Eichwald, E., and Keating, M. T. (1998a). Elastin is an essential determinant of arterial morphogenesis. *Nature* **393,** 276–280.

Li, D. Y., Faury, G., Taylor, D. G., Davis, E. C., Boyle, W. A., Mecham, R. P., Stenzel, P., Boak, B., and Keating, M. T. (1998b). Novel arterial pathology in mice and humans hemizygous for elastin. *J. Clin. Invest.* **102,** 1783–1787.

Li, D. Y., Toland, A. E., Boak, B. B., Atkinson, D. L., Ensing, G. J., Morris, C. A., and Keating, M. T. (1997a). Elastin point mutations cause an obstructive vascular disease, supravalvular aortic stenosis. *Hum. Mol. Genet.* **6,** 1021–1028.

Li, L., Krantz, I. D., Deng, Y., Genin, A., Banta, A. B., Collins, C. C., Qi, M., Trask, B. J., Kuo, W. L., Cochran, J., Costa, T., Pierpont, M. E., *et al.* (1997b). Alagille syndrome is caused by mutations in human Jagged1, which encodes a ligand for Notch1. *Nat. Genet.* **16,** 243–251.

Li, Q. Y., Newbury-Ecob, R. A., Terrett, J. A., Wilson, D. I., Curtis, A. R., Yi, C. H., Gebuhr, T., Bullen, P. J., Robson, S. C., Strachan, T., Bonnet, D., Lyonnet, S., *et al.* (1997c). Holt-Oram syndrome is caused by mutations in TBX5, a member of the Brachyury (T) gene family. *Nat. Genet.* **15,** 21–29.

Li, W. E., Waldo, K., Linask, K. L., Chen, T., Wessels, A., Parmacek, M. S., Kirby, M. L., and Lo, C. W. (2002). An essential role for connexin43 gap junctions in mouse coronary artery development. *Development* **129,** 2031–2042.

Liao, J., Kochilas, L., Nowotschin, S., Arnold, J. S., Aggarwal, V. S., Epstein, J. A., Brown, M. C., Adams, J., and Morrow, B. E. (2004). Full spectrum of malformations in velo-cardio-facial syndrome/DiGeorge syndrome mouse models by altering Tbx1 dosage. *Hum. Mol. Genet.* **13,** 1577–1585.

Lin, C. R., Kioussi, C., O'Connell, S., Briata, P., Szeto, D., Liu, F., Izpisua-Belmonte, J. C., and Rosenfeld, M. G. (1999). Pitx2 regulates lung asymmetry, cardiac positioning and pituitary and tooth morphogenesis. *Nature* **401,** 279–282.

Lin, L., Cui, L., Zhou, W., Dufort, D., Zhang, X., Cai, C. L., Bu, L., Yang, L., Martin, J., Kemler, R., Rosenfeld, M. G., Chen, J., *et al.* (2007). Beta-catenin directly regulates Islet1 expression in cardiovascular progenitors and is required for multiple aspects of cardiogenesis. *Proc. Natl. Acad Sci. USA* **104,** 9313–9318.

Lin, Q., Schwarz, J., Bucana, C., and Olson, E. N. (1997). Control of mouse cardiac morphogenesis and myogenesis by transcription factor MEF2C. *Science* **276,** 1404–1407.

Lincoln, J., Alfieri, C. M., and Yutzey, K. E. (2004). Development of heart valve leaflets and supporting apparatus in chicken and mouse embryos. *Dev. Dyn.* **230,** 239–250.

Lincoln, J., Florer, J. B., Deutsch, G. H., Wenstrup, R. J., and Yutzey, K. E. (2006a). ColVa1 and ColXIa1 are required for myocardial morphogenesis and heart valve development. *Dev. Dyn.* **235,** 3295–3305.

Lincoln, J., Lange, A. W., and Yutzey, K. E. (2006). Hearts and bones: Shared regulatory mechanisms in heart valve, cartilage, tendon, and bone development. *Dev. Biol.* **294,** 292–302.

Lincoln, J., Kist, R., Scherer, G., and Yutzey, K. E. (2007). Sox9 is required for precursor cell expansion and extracellular matrix organization during mouse heart valve development. *Dev. Biol.* **305,** 120–132.

Lindsay, E. A. (2001). Chromosomal microdeletions: Dissecting del22q11 syndrome. *Nat. Rev. Genet.* **2,** 858–868.

Lindsay, E. A., Botta, A., Jurecic, V., Carattini-Rivera, S., Cheah, Y. C., Rosenblatt, H. M., Bradley, A., and Baldini, A. (1999). Congenital heart disease in mice deficient for the DiGeorge syndrome region. *Nature* **401,** 379–383.

Lindsay, E. A., Vitelli, F., Su, H., Morishima, M., Huynh, T., Pramparo, T., Jurecic, V., Ogunrinu, G., Sutherland, H. F., Scambler, P. J., Bradley, A., and Baldini, A. (2001). Tbx1 haploinsufficiency in the DiGeorge syndrome region causes aortic arch defects in mice. *Nature* **410,** 97–101.

Liu, C., Liu, W., Palie, J., Lu, M. F., Brown, N. A., and Martin, J. F. (2002). Pitx2c patterns anterior myocardium and aortic arch vessels and is required for local cell movement into atrioventricular cushions. *Development* **129,** 5081–5091.

Liu, W., Selever, J., Wang, D., Lu, M. F., Moses, K. A., Schwartz, R. J., and Martin, J. F. (2004). Bmp4 signaling is required for outflow-tract septation and branchial-arch artery remodeling. *Proc. Natl. Acad Sci. USA* **101,** 4489–4494.

Loddenkemper, T., Grote, K., Evers, S., Oelerich, M., and Stogbauer, F. (2002). Neurological manifestations of the oculodentodigital dysplasia syndrome. *J. Neurol.* **249,** 584–595.

Loeys, B. L., Chen, J., Neptune, E. R., Judge, D. P., Podowski, M., Holm, T., Meyers, J., Leitch, C. C., Katsanis, N., Sharifi, N., Xu, F. L., Myers, L. A., *et al.* (2005). A syndrome of altered cardiovascular, craniofacial, neurocognitive and skeletal development caused by mutations in TGFBR1 or TGFBR2. *Nat. Genet.* **37,** 275–281.

Logan, M., Pagan-Westphal, S. M., Smith, D. M., Paganessi, L., and Tabin, C. J. (1998). The transcription factor Pitx2 mediates situs-specific morphogenesis in response to left-right asymmetric signals. *Cell* **94,** 307–317.

Lowe, L. A., Supp, D. M., Sampath, K., Yokoyama, T., Wright, C. V., Potter, S. S., Overbeek, P., and Kuehn, M. R. (1996). Conserved left-right asymmetry of nodal expression and alterations in murine situs inversus. *Nature* **381,** 158–161.

Lu, M. F., Pressman, C., Dyer, R., Johnson, R. L., and Martin, J. F. (1999). Function of Rieger syndrome gene in left-right asymmetry and craniofacial development. *Nature* **401,** 276–278.

Ma, L., Lu, M. F., Schwartz, R. J., and Martin, J. F. (2005). Bmp2 is essential for cardiac cushion epithelial-mesenchymal transition and myocardial patterning. *Development* **132,** 5601–5611.

Maass, A., and Leinwand, L. A. (2000). Animal models of hypertrophic cardiomyopathy. *Curr. Opin. Cardiol.* **15,** 189–196.

Macatee, T. L., Hammond, B. P., Arenkiel, B. R., Francis, L., Frank, D. U., and Moon, A. M. (2003). Ablation of specific expression domains reveals discrete functions of ectoderm- and endoderm-derived FGF8 during cardiovascular and pharyngeal development. *Development* **130,** 6361–6374.

Macri, L., Silverstein, D., and Clark, R. A. (2007). Growth factor binding to the pericellular matrix and its importance in tissue engineering. *Adv. Drug Deliv. Rev.* **59,** 1366–1381.

Mahtab, E. A., Wijffels, M. C., Van Den Akker, N. M., Hahurij, N. D., Lie-Venema, H., Wisse, L. J., Deruiter, M. C., Uhrin, P., Zaujec, J., Binder, B. R., Schalij, M. J., Poelmann, R. E., *et al.* (2008). Cardiac malformations and myocardial abnormalities in podoplanin knockout mouse embryos: Correlation with abnormal epicardial development. *Dev. Dyn.* **237,** 847–857.

Manner, J., Perez-Pomares, J. M., Macias, D., and Munoz-Chapuli, R. (2001). The origin, formation and developmental significance of the epicardium: A review. *Cells Tissues Organs* **169,** 89–103.

Marszalek, J. R., Ruiz-Lozano, P., Roberts, E., Chien, K. R., and Goldstein, L. S. (1999). Situs inversus and embryonic ciliary morphogenesis defects in mouse mutants lacking the KIF3A subunit of kinesin-II. *Proc. Natl. Acad Sci. USA* **96,** 5043–5048.

Martin, J. F. (2007). Left right asymmetry, the pulmonary vein, and a-fib. *Circ. Res.* **101,** 853–855.

Maslen, C. L., Babcock, D., Robinson, S. W., Bean, L. J., Dooley, K. J., Willour, V. L., and Sherman, S. L. (2006). CRELD1 mutations contribute to the occurrence of cardiac atrioventricular septal defects in Down syndrome. *Am. J. Med. Genet. A* **140,** 2501–2505.

McCright, B., Lozier, J., and Gridley, T. (2002). A mouse model of Alagille syndrome: Notch2 as a genetic modifier of Jag1 haploinsufficiency. *Development* **129,** 1075–1082.

McDaniell, R., Warthen, D. M., Sanchez-Lara, P. A., Pai, A., Krantz, I. D., Piccoli, D. A., and Spinner, N. B. (2006). NOTCH2 mutations cause Alagille syndrome, a heterogeneous disorder of the notch signaling pathway. *Am. J. Hum. Genet.* **79,** 169–163.

McElhinney, D. B., Geiger, E., Blinder, J., Benson, D. W., and Goldmuntz, E. (2003). NKX2.5 mutations in patients with congenital heart disease. *J. Am. Coll. Cardiol.* **42,** 1650–1655.

McFadden, D. G., Barbosa, A. C., Richardson, J. A., Schneider, M. D., Srivastava, D., and Olson, E. N. (2005). The Hand1 and Hand2 transcription factors regulate expansion of the embryonic cardiac ventricles in a gene dosage-dependent manner. *Development* **132,** 189–201.

McFadden, D. G., Charite, J., Richardson, J. A., Srivastava, D., Firulli, A. B., and Olson, E. N. (2000). A GATA-dependent right ventricular enhancer controls dHAND transcription in the developing heart. *Development* **127,** 5331–5341.

Meilhac, S. M., Esner, M., Kelly, R. G., Nicolas, J. F., and Buckingham, M. E. (2004). The clonal origin of myocardial cells in different regions of the embryonic mouse heart. *Dev. Cell.* **6,** 685–698.

Meno, C., Shimono, A., Saijoh, Y., Yashiro, K., Mochida, K., Ohishi, S., Noji, S., Kondo, H., and Hamada, H. (1998). Lefty-1 is required for left-right determination as a regulator of lefty-2 and nodal. *Cell* **94,** 287–297.

Mercado-Pimentel, M. E., and Runyan, R. B. (2007). Multiple transforming growth factor-beta isoforms and receptors function during epithelial-mesenchymal cell transformation in the embryonic heart. *Cells Tissues Organs* **185,** 146–156.

Merki, E., Zamora, M., Raya, A., Kawakami, Y., Wang, J., Zhang, X., Burch, J., Kubalak, S. W., Kaliman, P., Belmonte, J. C., Chien, K. R., and Ruiz-Lozano, P. (2005). Epicardial retinoid X receptor alpha is required for myocardial growth and coronary artery formation. *Proc. Natl. Acad Sci. USA* **102,** 18455–18460.

Merscher, S., Funke, B., Epstein, J. A., Heyer, J., Puech, A., Lu, M. M., Xavier, R. J., Demay, M. B., Russell, R. G., Factor, S., Tokooya, K., Jore, B. S., *et al.* (2001). TBX1 is responsible for cardiovascular defects in velo-cardio-facial/DiGeorge syndrome. *Cell* **104,** 619–629.

Meyer, D., and Birchmeier, C. (1995). Multiple essential functions of neuregulin in development. *Nature* **378,** 386–390.

Meyers, E., and Martin, G. R. (1999). Differences in Left-Right axis pathways in the mouse and chick: Functions of FGF8 and SHH. *Science* **285,** 403–406.

Mikawa, T., and Gourdie, R. G. (1996). Pericardial mesoderm generates a population of coronary smooth muscle cells migrating into the heart along with ingrowth of the epicardial organ. *Dev. Biol.* **174,** 221–232.

Minamisawa, S., Hoshijima, M., Chu, G., Ward, C. A., Frank, K., Gu, Y., Martone, M. E., Wang, Y., Ross, J., Jr., Kranias, E. G., Giles, W. R., and Chien, K. R. (1999). Chronic phospholamban-sarcoplasmic reticulum calcium ATPase interaction is the critical calcium cycling defect in dilated cardiomyopathy. *Cell* **99,** 313–322.

Miyabara, S., Gropp, A., and Winking, H. (1982). Trisomy 16 in the mouse fetus associated with generalized edema and cardiovascular and urinary tract anomalies. *Teratology* **25,** 369–380.

Mizugishi, K., Aruga, J., Nakata, K., and Mikoshiba, K. (2001). Molecular properties of Zic proteins as transcriptional regulators and their relationship to GLI proteins. *J. Biol. Chem.* **276,** 2180–2188.

Mjaatvedt, C. H., Nakaoka, T., Moreno-Rodriguez, R., Norris, R. A., Kern, M. J., Eisenberg, C. A., Turner, D., and Markwald, R. R. (2001). The outflow tract of the heart is recruited from a novel heart-forming field. *Dev. Biol.* **238,** 97–109.

Mjaatvedt, C. H., Yamamura, H., Capehart, A. A., Turner, D., and Markwald, R. R. (1998). The Cspg2 gene, disrupted in the hdf mutant, is required for right cardiac chamber and endocardial cushion formation. *Dev. Biol.* **202,** 56–66.

Mohamed, S. A., Aherrahrou, Z., Liptau, H., Erasmi, A. W., Hagemann, C., Wrobel, S., Borzym, K., Schunkert, H., Sievers, H. H., and Erdmann, J. (2006). Novel missense mutations (p.T596M and p.P1797H) in NOTCH1 in patients with bicuspid aortic valve. *Biochem. Biophys. Res. Commun.* **345,** 1460–1465.

Molkentin, J. D., Lin, Q., Duncan, S. A., and Olson, E. N. (1997). Requirement of the transcription factor GATA4 for heart tube formation and ventral morphogenesis. *Genes Dev.* **11,** 1061–1072.

Molkentin, J. D., Lu, J. R., Antos, C. L., Markham, B., Richardson, J., Robbins, J., Grant, S. R., and Olson, E. N. (1998). A calcineurin-dependent transcriptional pathway for cardiac hypertrophy. *Cell* **93,** 215–228.

Mommersteeg, M. T., Brown, N. A., Prall, O. W., de Gier-de Vries, C., Harvey, R. P., Moorman, A. F., and Christoffels, V. M. (2007a). Pitx2c and Nkx2-5 are required for the formation and identity of the pulmonary myocardium. *Circ. Res.* **101,** 902–909.

Mommersteeg, M. T., Hoogaars, W. M., Prall, O. W., de Gier-de Vries, C., Wiese, C., Clout, D. E., Papaioannou, V. E., Brown, N. A., Harvey, R. P., Moorman, A. F., and Christoffels, V. M. (2007b). Molecular pathway for the localized formation of the sinoatrial node. *Circ. Res.* **100,** 354–362.

Moon, A. M., Guris, D. L., Seo, J. H., Li, L., Hammond, J., Talbot, A., and Imamoto, A. (2006). Crkl deficiency disrupts fgf8 signaling in a mouse model of 22q11 deletion syndromes. *Dev. Cell* **10,** 71–80.

Moore, C. S., and Roper, R. J. (2007). The power of comparative and developmental studies for mouse models of Down syndrome. *Mamm. Genome.* **18,** 431–443.

Moorman, A. F., de Jong, F., Denyn, M. M., and Lamers, W. H. (1998). Development of the cardiac conduction system. *Circ. Res.* **82,** 629–644.

Mori, A. D., and Bruneau, B. G. (2004). TBX5 mutations and congenital heart disease: Holt-Oram syndrome revealed. *Curr. Opin. Cardiol.* **19,** 211–215.

Moric-Janiszewska, E., and Markiewicz-Loskot, G. (2008). Genetic Heterogeneity of Left-ventricular Noncompaction Cardiomyopathy. *Clin. Cardiol.* **31,** 201–204.

Morita, H., Rehm, H. L., Menesses, A., McDonough, B., Roberts, A. E., Kucherlapati, R., Towbin, J. A., Seidman, J. G., and Seidman, C. E. (2008). Shared genetic causes of cardiac hypertrophy in children and adults. *N. Engl. J. Med.* **358,** 1899–1908.

Morrison-Graham, K., Schatteman, G. C., Bork, T., Bowen-Pope, D. F., and Weston, J. A. (1992). A PDGF receptor mutation in the mouse (Patch) perturbs the development of a non-neuronal subset of neural crest-derived cells. *Development* **115,** 133–142.

Moskowitz, I. P., Kim, J. B., Moore, M. L., Wolf, C. M., Peterson, M. A., Shendure, J., Nobrega, M. A., Yokota, Y., Berul, C., Izumo, S., Seidman, J. G., and Seidman, C. E. (2007). A molecular pathway including Id2, Tbx5, and Nkx2-5 required for cardiac conduction system development. *Cell* **129,** 1365–1376.

Moskowitz, I. P., Pizard, A., Patel, V. V., Bruneau, B. G., Kim, J. B., Kupershmidt, S., Roden, D., Berul, C. I., Seidman, C. E., and Seidman, J. G. (2004). The T-Box transcription factor Tbx5 is required for the patterning and maturation of the murine cardiac conduction system. *Development* **131,** 4107–4116.

Mounkes, L. C., Kozlov, S., Hernandez, L., Sullivan, T., and Stewart, C. L. (2003). A progeroid syndrome in mice is caused by defects in A-type lamins. *Nature* **423,** 298–301.

Murcia, N. S., Richards, W. G., Yoder, B. K., Mucenski, M. L., Dunlap, J. R., and Woychik, R. P. (2000). The Oak Ridge Polycystic Kidney (orpk) disease gene is required for left-right axis determination. *Development* **127,** 2347–2355.

Naito, T., Masaki, T., Nikolic-Paterson, D. J., Tanji, C., Yorioka, N., and Kohno, N. (2004). Angiotensin II induces thrombospondin-1 production in human mesangial cells via p38 MAPK and JNK: A mechanism for activation of latent TGF-beta1. *Am. J. Physiol. Renal Physiol.* **286,** F278–F287.

Nakamura, T., Colbert, M., Krenz, M., Molkentin, J. D., Hahn, H. S., Dorn, G. W., 2nd, and Robbins, J. (2007). Mediating ERK 1/2 signaling rescues congenital heart defects in a mouse model of Noonan syndrome. *J. Clin. Invest.* **117,** 2123–2132.

Neptune, E. R., Frischmeyer, P. A., Arking, D. E., Myers, L., Bunton, T. E., Gayraud, B., Ramirez, F., Sakai, L. Y., and Dietz, H. C. (2003). Dysregulation of TGF-beta activation contributes to pathogenesis in Marfan syndrome. *Nat. Genet.* **33,** 407–411.

Ng, C. M., Cheng, A., Myers, L. A., Martinez-Murillo, F., Jie, C., Bedja, D., Gabrielson, K. L., Hausladen, J. M., Mecham, R. P., Judge, D. P., and Dietz, H. C. (2004). TGF-beta-dependent pathogenesis of mitral valve prolapse in a mouse model of Marfan syndrome. *J. Clin. Invest.* **114,** 1586–1592.

Nilles, K. M., and London, B. (2007). Knockin animal models of inherited arrhythmogenic diseases: What have we learned from them? *J. Cardiovasc. Electrophysiol.* **18,** 1117–1125.

Niu, Z., Yu, W., Zhang, S. X., Barron, M., Belaguli, N. S., Schneider, M. D., Parmacek, M., Nordheim, A., and Schwartz, R. J. (2005). Conditional mutagenesis of the murine serum response factor gene blocks cardiogenesis and the transcription of downstream gene targets. *J. Biol. Chem.* **280,** 32531–32538.

Nonaka, S., Shiratori, H., Saijoh, Y., and Hamada, H. (2002). Determination of left-right patterning of the mouse embryo by artificial nodal flow. *Nature* **418,** 96–99.

Nonaka, S., Tanaka, Y., Okada, Y., Takeda, S., Harada, A., Kanai, Y., Kido, M., and Hirokawa, N. (1998). Randomization of left-right asymmetry due to loss of nodal cilia generating leftward flow of extraembryonic fluid in mice lacking KIF3B motor protein. *Cell* **95,** 829–837.

Norris, R. A., Moreno-Rodriguez, R. A., Sugi, Y., Hoffman, S., Amos, J., Hart, M. M., Potts, J. D., Goodwin, R. L., and Markwald, R. R. (2008). Periostin regulates atrioventricular valve maturation. *Dev. Biol.* **316,** 200–213.

Oda, T., Elkahloun, A. G., Pike, B. L., Okajima, K., Krantz, I. D., Genin, A., Piccoli, D. A., Meltzer, P. S., Spinner, N. B., Collins, F. S., and Chandrasekharappa, S. C. (1997). Mutations in the human Jagged1 gene are responsible for Alagille syndrome. *Nat. Genet.* **16,** 235–242.

Oh, S. P., and Li, E. (1997). The signaling pathway mediated by the type IIB activin receptor controls axial patterning and lateral asymmetry in the mouse. *Genes Dev.* **11,** 1812–1826.

Okada, Y., Nonaka, S., Tanaka, Y., Saijoh, Y., Hamada, H., and Hirokawa, N. (1999). Abnormal nodal flow precedes situs inversus in iv and inv mice. *Mol. Cell* **4,** 459–468.

Park, C., Lavine, K., Mishina, Y., Deng, C. X., Ornitz, D. M., and Choi, K. (2006a). Bone morphogenetic protein receptor 1A signaling is dispensable for hematopoietic development but essential for vessel and atrioventricular endocardial cushion formation. *Development* **133,** 3473–3484.

Park, E. J., Ogden, L. A., Talbot, A., Evans, S., Cai, C. L., Black, B. L., Frank, D. U., and Moon, A. M. (2006b). Required, tissue-specific roles for Fgf8 in outflow tract formation and remodeling. *Development* **133,** 2419–2433.

Park, E. J., Watanabe, Y., Smyth, G., Miyagawa-Tomita, S., Meyers, E., Klingensmith, J., Camenisch, T., Buckingham, M., and Moon, A. M. (2008). An FGF autocrine loop initiated in second heart field mesoderm regulates morphogenesis at the arterial pole of the heart. *Development* **135,** 3599–3610.

Parlakian, A., Tuil, D., Hamard, G., Tavernier, G., Hentzen, D., Concordet, J. P., Paulin, D., Li, Z., and Daegelen, D. (2004). Targeted inactivation of serum response factor in the developing heart results in myocardial defects and embryonic lethality. *Mol Cell Biol.* **24,** 5281–5289.

Pashmforoush, M., Lu, J. T., Chen, H., Amand, T. S., Kondo, R., Pradervand, S., Evans, S. M., Clark, B., Feramisco, J. R., Giles, W., Ho, S. Y., Benson, D. W., *et al.* (2004). Nkx2-5 pathways and congenital heart disease; loss of ventricular myocyte lineage specification leads to progressive cardiomyopathy and complete heart block. *Cell* **117,** 373–386.

Paznekas, W. A., Boyadjiev, S. A., Shapiro, R. E., Daniels, O., Wollnik, B., Keegan, C. E., Innis, J. W., Dinulos, M. B., Christian, C., Hannibal, M. C., and Jabs, E. W. (2003). Connexin 43 (GJA1) mutations cause the pleiotropic phenotype of oculodentodigital dysplasia. *Am. J. Hum. Genet.* **72,** 408–418.

Pereira, L., Andrikopoulos, K., Tian, J., Lee, S. Y., Keene, D. R., Ono, R., Reinhardt, D. P., Sakai, L. Y., Biery, N. J., Bunton, T., Dietz, H. C., and Ramirez, F. (1997). Targetting of the gene encoding fibrillin-1 recapitulates the vascular aspect of Marfan syndrome. *Nat. Genet.* **17,** 218–222.

Pereira, L., Lee, S. Y., Gayraud, B., Andrikopoulos, K., Shapiro, S. D., Bunton, T., Biery, N. J., Dietz, H. C., Sakai, L. Y., and Ramirez, F. (1999). Pathogenetic sequence for aneurysm revealed in mice underexpressing fibrillin-1. *Proc. Natl. Acad Sci. USA* **96,** 3819–3823.

Person, A. D., Klewer, S. E., and Runyan, R. B. (2005). Cell biology of cardiac cushion development. *Int. Rev. Cytol.* **243,** 287–335.

Peterkin, T., Gibson, A., Loose, M., and Patient, R. (2005). The roles of GATA-4, -5 and -6 in vertebrate heart development. *Semin. Cell. Dev. Biol.* **16,** 83–94.

Piedra, M. E., Icardo, J. M., Albajar, M., Rodriguez-Rey, J. C., and Ros, M. A. (1998). Pitx2 participates in the late phase of the pathway controlling left-right asymmetry. *Cell* **94,** 319–324.

Pinto, A. C., and Faria, M. (1985). Hypertrophic ileo-caecal tuberculosis with complete situs inversus (a case report). *J. Postgrad. Med.* **31,** 175–176.

Pizard, A., Burgon, P. G., Paul, D. L., Bruneau, B. G., Seidman, C. E., and Seidman, J. G. (2005). Connexin 40, a target of transcription factor Tbx5, patterns wrist, digits, and sternum. *Mol. Cell. Biol.* **25,** 5073–5083.

Prall, O. W., Elliott, D. A., and Harvey, R. P. (2002). Developmental paradigms in heart disease: Insights from tinman. *Ann. Med.* **34,** 148–156.
Prall, O. W., Menon, M. K., Solloway, M. J., Watanabe, Y., Zaffran, S., Bajolle, F., Biben, C., McBride, J. J., Robertson, B. R., Chaulet, H., Stennard, F. A., Wise, N., *et al.* (2007). An Nkx2-5/Bmp2/Smad1 negative feedback loop controls heart progenitor specification and proliferation. *Cell* **128,** 947–959.
Pu, W. T., Ishiwata, T., Juraszek, A. L., Ma, Q., and Izumo, S. (2004). GATA4 is a dosage-sensitive regulator of cardiac morphogenesis. *Dev. Biol.* **275,** 235–244.
Puech, A., Saint-Jore, B., Merscher, S., Russell, R. G., Cherif, D., Sirotkin, H., Xu, H., Factor, S., Kucherlapati, R., and Skoultchi, A. I. (2000). Normal cardiovascular development in mice deficient for 16 genes in 550 kb of the velocardiofacial/DiGeorge syndrome region. *Proc. Natl. Acad. Sci. USA* **97,** 10090–10095.
Purandare, S. M., Ware, S. M., Kwan, K. M., Gebbia, M., Bassi, M. T., Deng, J. M., Vogel, H., Behringer, R. R., Belmont, J. W., and Casey, B. (2002). A complex syndrome of left-right axis, central nervous system and axial skeleton defects in Zic3 mutant mice. *Development* **129,** 2293–2302.
Ramsdell, A. F. (2005). Left-right asymmetry and congenital cardiac defects: Getting to the heart of the matter in vertebrate left-right axis determination. *Dev. Biol.* **288,** 1–20.
Ranger, A. M., Grusby, M. J., Hodge, M. R., Gravallese, E. M., de la Brousse, F. C., Hoey, T., Mickanin, C., Baldwin, H. S., and Glimcher, L. H. (1998). The transcription factor NF-ATc is essential for cardiac valve formation. *Nature* **392,** 186–190.
Reamon-Buettner, S. M., Ciribilli, Y., Inga, A., and Borlak, J. (2008). A loss-of-function mutation in the binding domain of HAND1 predicts hypoplasia of the human hearts. *Hum. Mol. Genet.* **17,** 1397–1405.
Reaume, A. G., de Sousa, P. A., Kulkarni, S., Langille, B. L., Zhu, D., Davies, T. C., Juneja, S. C., Kidder, G. M., and Rossant, J. (1995). Cardiac malformation in neonatal mice lacking connexin43. *Science* **267,** 1831–1834.
Reeves, R. H. (2006). Down syndrome mouse models are looking up. *Trends Mol. Med.* **12,** 237–240.
Rene, C., Taulan, M., Iral, F., Doudement, J., L'Honore, A., Gerbon, C., Demaille, J., Claustres, M., and Romey, M. C. (2005). Binding of serum response factor to cystic fibrosis transmembrane conductance regulator CArG-like elements, as a new potential CFTR transcriptional regulation pathway. *Nucleic Acids Res.* **33,** 5271–5290.
Richarte, A. M., Mead, H. B., and Tallquist, M. D. (2007). Cooperation between the PDGF receptors in cardiac neural crest cell migration. *Dev. Biol.* **306,** 785–796.
Riley, P. R., Gertsenstein, M., Dawson, K., and Cross, J. C. (2000). Early exclusion of hand1-deficient cells from distinct regions of the left ventricular myocardium in chimeric mouse embryos. *Dev. Biol.* **227,** 156–168.
Risebro, C. A., and Riley, P. R. (2006). Formation of the ventricles. *Scientific World Journal* **6,** 1862–1880.
Robbins, J. R., McGuire, P. G., Wehrle-Haller, B., and Rogers, S. L. (1999). Diminished matrix metalloproteinase 2 (MMP-2) in ectomesenchyme-derived tissues of the Patch mutant mouse: Regulation of MMP-2 by PDGF and effects on mesenchymal cell migration. *Dev. Biol.* **212,** 255–263.
Robinson, H. B., Jr. (1975). DiGeorge's or the III-IV pharyngeal pouch syndrome: pathology and a theory of pathogenesis. *Perspect. Pediatr. Pathol.* **2,** 173–206.
Rodgers, L. S., Lalani, S., Runyan, R. B., and Camenisch, T. D. (2008). Differential growth and multicellular villi direct proepicardial translocation to the developing mouse heart. *Dev. Dyn.* **237,** 145–152.
Roessler, E., Ouspenskaia, M. V., Karkera, J. D., Velez, J. I., Kantipong, A., Lacbawan, F., Bowers, P., Belmont, J. W., Towbin, J. A., Goldmuntz, E., Feldman, B., and Muenke, M. (2008). Reduced NODAL Signaling Strength via Mutation of Several

Pathway Members Including FOXH1 Is Linked to Human Heart Defects and Holoprosencephaly. *Am. J. Hum. Genet.*

Rosenthal, N., and Harvey, R. (Eds.), Heart Development. Academic Press San Diego, California.

Rothermel, B., Vega, R. B., Yang, J., Wu, H., Bassel-Duby, R., and Williams, R. S. (2000). A protein encoded within the Down syndrome critical region is enriched in striated muscles and inhibits calcineurin signaling. *J. Biol. Chem.* **275,** 8719–8725.

Rothermel, B. A., McKinsey, T. A., Vega, R. B., Nicol, R. L., Mammen, P., Yang, J., Antos, C. L., Shelton, J. M., Bassel-Duby, R., Olson, E. N., and Williams, R. S. (2001). Myocyte-enriched calcineurin-interacting protein, MCIP1, inhibits cardiac hypertrophy in vivo. *Proc. Natl. Acad. Sci. USA* **98,** 3328–3333.

Ryan, A. K., Blumberg, B., Rodriguez-Esteban, C., Yonei-Tamura, S., Tamura, K., Tsukui, T., de la Pena, J., Sabbagh, W., Greenwald, J., Choe, S., Norris, D. P., Robertson, E. J., *et al.* (1998). Pitx2 determines left-right asymmetry of internal organs in vertebrates. *Nature* **394,** 545–551.

Saint-Jore, B., Puech, A., Heyer, J., Lin, Q., Raine, C., Kucherlapati, R., and Skoultchi, A. I. (1998). Goosecoid-like (Gscl), a candidate gene for velocardiofacial syndrome, is not essential for normal mouse development. *Hum. Mol. Genet.* **7,** 1841–1849.

Salama, G., and London, B. (2007). Mouse models of long QT syndrome. *J. Physiol.* **578,** 43–53.

Sanford, L. P., Ormsby, I., Gittenberger-de Groot, A. C., Sariola, H., Friedman, R., Boivin, G. P., Cardell, E. L., and Doetschman, T. (1997). TGFbeta2 knockout mice have multiple developmental defects that are non-overlapping with other TGFbeta knockout phenotypes. *Development* **124,** 2659–2670.

Sanna, B., Brandt, E. B., Kaiser, R. A., Pfluger, P., Witt, S. A., Kimball, T. R., van Rooij, E., De Windt, L. J., Rothenberg, M. E., Tschop, M. H., Benoit, S. C., and Molkentin, J. D. (2006). Modulatory calcineurin-interacting proteins 1 and 2 function as calcineurin facilitators *in vivo. Proc. Natl. Acad Sci. USA* **103,** 7327–7332.

Scambler, P. J. (2000). The 22q11 deletion syndromes. *Hum Mol Genet* **9,** 2421–2426.

Schlame, M., and Ren, M. (2006). Barth syndrome, a human disorder of cardiolipin metabolism. *FEBS Lett.* **580,** 5450–5455.

Schmitt, J. P., Kamisago, M., Asahi, M., Li, G. H., Ahmad, F., Mende, U., Kranias, E. G., MacLennan, D. H., Seidman, J. G., and Seidman, C. E. (2003). Dilated cardiomyopathy and heart failure caused by a mutation in phospholamban. *Science* **299,** 1410–1413.

Schneider, H., and Brueckner, M. (2000). Of mice and men: dissecting the genetic pathway that controls left-right asymmetry in mice and humans. *Am. J. Med. Genet.* **97,** 258–270.

Schott, J. J., Benson, D. W., Basson, C. T., Pease, W., Silberbach, G. M., Moak, J. P., Maron, B. J., Seidman, C. E., and Seidman, J. G. (1998). Congenital heart disease caused by mutations in the transcription factor NKX2-5. *Science* **281,** 108–111.

Schroeder, J. A., Jackson, L. F., Lee, D. C., and Camenisch, T. D. (2003). Form and function of developing heart valves: Coordination by extracellular matrix and growth factor signaling. *J. Mol. Med.* **81,** 392–403.

Semina, E. V., Reiter, R., Leysens, N. J., Alward, W. L., Small, K. W., Datson, N. A., Siegel-Bartelt, J., Bierke-Nelson, D., Bitoun, P., Zabel, B. U., Carey, J. C., and Murray, J. C. (1996). Cloning and characterization of a novel bicoid-related homeobox transcription factor gene, RIEG, involved in Rieger syndrome. *Nat Genet* **14,** 392–399.

Seo, S., and Kume, T. (2006). Forkhead transcription factors, Foxc1 and Foxc2, are required for the morphogenesis of the cardiac outflow tract. *Dev. Biol.* **296,** 421–436.

Sepulveda, J. L., Vlahopoulos, S., Iyer, D., Belaguli, N., and Schwartz, R. J. (2002). Combinatorial expression of GATA4, Nkx2-5, and serum response factor directs early cardiac gene activity. *J. Biol. Chem.* **277,** 25775–25782.

Severs, N. J., Dupont, E., Thomas, N., Kaba, R., Rothery, S., Jain, R., Sharpey, K., and Fry, C. H. (2006). Alterations in cardiac connexin expression in cardiomyopathies. *Adv. Cardiol.* **42,** 228–242.

Sheikh, F., Bang, M. L., Lange, S., and Chen, J. (2007). "Z"eroing in on the role of Cypher in striated muscle function, signaling, and human disease. *Trends Cardiovasc. Med.* **17,** 258–262.

Shiratori, H., and Hamada, H. (2006). The left-right axis in the mouse: From origin to morphology. *Development* **133,** 2095–2104.

Shou, W., Aghdasi, B., Armstrong, D. L., Guo, Q., Bao, S., Charng, M. J., Mathews, L. M., Schneider, M. D., Hamilton, S. L., and Matzuk, M. M. (1998). Cardiac defects and altered ryanodine receptor function in mice lacking FKBP12. *Nature* **391,** 489–492.

Shprintzen, R. J., Goldberg, R. B., Lewin, M. L., Sidoti, E. J., Berkman, M. D., Argamaso, R. V., and Young, D. (1978). A new syndrome involving cleft palate, cardiac anomalies, typical facies, and learning disabilities: velo-cardio-facial syndrome. *Cleft Palate J.* **15,** 56–62.

Simon, A. M., Goodenough, D. A., and Paul, D. L. (1998). Mice lacking connexin40 have cardiac conduction abnormalities characteristic of atrioventricular block and bundle branch block. *Curr. Biol.* **8,** 295–298.

Singh, M. K., Christoffels, V. M., Dias, J. M., Trowe, M. O., Petry, M., Schuster-Gossler, K., Burger, A., Ericson, J., and Kispert, A. (2005). Tbx20 is essential for cardiac chamber differentiation and repression of Tbx2. *Development* **132,** 2697–2707.

Smith, T. K., and Bader, D. M. (2007). Signals from both sides: Control of cardiac development by the endocardium and epicardium. *Semin. Cell Dev. Biol.* **18,** 84–89.

Snarr, B. S., O'Neal, J. L., Chintalapudi, M. R., Wirrig, E. E., Phelps, A. L., Kubalak, S. W., and Wessels, A. (2007). Isl1 expression at the venous pole identifies a novel role for the second heart field in cardiac development. *Circ. Res.* **101,** 971–974.

Snider, P., Hinton, R. B., Moreno-Rodriguez, R. A., Wang, J., Rogers, R., Lindsley, A., Li, F., Ingram, D. A., Menick, D., Field, L., Firulli, A. B., Molkentin, J. D., *et al.* (2008). Periostin is required for maturation and extracellular matrix stabilization of noncardiomyocyte lineages of the heart. *Circ. Res.* **102,** 752–760.

Snider, P., Olaopa, M., Firulli, A. B., and Conway, S. J. (2007a). Cardiovascular development and the colonizing cardiac neural crest lineage. *Scientific World Journal* **7,** 1090–1113.

Solloway, M. J., and Robertson, E. J. (1999). Early embryonic lethality in Bmp5;Bmp7 double mutant mice suggests functional redundancy within the 60A subgroup. *Development* **126,** 1753–1768.

Song, L., Fassler, R., Mishina, Y., Jiao, K., and Baldwin, H. S. (2007a). Essential functions of Alk3 during AV cushion morphogenesis in mouse embryonic hearts. *Dev. Biol.* **301,** 276–286.

Song, L., Yan, W., Chen, X., Deng, C. X., Wang, Q., and Jiao, K. (2007b). Myocardial smad4 is essential for cardiogenesis in mouse embryos. *Circ. Res.* **101,** 277–285.

Song, W., Jackson, K., and McGuire, P. G. (2000). Degradation of type IV collagen by matrix metalloproteinases is an important step in the epithelial-mesenchymal transformation of the endocardial cushions. *Dev. Biol.* **227,** 606–617.

Soriano, P. (1997). The PDGF alpha receptor is required for neural crest cell development and for normal patterning of the somites. *Development* **124,** p2691–p2700.

Srivastava, D., and Olson, E. N. (2000). A genetic blueprint for cardiac development. *Nature* **407,** 221–226.

Stalmans, I., Lambrechts, D., De Smet, F., Jansen, S., Wang, J., Maity, S., Kneer, P., von der Ohe, M., Swillen, A., Maes, C., Gewillig, M., Molin, D. G., *et al.* (2003). VEGF: A modifier of the del22q11 (DiGeorge) syndrome? *Nat. Med.* **9,** 173–182.

Stennard, F. A., Costa, M. W., Elliott, D. A., Rankin, S., Haast, S. J., Lai, D., McDonald, L. P., Niederreither, K., Dolle, P., Bruneau, B. G., Zorn, A. M., and

Harvey, R. P. (2003). Cardiac T-box factor Tbx20 directly interacts with Nkx2-5, GATA4, and GATA5 in regulation of gene expression in the developing heart. *Dev. Biol.* **262,** 206–224.
Stennard, F. A., Costa, M. W., Lai, D., Biben, C., Furtado, M. B., Solloway, M. J., McCulley, D. J., Leimena, C., Preis, J. I., Dunwoodie, S. L., Elliott, D. E., Prall, O. W., *et al.* (2005). Murine T-box transcription factor Tbx20 acts as a repressor during heart development, and is essential for adult heart integrity, function and adaptation. *Development* **132,** 2451–2462.
Stennard, F. A., and Harvey, R. P. (2005). T-box transcription factors and their roles in regulatory hierarchies in the developing heart. *Development* **132,** 4897–4910.
Stephenson, D. A., Mercola, M., Anderson, E., Wang, C. Y., Stiles, C. D., Bowen-Pope, D. F., and Chapman, V. M. (1991). Platelet-derived growth factor receptor alpha-subunit gene (Pdgfra) is deleted in the mouse patch (Ph) mutation. *Proc. Natl. Acad. Sci. USA* **88,** 6–10.
Stevens, C. A., and Bhakta, M. G. (1995). Cardiac abnormalities in the Rubinstein-Taybi syndrome. *Am. J. Med. Genet.* **59,** 346–348.
Stottmann, R. W., Choi, M., Mishina, Y., Meyers, E. N., and Klingensmith, J. (2004). BMP receptor IA is required in mammalian neural crest cells for development of the cardiac outflow tract and ventricular myocardium. *Development* **131,** 2205–2218.
Stuart, A. G., and Williams, A. (2007). Marfan's syndrome and the heart. *Arch. Dis. Child.* **92,** 351–356.
Sugishita, Y., Watanabe, M., and Fisher, S. A. (2004). The development of the embryonic outflow tract provides novel insights into cardiac differentiation and remodeling. *Trends Cardiovasc. Med.* **14,** 235–241.
Supp, D. M., Brueckner, M., Kuehn, M. R., Witte, D. P., Lowe, L. A., McGrath, J., Corrales, J., and Potter, S. S. (1999). Targeted deletion of the ATP binding domain of left-right dynein confirms its role in specifying development of left-right asymmetries. *Development* **126,** 5495–5504.
Supp, D. M., Witte, D. P., Potter, S. S., and Brueckner, M. (1997). Mutation of an axonemal dynein affects left-right asymmetry in inversus viscerum mice. *Nature* **389,** 963–966.
Sussman, M. A., Lim, H. W., Gude, N., Taigen, T., Olson, E. N., Robbins, J., Colbert, M. C., Gualberto, A., Wieczorek, D. F., and Molkentin, J. D. (1998). Prevention of cardiac hypertrophy in mice by calcineurin inhibition. *Science* **281,** 1690–1693.
Tadano, M., Edamatsu, H., Minamisawa, S., Yokoyama, U., Ishikawa, Y., Suzuki, N., Saito, H., Wu, D., Masago-Toda, M., Yamawaki-Kataoka, Y., Setsu, T., Terashima, T., *et al.* (2005). Congenital semilunar valvulogenesis defect in mice deficient in phospholipase C epsilon. *Mol. Cell. Biol.* **25,** 2191–2199.
Tallquist, M. D., and Soriano, P. (2003). Cell autonomous requirement for PDGFRalpha in populations of cranial and cardiac neural crest cells. *Development* **130,** 507–518.
Tanaka, M., Chen, Z., Bartunkova, S., Yamasaki, N., and Izumo, S. (1999). The cardiac homeobox gene Csx/Nkx2.5 lies genetically upstream of multiple genes essential for heart development. *Development* **126,** 1269–1280.
Tartaglia, M., Mehler, E. L., Goldberg, R., Zampino, G., Brunner, H. G., Kremer, H., van der Burgt, I., Crosby, A. H., Ion, A., Jeffery, S., Kalidas, K., Patton, M. A., *et al.* (2001). Mutations in PTPN11, encoding the protein tyrosine phosphatase SHP-2, cause Noonan syndrome. *Nat. Genet.* **29,** 465–468.
Tepper, S. J., Sheftell, F. D., and Bigal, M. E. (2007). The patent foramen ovale-migraine question. *Neurol. Sci.* **28**(Suppl 2), S118–S123.
Tevosian, S. G., Deconinck, A. E., Tanaka, M., Schinke, M., Litovsky, S. H., Izumo, S., Fujiwara, Y., and Orkin, S. H. (2000). FOG-2, a cofactor for GATA transcription

factors, is essential for heart morphogenesis and development of coronary vessels from epicardium. *Cell* **101,** 729–739.

Timmerman, L. A., Grego-Bessa, J., Raya, A., Bertran, E., Perez-Pomares, J. M., Diez, J., Aranda, S., Palomo, S., McCormick, F., Izpisua-Belmonte, J. C., and de la Pompa, J. L. (2004). Notch promotes epithelial-mesenchymal transition during cardiac development and oncogenic transformation. *Genes Dev.* **18,** 99–115.

Todorovic, V., Frendewey, D., Gutstein, D. E., Chen, Y., Freyer, L., Finnegan, E., Liu, F., Murphy, A., Valenzuela, D., Yancopoulos, G., and Rifkin, D. B. (2007). Long form of latent TGF-beta binding protein 1 (Ltbp1L) is essential for cardiac outflow tract septation and remodeling. *Development* **134,** 3723–3732.

Tran, T. S., Kolodkin, A. L., and Bharadwaj, R. (2007). Semaphorin regulation of cellular morphology. *Annu. Rev. Cell. Dev. Biol.* **23,** 263–292.

van Mierop, L. H., and Kutsche, (1986). Cardiovascular anomalies in DiGeorge syndrome and importance of neural crest asp a possible pathogenetic factor. *American Journal of Cardiology* **58,** 133–137.

Vikstrom, K. L., Seiler, S. H., Sohn, R. L., Strauss, M., Weiss, A., Welikson, R. E., and Leinwand, L. A. (1997). The vertebrate myosin heavy chain: genetics and assembly properties. *Cell Struct. Funct.* **22,** 123–129.

Vitelli, F., Taddei, I., Morishima, M., Meyers, E. N., Lindsay, E. A., and Baldini, A. (2002). A genetic link between Tbx1 and fibroblast growth factor signaling. *Development* **129,** 4605–4611.

von Both, I., Silvestri, C., Erdemir, T., Lickert, H., Walls, J. R., Henkelman, R. M., Rossant, J., Harvey, R. P., Attisano, L., and Wrana, J. L. (2004). Foxh1 is essential for development of the anterior heart field. *Dev. Cell* **7,** 331–345.

Vyas, P., and Crispino, J. D. (2007). Molecular insights into Down syndrome-associated leukemia. *Curr. Opin. Pediatr.* **19,** 9–14.

Wagner, T., Wirth, J., Meyer, J., Zabel, B., Held, M., Zimmer, J., Pasantes, J., Bricarelli, F. D., Keutel, J., Hustert, E., *et al.* (1994). Autosomal sex reversal and campomelic dysplasia are caused by mutations in and around the SRY-related gene SOX9. *Cell* **79,** 1111–1120.

Waldo, K. L., Kumiski, D. H., Wallis, K. T., Stadt, H. A., Hutson, M. R., Platt, D. H., and Kirby, M. L. (2001). Contoruncal myocardium arises from a secondary heart field. *Development* **128,** 3179–3188.

Wang, J., Nagy, A., Larsson, J., Dudas, M., Sucov, H. M., and Kaartinen, V. (2006). Defective ALK5 signaling in the neural crest leads to increased postmigratory neural crest cell apoptosis and severe outflow tract defects. *BMC Dev. Biol.* **6,** 51.

Wang, J., Sridurongrit, S., Dudas, M., Thomas, P., Nagy, A., Schneider, M. D., Epstein, J. A., and Kaartinen, V. (2005). Atrioventricular cushion transformation is mediated by ALK2 in the developing mouse heart. *Dev. Biol.* **286,** 299–310.

Ware, S. M. (2006). DNA mutation analysis in heterotaxy. *Methods Mol Med* **126,** 247–256.

Ware, S. M., Harutyunyan, K. G., and Belmont, J. W. (2006). Heart defects in X-linked heterotaxy: evidence for a genetic interaction of Zic3 with the nodal signaling pathway. *Dev. Dyn.* **235,** 1631–1637.

Washington Smoak, I., Byrd, N. A., Abu-Issa, R., Goddeeris, M. M., Anderson, R., Morris, J., Yamamura, K., Klingensmith, J., and Meyers, E. N. (2005). Sonic hedgehog is required for cardiac outflow tract and neural crest cell development. *Dev. Biol.* **283,** 357–372.

Watanabe, D., Saijoh, Y., Nonaka, S., Sasaki, G., Ikawa, Y., Yokoyama, T., and Hamada, H. (2003). The left-right determinant Inversin is a component of node monocilia and other 9+0 cilia. *Development* **130,** 1725–1734.

Watt, A. J., Battle, M. A., Li, J., and Duncan, S. A. (2004). GATA4 is essential for formation of the proepicardium and regulates cardiogenesis. *Proc. Natl. Acad Sci. USA* **101,** 12573–12578.

Wenstrup, R. J., Florer, J. B., Davidson, J. M., Phillips, C. L., Pfeiffer, B. J., Menezes, D. W., Chervoneva, I., and Birk, D. E. (2006). Murine model of the Ehlers-Danlos syndrome. col5a1 haploinsufficiency disrupts collagen fibril assembly at multiple stages. *J. Biol. Chem.* **281,** 12888–12895.

Williams, A. D., Mjaatvedt, C. H., and Moore, C. S. (2008). Characterization of the cardiac phenotype in neonatal Ts65Dn mice. *Dev. Dyn.* **237,** 426–435.

Winnier, G. E., Kume, T., Deng, K., Rogers, R., Bundy, J., Raines, C., Walter, M. A., Hogan, B. L., and Conway, S. J. (1999). Roles for the winged helix transcription factors MF1 and MFH1 in cardiovascular development revealed by nonallelic noncomplementation of null alleles. *Dev. Biol.* **213,** 418–431.

Wu, H., Lee, S. H., Gao, J., Liu, X., and Iruela-Arispe, M. L. (1999). Inactivation of erythropoietin leads to defects in cardiac morphogenesis. *Development* **126,** 3597–3605.

Xu, H., and Baldini, A. (2007). Genetic pathways to mammalian heart development: Recent progress from manipulation of the mouse genome. *Semin. Cell. Dev. Biol.* **18,** 77–83.

Xu, H., Morishima, M., Wylie, J. N., Schwartz, R. J., Bruneau, B. G., Lindsay, E. A., and Baldini, A. (2004). Tbx1 has a dual role in the morphogenesis of the cardiac outflow tract. *Development* **131,** 3217–3227.

Xue, Y., Gao, X., Lindsell, C. E., Norton, C. R., Chang, B., Hicks, C., Gendron-Maguire, M., Rand, E. B., Weinmaster, G., and Gridley, T. (1999). Embryonic lethality and vascular defects in mice lacking the Notch ligand Jagged1. *Hum. Mol. Genet.* **8,** 723–730.

Yagi, H., Furutani, Y., Hamada, H., Sasaki, T., Asakawa, S., Minoshima, S., Ichida, F., Joo, K., Kimura, M., Imamura, S., Kamatani, N., Momma, K., *et al.* (2003). Role of TBX1 in human del22q11.2 syndrome. *Lancet* **362,** 1366–1373.

Yamagishi, H., Maeda, J., Hu, T., McAnally, J., Conway, S. J., Kume, T., Meyers, E. N., Yamagishi, C., and Srivastava, D. (2003). Tbx1 is regulated by tissue-specific forkhead proteins through a common Sonic hedgehog-responsive enhancer. *Genes Dev.* **17,** 269–281.

Yanagisawa, H., Hammer, R. E., Richardson, J. A., Williams, S. C., Clouthier, D. E., and Yanagisawa, M. (1998). Role of Endothelin-1/Endothelin-A receptor-mediated signaling pathway in the aortic arch patterning in mice. *J. Clin. Invest.* **102,** 22–33.

Yang, J. T., Rayburn, H., and Hynes, R. O. (1994). Cell adhesion events mediated by a4 integrins are essential in placental and cardiac development. *Development.* in press.

Yin, Z., Haynie, J., Yang, X., Han, B., Kiatchoosakun, S., Restivo, J., Yuan, S., Prabhakar, N. R., Herrup, K., Conlon, R. A., Hoit, B. D., Watanabe, M., *et al.* (2002). The essential role of Cited2, a negative regulator for HIF-1alpha, in heart development and neurulation. *Proc. Natl. Acad Sci. USA* **99,** 10488–10493.

Yokoyama, T., Copeland, N. G., Jenkins, N. A., Montgomery, C. A., Elder, F. F., and Overbeek, P. A. (1993). Reversal of left-right asymmetry: a situs inversus mutation. *Science* **260,** 679–682.

Yoshioka, H., Meno, C., Koshiba, K., Sugihara, M., Itoh, H., Ishimaru, Y., Inoue, T., Ohuchi, H., Semina, E. V., Murray, J. C., Hamada, H., and Noji, S. (1998). Pitx2, a bicoid-type homeobox gene, is involved in a lefty-signaling pathway in determination of left-right asymmetry. *Cell* **94,** 299–305.

Zariwala, M. A., Knowles, M. R., and Omran, H. (2007). Genetic defects in ciliary structure and function. *Annu. Rev. Physiol.* **69,** 423–450.

Zeisberg, E. M., Ma, Q., Juraszek, A. L., Moses, K., Schwartz, R. J., Izumo, S., and Pu, W. T. (2005). Morphogenesis of the right ventricle requires myocardial expression of Gata4. *J. Clin. Invest.* **115,** 1522–1531.

Zhang, J., Lin, Y., Zhang, Y., Lan, Y., Lin, C., Moon, A. M., Schwartz, R. J., Martin, J. F., and Wang, F. (2008). Frs2{alpha}-deficiency in cardiac progenitors disrupts a subset of FGF signals required for outflow tract morphogenesis. *Development* **135,** 3611–3622.

Zhang, Z., and Baldini, A. (2008). *In vivo* response to high-resolution variation of Tbx1 mRNA dosage. *Hum. Mol. Genet.* **17,** 150–157.

Zhang, Z., Huynh, T., and Baldini, A. (2006). Mesodermal expression of Tbx1 is necessary and sufficient for pharyngeal arch and cardiac outflow tract development. *Development* **133,** 3587–3595.

Zhou, B., Ma, Q., Rajagopal, S., Wu, S. M., Domian, I., Rivera-Feliciano, J., Jiang, D., von Gise, A., Ikeda, S., Chien, K. R., and Pu, W. T. (2008). Epicardial progenitors contribute to the cardiomyocyte lineage in the developing heart. *Nature.*

Zhou, Q., Chu, P. H., Huang, C., Cheng, C. F., Martone, M. E., Knoll, G., Shelton, G. D., Evans, S., and Chen, J. (2001). Ablation of Cypher, a PDZ-LIM domain Z-line protein, causes a severe form of congenital myopathy. *J. Cell Biol.* **155,** 605–612.

Zweier, C., Sticht, H., Aydin-Yaylagul, I., Campbell, C. E., and Rauch, A. (2007). Human TBX1 missense mutations cause gain of function resulting in the same phenotype as 22q11.2 deletions. *Am. J. Hum. Genet.* **80,** 510–517.

CHAPTER FIVE

Modeling Ciliopathies: Primary Cilia in Development and Disease

Robyn J. Quinlan, Jonathan L. Tobin, *and* Philip L. Beales

Contents

Abstract

Primary (nonmotile) cilia are currently enjoying a renaissance in light of novel ascribed functions ranging from mechanosensory to signal transduction. Their importance for key developmental pathways such as Sonic Hedgehog (Shh) and Wnt is beginning to emerge. The function of nodal cilia, for example, is vital for

Molecular Medicine Unit, Institute of Child Health, University College London, London, WC1N1EH, United Kingdom

Current Topics in Developmental Biology, Volume 84
ISSN 0070-2153, DOI: 10.1016/S0070-2153(08)00605-4

breaking early embryonic symmetry, Shh signaling is important for tissue morphogenesis and successful Wnt signaling for organ growth and differentiation. When ciliary function is perturbed, photoreceptors may die, kidney tubules develop cysts, limb digits multiply and brains form improperly. The etiology of several uncommon disorders has recently been associated with cilia dysfunction. The causative genes are often similar and their cognate proteins certainly share cellular locations and/or pathways. Animal models of ciliary gene ablation such as *Ift88*, *Kif3a*, and *Bbs* have been invaluable for understanding the broad function of the cilium. Herein, we describe the wealth of information derived from the study of the ciliopathies and their animal models.

"This world, after all our science and sciences, is still a miracle; wonderful, inscrutable, magical and more, to whosoever will think of it." (Thomas Carlyle)

1. The Human Ciliopathies

The term "ciliopathy" has been coined to describe a class of rare human genetic diseases whose etiologies lie in defective cilia. In vertebrates, cilia are present in nearly all organs and cell types; however, amongst invertebrates they are confined to the sensory neurons that sense chemical stimuli, changes in environment and even vibration (Evans *et al.*, 2006). Cilia fall into two broad categories: those that are motile and those that are not. As the name would suggest, motile cilia and flagella (which share structural identity) are important for cell motility or for establishing fluid flow across their surface. For example, cells lining the respiratory tract, oviducts, epididymis, and ependymal surface of the brain display large clusters of motile cilia, which beat in concert to generate a wave-like motion. In contrast, the ubiquitous immotile or "primary" cilium, long regarded as a vestigial remnant of its motile cousin, is sessile in nature and present as a solitary extension of the plasma membrane. It is now accepted that primary cilia serve a broad sensory purpose in transducing extracellular information to the cell interior. Central to this is their role in several signal transduction pathways including the noncanonical Wnt/planar cell polarity (PCP) pathway and the Hedgehog (HH) pathway as well as for regulation of intracellular calcium concentration.

In the past few years, appreciation of ciliary function has led to reanalysis of a number of human syndromes. Because of a growing wealth of cellular and molecular evidence for the role of cilia in development and normal human physiology, a number of diseases that had previously been loosely associated due to shared clinical features (such as Bardet-Biedl and Alström syndromes) can now confidently be grouped together under the classification of a ciliopathy. This progress has also revealed some surprising new interrelationships amongst diseases (such as Bardet-Biedl and Meckel syndromes). What now links these individual disorders is the findings that the gene products, known to cause the disease if mutated, are localized to the cilia or their anchoring structure, the basal body, and play some role in its function. Consistent with their broad tissue and cellular distribution, it is now recognized that defects in cilia give rise to an eqaully broad but consistent range of phenotypes in mammals that are associated with organ-specific disorders such as polycystic kidney disease (PKD) as well as broad, pleiotropic syndromes. These syndromes and their main clinical features are summarized in Table 5.1. This table encapsulates most considerable clinical overlap between these syndromes; for instance most display renal cystic and hepatobiliary disease, some have laterality defects and retinal degeneration and several have polydactyly. It is also possible to further subdivide these disorders into those with skeletal involvement (JATD, OFD1, EVC) and those without (BBS, NPHP, MKS, JBTS, ALMS) as illustrated in Fig. 5.1. Although the individual disorders that make up the ciliopathies are thought of as rare genetic diseases, when viewed collectively, their prevalence rate could be as high as ~1 in 2000 [based on three common disease traits: renal cysts (1 in 500 adults), retinal degeneration (1 in 3000), and polydactyly (1 in 500)].

Although we depend on experimental models for our understanding of the biological processes underpinning disease, perhaps the most informative and diverse of all disease models are *Homo sapiens* themselves. Thus, this review will discuss the role of primary cilia in relation to human health and disease (with only limited reference to motile cilia defects) and with reference to relevant murine and other animal models that have informed and helped us to understand their etiology. The ciliopathies and the loci/genes that underlie their manifestation are summarized in Table 5.2.

2. Bardet-Biedl Syndrome

Bardet-Biedl syndrome (BBS) is a heterogeneous pleiotropic disorder inherited in a mainly recessive manner. Despite the growing list of syndromes now classified as a ciliopathy, it has been BBS that has led the quest to understand the pathomechanisms of this class of disease. Its clinical features include retinal degeneration, cognitive impairment, obesity, renal cystic disease, polydactyly, and occasionally *situs inversus*. It was these observations, in light of emerging evidence of the etiology of left–right (LR) asymmetry (discussed further below) that implicated cilia in the pathogenesis of BBS (Ansley *et al.*, 2003). Currently, twelve genes (*BBS1–12*) have been identified (summarized in Table 5.2) although, to date, no consistent phenotype–genotype correlations have been established (Ansley *et al.*, 2003; Badano *et al.*, 2003; Chiang *et al.*, 2004, 2006; Fan *et al.*, 2004;

Table 5.1 Common clinical features of the ciliopathies

Feature	BBS	MKS	JBTS	NPHP	SLSN	JATD	OFD1	EVC	ALMS	PKD
Renal cysts	✓	✓	✓	✓	✓	✓	✓		✓	✓
Hepatobiliary disease	✓	✓	✓	✓	✓	✓	✓		✓	✓
Laterality defects	✓	✓		✓		✓				
Polydactyly	✓	✓	✓			✓	✓	✓		
Agenesis of corpus callosum	✓	✓	✓			✓	✓		✓	
Cognitive impairment	✓	✓	✓			✓	✓	✓		
Retinal degeneration	✓	✓	✓		✓	✓			✓	
Posterior fossa defects/ encephalocoele	✓	✓	✓			✓		✓		
Skeletal bone defects						✓	✓	✓		
Obesity	✓								✓	

BBS: Bardet-Biedl syndrome; MKS: Meckel syndrome; JBTS: Joubert syndrome; NPHP: Nephrophthisis; SLSN: Senior-Løken syndrome; JATD: Jeune syndrome; OFD1: Oro-facial-digital syndrome type 1; EVC: Ellis van Creveld syndrome; ALMS: Alström Syndrome; PKD: Polycystic kidney disease

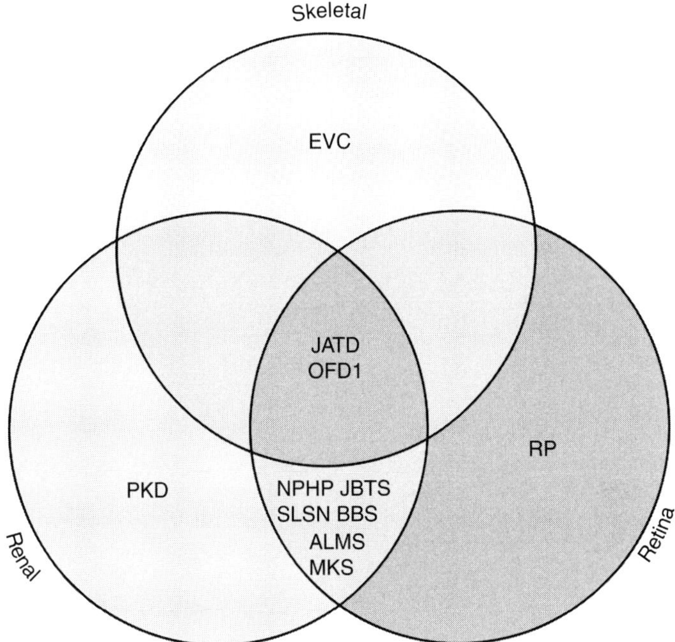

Figure 5.1 Clinical overlap between the ciliopathies. As outlined in Table 5.1, the ciliopathies can display a broad range of clinical features. Sorting the individual syndromes according to the occurrence or not of three features (renal cysts = green; retinal degeneration = blue; skeletal bone defects = yellow) helps to illustrate the commonalities and differences between these disorders. *Abbreviations*: BBS, Bardet-Biedl syndrome; MKS, Meckel syndrome; JBTS, Joubert syndrome; NPHP, nephrophthisis; SLSN, Senior-Løken syndrome; JATD, Jeune syndrome; OFD1, Orofacial–digital syndrome type 1; EVC, Ellis-van Creveld syndrome; ALMS, Alström syndrome; RP, retinitis pigmentosa; PKD, polycystic kidney disease.

Katsanis *et al.*, 2000; Li *et al.*, 2004; Mykytyn *et al.*, 2001, 2002; Nishimura *et al.*, 2001, 2005; Slavotinek *et al.*, 2000; Stoetzel *et al.*, 2006). The link with cilia was not at first apparent owing to a lack of sequence-based information for the early BBS genes. For example, the amino acid sequence of *BBS6/MKKS*, *BBS10*, and *BBS12* have strong homology with the type 2 group of chaperones, which initially suggested a role in protein folding but gave little insight into their actual, though still not clearly determined, function (Kim *et al.*, 2005; Stoetzel *et al.*, 2006, 2007). Of the remaining genes, only *BBS3/ARL6* (a member of the Ras superfamily of small GTP-binding proteins) and *BBS11/TRIM32* (an E3 ubiquitin ligase) encode proteins of known function. As indicated in Table 5.2, knockout and conditionally mutant mice have now been generated for a number of BBS genes. An interesting aspect of these mutants is that they do not fully phenocopy the cardinal features seen in BBS patients. For instance,

Table 5.2 Summary of ciliopathy genes associated with human syndromes

Human Syndrome	HUMAN Gene	*also known as*	Associated Ciliopathy	Locus and OMIM	Protein	Murine model	Key References
Alström Syndrome OMIM 203800	***ALMS1***		**Type 2 diabetes**	*2p13* **606844*	**Alström syndrome protein 1**	*Foz* *Alms null*	Cano *et al.* (2004, 2006), Caridi *et al.* (1998), Caspary *et al.* (2007), Chang *et al.* (2006), Porter *et al.* (1999)
Bardet-Biedl Syndrome OMIM 209900	***BBS1***			*11q13* **209901*	**Bardet-Biedl syndrome 1 protein**	*Bbs1 (M390R)* *Bbs1 null*	Atala *et al.* (1993), Bergmann *et al.* (2008), Mollet *et al.* (2005), Qian *et al.* (1997)
	BBS2			*16q21* **606151*	**Bardet-Biedl syndrome 2 protein**	*Bbs2 null*	Arts *et al.* (2007), Ibraghimov-Beskrovnaya *et al.* (2008), Qin *et al.* (2001)
	ARL6	*BBS3*		*3p12-q13* **608845*	**ADP-ribosylation factor-like protein 6** Bardet-Biedl syndrome 3 protein		Barr *et al.* (1999), Beales *et al.* (2007)
	BBS4			*15q22.3* **600374*	**Bardet-Biedl syndrome 4 protein**	Bbs4 null	Attanasio *et al.* (2007), Bergmann *et al.* (2008), Huangfu *et al.* (2003), Hughes *et al.* (1995), Ingham *et al.* (2008), Qin *et al.* (2001), Mollet *et al.* (2005), Rahmouni *et al.* (2008)

	BBS5			*2q31* *⋆603650*	**Bardet-Biedl syndrome 5 protein**		Baala *et al.* (2007)
	MKKS	*BBS6*	**McKusick-Kaufman Syndrome** OMIM 236700	*20p12* *⋆604896*	**Bardet-Biedl syndrome 6 protein** McKusick-Kaufman/Bardet-Biedl syndromes putative chaperonin	Bbs6 null Mkks null	Badano *et al.* (2003, 2006), Benzing *et al.* (2001), Huangfu *et al.* (2003), Mollet *et al.* (2005)
	BBS7			*4q27* *⋆607590*	**Bardet-Biedl syndrome 7 protein**		Beales *et al.* (1999)
	TTC8	*BBS8*		*14q32.11* *⋆608132*	**Bardet-Biedl syndrome 8 protein** Tetratricopeptide repeat protein 8		Alberts *et al.* (2008)
	PTHB1	*BBS9*		*7p14* *⋆607968*	**Bardet-Biedl syndrome 9 protein** Parathyroid hormone-responsive B1 gene protein		Arsov *et al.* (2006)
	BBS10			*12q* *⋆610148*	**Bardet-Biedl syndrome 10 protein**		Ansley *et al.* (2003), Rana *et al.* (2004)
	TRIM32	*BBS11*		*9q33.1* *⋆602290*	**Tripartite motif-containing** Bardet-Biedl syndrome 11 protein		Andersen *et al.* (2003)
	BBS12			*4q27* *⋆610683*	**Bardet-Biedl syndrome 12 protein**		Berbari *et al.* (2008)
Modifies BBS phenotypes	***CCDC28B***			*⋆610162*			Peters *et al.* (1999)

(continued)

Table 5.2 (*continued*)

Human Syndrome	HUMAN Gene	*also known as*	Associated Ciliopathy	Locus and OMIM	Protein	Murine model	Key References
					Coiled-coil domain-containing protein 28B		
Ellis van Creveld syndrome OMIM 225500	***EVC***			*14p16* **604831*	**Ellis-van Creveld syndrome protein**	*Evc null*	Robert *et al.* (2007), Roepman *et al.* (2005)
	LBN	*EVC2*		*14p16* **607261*	**Limbin**		Rohatgi *et al.* (2007), Ross *et al.* (2005)
Jeune syndrome OMIM 208500	***unknown***	*JATD1*		*15q13*	**n/a**		Ruiz-Perez *et al.* (2007)
Jeune syndrome OMIM 611263	***IFT80***	*JATD2*		*3q24-26* **611177*	**Intraflagellar transport protein 80 homolog**		Ferrante *et al.* (2006), Guay-Woodford *et al.* (1995)
Joubert syndrome OMIM 213300 **including Type B/ CORS and JSRD**	***unknown***	*JBTS1*		*9q34.3*			Ruiz-Perez *et al.* (2000)
	unknown	*JBTS2*		*11p12-q13.3* *608091			Galdzicka *et al.* (2002), Saar *et al.* (1999)
	AHI1	*JBTS3*		*6q23.3* **608894*	**Abelson helper integration site 1 protein homolog** Jouberin		Garcia-Garcia *et al.* (2005), Gattone *et al.* (1996), Gerdes *et al.* (2007), Germino and Somlo (1993), Giorgio *et al.* (2007), Giusto and Sciubba (2004), Saburi *et al.* (2008), Sage (1991)

	RPGRIP1L	*FTM* *JBTS7* *NPHP8* *MKS5*	**MKS Type 5** OMIM 611561 **NPHP Type 8** OMIM n/a	*16q12.2* *⋆610937*	**RPGR-interacting protein 1-like protein** Protein fantom	*Ft*	Eggenschwiler and Anderson (2007), Eggenschwiler *et al.* (2006), Eichers *et al.* (2006), Eley *et al.* (2008), Saalonen and Paavola (1998)
	ARL13B	*JBTS8*		3p12.3-q12.3 608922⋆	**ADP-ribosylation factor-like protein 13B**	*hnn* *scorpion (zebrafish)*	Grace *et al.* (2003), Green *et al.* (1989), Gretz *et al.* (1996), Guay-Woodford *et al.* (1996)
Meckel syndrome OMIM 249000	***MKS1***	*BBS13*	**BBS Type 13** OMIM 609883	*17q23* *⋆609883*	**Meckel syndrome type 1 protein**		Salonen (1984)
	TMEM67	*MKS3* *JBTS6*	Originally used as model for human **ARPKD** OMIM 263200 **JBTS Type 6** OMIM 609884	*Chr. 8q* *⋆609884*	**Meckelin** Meckel syndrome Type 3 protein Transmembrane protein 67	*wpk* (RAT)	Ferland *et al.* (2004), Ferrante *et al.* (2001), Gonzalez-Perrett *et al.* (2001), Sanzen *et al.* (2001)
	CC2D2A	*MKS6*		*4p15.3* *⋆612013*	**Coiled-coil and C2 domain-containing protein 2A**		Ferrante *et al.* (2006), Flaherty *et al.* (1995), Fliegauf *et al.* (2003)
Nephronophthisis (Juvenile) OMIM 256100 **(Infantile)**	***NPHP1***	*JBTS4*	**JBTS Type 4** OMIM 609583 **SLSN Type 1** OMIM 266900 **NPHP Type 1** OMIM 607100	2q13 ⋆607100	**Nephrocystin-1** Juvenile nephronophthisis 1 protein	*Nphp1 null*	Sayer *et al.* (2006), Schafer *et al.* (1994), Schneider *et al.* (1996), Schrick *et al.* (1995)

(continued)

Table 5.2 (*continued*)

Human Syndrome	HUMAN Gene	*also known as*	Associated Ciliopathy	Locus and OMIM	Protein	Murine model	Key References
OMIM 602088	***INVS***	*NPHP2*		9q31 *243305	**Inversin** Nephrocystin-2	*inv*	Davenport *et al.* (2007), Kim *et al.* (2004), Marszalek *et al.* (1999), Schuermann *et al.* (2002), Shah *et al.* (2008)
	NPHP3		Originally used as model for human **ADPKD** OMIM 173900 **SLSN Type 3** OMIM 606995	3q22 *608002	**Nephrocystin-3**	*pcy* *Nphp3 null*	Dawe *et al.* (2007), Deane *et al.* (2001), Sheng *et al.* (2008), Shillingford *et al.* (2006)
	NPHP4		**SLSN Type 4** OMIM 606996	1p36 *607215	**Nephrocystin-4** Nephroretinin		Doudney and Stanier (2005), Signor *et al.* (1999), Simons *et al.* (2005), Slavotinek *et al.* (2000), Smith *et al.* (2006)
	IQCB1	*NPHP5*	**SLSN Type 5** OMIM 609254	3q21.1 *609237	**IQ calmodulin-binding motif-containing protein 1** Nephrocystin-5	*n/a*	Drummond *et al.* (1998)
	CEP290	*MKS4* *JBTS5* *LCA10* *BBS14* *NPHP6* *SLSN6*	**SLSN Type 6** OMIM 610189 **MKS Type 4** OMIM 611134 **LCA Type 10** OMIM 611755 **BBS Type 14** OMIM 209900 **JBTS Type 5** OMIM 610188	12q21.3 *610142	**Centrosomal protein of 290 kDa** Nephrocystin-6	*rd16*	Drummond (2005), Eberhart *et al.* (2006), Edelstein (2008), Sohara *et al.* (2008)

	GLIS2	*NPHP7* *NKL*		16p13.3 ⋆608539	**Zinc finger protein GLIS2 (GLI-similar 2)** Neuronal Krueppel-like protein	*Glis2 null*	Edelstein (2008), Somlo and Ehrlich (2001), Spassky *et al.* (2008)
	NEK8	*JCK* *NPHP9*	originally used for model of human **ARPKD** OMIM 263200 **NPHP Type 9** OMIM 256100	17q11.1 ⋆609799	**Serine/threonine-protein kinase Nek8** Never in mitosis A-related kinase 8	*jck*	Eley *et al.* (2005), Nonaka *et al.* (1998), Patel *et al.* (2008), Stamataki *et al.* (2005), Starich *et al.* (1995)
Obesity	***TUB***		maturity-onset obesity and insulin resistence	11p15.5 ⋆601197	**Tubby protein homolog**	*tub* *Rd5 gene in mouse*	Stoetzel *et al.* (2006, 2007)
	FTO		Type II diabetes obesity	16q12.2 ⋆610966	**Protein fatso** Fat mass and obesity-associated protein	*Ft*	Sulik *et al.* (1994), Sullivan-Brown *et al.* (2008)
Oro-facial-digital syndrome OMIM 311200	***OFD1***			Xp22.3-p22.2 ⋆300170	**Oral-facial-digital syndrome 1 protein**	Ofd1 null and conditional alleles	Guay-Woodford (2003), Habas *et al.* (2003), Han *et al.* (2008)
***Polycystic Kidney Disease* (autosomal dominant)** OMIM 173900	***PKD1***			16p13.3-p13.12 ⋆601313	**Polycystin-1** Autosomal dominant polycystic kidney disease protein 1	*Pkd null and various conditional alleles*	Chiang *et al.* (2004), Nauta *et al.* (1993), Sun *et al.* (2004)
	PKD2			4q21-q23 ⋆173910	**Polycystin-2** Autosomal dominant polycystic kidney disease type II protein Polycystwin	*Pkd null and various conditional alleles* *Pkd2$^{WS25/-}$*	Chiang *et al.* (2004), Chizhikov *et al.* (2007), Nauli *et al.* (2003), Nauta *et al.* (1993), Ocbina and Anderson (2008), Supp *et al.* (1997, 1999), Takahashi *et al.* (1991)

(*continued*)

Table 5.2 *(continued)*

Human Syndrome	HUMAN Gene	*also known as*	Associated Ciliopathy	Locus and OMIM	Protein	Murine model	Key References
***Polycystic Kidney Disease* (autosomal recessive)** OMIM 263200	***PKHD1***		initially proposed as model for **ADPKD** OMIM 173900 because of late onset of disease in rat.	6p21.1-p12 *606702	**Fibrocystin** Polycystic kidney and hepatic disease 1 protein Polyductin	*pck (rat)*	Cole *et al.* (1998), Nauta *et al.* (1997), Takahashi *et al.* (1986), Takeda *et al.* (1999, 2002)
Primary Ciliary Dyskinesia OMIM 242650	***DNAHC11***	*LRD*		7p21 *603339	**Axonemal dynein heavy chain isotype 11** left/right-dynein	*iv*	Kaspareit-Rittinghausen *et al.* (1990), Keeler (1931), Tallila *et al.* (2008), Taulman *et al.* (2001), Tobin and Beales (2007)
used as a model for **ADPKD** OMIM 173900	***unknown***			*n/a*	n/a	Han:SPRD-Cy *(rat)*	Olbrich *et al.* (2003), Omran *et al.* (2001), Tobin and Beales (2008), Tobin *et al.* (2008)
bpk used as model for human **ARPKD** OMIM 263200 *jcpk* used as model for human **ADPKD** OMIM 173900	***BICC1***			*10q21.1*	**Protein bicaudal C homolog 1**	*bpk* *jcpk*	Torres *et al.* (2004), Tran *et al.* (2008), Tsiokas *et al.* (1997), Upadhya *et al.* (1999)

used as model for human **ARPKD** OMIM 263200	***CYS1***			*2p25.1*	**Cystin-1** Cilia-associated protein	*cpk*	Nauta *et al.* (1993), Nonaka *et al.* (2002), Upadhya *et al.* (2000), Utsch *et al.* (2006)
no Human syndrome	***KIF3A***			*5q31* **604683*	**Kinesin-like protein KIF3A** Microtubule plus end-directed kinesin motor 3A	KIF3A null and conditional alleles	Kimberling *et al.* (1993), Koyama *et al.*(2007), Onuchic *et al.* (2002), Valente *et al.* (2003, 2005, 2006), Vierkotten *et al.* (2007), Vogler *et al.* (1999), Ward *et al.* (2002, 2003), Watanabe *et al.* (2003), Watnick and Germino (2003), White *et al.* (2007), Whitehead *et al.* (1999), Wilson *et al.* (2006)
no Human syndrome	***IFT52***				**Intraflagellar transport protein 52 homolog**	*Ift52*^{hypo}	Wolf *et al.* (2007), Wong *et al.* (2003)
no Human syndrome	***IFT88***	*TG737*	Used as model for human **ARPKD** OMIM #263200	*13q12.1* **600595*	**Intraflagellar transport protein 88 homolog** Recessive polycystic kidney disease protein Tg737 homolog	*Orpk/polaris fxo*	Janaswami *et al.* (1997), Koyama *et al.* (2007), Marszalek *et al.* (2000), Morgan *et al.* (2002, 2003), Nauta *et al.* (1993), Nauli *et al.* (2003),

(*continued*)

Table 5.2 (*continued*)

Human Syndrome	HUMAN Gene	*also known as*	Associated Ciliopathy	Locus and OMIM	Protein	Murine model	Key References
							Nishimura *et al.* (2001), White *et al.* (2008), Wong *et al.* (2000, 2003), Wu *et al.* (1997, 1998), Xiong *et al.* (2002), Yodey *et al.* (1995, 2002)
no Human syndrome	***NEK1***		used as a model for human **ADPKD** OMIM 173900	*4q33* **604588*	**Serine/threonine-protein kinase Nek1** NimA-related protein kinase 1	*kat* *kat*^{2J}	Collin *et al.* (2002), Yodey *et al.* (2002); Zhang *et al.* (2001, 2002, 2003).

polydactyly is often seen in human BBS patients but has not yet been recorded in mutant mice. However, in other ways these models have been informative for clinicians; for instance, us and others recently reported anosmia in *Bbs1* and *Bbs4* mutant mice, which arises from a depletion of olfactory proteins in the ciliary layer of olfactory neurons (Kulaga *et al.*, 2004) and the presence of anosmia in human BBS patients, which is now established as a novel feature of the syndrome (Iannaccone *et al.*, 2005; Kulaga *et al.*, 2004).

Recently a study of craniofacial dysmorphology in BBS patients utilized no fewer than three organisms to delineate its underlying cause. Using three-dimensional (3D) facial imaging combined with dense surface morphometry and principle component analysis, Tobin *et al.* (2008) characterized the predominantly midfacial defects in BBS patients and demonstrated that these were also present in *Bbs4* and *Bbs6* mouse mutants. On the basis that human, mouse, and fish share similar embryonic craniofacial developmental processes and that most of the anterior head is derived from migrating cranial neural crest cells (NCC), zebrafish were used to demonstrate aberrant NCC migration when BBS genes are abrogated (Tobin and Beales, 2007; Tobin *et al.*, 2008). This study provides a quintessential example of the power of diverse model integration, something that is increasingly becoming commonplace within biomedical research.

3. Alström Syndrome

Alström syndrome (ALMS) is a rare recessive disorder that shows strong resemblance to BBS. It is characterized by cardinal features such as cone-rod dystrophy, neurosensory hearing loss, early-onset obesity, and insulin resistance leading to type 2 diabetes. A number of other features may also be present such as dilated cardiomyopathy, hepatic and urinary dysfunction, short stature, and male hypogonadism (Joy *et al.*, 2007). It has been demonstrated that variable sensorineural hearing loss in ALMS patients results from cochlear neuronal degeneration (Marshall *et al.*, 2005). Commonly, infants display nystagmus and photophobia, progressing to cone and rod photoreceptor degeneration, which gives a key differential diagnosis from BBS (Marshall *et al.*, 2005; Russell-Eggitt *et al.*, 1998).

Mutations in *ALMS1* account for all cases of this syndrome; again, there appears to be no phenotype–genotype correlations (Collin *et al.*, 2002; Hearn *et al.*, 2002). ALMS1 is ubiquitously expressed throughout all organ tissues (Collin *et al.*, 2002); it was first identified in a proteomic study of the centrosome (Andersen *et al.*, 2003) where it is localized to the centrosome and basal body (Hearn *et al.*, 2005), which suggested an involvement in ciliary function and a possible explanation for the phenotypic overlap seen with BBS (Hearn *et al.*, 2002). Although there are several conserved sequence motifs within ALMS1, albeit of limited functional significance, its predicted sequence displays no similarity to any other known gene/protein. Studies in *foz* (*fat aussie*) mutants that carry a mutation in *Alms1*, as well as an *Alms1* knockout mouse, suggest that ALMS1 has a role in intracellular trafficking (Arsov *et al.*, 2006; Collin *et al.*, 2005). These mice develop features similar to ALMS patients such as obesity, hypogonadism, hyperinsulinemia, retinal degeneration, late-onset hearing loss and males, like in BBS null mice, are infertile due to aflagellate spermatozoa making this mouse an important model for further investigating disease manifestation in patients.

4. Polycystic Kidney Disease

The most commonly inherited PKDs are transmitted as autosomal-dominant (ADPKD) or autosomal-recessive (ARPKD) traits (Ibraghimov-Beskrovnaya and Bukanov, 2008). Although both forms of PKD initiate early in life, ARPKD rapidly progresses to kidney failure shortly after birth whereas ADPKD can take many years to reach end-stage renal disease (ESRD); typically by the fifth decade. Although both forms of the disease are characterized by the development and expansion of numerous fluid-filled cysts in the kidney as well as defects in multiple other tissues, the distribution of the renal cysts and the other body organs affected vary. Significantly, ADPKD is one of the most common human monogenic diseases with an incidence of 1:1000 and accounting for up to 10% of all end-stage renal patients; it is a systemic disorder, characterized by fluid-filled cysts not only in the kidneys but also in liver, pancreas, and other organs as well as cardiovascular defects and aneurysm. Although the incidence of ARPKD is lower at 1:20,000 it is associated with a high level of mortality in affected newborns. Most cases manifest *in utero* or at birth with renal enlargement and biliary dysgenesis. A number of well-characterized murine models, many spontaneous mutants, have been utilized to study the pathology of PKD (Guay-Woodford, 2003). By the mid-1990s, the genes mutated in the different types of human PKD were identified using positional cloning strategies. In the case of ADPKD, the causative genes are *PKD1* and *PKD2* that encode polycystin-1 and polycystin-2, respectively (Consortium, 1994; Mochizuki *et al.*, 1996), whereas mutations in a single gene, *PKHD1*, is sufficient to cause ARPKD (Onuchic *et al.*, 2002; Ward *et al.*, 2002; Xiong *et al.*, 2002) (Guay-Woodford, 2003). Subsequently, it was established that the disease-causing mutations in the murine models were not allelic to the genes known to cause PKD in humans. Additionally, there have been a number of reports demonstrating the profound effects of the genetic background of these animals on the disease phenotype (Gattone *et al.*, 1996;

Janaswami *et al.*, 1997). Despite this, these models have provided much insight into disease mechanisms and they have been instrumental in revealing a role for cilia in the mechanisms of cystogenesis; this will be discussed later.

5. Nephronophthisis

Nephronophthisis (NPHP) is an autosomal-recessive cystic renal condition characterized by corticomedullary clustering of cysts and tubulointerstitial fibrosis. In contrast to PKD where enlarged kidneys are a common diagnostic feature, the overall size of the kidney in NPHP is normal or diminished. Although, strictly speaking, NPHP describes a renal histopathology, ~10% of cases also present with extrarenal manifestations that can be associated with other syndromes such as retinitis pigmentosa (SLSN), cerebellar vermis hypoplasia (JBTS), ocular motor apraxia (Cogan type), cognitive impairment, hepatic fibrosis, phalangeal cone-shaped epiphyses (Mainzer-Saldino), and *situs inversus*. NPHP has also been described in cases of BBS, EVC, JATD, AS, and MKS (Hildebrandt and Zhou, 2007). NPHP presents as three forms characterized by the time of onset of ESRD: infantile, juvenile, and adolescent. Collectively, they constitute the most frequent genetic cause of end-stage renal failure in the young. The earliest presenting features are polyuria and polydipsia accompanied by renal concentrating defects. Eight causative genes have now been identified (summarized in Table 5.2) and the analysis of their protein products has provided a strong link between ciliary function and the pathogenesis of this disease. The time of disease onset is dependent on the variant form of NPHP involved such that *NPHP1, 3–9* give rise to the juvenile and adolescent forms and *NPHP2* is responsible for the more severe infantile form. *NPHP1* encodes nephrocystin-1 a protein that interacts with other syndrome-related proteins: NPHP2, NPHP3, and NPHP4. Nephrocystin-1 localizes to adherens junctions and interacts with focal adhesion proteins within the renal epithelia with a likely role in cell polarity (Benzing *et al.*, 2001; Donaldson *et al.*, 2002; Nurnberger *et al.*, 2002). *NPHP2* was identified as the human orthologue of the murine *Inv* (*inversin*) mutant that encodes for the Inversin protein. Inversin interacts with β-tubulin, thus linking it closely to cilium structures (Otto, 2003; Watnick and Germino, 2003). *NPHP3* seems to be responsible for an adolescent form of the disease. The *pcy* (*polycystic kidney disease*) mouse was first described by (Takahashi *et al.*, 1986) and, at that time, identified as a model for ADPKD (Nagao *et al.*, 1991; Takahashi *et al.*, 1991). However, it was later revealed that the *pcy* mutation generated a hypomorphic allele for *Nphp3*, which itself is a cause of nephronophthisis in humans (Olbrich *et al.*, 2003; Omran *et al.*, 2001); this animal provided the first example of synteny between a human and a spontaneous murine renal cystic disease. *NPHP4* encodes nephroretinin, which forms a complex with cell adhesion and actin cytoskeletal proteins (Mollet *et al.*, 2005). Mutations in *NPHP5/IQCB1* are associated with SLSN and NPHP5 interacts directly with RPGR (retinitis pigmentosa GTPase regulator) encoded by an X-linked gene which when mutated causes retinitis pigmentosa in males. Both localize to the connecting cilium of the photoreceptor and primary cilium of renal epithelial cells (Otto *et al.*, 2005). Perhaps the most important gene with respect to unifying the ciliopathies is *NPHP6/Cep290.* Cep290 is a novel centrosomal protein, the deficiency of which causes NPHP (type 6), JBTS (type 5), MKS (type 4), and BBS (type 14) (Valente *et al.*, 2006b). *Nphp6* knockdown in zebrafish leads to convergent extension defects, pronephric cysts, retinal degeneration, and hindbrain defects, nicely recapitulating the human disease (Sayer *et al.*, 2006). Furthermore, from the same study, *Nphp6* appears to regulate cell size and morphogenesis in the tunicate (sea squirt), *Ciona intestinalis. NPHP6* mutations cause Leber's congenital amaurosis (that affects the retina only) in 20% of patients diagnosed with NPHP (Chang *et al.*, 2006), which is supported by the finding of an in-frame deletion of *Nphp6/Cep290* in a *rds16* (*retinal degeneration, slow*) mouse, which is a model for retinitis pigmentosa (RP). Although the severity varies, RP is associated with mutations in most NPHP genes, except NPHP7. Mutations in *NPHP7/GLIS2* have only been reported in a single Oji-Cree Canadian family with disease confined to the kidney (Attanasio *et al.*, 2007). In support of this is the *Glis2* mutant mouse that develops, from 2 months postnatal, renal atrophy with fibrosis and increased apoptosis in renal tubular cells. Mutations in the novel *NPHP8/RPGRIP1L* gene causes MKS (type 5) and JBTS (type 7). Two truncating mutations were reported by Delous *et al.* (2007) in MKS fetuses whereas their JBTS patients carried missense and/or one truncating mutation, perhaps revealing a phenotypic spectrum of severity based on number and type of mutation. RPGRIP1L interacts with and colocalizes at the basal bodies, centrosomes, and primary cilia in renal tubular cells with nephrocystin-4 and nephrocystin-6 (Arts *et al.*, 2007). *RPGRIP1L* missense mutations were demonstrated to decrease the protein interaction with nephrocystin-4. *Rpgrip1l* is also one of the genes deleted in the (*Ft*) *fused-toe* mouse where the gene is referred to as (*Ftm*) *Fantom*. Mutations in *Ftm* have been linked to the renal and *situs inversus* phenotype in the *Ft* mouse. Rpgrip1l has recently been shown to participate in Sonic Hedgehog (Shh) signaling through the cilium thereby affecting patterning of the developing neural tube and limb (Vierkotten *et al.*, 2007). Intriguingly, further extrarenal manifestations have been reported in patients carrying *RPGRIP1L* mutations such as liver fibrosis, postaxial polydactyly, pituitary agenesis, and partial growth hormone deficiency (Wolf *et al.*, 2007), making this gene a plausible candidate for the rare RHYNS syndrome (retinitis pigmentosa, hypopituitarism, and skeletal

Table 5.3 Genetic and allelic heterogeneity of nephronophthisis (NPHP) and overlap with other ciliopathies, including: BBS – Bardet-Biedl syndrome; MKS: Meckel syndrome; JBTS: Joubert syndrome; NPHP: Nephrophthisis; SLSN: Senior-Løken syndrome.

Gene	NPHP	JBTS	MKS	BBS	SLSN
NPHP1	NPHP1	JBTS4			SLSN1
INVS	NPHP2				
NPHP3	NPHP3				SLSN3
NPHP4	NPHP4				SLSN4
IQCB1	NPHP5				SLSN5
CEP290	NPHP6	JBTS5	MKS4	BBS14	SLSN6
GLIS2	NPHP7				
RPGRIP1L/Ftm	NPHP8	JBTS7	MKS5		
NEK8	NPHP9				

dysplasia). Furthermore, the analysis of *jck* (*juvenile cystic kidney*) mice that carry mutations in *Nek8* (*never in mitosis gene A-related kinase 8*) led to the evaluation of *NEK8* in 588 NPHP patients revealing three missense mutations linked to a ninth NPHP locus (*NPHP9*) (Liu *et al.*, 2002). One patient had an additional mutation in the *NPHP5* gene raising the possibility that oligenic inheritance may be operating in some cases of NPHP. Nephronophthisis offers a clear example of the genetic and allelic heterogeneity that can be features of the ciliopathies; this is illustrated in Table 5.3.

6. Meckel Syndrome

The lethal autosomal-recessive Meckel syndrome (MKS) is characterized by renal cystic dysplasia with fibrotic changes in the liver, occipital encephalocoele or some other central nervous system malformation. Additionally, polydactyly is frequently reported and some patients have cleft palate, cardiac abnormalities, and incomplete development of genitalia and gonads (Dawe *et al.*, 2007; Paavola *et al.*, 1997; Salonen, 1984; Salonen and Paavola, 1998). Patients with MKS invariably die from respiratory and renal failure. Thus far, MKS has been linked to six loci of which five genes have so far been identified: MKS1, MKS3–5 (summarized in Table 5.2). All identified gene products are associated with ciliary functions. For instance, the products of *MKS1* and *MKS3/TMEM67* (meckelin) interact with each other and are required for centriole migration to the apical membrane and consequent formation of the primary cilium (Dawe *et al.*, 2007). *MKS3/TMEM67* has now been mapped to a region syntenic to the spontaneous *wpk* (*wistar polycystic kidneys*) locus in rat, which has polycystic kidney disease (similar to ARPKD), agenesis of the corpus callosum, and hydrocephalus (Nauta *et al.*, 2000; Smith *et al.*, 2006). As described above, mutations in NPHP genes have been found in patients with MKS as well as JBTS suggesting that these conditions represent a spectrum of the same underlying disorder; these include *MKS4/CEP290/NPHP6* and *MKS5/RPGRIP1L/NPHP8*. Recently, a sixth locus for MKS was identified and the gene *MKS6/CC2D2A* reported; (Nagase, 2000) although the biological function of CC2D2A is uncharacterized, the identification of a calcium-binding domain as well as the fact that patient-derived fibroblasts lack cilia indicate a critical role for CC2D2A in cilia function (Noor *et al.*, 2008; Tallila *et al.*, 2008).

7. Joubert Syndrome

Joubert syndrome (JBTS) is an autosomal-recessive condition characterized by hypotonia, ataxia, severe psychomotor delay, oculomotor apraxia, and episodes of rapid breathing. Diagnosis may be supported by the neuroradiological hallmark referred to as the "molar tooth sign" (MTS), owing to horizontally oriented and thickened superior cerebellar peduncles and a deepened interpeduncular fossa combined with cerebellar vermis hypoplasia (Louie and Gleeson, 2005). The MTS has greatly enhanced the diagnosis of JBTS and with this has identified a group of Joubert syndrome-related disorders (JSRD) with involvement of other organs. For instance, approximately a quarter of patients develop juvenile nephronophthisis and retinal dystrophy, termed cerebello-oculo-renal syndrome (CORS) or Joubert syndrome type B (Valente *et al.*, 2003). Several additional clinical features have been reported including occipital encephalocele, polymicrogyria, cystic kidneys, polydactyly, hepatic fibrosis, and ocular coloboma, thus overlapping with the lethal, recessive disorder MKS and SLSN. Six of the eight JBTS loci have now been described (summarized in Table 5.2); again a clear overlap with other ciliopathies is apparent. The underlying gene mutations for *JBTS1* and *JBTS2* loci are not yet identified; however, *JBTS3* has been shown to be caused by mutations in the *AHI1* gene (Dixon-Salazar *et al.*, 2004; Ferland *et al.*, 2004) that accounts for between 7% and 11% of JBTS cases most of which are accompanied by retinopathy (Parisi *et al.*, 2006; Utsch *et al.*, 2006; Valente *et al.*, 2006a). The *AHI1* gene product is Jouberin; it is expressed in the brain and kidney and has been shown to interact with nephrocystin-1. In a mouse kidney cell line, AHI1 was shown to localize at adherens junctions, the primary cilia and basal body, which is consistent with a role in nephronophthisis (Baala 2007; Eley *et al.*, 2008). *JBTS6/TMEM67/MKS3* was identified as the sixth locus and linked to MKS. JBTS patients with associations with NPHP have mutations in *JBTS4/NPHP1*, *JBTS5/*

NPHP6/CEP290, and *JBTS7/NPHP8/RPGRIP1L* genes; the presence of RP and its severity is variable in these cases. *JBTS8/ARL13B* mutations have recently been identified in families with a classical form of JBTS (Cantagrel *et al.*, 2008). ARL13B belongs to the Ras-GTPase family and in other species is required for ciliogenesis, body axis formation, and renal function. Two animal models have been reported for this gene: the lethal *hnn* (*hennin*) mouse that has a coupled defect in cilia structure and Shh signaling and the zebrafish *scorpion* mutant that displays renal cysts and a curved tail, both of which are phenotypes common in morphants with cilia dysregulation (Cantagrel *et al.*, 2008; Caspary *et al.*, 2007; Garcia-Garcia *et al.*, 2005; Sun *et al.*, 2004).

8. Jeune Syndrome

Jeune asphyxiating thoracic dystrophy (JATD) is an autosomal-recessive chondrodysplasia. Affected children often die in the perinatal period owing to respiratory insufficiency that is a consequence of narrow and slender ribs and abnormal cage formation. Radiographical analysis can also indicate a shortening of the long bones and changes of the pelvic bones and the phalanges. There can be multiorgan involvement such as biliary dysgenesis with portal fibrosis and bile duct proliferation, renal cystogenesis and failure, polydactyly, and retinal degeneration.

These phenotypic clues led to the identification of mutations in *IFT80* in a subgroup of patients presenting with milder disease with no renal, liver, pancreatic, or retinal features (Beales *et al.*, 2007). There is, however, genetic heterogeneity with another, as yet unidentified locus on 15q. IFT80 is a member of the intraflagellar transport (IFT) complex B proteins that are important for cilia structure and function and which will be discussed further below. *Ift80* knockdown in the multiciliate protozoan, *Tetrahymena*, resulted in fewer cilia and nuclear duplication (Beales *et al.*, 2007). In *zebrafish*, silencing *ift80* results in convergent extension defects, cystic pronephros, and cardiac edema whereas knockout mice all display early embryonic lethality (unpublished observations).

9. Oral–Facial–Digital Syndrome

Oral–facial–digital (OFD) type 1 syndrome is an X-linked-dominant disease characterized by malformations of oral cavity, face, and digits and by cystic kidneys. Facial features include hypertelorism, broad nasal bridge, buccal frenula, cleft palate, lobulated tongue, lingual hamartomas; in the hands and feet, brachydactyly, and polydactyly may be present. PKD is common and central nervous system malformations include corpus callosum agenesis, cerebellar abnormalities, and hydrocephalus, with accompanying mental retardation. It is presumed lethal in males. Mutations in the novel gene *OFD1* are causative and the OFD1 protein localizes both to the primary cilium and to the nucleus (Ferrante *et al.*, 2001; Giorgio *et al.*, 2007). *OFD1* has phenotypic similarities with 11 other forms of OFD syndrome, however, the underlying genes for these other disease forms remain to be identified. Franco and colleagues knocked out *Ofd1* in mice and recapitulated the human phenotype albeit with increased severity, possibly owing to differences of X inactivation patterns between species (Ferrante *et al.*, 2006). They also showed a failure of left–right axis specification in mutant male embryos, a lack of cilia in the embryonic node, mispatterning of the neural tube, and altered expression of *Hox* genes in the limb buds, all of which are indicative of cilia defects and demonstrate that Ofd1 plays a role in ciliogenesis.

10. The Structure and Function of the Cilium

Historically, much of what we know of the structure and workings of the cilium/flagellum has come from a number of model organisms whose strengths are include both that they are amenable to genetic manipulation and relatively simple organisms, compared to mammalian models making them ideal for microscopic analysis. These include the unicellular green algae flagellate, *Chlamydomonas reinhardtii*, the ciliated protozoan, *Tetrahymena*, and the nematode worm, *Caenorhabditis elegans*. The finding that the cilium and flagellum are structures that are evolutionarily conserved throughout nature means that what we learn from our more simple experimental models can be transferred to vertebrates, indeed humans. This has led to a rapid increase in our understanding of the biology and the consequences to development and disease when cilia are defective.

10.1. The structure of primary cilia

Cilia typically project from the apical surface of cells and are composed of a microtubular (MT) backbone (axoneme), nucleating from the basal body and ending at the tip complex, ensheathed by membrane contiguous with the plasma membrane. This generic structure is illustrated in Fig. 5.2; it indicates that the two types of cilia, motile and nonmotile, differ slightly in their structure. In general, motile cilia consist of an axoneme of nine microtubule doublets arranged in a ring with a central doublet pair (termed

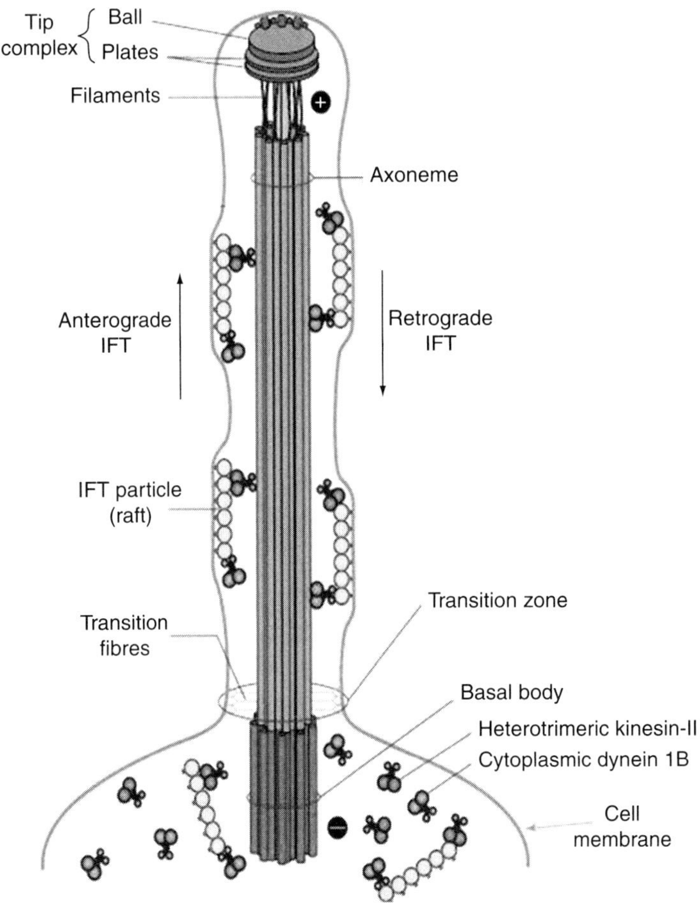

Figure 5.2 Structure of the cilium illustrating intraflagellar transport. Adapted from Eley *et al.* (2005) (See Color Insert.).

"9 + 2"). They tend to form in multiples on the cell surface and beat in concert with each other to provide unidirectional fluid flow. Each MT doublet consists of an "A" strand comprising 13-tubulin protofilaments, and a "B" strand made from 10 protofilaments (Alberts *et al.*, 2008). In motile 9 + 2 cilia, dynein motor proteins crosslink and enable doublets to slide against each other (Johnson and Gilbert, 1995; Porter and Sale, 2000). The outer and inner doublets are connected by radial spokes that serve to convert the sliding motion of the MT into a rhythmical beat; however, it is unknown how this beating is coordinated amongst cells. Nonmotile or "primary" cilia have a simpler 9 + 0 microtubule configuration, lacking the central doublet pair, dynein arms, and radial spokes. They are usually present as single apical membrane extensions.

10.2. Intraflagellar transport

Intracellular transport makes use of cytoskeletal microtubules and molecular motors to move membrane-bound vesicular cargo around the cell. Within the eukaryotic cilium/flagellum, cargo is also carried along the microtubular axoneme, in association with nonmembrane-bound protein complexes, in a highly conserved process termed, intraflagellar transport (IFT). As no protein synthesis apparatus exists within the cilium, one function of the IFT particle, is to carry cargo from the cell body, such as receptors and structural proteins into the cilia, as well as to deliver signals that emanate from the cilia into the cytoplasm in response to external environmental stimuli. Cargo is first translocated to a "loading dock" at the base of the cilium, which consists of transitional fibers that extend out from the basal body to the cell membrane (Deane *et al.*, 2001). Here a molecular motor–cargo complex is assembled and takes up its position at the proximal end (minus-end) of the axoneme in preparation for transport. The anterograde movement of the complex is facilitated by the heterotrimeric kinesin-II complex (Kif3a, Kif3b, and KAP) along the microtubules. At the tip (plus-end), the complex is remodeled and kinesin itself becomes cargo for retrograde transport back down the axoneme. This is accomplished by another motor, IFT-dynein (Pazour *et al.*, 1999; Porter *et al.*, 1999; Signor *et al.*, 1999). This switching of transport direction takes place only at the tip and basal body (Kozminski *et al.*, 1993). Particles undergoing IFT were first visualized on the axonemes of mutant *Chlamydomonas* with paralyzed flagella (Kozminski *et al.*, 1993, 1995). Interestingly, disruption of dynein-dependent retrograde transport can result in swelling of the cilium tip owing presumably to backup of cargo and highlighting the importance of this intricate but coordinated transport. Studies, in which IFT particles were isolated from *Chlamydomonas* flagella, identified a complex of 15 (now 17) polypeptides (Cole *et al.*, 1998; Piperno and Mead, 1997). This assembly was further resolved into two subcomplexes, A (550 kDa) and B (710–760 kDa) (Cole *et al.*, 1998). Sequence identity of these polypeptides revealed they are homologous to proteins required for sensory cilia assembly in *C. elegans* (Cole *et al.*, 1998; Perkins *et al.*, 1986; Starich *et al.*, 1995).

10.3. IFT and ciliopathies

The cilium is a highly regulated organelle, projecting from the surface of the cell, its membrane continuous with the plasma membrane of the cell and with a transport system that moves cargo between it and the cytoplasm of the cell. Evidence indicates that many proteins are seen to localize to the cilium. Hence the cilium can be described as a functional compartment, not only of the cytoplasm, but also of the plasma membrane itself with IFT acting as a coordinated link between them. Studies into the pathogenesis of BBS in a number of different models have provided evidence for a functional link between the cilium and ciliopathies.

Most BBS proteins studied thus far localize to the cilium/basal body/centrosome complex. In mammalian cultured cells BBS1, BBS4, BBS5, BBS6, and BBS8 localize to the basal body and pericentriolar region whereas BBS4 and BBS5 have been observed in the ciliary axoneme (Ansley *et al.*, 2003; Kim *et al.*, 2004, 2005; Li *et al.*, 2004; Ross *et al.*, 2005) (and unpublished data). In *C. elegans*, BBS1::GFP, BBS7::GFP, and BBS8::GFP fusion proteins have been localized to the cilia and mutations for BBS7 and BBS8 show defective chemotaxis, behavior dependent upon functional ciliated neurons; thus, not only are the proteins localized to the cilia but are important for their proper function. Indeed, time-lapse microscopy studies in worms have shown that GFP::BBS7 and BBS8 fusion proteins, localized at the base of the cilium in ciliated neurons, are participating in IFT (Blacque and Leroux, 2006). This is further supported by evidence in *bbs7* and *bbs8* mutant mice where GFP-tagged IFT proteins fail to localize and move properly along the cilium suggesting that these proteins facilitate the incorporation of the IFT particle into the motor protein complex and are selective for particular cargoes. As indicated in Table 5.1, a common feature of the ciliopathies is a number of retinal pathologies. In mammalian photoreceptors, Rhodopsin relies on IFT for transport to the outer segment (a specialized cilia structure of the retina) and a feature of mouse *Bbs1*, *Bbs2*, *Bbs4*, and *Bbs6* mutants is the development of severe retinal degeneration akin to BBS patients (Fath *et al.*, 2005; Kulaga *et al.*, 2004; Mykytyn *et al.*, 2004; Nishimura *et al.*, 2004; Ross *et al.*, 2005). In these mouse mutants, degeneration is linked to failed Rhodopsin transport that subsequently accumulates in the cell body and triggers apoptosis (Nishimura *et al.*, 2004; Ross *et al.*, 2005). The development of a similar retinal degeneration in IFT mouse mutants supports the relationship between BBS and IFT (Pazour *et al.*, 2002a). It was later determined in *C. elegans* that BBS7 and BBS8 coordinate the movement of IFT subcomplexes A and B (Blacque *et al.*, 2004). In the absence of BBS proteins, the A and B subcomplexes move separately and at different rates, most likely because each is moved by a different type of kinesin. The authors proposed a model whereby BBS7 and BBS8 act as a bridge between the two IFT subcomplexes, stabilizing and coordinating transport, which may explain why BBS phenotypes are less severe than IFT knockouts. Figure 5.3 shows how BBS7 and BBS8 act together to coordinate IFT in *C. elegans*. Further evidence for the role of BBS proteins in IFT in mammalian cells comes from two-hybrid screens in which p150^{glued} (a component of dynactin and important in retrograde motor function) and PCM1 (pericentriolar material 1) were identified as proteins that interact with BBS4. Silencing of *Bbs4* results in the abrogation of PCM1 recruitment to the pericentriolar

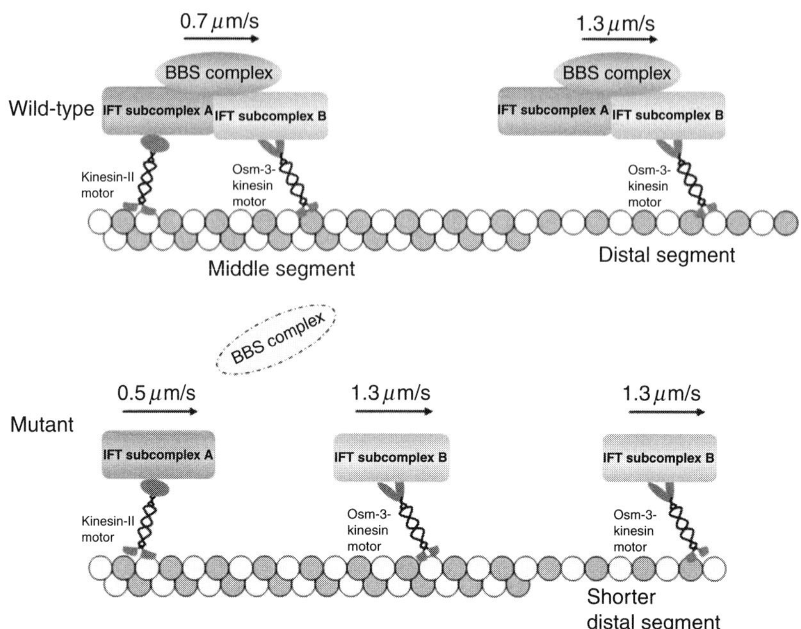

Figure 5.3 BBS7 and BBS8 act together to coordinate IFT in *C. elegans*. Upper panel. In wild-type worms, BBS proteins are transported along the ciliary midsegment by kinesin-II and OSM-3-kinesin at an "intermediate" rate (<0.7 μm/s). They travel faster (<1.3 μm/s) when reaching the distal segments propelled by OSM-3-kinesin alone. Lower panel. Loss of BBS protein function causes the separation of both kinesin-II motors and IFT particle subcomplexes A and B, resulting in subcomplex A moving at a slow rate (<0.5 μm/s) with kinesin-II, and subcomplex B moving at a fast rate (<1.3 μm/s) with OSM-3-kinesin. Redrawn with permission from Blacque and Leroux (2006).

satellites, deanchoring of microtubules from the centrosome, nuclear duplication, and blocked cytokinesis and apoptosis (Kim *et al.*, 2004). Thus, it has been proposed that BBS4 functions as an adaptor protein that is important for loading cargo (e.g., PCM1) onto IFT particles in preparation for its transport along the axoneme. Figure 5.4 illustrates how BBS4 modulates retrograde IFT.

Further insight into how BBS proteins behave, at the molecular level, was recently provided by (Nachury and Colleagues, 2007). Tandem affinity purification (TAP) has demonstrated that the BBS proteins form a complex termed the "BBSome." This 438 kDa complex consists of stoichiometric ratios of BBS1, 2, 4, 5, 7, 8, and 9. BBS9 is observed to interact with all other subunits and is therefore thought to act as a central organizer of the BBSome. The most marked defects in ciliation are found when cells are depleted of BBS1 and BBS5. One theory suggests that the BBSome is transported to the basal body by the centriolar satellites where it associates

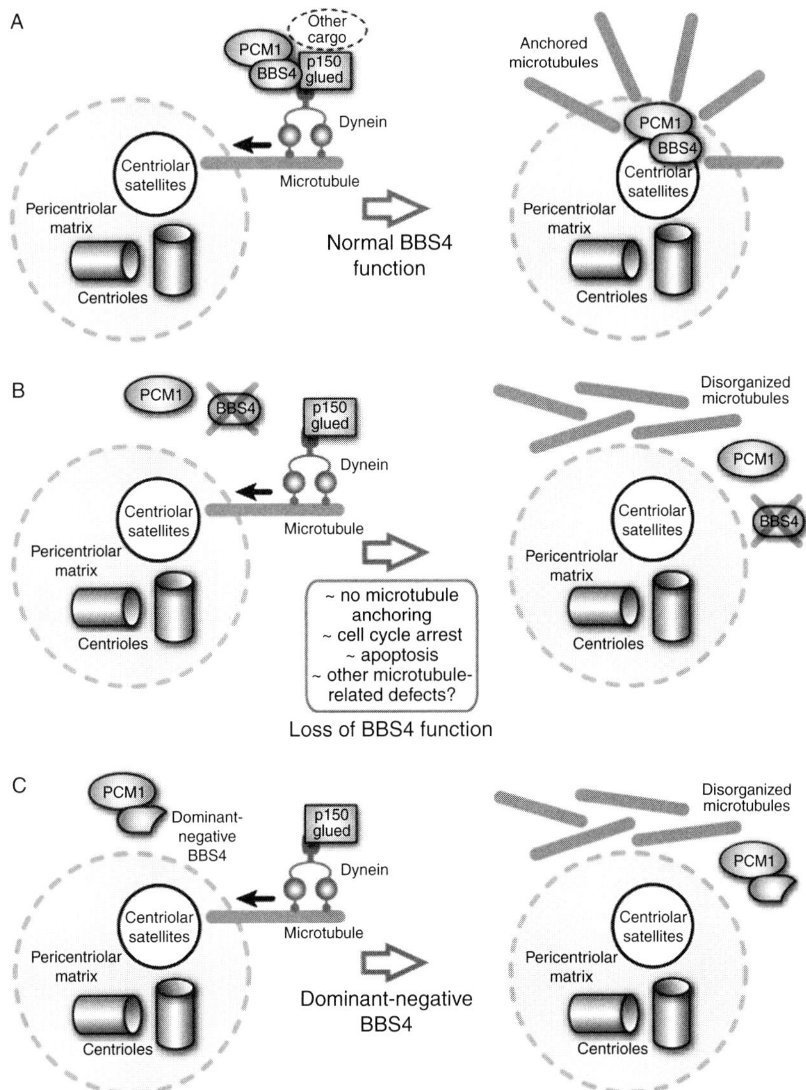

Figure 5.4 Model of BBS4 (dys)function. (A) In normal BBS4 function, BBS4 recruits PCM1 and associated cargo to the centriolar satellites in a dynein-dependent manner, likely through its interaction with $p150^{glued}$. (B) In the absence of BBS4, PCM1 and possibly its cargo do not localize to the centriolar satellites, leading to several cellular phenotypes. (C) Some BBS4 truncation mutants have an equally detrimental effect by inhibiting PCM1 localization to the centriolar satellites. Reproduced from Kim *et al.* (2004).

with the ciliary membrane. In addition, the BBSome interacts with Rabin8, a guanosyl exchange factor for Rab8 that is a small GTPase involved in regulating trafficking from post-Golgi vesicles. Rab8 itself enters the ciliary membrane and promotes ciliogenesis, therefore, it has been proposed that pathogenesis in BBS might be caused by defects in vesicular transport to the cilium.

These data sets generate two general, nonexclusive models for how BBS proteins might function in relation to IFT. One suggests that at least BBS4, 7, and 8 are necessary for microtubule-based transport along the cilium, the other that BBS proteins function as a complex mediating vesicle transport from the cytoplasm to the cilium. As such, no one clear model for the function of BBS proteins has been described and the proteins may function in numerous roles, including cytoplasmic transport.

11. Cilia and Development

11.1. Left–right determination

A fundamental question perplexing developmental biologists for decades concerned the mechanism that establishes LR asymmetry, almost universal amongst vertebrates. A failure to break symmetry during early development can lead to a randomization of subsequent body organs, referred to as *situs inversus*, which is a common ciliopathy phenotype. LR asymmetry was first associated with cilia motility following observations of patients with a form of primary ciliary dyskinesia (PCD) known as Kartagener's syndrome that display increased pulmonary infections, male sterility, absence of sinuses/sinusitis and of a left–right organ reversal (Afzelius, 1976). Such disorders are associated with motile cilia dysfunction. Additionally, Afzelius accurately predicted that motile cilia are also important for cardiac situs during early embryogenesis. Since then mutations in genes encoding dynein heavy and light chains in the ciliary axoneme are responsible for some cases of Kartagener's syndrome (Giusto and Sciubba, 2004). Whilst studying the morphogenesis of the mouse embryonic notochordal plate as well as the node, situated at the rostral end of the primitive streak, Sulik *et al.* (1994) remarked upon the presence of a single, central cilium on each cell. Although they observed that cilia of the node and of the prechordal/notochordal plates were motile, they were unaware of the "potential significance of this motile behavior." It was later, in a landmark study, that Nonaka *et al.* (1998) demonstrated a clockwise rotation in the cilia lining the embryonic node that generated a leftward flow across the embryonic pole. Subsequent studies in the mouse mutant, *iv* (inversus viscerum) which results from mutations in the *lrd* (*left–right dynein*) gene showed a number of laterality defects as a result of abnormal nodal cilia and consequently fluid flow disruption (Okada *et al.*, 1999; Supp *et al.*, 1999). In fact, microfluidic experiments in which the fluid flow is reversed in the developing node produced *situs inversus totalis* in wild-type embryos thus establishing the "morphogen flow" hypothesis (Nonaka *et al.*, 2002). This model proposes that the net leftward flow establishes a concentration gradient of signaling molecules such as Nodal, Fgf, or Shh across the node resulting in asymmetric gene expression and preferential activation of left-sided downstream pathways. Although there is undoubtedly leftward fluid flow, no left-sided accumulation of signaling proteins has been demonstrated. To account for this, a second hypothesis, or rather modification of the first has been proposed. The "two cilia" model is supported by the observation that two populations of primary cilia are found in the node (McGrath *et al.*, 2003). It is proposed that a centrally placed "rotating" population of cilia are surrounded by a population of nonmotile sensory cilia that "reads" the leftward flow by differential deflection and subsequent calcium influx and signaling. In support of this hypothesis, *lrd* mutant mice in which nodal cilia are paralyzed do not display the expected left-sided nodal calcium signal (McGrath *et al.*, 2003). Calcium signaling is likely triggered in response to cilia deflection during urine flow and it is known that it depends on polycystin-2, a major cause of adult PKD when mutated. Thus it is remarkable that *Pkd2* mice develop *situs inversus* in the presence of morphologically normal and motile nodal cilia (Pennekamp *et al.*, 2002). Consistent with the overall flow hypothesis are reports of cilia structural defects and laterality disorders in mouse mutants of the kinesin-II complex proteins (*Kif3a* and *Kif3b*) as well as *Invs*, *polaris* (*Tg737/Ift88*), *wimple* (*Ift172*), and *Rfx3* (Bonnafe *et al.*, 2004; Marszalek *et al.*, 1999; Mochizuki *et al.*, 1998; Murcia *et al.*, 2000; Nonaka *et al.*, 1998; Pennekamp *et al.*, 2002; Rana *et al.*, 2004) and in zebrafish embryos in which similar proteins such as *polaris*, *invs*, *lrd*, or *pc2* are abrogated (Bisgrove *et al.*, 2005; Essner *et al.*, 2005; Kramer-Zucker *et al.*, 2005b; Otto *et al.*, 2003; Sun *et al.*, 2004).

11.2. Hedgehog signaling in ciliopathies

An exciting development of the last 5 years has been the finding that cilia and IFT are essential for the activity of the HH pathway (Fig. 5.5) (Eggenschwiler and Anderson, 2007). HH signaling has a role in the development of nearly every organ in the body. Certainly, a role for cilia in the transduction of this pathway provides a valuable insight into mechanisms that may be underlying certain phenotypes common to the ciliopathies. In particular, we highlight phenotypes that arise in tissue where Shh signaling is known to play an important role during embryonic development in mammals; phenotypes such as postaxial polydactyly, external genitalia anomalies, and craniofacial defects.

Arguably, the best-known ligand for the HH pathway is Shh (Ingham, 2008). In addition to Shh, there are other Hedgehog ligand subgroups in vertebrates, such as Desert Hedgehog (Dhh) and Indian Hedgehog (Ihh). The downstream components of the pathway are the same, however, the

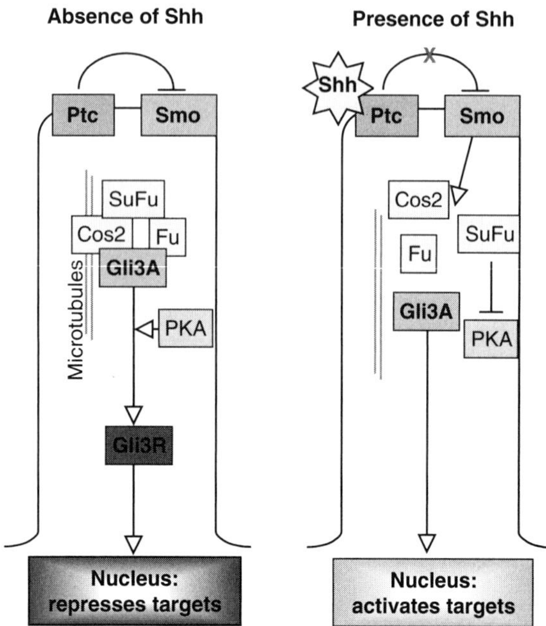

Figure 5.5 Mechanism of Shh transduction. In the absence of Shh, Ptc inhibits Smo. In the cytoplasm, the tetrameric complex of Su(Fu), Costal2, Fused, and Gli3 associates with microtubules and Gli3 is cleaved into its repressor form by PKA. In the presence of Shh, the inhibition of Smo by Ptc is relieved. Smo acts to disassemble the tetrameric complex, preventing the action of PKA and leaving active Gli3 free to enter the nucleus to induce target genes.

posttranslational modifications that effect diffusion kinetics vary between the different ligands types. Activation of the pathway relies on the export and extracellular diffusion of its ligand, which in the case of Shh is a cleaved 20 kDa N-terminal isoform called Shh-N, that binds to its membrane-bound receptor, Patched-1 (Ptc1). In the absence of bound ligand, Ptc1 inhibits Smoothened (Smo), a member of the G protein-coupled receptor family. Binding of Shh to Ptc1 relieves this inhibition on Smo, allowing the downstream transduction of the pathway signal that results in the transcriptional activation of target genes by the transcription factor, Gli3. A fundamental aspect of the pathway is that varying concentrations of Shh generate a graded response through the differential transcription of target genes. This morphogenic effect of Shh is encoded by a gradient of Gli transcriptional activity; in this way, Gli activity can be seen as the intracellular correlate of the extracellular Shh concentrations (Stamataki *et al.*, 2005). Hence, differential cell fate is determined by varying levels of Shh signaling, which in turn is dependent on the distance of the responding cells from the original source of secreted Shh. Ptc1 is an early response gene of the pathway, which acts itself to restrict the diffusion of the ligand.

The first link between IFT and the HH pathway was reported in embryonic patterning mutants from an ethyl nitrosourea (ENU)-based mutagenesis screen. Huangfu *et al.* (2003) identified two IFT mutants, called *wim* (*Ift172*) and *flexo* (*Ift88*) (refer Table 5.2); although these embryos expressed Shh normally, their *Ptc1* expression levels are markedly down-regulated indicative of a loss of pathway activity. Further analysis in the neural tube showed that ventral cell types are absent while dorsal cell types, usually repressed by Shh, are expanded. Cilia are present on neural tube cells and the cilia in these embryos are absent or stunted. Subsequently, many of the downstream components of the pathway, at some stage in their activity, show subcellular localization to the cilium. These include Smo, Ptc1, and Gli2/Gli3, as well as a number of proteins that interact with Gli in a tetrameric complex that regulates the proteolytic conversion from an active isoform (Gli3A) to a repressor isoform (Gli3R); this includes Suppressor of Fused (Su(Fu), a negative regulator of the pathway and Costal2 (Cos2), a microtubule-associated kinesin-like protein. (Corbit *et al.*, 2005; Haycraft *et al.*, 2005; Ocbina and Anderson, 2008; Rohatgi *et al.*, 2007). Given these insights, it is perhaps unsurprising that IFT plays an essential role in Shh signaling (Haycraft *et al.*, 2005; May *et al.*, 2005). Several other ciliary proteins have since been shown to be essential for Shh signaling, including THM1 and RPGRIP1L/Ftm/JBTS7/MKS5 (Tran *et al.*, 2008; Vierkotten *et al.*, 2007).

In the case of embryonic limb bud development, Shh is known to play an important role in anteroposterior patterning. Shh ligand itself originates from a source at the posterior margin of the limb bud in a region called the zone of polarizing activity (ZPA). Ihh is expressed in chondrocytes and acts to signal to the growth plate to induce cell division essential for bone growth. Interestingly, mutations in *IHH* in humans are associated with bone growth defects such as brachydactyly (short fingers) and dysplasia of the long bones such as the femur (Hellemans *et al.*, 2003). Loss-of-function mutations to *SHH* results in a loss of digits, while mutations in *GLI3* cause polydactyly (McGlinn and Tabin, 2006), a common feature of ciliopathies in general and in BBS, present in at least 70% of patients (Tobin and Beales, 2007). It is interesting, as mentioned above, that polydactyly is not, however, a feature of mouse BBS mutants.

Beales *et al.* (1999) have reported that 98% of BBS patients have malformed external genitalia. In males this includes micropenis, small and undescended testes, and hypospadias (mispositioning of the urethral meatus). This is recapitulated in *Shh* null mice that fail to grow out a genital tubercle, the precursor to the penis (Haraguchi *et al.*, 2001). Perriton *et al.* (2002) showed that transplanting the normal urethral epithelium to the limb bud in chick embryos was sufficient to induce mirror-image duplication of the digits, implying that Shh is acting to also pattern genital development. Additionally, Dhh is expressed in the testis and involved in spermatogenesis.

Thus, the genital and limb phenotypes seen in ciliopathies, such as in BBS and MKS, may share an etiology underpinned by aberrant Shh signaling. In support of this hypothesis, it is interesting to note that ALMS is often misdiagnosed as BBS; patients present similar features: obesity, blindness, deafness, and kidney disease. However, ALMS patients do not develop polydactyly or genital anomalies, features likely attributable, at least in part, to faulty Shh signaling.

In the skull, Shh signaling is essential for normal craniofacial development; in particular it is essential for development of the frontonasal and maxillary processes of the upper and midface (Hu and Helms, 1999). Mutations in *Shh* in mice and humans cause holoprosencephaly (HPE) where the lobes of the forebrain fail to split, leading to a single central eye (cyclopia) and the absence of a nose, but with a less severe effect in the lower face. During embryogenesis, Shh is expressed in a rostral region of the forebrain as well as within the oral ectoderm and it has been demonstrated that these sources of Shh are important for the correct migration of cranial NCC that contribute to the bony features of the face (Eberhart *et al.*, 2006). Interestingly, craniofacial dysmorphology is a notable feature of BBS and it is likely that this may reflect, in part, defects in Shh signaling (Tobin *et al.*, 2008). It is noteworthy that homozygous mutations in *Rab23* have been shown to cause Carpenter syndrome characterized by craniosynostosis, polysyndactyly, obesity, and cardiac defects. (Jenkins *et al.*, 2007) Rab23 is a small GTPase involved in the endocytosis or targeting of vesicles containing Shh, Ptc, or Smo (Eggenschwiler *et al.*, 2006).

11.3. Cilia and polarity

Cells within an epithelium often display a polarity across the plane of cells known as epithelial PCP. Recently, PCP genes have also been shown to have a role in ciliogenesis and, conversely, genes previously known to be required for ciliogenesis have a role in PCP (Jones and Chen, 2007). Examples of PCP in vertebrates include the uniform orientation of stereociliary bundles on the apical surfaces of sensory hair cells in the organ of Corti of the mammalian auditory sensory organ (Kelly and Chen, 2007), as well as convergent extension (CE) movements that involve a polarized cellular rearrangement that leads to the narrowing and concomitant lengthening of a tissue along two perpendicular axes (Jessen *et al.*, 2002), both linked to cilia. Vertebrate PCP signaling utilizes some of the components of the canonical Wnt/β-catenin signaling pathway. Wnt signaling is most well known for its role in embryogenesis and cancer but is also important for a number of normal physiological processes in adult animals. The central purpose of the Wnt/β-catenin pathway is to stabilize β-catenin by preventing its phosphorylation-dependent degradation. This involves the binding of extracellular Wnt ligands to receptors of the Frizzled (Fz) and

LRP families on the cell surface causing the activation of Disheveled (Dsh or Dvl) family proteins. Dsh inhibits a second complex of proteins, Axin, GSK3, and APC, which normally promotes the proteolytic destruction of β-catenin. Consequently, the inhibition of the "β-catenin destruction complex" affects the amount of β-catenin that reaches the nucleus to interact with TCF/LEF family transcription factors to promote specific gene expression.

Vertebrate PCP signaling also requires the binding of Wnt ligand to its Fz receptor. Because extracellular ligands and membrane receptors constitute the first signaling modalities in a pathway, it is often tempting to think that specific Wnt ligand and/or ligand/receptor combinations dictates the strict spatiotemporal specificity of pathway activation, as well as the outcome of downstream signaling events. However, many examples of ambiguity are now emerging that caution taking too simplistic a view. Additionally, the subcellular localization of Dsh is also now known to have an important impact on the switch between the different branches of the pathway (Wnt/β-catenin or PCP). In PCP signaling, Dsh localizes to the plasma membrane via its DEP domain where it is activated by Fz in combination with a coreceptor, Knypek (Wong *et al.*, 2000, 2003). Dsh activation does not involve β-catenin or the destruction complex in PCP signaling but instead involves a number of proteins specific to PCP signaling. For instance, the nonclassical cadherins Fat, Dachsous and Flamingo, and other proteins, including Prickle and Strabismus act downstream of Fz and Dsh to regulate the cytoskeleton through RhoA and Rho-kinase (Habas *et al.*, 2003). Figure 5.6 outlines the main steps in canonical Wnt and PCP pathways.

Konig and Hausen (1993) were the first to implicate cilia in planar polarity events, although at the time they did not know that this involved PCP signaling. They described the polarity of epithelial cells in the embryos of the African clawed frog, *Xenopus laevis*, aligned with the direction of ciliary beating. Cano *et al.* (2004) subsequently demonstrated that the subcellular localization of β-catenin was altered in *Ift88* mutant mice and that this affected the organization of pancreatic tissue. Perhaps the most significant finding to link PCP signaling to the cilium and consequently to ciliopathies, was that of (Simons and colleagues 2005) that showed Inversin (NPHP2) localizes to the cilium and acts as a molecular switch between the Wnt/β-catenin and PCP branches of the pathway. This is achieved through the degradation of cytoplasmic pools of Dsh, in contrast to membrane-bound Dsh that is linked to PCP signaling (Simons *et al.*, 2005). Interestingly, fluid flow such as the flow of urine in the developing renal tubules, may act to increase the level of Inversin in cultured kidney cells, which has been speculated to favor the patterning of renal epithelia over cellular proliferation. In ciliopathy mutants, where PCP signaling is suppressed, the balance of proliferation and differentiation appears to be lost, leading to cyst formation. Further evidence to support the role of PCP signaling in

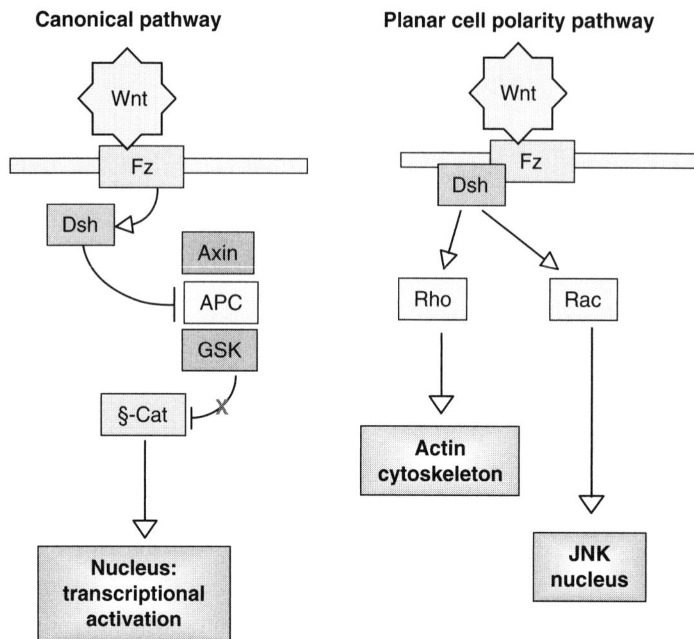

Figure 5.6 Wnt signaling pathways. In the canonical pathway, Wnt binds to its receptor Frizzled (Fz) which activates cytoplasmic Disheveled (Dsh). This inhibits the destruction complex consisting of Axin, APC, and GSK3β, preventing the destruction of cytoplasmic β-catenin. β-catenin then enters the nucleus and activates transcriptional targets by association with TCF and LEF. In the PCP pathway, Wnt binds to Fz which recruits Dsh to the plasma membrane. This results in RhoA activation and consequently remodeling of the cytoskeleton. Dsh can also activate the JNK pathway, through Rac1, to affect the transcription of downstream targets of this pathway.

normal renal development comes from a mouse mutant for the PCP gene *Fat4*, which develops cystic kidneys as a result of disoriented cell divisions and tubule elongation (Saburi *et al.*, 2008).

Mouse PCP mutants such as *circle tail*, *loop tail*, *crash*, and *Dvl1* and *2*, display neural tube defects (NTD) such as spina bifida (failure of posterior neural tube closure), craniorachischisis (failure of the entire neural tube to close), and exencephaly (failure of rostral neural tube closure) (Doudney and Stanier, 2005). These gene knockout models tend to be embryonic or perinatal lethal. In humans, missense mutations in *VANGL1*, the mouse orthologue of the gene mutated in *loop tail*, causes NTD such as spina bifida (Kibar *et al.*, 2007). Ross and colleagues demonstrated that BBS proteins are involved in PCP signaling and that a proportion of *Bbs4* null mice phenocopied PCP mutants such as *loop tail* (Eley *et al.*, 2005). They had open eyelids at birth, exencephaly, and misoriented stereociliary bundles. Compound *Bbs4/Lt* heterozygotes recapitulated the phenotype, whereas single heterozygotes for each allele was not phenotypic, implying a genetic interaction between the two genes. Vangl2, a membrane-bound receptor involved in PCP signaling, localizes to the cilium. Analysis of downstream effector proteins for the PCP pathway have demonstrated that mutations in *inturned* and *fuzzy* abrogate PCP signaling and cause indirect Shh-deficient phenotypes as a result of defective ciliogenesis (Park *et al.*, 2006). It is likely that affecting the structure and/or function of the cilia and its associated proteins acts to disrupt the balance between the different branches of the Wnt signaling pathway. Corbit *et al.* (2008) showed that the Wnt pathway components β-catenin and APC are also localized to the cilium. Additionally, *Ofd1* null cells that lack cilia are five times more sensitive to stimulation with exogenous Wnt ligand compared to wild-type cells, probably because the cilium in normal cells act to impose a brake on the Wnt/β-catenin pathway. In support of this, Gerdes *et al.* (2007) showed that suppressing expression of *bbs1*, *bbs4*, *bbs6*, and *Kif3a* affects the stabilization of β-catenin and consequently the upregulation of TCF-mediated transcriptional activity, which is indicative of overactive Wnt/β-catenin signaling.

11.4. Cilia and cystogenesis in mammals

Although *C. elegans* has no kidney *per se*, the worm has provided many seminal moments in renal-related research, beginning with the initial findings that *C. elegans* harbor orthologues of the human polycystins, known to be one of the primary causes of PKD in humans (Mochizuki *et al.*, 1996) and that these proteins (called LOV1 and PKD2) are found to be involved in mating behavioral responses mediated by sensory neurons that require cilia function (Barr and Sternberg, 1999). Subsequently, IFT88 was identified in algae and worms (where it is called Osm-5) and shown to be the underlying mutation in the important *orpk* (*oak ridge polycystic kidney*) mouse that models ARPKD (Moyer *et al.*, 1994). This work demonstrated that IFT is a conserved process common to all ciliated or flagellated eukaryotic cells and that the IFT protein, *Polaris/Tg373*, is important for ciliogenesis in mammals, thus, providing the first evidence of a link between cilia dysfunction and renal cyst formation (Haycraft *et al.*, 2001; Pazour *et al.*, 2000, 2002a; Qin *et al.*, 2001; Taulman *et al.*, 2001). In the mammalian kidney, primary cilia extend from the apical surface of renal epithelial cells into the tubule lumen as well as from cells of the nephron (with the exception of intercalated cells) and it is now known that proteins whose functions are disrupted in cystic diseases have all been localized to the cilium or the basal body. The model currently held for how primary cilia function in the renal epithelium is that they act as mechanosensors of extracellular fluid flow through the lumen of renal tubules to regulate cell growth and differentiation (Nauli *et al.*, 2003). Invariably, a failure in this balance between proliferation and differentiation seems to be the underlying cause of abnormal cell proliferation leading to the production of renal cysts (Lina and

Satlinb, 2004) and it seems that deflection of the axoneme due to flow is the initiator of an increase in intracellular calcium signaling that results in altered gene expression (Yoder, 2007). Polycystin-1 and polycystin-2 have been shown to interact with each other (Qian *et al.*, 1997; Tsiokas *et al.*, 1997) and are thought to be part of a calcium channel localized in the primary cilium of renal epithelial cells (Gonzalez-Perrett *et al.*, 2001; Hanaoka *et al.*, 2000; Pazour *et al.*, 2002b; Somlo and Ehrlich, 2001; Yoder *et al.*, 2002). Furthermore, *PKHD1* (the underlying cause of ARPKD) that encodes polyductin localizes to primary cilia and has been shown to affect renal duct differentiation (Onuchic *et al.*, 2002; Ward *et al.*, 2002, 2003; Xiong *et al.*, 2002). However, other findings show that polycystin-1 can regulate the expression of p21, a tumor suppressor that inhibits cyclin-dependent kinases (CDK) leading to cell-cycle arrest (Bhunia *et al.*, 2002) and that IFT88, required for cilia assembly, has an additional role in regulating cell-cycle progression that is independent of its function in ciliogenesis (Robert *et al.*, 2007). This highlights the need for a better understanding of the relationship between the cilium, cell-cycle, and cilia-mediated mechanosensation and signaling activity.

Certainly, there are now many well-characterized murine models for studying human renal cystic diseases. The hypomorphic nature of the *orpk* allele (originally named $Tg737^{orpk}$ but now designated as $Ift88^{Tg737Rpw}$) allows this animal to survive into young adulthood, making this a valuable model to study cystic renal phenotypes (Moyer *et al.*, 1994). The early onset of the renal phenotype and the subsequent distribution of cysts in the kidney has made this animal a good model for human ARPKD. *Chlamydomonas* mutants for Ift88 have a complete absence of flagella, the *orpk* mutants have stunted and malformed cilia, but are not completely abolished. Given the localization of IFT88 in the cilia (Haycraft *et al.*, 2001; Moyer *et al.*, 1994; Pazour *et al.*, 2000; Taulman *et al.*, 2001) and its requirement for ciliogenesis, it is not surprising that abnormal cilia are observed in the renal epithelium of *orpk* mice. The original gross phenotype of the *orpk* mouse described scruffy fur, severe growth retardation and preaxial polydactyly on all limbs (Moyer *et al.*, 1994). Subsequent work has revealed *orpk* mice to also have random LR axis specification, liver and pancreatic defects, hydrocephalus, and skeletal patterning abnormalities in addition to PKD. A role supported by the loss of cilia on the ependymal cell layer in brain ventricles and by the lack of node cilia in the Ift88 mutants (Taulman *et al.*, 2001). Additionally, the follicular dysplasia that gives rise to the disordered and scruffy fur may suggest a role for primary cilia in the skin and hair follicle and provide insight into ectodermal dysplasias seen in the human syndromes such as Ellis-van Creveld and Sensenbrenner's syndromes (Lehman *et al.*, 2008). Thus, this model continues to provide insights into the role of cilia in multiple tissues.

Mandell and colleagues first described the *cpk* (*congenital polycystic kidney*) mouse for which the underlying mutation was subsequently found to be in the *Cys1* gene (Hou *et al.*, 2002; Mandell *et al.*, 1983); the finding that its gene product, cystin, localizes to the cilium along with polycystins (Yoder *et al.*, 2002) means this mouse continues to be a useful model for studying a recessively transmitted form of PKD. The *jck* (*juvenile cystic kidneys*) mouse that carries a recessive mutation has an intermediate phenotype relative to *cpk* and *pcy* with mutants being fertile and surviving into adulthood (Atala *et al.*, 1993).

ADPKD has also been difficult to model. Although probably not an entirely satisfying theory, it has long been proposed that there is a two-hit mechanism required for cyst formation, consisting of a germ line mutation to one allele and a somatic mutation to the other (Ong and Harris, 2005). Embryonic renal cyst development in homozygous knockout *Pkd1* or *Pkd2* animals and the progressive cystic disease associated with the $Pkd2^{WS25}$ mutant, which has a *Pkd2* allele prone to inactivation by somatic mutation, are consistent with a "two-hit model" of cyst development (Lu *et al.*, 1997; Wu *et al.*, 1998). In all homozygous animals, renal development appears to be normal until embryonic day (E) 14.5 after which renal cysts start to develop, suggesting that the polycystins are not required for nephronic induction. But these mutations are generally embryonic lethal due to systemic effects. In contrast to the human condition, the phenotype seen in heterozygote mice is generally mild and variable probably due to the late onset of this disorder, reflecting the difference in lifespans between humans and mice. Transheterozygotes are somewhat more severe. This and a number of conditional knockouts of *Pkd1* that are emerging as orthologous models of ADPKD suggest a productive future for therapeutic testing (Lantinga-van Leeuwen *et al.*, 2004; Wu *et al.*, 1998).

A number of PKD models with an autosomal-dominant inheritance pattern are also available; for instance, the Han:SPRD rat has been well documented with several features which resemble human ADPKD (Cowley *et al.*, 1993; Gretz *et al.*, 1996) despite the fact that the disease in these animals is not linked to the rat orthologue of the human PKD1 gene (Nauta *et al.*, 1997).

Interestingly, the genetic background in these murine models has a strong influence on the expression of the mutated gene and consequently how comparable the disease manifests between mice/humans. For instance the *cpk* mouse has been extensively studied in the C57BL strain but it was only when the *cpk* gene was bred onto a CD1 background (Gattone *et al.*, 1996) that the extrarenal pathology, similar to that seen in the human infantile PKD, was penetrative. Additionally, Janaswami *et al.* (1997) reported that the genetic background had profound effects on the disease phenotype in the *kat(2j)* (*kidney, anemia, testes*) mouse. These observations have also been reported in most other murine models.

11.5. Cilia and obesity

Obesity is a cardinal feature of two ciliopathies whose diagnosis is often confused: BBS and ALMS. In BBS, the frequency of overweight varies by from 72% (Beales *et al.*, 1999) to 91% (Green *et al.*, 1989) whereas childhood obesity is present in 95% of ALMS patients (Joy *et al.*, 2007). Although obesity is a common thread for certain ciliopathies, its precise cause is not yet clear. Nevertheless, this does raise the question as to whether cilia function is required for body mass homeostasis. In BBS patients, it has been proposed that obesity arises from hyperphagia (overeating) and that this feeding behavior is perhaps a result of hypothalamic defects impacting on the satiety centre where ciliated neurons regulate hunger and satiation (Davenport *et al.*, 2007). Others have found that BBS patients tend to be less physically active, albeit with no significant difference in their energy metabolism compared to BMI matched controls (Grace *et al.*, 2003). In support of findings in human patients, mouse BBS mutants have been shown to consistently eat more than their wild-type littermates and go on to display early-onset obesity. Additionally, Rahmouni *et al.* (2008) found BBS2, 4, and 6 null mice have combination of low locomotor activity. These mice were also shown to have and increased levels of circulating leptin, a hormone that regulates satiety. Whereas the administration of exogenous leptin to wild-type mice reduced bodyweight by around 10% in a 4-day period, no such effect was seen in the BBS mutants, regardless of the underlying mutation. This suggests that hypothalamic neurons in the BBS mutants are unresponsive to leptin, which results in changes in their feeding behavior. Hypertension is a common feature of BBS patients; it is interesting that the mouse BBS mutants also develop hypertension most likely as a result of this resistance to leptin signaling. As mentioned above, the *foz* mutant mouse, which carries a mutation in Alms1, also develops hyperphagia and obesity as well as insulin tolerance progressing to type 2 diabetes (Arsov *et al.*, 2006). Although fibroblasts taken from ALMS patients do not indicate a role for ALMS1 in ciliogenesis, the protein itself localizes to the cilium and basal body (Hearn *et al.*, 2005). Other proteins involved in the regulation of feeding behavior also show a distribution to the cilium, such as the G protein-coupled receptor Melanin concentrating hormone receptor-1 (Mchr1), which is abrogated in BBS mutant mice (Berbari *et al.*, 2008). Davenport *et al.* (2007) further provided the first insight into a mechanism for obesity. They conditionally ablated Kif3a, therefore blocking ciliogenesis, in POMC (pro-opiomelanocortin) expressing hypothalamic neurons involved in negatively regulating appetite. These mice most notably display an increase in feeding behavior that progressively leads to obesity. These data have provided important evidence for a role of cilia in regulating appetite and indirectly suggests that BBS proteins and ALMS1 play a role in appetite homeostasis.

12. Therapies for Cystic Disease

Of all phenotypes seen within the ciliopathies, it is the renal pathologies that are the greatest contributor to morbidity and mortality in affected patients. As an example, in ADPKD, cyst growth can lead to dramatic increases in overall kidney mass; at times up to 20 kg (Germino and Somlo, 1993). Eventually, the size and number of cysts replace functional tissue and normal kidney filtration is impaired, resulting in ESRD in approximately 50% of patients. The only effective treatment currently available to PKD patients is transplantation, which emphasizes the need to develop potential therapeutic agents for clinical use. Because the mode of cystogenesis is probably similar irrespective of the gene mutated, there are opportunities for treatment using small molecules/drugs to treat the cystic diseases seen in the ciliopathies. A number of drugs are currently under investigation for their ability to slow cyst development and these represent significant hope as a treatment strategy (summarized in Table 5.4). Of particular mention are Rapamycin and Roscovitine, both of which have well-characterized modes of action, and here we will highlight findings from their use in cystic kidney models.

Rapamycin (Sirolimus) is an inhibitor of mTOR pathway that is an important regulator of cell proliferation; it is currently being used in clinical trials in humans. It has been proposed that the inappropriate activation of mTOR is a common feature of all forms of cystic kidney disease and may indeed be the predominant cause of PKD (Edelstein, 2008a,b). Interestingly, mTOR interacts with the cytoplasmic tail of polycystin-1, indicating that it functions, at least in part, within the cilium (Shillingford *et al.*, 2006). Certainly, in support of this, Rapamycin treatment alone is able to rescue kidney phenotypes in several different mouse models. For instance, Rapamycin treatment in mice with cystic kidneys increases apoptosis and shedding of cystic cells, resulting in an overall reduction in the size of the kidneys from 16% of total body weight to 4%, compared to 0.7% in wild-types (Shillingford *et al.*, 2006). In these mice, renal function, as measured by blood urea nitrogen concentration, was completely restored. Upstream of mTOR are two centrosomal proteins, TSC1 and TSC2 that, when mutated, cause the Tuberous sclerosis complex with renal cysts (Henske, 2005; Wilson *et al.*, 2006).

Roscovitine is also currently in clinical trials for cancer treatment. It is an inhibitor of cell cycle, acting through CDK. Cell-cycle regulation, disrupted in PKD through a ciliary mechanism, is a plausible target for preventing cyst growth (Ibraghimov-Beskrovnaya and Bukanov, 2008). Both in humans and in the *jck* mouse, an increase in proliferating cells lining the renal tubules has been reported, which is reduced with Roscovitine treatment (Bukanov *et al.*, 2006). The increased proliferation seen within diseased tubules is attributed to an upregulation in the activity of the Ras

has all the attributes of the human ALMS and some of the BBS but no mutations have yet been found.

REFERENCES

Afzelius, B. A. (1976). A human syndrome caused by immotile cilia. *Science* **193,** 317–319.

Alberts, B., Johnson, A., Lewis, J., Raff, M., Roberts, K., and Walter, P. (2008). "Molecular Biology of the Cell," 5th edn. Garland Science, New York.

Andersen, J. S., Wilkinson, C. J., Mayor, T., Mortensen, P., Nigg, E. A., and Mann, M. (2003). Proteomic characterization of the human centrosome by protein correlation profiling. *Nature* **426,** 570–574.

Ansley, S. J., Badano, J. L., Blacque, O. E., Hill, J., Hoskins, B. E., Leitch, C. C., Kim, J. C., Ross, A. J., Eichers, E. R., Teslovich, T. M., *et al.* (2003). Basal body dysfunction is a likely cause of pleiotropic Bardet-Biedl syndrome. *Nature* **425,** 628–633.

Arsov, T., Silva, D. G., O'Bryan, M. K., Sainsbury, A., Lee, N. J., Kennedy, C., Manji, S. S., Nelms, K., Liu, C., Vinuesa, C. G., *et al.* (2006). Fat aussie—A new Alstrom syndrome mouse showing a critical role for ALMS1 in obesity, diabetes, and spermatogenesis. *Mol. Endocrinol.* **20,** 1610–1622.

Arts, H. H., Doherty, D., van Beersum, S. E., Parisi, M. A., Letteboer, S. J., Gorden, N. T., Peters, T. A., Marker, T., Voesenek, K., Kartono, A., *et al.* (2007). Mutations in the gene encoding the basal body protein RPGRIP1L, a nephrocystin-4 interactor, cause Joubert syndrome. *Nat. Genet.* **39,** 882–888.

Atala, A., Freeman, M. R., Mandell, J., and Beier, D. R. (1993). Juvenile cystic kidneys (jck): A new mouse mutation which causes polycystic kidneys. *Kidney Int.* **43,** 1081–1085.

Attanasio, M., Uhlenhaut, N. H., Sousa, V. H., O'Toole, J. F., Otto, E., Anlag, K., Klugmann, C., Treier, A. C., Helou, J., Sayer, J. A., *et al.* (2007). Loss of GLIS2 causes nephronophthisis in humans and mice by increased apoptosis and fibrosis. *Nat. Genet.* **39,** 1018–1024.

Baala, L., Romano, S., Khaddour, R., Saunier, S., Smith, U. M., Audollent, S., Ozilou, C., Faivre, L., Laurent, N., Foliguet, B., *et al.* (2007). The Meckel-Gruber syndrome gene, MKS3, is mutated in Joubert syndrome. *Am. J. Hum. Genet.* **80,** 186–194.

Badano, J. L., Ansley, S. J., Leitch, C. C., Lewis, R. A., Lupski, J. R., and Katsanis, N. (2003). Identification of a novel Bardet-Biedl syndrome protein, BBS7, that shares structural features with BBS1 and BBS2. *Am. J. Hum. Genet.* **72,** 650–658.

Badano, J. L., Leitch, C. C., Ansley, S. J., May-Simera, H., Lawson, S., Lewis, R. A., Beales, P. L., Dietz, H. C., Fisher, S., and Katsanis, N. (2006). Dissection of epistasis in oligogenic Bardet-Biedl syndrome. *Nature* **439,** 326–330.

Barr, M. M., and Sternberg, P. W. (1999). A polycystic kidney-disease gene homologue required for male mating behaviour in *C. elegans*. *Nature* **401,** 386–389.

Beales, P. L., Elcioglu, N., Woolf, A. S., Parker, D., and Flinter, F. A. (1999). New criteria for improved diagnosis of Bardet-Biedl syndrome: Results of a population survey. *J. Med. Genet.* **36,** 437–446.

Beales, P. L., Bland, E., Tobin, J. L., Bacchelli, C., Tuysuz, B., Hill, J., Rix, S., Pearson, C. G., Kai, M., Hartley, J., *et al.* (2007). IFT80, which encodes a conserved intraflagellar transport protein, is mutated in Jeune asphyxiating thoracic dystrophy. *Nat. Genet.* **39,** 727–729.

Benzing, T., Gerke, P., Hopker, K., Hildebrandt, F., Kim, E., and Walz, G. (2001). Nephrocystin interacts with Pyk2, p130(Cas), and tensin and triggers phosphorylation of Pyk2. *Proc. Natl Acad. Sci. USA* **98,** 9784–9789.

Berbari, N. F., Lewis, J. S., Bishop, G. A., Askwith, C. C., and Mykytyn, K. (2008). Bardet-Biedl syndrome proteins are required for the localization of G protein-coupled receptors to primary cilia. *Proc. Natl Acad. Sci. USA* **105,** 4242–4246.

Bergmann, C., Fliegauf, M., Bruchle, N. O., Frank, V., Olbrich, H., Kirschner, J., Schermer, B., Schmedding, I., Kispert, A., Kranzlin, B., *et al.* (2008). Loss of nephrocystin-3 function can cause embryonic lethality, Meckel-Gruber-like syndrome, situs inversus, and renal–hepatic–pancreatic dysplasia. *Am. J. Hum. Genet.* **82,** 959–970.

Bhunia, A. K., Piontek, K., Boletta, A., Liu, L., Qian, F., Xu, P. N., Germino, F. J., and Germino, G. G. (2002). PKD1 induces p21(waf1) and regulation of the cell cycle via direct activation of the JAK–STAT signaling pathway in a process requiring PKD2. *Cell* **109,** 157–168.

Bisgrove, B. W., Snarr, B. S., Emrazian, A., and Yost, H. J. (2005). Polaris and Polycystin-2 in dorsal forerunner cells and Kupffer's vesicle are required for specification of the zebrafish left–right axis. *Dev. Biol.* **287,** 274–288.

Blacque, O. E., and Leroux, M. R. (2006). Bardet-Biedl syndrome: An emerging pathomechanism of intracellular transport. *Cell. Mol. Life Sci.* **63,** 2145–2161.

Bonnafe, E., Touka, M., AitLounis, A., Baas, D., Barras, E., Ucla, C., Moreau, A., Flamant, F., Dubruille, R., Couble, P., *et al.* (2004). The transcription factor RFX3 directs nodal cilium development and left–right asymmetry specification. *Mol. Cell. Biol.* **24,** 4417–4427.

Bowers, A. J., and Boylan, J. F. (2004). *Nek8*, a NIMA family kinase member, is overexpressed in primary human breast tumors. *Gene* **328,** 135–142.

Bukanov, N. O., Smith, L. A., Klinger, K. W., Ledbetter, S. R., and Ibraghimov-Beskrovnaya, O. (2006). Long-lasting arrest of murine polycystic kidney disease with CDK inhibitor roscovitine. *Nature* **444,** 949–952.

Cano, D. A., Murcia, N. S., Pazour, G. J., and Hebrok, M. (2004). Orpk mouse model of polycystic kidney disease reveals essential role of primary cilia in pancreatic tissue organization. *Development* **131,** 3457–3467.

Cano, D. A., Sekine, S., and Hebrok, M. (2006). Primary cilia deletion in pancreatic epithelial cells results in cyst formation and pancreatitis. *Gastroenterology* **131,** 1856–1869.

Cantagrel, V., Silhavy, J. L., Bielas, S. L., Swistun, D., Marsh, S. E., Bertrand, J. Y., Audollent, S., Attie-Bitach, T., Holden, K. R., Dobyns, W. B., *et al.* (2008). Mutations in the cilia gene ARL13B lead to the classical form of Joubert syndrome. *Am. J. Hum. Genet.* **83,** 170–179.

Caridi, G., Murer, L., Bellantuono, R., Sorino, P., Caringella, D. A., Gusmano, R., and Ghiggeri, G. M. (1998). Renal–retinal syndromes: Association of retinal anomalies and recessive nephronophthisis in patients with homozygous deletion of the NPH1 locus. *Am. J. Kidney Dis.* **32,** 1059–1062.

Caspary, T., Larkins, C. E., and Anderson, K. V. (2007). The graded response to Sonic Hedgehog depends on cilia architecture. *Dev. Cell* **12,** 767–778.

Chang, B., Khanna, H., Hawes, N., Jimeno, D., He, S., Lillo, C., Parapuram, S. K., Cheng, H., Scott, A., Hurd, R. E., *et al.* (2006). In-frame deletion in a novel centrosomal/ciliary protein CEP290/NPHP6 perturbs its interaction with RPGR and results in early-onset retinal degeneration in the rd16 mouse. *Hum. Mol. Genet.* **15,** 1847–1857.

Chapelin, C., Duriez, B., Magnino, F., Goossens, M., Escudier, E., and Amselem, S. (1997). Isolation of several human axonemal dynein heavy chain genes: Genomic structure of the catalytic site, phylogenetic analysis and chromosomal assignment. *FEBS Lett.* **412,** 325–330.

Chiang, A. P., Nishimura, D., Searby, C., Elbedour, K., Carmi, R., Ferguson, A. L., Secrist, J., Braun, T., Casavant, T., Stone, E. M., *et al.* (2004). Comparative genomic analysis identifies an ADP-ribosylation factor-like gene as the cause of Bardet-Biedl syndrome (BBS3). *Am. J. Hum. Genet.* **75,** 475–484.

Chiang, A. P., Beck, J. S., Yen, H. J., Tayeh, M. K., Scheetz, T. E., Swiderski, R. E., Nishimura, D. Y., Braun, T. A., Kim, K. Y., Huang, J., *et al.* (2006). Homozygosity mapping with SNP arrays identifies TRIM32, an E3 ubiquitin ligase, as a Bardet-Biedl syndrome gene (BBS11). *Proc. Natl Acad. Sci. USA* **103,** 6287–6292.

Chizhikov, V. V., Davenport, J., Zhang, Q., Shih, E. K., Cabello, O. A., Fuchs, J. L., Yoder, B. K., and Millen, K. J. (2007). Cilia proteins control cerebellar morphogenesis by promoting expansion of the granule progenitor pool. *J. Neurosci.* **27,** 9780–9789.

Cogswell, C., Price, S. J., Hou, X., Guay-Woodford, L. M., Flaherty, L., and Bryda, E. C. (2003). Positional cloning of jcpk/bpk locus of the mouse. *Mamm. Genome.* **14,** 242–249.

Cole, D. G., Diener, D. R., Himelblau, A. L., Beech, P. L., Fuster, J. C., and Rosenbaum, J. L. (1998). Chlamydomonas kinesin-II-dependent intraflagellar transport (IFT): IFT particles contain proteins required for ciliary assembly in *Caenorhabditis elegans* sensory neurons. *J. Cell Biol.* **141,** 993–1008.

Coleman, D. L., and Eicher, E. M. (1990). Fat (fat) and tubby (tub): Two autosomal recessive mutations causing obesity syndromes in the mouse. *J. Hered.* **81,** 424–427.

Collin, G. B., Marshall, J. D., Ikeda, A., So, W. V., Russell-Eggitt, I., Maffei, P., Beck, S., Boerkoel, C. F., Sicolo, N., Martin, M., *et al.* (2002). Mutations in ALMS1 cause obesity, type 2 diabetes and neurosensory degeneration in Alstrom syndrome. *Nat. Genet.* **31,** 74–78.

Collin, G. B., Cyr, E., Bronson, R., Marshall, J. D., Gifford, E. J., Hicks, W., Murray, S. A., Zheng, Q. Y., Smith, R. S., Nishina, P. M., *et al.* (2005). Alms1-disrupted mice recapitulate human Alstrom syndrome. *Hum. Mol. Genet.* **14,** 2323–2333.

Consortium, T. E. P. K. D. (1994). The polycystic kidney disease 1 gene encodes a 14 kb transcript and lies within a duplicated region on chromosome 16. The European Polycystic Kidney Disease Consortium. *Cell* **78,** 725.

Corbit, K. C., Aanstad, P., Singla, V., Norman, A. R., Stainier, D. Y., and Reiter, J. F. (2005). Vertebrate smoothened functions at the primary cilium. *Nature* **437,** 1018–1021.

Corbit, K. C., Shyer, A. E., Dowdle, W. E., Gaulden, J., Singla, V., Chen, M. H., Chuang, P. T., and Reiter, J. F. (2008). Kif3a constrains beta-catenin-dependent Wnt signalling through dual ciliary and non-ciliary mechanisms. *Nat. Cell Biol.* **10,** 70–76.

Cowley, B. D., Jr., Gudapaty, S., Kraybill, A. L., Barash, B. D., Harding, M. A., Calvet, J. P., and Gattone, V. H., II (1993). Autosomal-dominant polycystic kidney disease in the rat. *Kidney Int.* **43,** 522–534.

Davenport, J. R., Watts, A. J., Roper, V. C., Croyle, M. J., van Groen, T., Wyss, J. M., Nagy, T. R., Kesterson, R. A., and Yoder, B. K. (2007). Disruption of intraflagellar transport in adult mice leads to obesity and slow-onset cystic kidney disease. *Curr. Biol.* **17,** 1586–1594.

Davis, R. E., Swiderski, R. E., Rahmouni, K., Nishimura, D. Y., Mullins, R. F., Agassandian, K., Philp, A. R., Searby, C. C., Andrews, M. P., Thompson, S., *et al.* (2007). A knockin mouse model of the Bardet-Biedl syndrome 1 M390R mutation has cilia defects, ventriculomegaly, retinopathy, and obesity. *Proc. Natl Acad. Sci. USA* **104,** 19422–19427.

Dawe, H. R., Smith, U. M., Cullinane, A. R., Gerrelli, D., Cox, P., Badano, J. L., Blair-Reid, S., Sriram, N., Katsanis, N., Attie-Bitach, T., *et al.* (2007). The Meckel-Gruber syndrome proteins MKS1 and meckelin interact and are required for primary cilium formation. *Hum. Mol. Genet.* **16,** 173–186.

Deane, J. A., Cole, D. G., Seeley, E. S., Diener, D. R., and Rosenbaum, J. L. (2001). Localization of intraflagellar transport protein IFT52 identifies basal body transitional fibers as the docking site for IFT particles. *Curr. Biol.* **11,** 1586–1590.

Delous, M., Baala, L., Salomon, R., Laclef, C., Vierkotten, J., Tory, K., Golzio, C., Lacoste, T., Besse, L., Ozilou, C., *et al.* (2007). The ciliary gene RPGRIP1L is mutated in cerebello-oculo-renal syndrome (Joubert syndrome type B) and Meckel syndrome. *Nat. Genet.* **39,** 875–881.

Dixon-Salazar, T., Silhavy, J. L., Marsh, S. E., Louie, C. M., Scott, L. C., Gururaj, A., Al-Gazali, L., Al-Tawari, A. A., Kayserili, H., Sztriha, L., *et al.* (2004). Mutations in the AHI1 gene, encoding jouberin, cause Joubert syndrome with cortical polymicrogyria. *Am. J. Hum. Genet.* **75,** 979–987.

Donaldson, J. C., Dise, R. S., Ritchie, M. D., and Hanks, S. K. (2002). Nephrocystin-conserved domains involved in targeting to epithelial cell–cell junctions, interaction with filamins, and establishing cell polarity. *J. Biol. Chem.* **277,** 29028–29035.

Doudney, K., and Stanier, P. (2005). Epithelial cell polarity genes are required for neural tube closure. *Am. J. Med. Genet. C Semin. Med. Genet.* **135,** 42–47.

Drummond, I. A. (2005). Kidney development and disease in the zebrafish. *J. Am. Soc. Nephrol.* **16,** 299–304.

Drummond, I. A., Majumdar, A., Hentschel, H., Elger, M., Solnica-Krezel, L., Schier, A. F., Neuhauss, S. C., Stemple, D. L., Zwartkruis, F., Rangini, Z., *et al.* (1998). Early development of the zebrafish pronephros and analysis of mutations affecting pronephric function. *Development* **125,** 4655–4667.

Eberhart, J. K., Swartz, M. E., Crump, J. G., and Kimmel, C. B. (2006). Early Hedgehog signaling from neural to oral epithelium organizes anterior craniofacial development. *Development* **133,** 1069–1077.

Edelstein, C. L. (2008a). Mammalian target of rapamycin and caspase inhibitors in polycystic kidney disease. *Clin. J. Am. Soc. Nephrol.* **3,** 1219–1226.

Edelstein, C. L. (2008b). Therapeutic interventions for autosomal dominant polycystic kidney disease. *Nephrol. News Issues* **22,** 25–26.

Eggenschwiler, J. T., and Anderson, K. V. (2007). Cilia and developmental signaling. *Annu. Rev. Cell Dev. Biol.* **23,** 345–373.

Eggenschwiler, J. T., Bulgakov, O. V., Qin, J., Li, T., and Anderson, K. V. (2006). Mouse Rab23 regulates hedgehog signaling from smoothened to Gli proteins. *Dev. Biol.* **290,** 1–12.

Eichers, E. R., Abd-El-Barr, M. M., Paylor, R., Lewis, R. A., Bi, W., Lin, X., Meehan, T. P., Stockton, D. W., Wu, S. M., Lindsay, E., *et al.* (2006). Phenotypic characterization of Bbs4 null mice reveals age-dependent penetrance and variable expressivity. *Hum. Genet.* **120,** 211–226.

Eley, L., Yates, L. M., and Goodship, J. A. (2005). Cilia and disease. *Curr. Opin. Genet. Dev.* **15,** 308–314.

Eley, L., Gabrielides, C., Adams, M., Johnson, C. A., Hildebrandt, F., and Sayer, J. A. (2008). Jouberin localizes to collecting ducts and interacts with nephrocystin-1. *Kidney Int* .

Essner, J. J., Amack, J. D., Nyholm, M. K., Harris, E. B., and Yost, H. J. (2005). Kupffer's vesicle is a ciliated organ of asymmetry in the zebrafish embryo that initiates left–right development of the brain, heart and gut. *Development* **132,** 1247–1260.

Evans, J. E., Snow, J. J., Gunnarson, A. L., Ou, G., Stahlberg, H., McDonald, K. L., and Scholey, J. M. (2006). Functional modulation of IFT kinesins extends the sensory repertoire of ciliated neurons in *Caenorhabditis elegans*. *J. Cell Biol.* **172,** 663–669.

Fan, Y., Esmail, M. A., Ansley, S. J., Blacque, O. E., Boroevich, K., Ross, A. J., Moore, S. J., Badano, J. L., May-Simera, H., Compton, D. S., *et al.* (2004). Mutations in a member of the Ras superfamily of small GTP-binding proteins causes Bardet-Biedl syndrome. *Nat. Genet.* **36,** 989–993.

Fath, M. A., Mullins, R. F., Searby, C., Nishimura, D. Y., Wei, J., Rahmouni, K., Davis, R. E., Tayeh, M. K., Andrews, M., Yang, B., *et al.* (2005). Mkks-null mice have a phenotype resembling Bardet-Biedl syndrome. *Hum. Mol. Genet.* **14,** 1109–1118.

Ferland, R. J., Eyaid, W., Collura, R. V., Tully, L. D., Hill, R. S., Al-Nouri, D., Al-Rumayyan, A., Topcu, M., Gascon, G., Bodell, A., *et al.* (2004). Abnormal cerebellar development and axonal decussation due to mutations in AHI1 in Joubert syndrome. *Nat. Genet.* **36,** 1008–1013.

Ferrante, M. I., Giorgio, G., Feather, S. A., Bulfone, A., Wright, V., Ghiani, M., Selicorni, A., Gammaro, L., Scolari, F., Woolf, A. S., *et al.* (2001). Identification of the gene for oral–facial–digital type I syndrome. *Am. J. Hum. Genet.* **68,** 569–576.

Ferrante, M. I., Zullo, A., Barra, A., Bimonte, S., Messaddeq, N., Studer, M., Dolle, P., and Franco, B. (2006). Oral–facial–digital type I protein is required for primary cilia formation and left–right axis specification. *Nat. Genet.* **38,** 112–117.

Flaherty, L., Bryda, E. C., Collins, D., Rudofsky, U., and Montogomery, J. C. (1995). New mouse model for polycystic kidney disease with both recessive and dominant gene effects. *Kidney Int.* **47,** 552–558.

Fliegauf, M., Frohlich, C., Horvath, J., Olbrich, H., Hildebrandt, F., and Omran, H. (2003). Identification of the human CYS1 gene and candidate gene analysis in Boichis disease. *Pediatr. Nephrol.* **18,** 498–505.

Fry, J. L., Jr., Koch, W. E., Jennette, J. C., McFarland, E., Fried, F. A., and Mandell, J. (1985). A genetically determined murine model of infantile polycystic kidney disease. *J. Urol.* **134,** 828–833.

Galdzicka, M., Patnala, S., Hirshman, M. G., Cai, J. F., Nitowsky, H., Egeland, J. A., and Ginns, E. I. (2002). A new gene, EVC2, is mutated in Ellis-van Creveld syndrome. *Mol. Genet. Metab.* **77,** 291–295.

Garcia-Garcia, M. J., Eggenschwiler, J. T., Caspary, T., Alcorn, H. L., Wyler, M. R., Huangfu, D., Rakeman, A. S., Lee, J. D., Feinberg, E. H., Timmer, J. R., *et al.* (2005). Analysis of mouse embryonic patterning and morphogenesis by forward genetics. *Proc. Natl Acad. Sci. USA* **102,** 5913–5919.

Gattone, V. H., II, Kuenstler, K. A., Lindemann, G. W., Lu, X., Cowley, B. D., Jr., Rankin, C. A., and Calvet, J. P. (1996). Renal expression of a transforming growth factor-alpha transgene accelerates the progression of inherited, slowly progressive polycystic kidney disease in the mouse. *J. Lab. Clin. Med.* **127,** 214–222.

Gerdes, J. M., Liu, Y., Zaghloul, N. A., Leitch, C. C., Lawson, S. S., Kato, M., Beachy, P. A., Beales, P. L., DeMartino, G. N., Fisher, S., *et al.* (2007). Disruption of the basal body compromises proteasomal function and perturbs intracellular Wnt response. *Nat. Genet.* **39,** 1350–1360.

Germino, G. G., and Somlo, S. (1993). Inherited diseases of the kidney. *Curr. Opin. Nephrol. Hypertens.* **2,** 430–440.

Giorgio, G., Alfieri, M., Prattichizzo, C., Zullo, A., Cairo, S., and Franco, B. (2007). Functional characterization of the OFD1 protein reveals a nuclear localization and physical interaction with subunits of a chromatin remodeling complex. *Mol. Biol. Cell* **18,** 4397–4404.

Giusto, T. J., and Sciubba, J. J. (2004). Kartagener's syndrome: A review of the literature and case report of oral findings in two siblings. *J. N. J. Dent. Assoc.* **75,** 30–32.

Gonzalez-Perrett, S., Kim, K., Ibarra, C., Damiano, A. E., Zotta, E., Batelli, M., Harris, P. C., Reisin, I. L., Arnaout, M. A., and Cantiello, H. F. (2001). Polycystin-2, the protein mutated in autosomal dominant polycystic kidney disease (ADPKD), is a Ca2+-permeable nonselective cation channel. *Proc. Natl Acad. Sci. USA* **98,** 1182–1187.

Grace, C., Beales, P., Summerbell, C., Jebb, S. A., Wright, A., Parker, D., and Kopelman, P. (2003). Energy metabolism in Bardet-Biedl syndrome. *Int. J. Obes. Relat. Metab. Disord.* **27,** 1319–1324.

Green, J. S., Parfrey, P. S., Harnett, J. D., Farid, N. R., Cramer, B. C., Johnson, G., Heath, O., McManamon, P. J., O'Leary, E., and Pryse-Phillips, W. (1989). The cardinal manifestations of Bardet-Biedl syndrome, a form of Laurence-Moon-Biedl syndrome. *N. Engl. J. Med.* **321,** 1002–1009.

Gretz, N., Kranzlin, B., Pey, R., Schieren, G., Bach, J., Obermuller, N., Ceccherini, I., Kloting, I., Rohmeiss, P., Bachmann, S., *et al.* (1996). Rat models of autosomal dominant polycystic kidney disease. *Nephrol. Dial. Transplant.* **11**(Suppl. 6), 46–51.

Guay-Woodford, L. M. (2003). Murine models of polycystic kidney disease: Molecular and therapeutic insights. *Am. J. Physiol. Renal Physiol.* **285,** F1034–F1049.

Guay-Woodford, L. M., Muecher, G., Hopkins, S. D., Avner, E. D., Germino, G. G., Guillot, A. P., Herrin, J., Holleman, R., Irons, D. A., Primack, W., *et al.* (1995). The severe perinatal form of autosomal recessive polycystic kidney disease maps to chromosome 6p21.1–p12: Implications for genetic counseling. *Am. J. Hum. Genet.* **56,** 1101–1107.

Guay-Woodford, L. M., Bryda, E. C., Christine, B., Lindsey, J. R., Collier, W. R., Avner, E. D., D'Eustachio, P., and Flaherty, L. (1996). Evidence that two phenotypically distinct mouse PKD mutations, bpk and jcpk, are allelic. *Kidney Int.* **50,** 1158–1165.

Habas, R., Dawid, I. B., and He, X. (2003). Coactivation of Rac and Rho by Wnt/Frizzled signaling is required for vertebrate gastrulation. *Genes Dev.* **17,** 295–309.

Han, Y. G., Spassky, N., Romaguera-Ros, M., Garcia-Verdugo, J. M., Aguilar, A., Schneider-Maunoury, S., and Alvarez-Buylla, A. (2008). Hedgehog signaling and primary cilia are required for the formation of adult neural stem cells. *Nat. Neurosci.* **11,** 277–284.

Hanaoka, K., Qian, F., Boletta, A., Bhunia, A. K., Piontek, K., Tsiokas, L., Sukhatme, V. P., Guggino, W. B., and Germino, G. G. (2000). Co-assembly of polycystin-1 and -2 produces unique cation-permeable currents. *Nature* **408,** 990–994.

Haraguchi, R., Mo, R., Hui, C., Motoyama, J., Makino, S., Shiroishi, T., Gaffield, W., and Yamada, G. (2001). Unique functions of Sonic hedgehog signaling during external genitalia development. *Development* **128,** 4241–4250.

Haycraft, C. J., Swoboda, P., Taulman, P. D., Thomas, J. H., and Yoder, B. K. (2001). The *C. elegans* homolog of the murine cystic kidney disease gene Tg737 functions in a ciliogenic pathway and is disrupted in osm-5 mutant worms. *Development* **128,** 1493–1505.

Haycraft, C. J., Banizs, B., Aydin-Son, Y., Zhang, Q., Michaud, E. J., and Yoder, B. K. (2005). Gli2 and Gli3 localize to cilia and require the intraflagellar transport protein polaris for processing and function. *PLoS Genet.* **1,** e53.

Hearn, T., Renforth, G. L., Spalluto, C., Hanley, N. A., Piper, K., Brickwood, S., White, C., Connolly, V., Taylor, J. F., Russell-Eggitt, I., *et al.* (2002). Mutation of ALMS1, a large gene with a tandem repeat encoding 47 amino acids, causes Alstrom syndrome. *Nat. Genet.* **31,** 79–83.

Hearn, T., Spalluto, C., Phillips, V. J., Renforth, G. L., Copin, N., Hanley, N. A., and Wilson, D. I. (2005). Subcellular localization of ALMS1 supports involvement of centrosome and basal body dysfunction in the pathogenesis of obesity, insulin resistance, and type 2 diabetes. *Diabetes* **54,** 1581–1587.

Hellemans, J., Coucke, P. J., Giedion, A., De Paepe, A., Kramer, P., Beemer, F., and Mortier, G. R. (2003). Homozygous mutations in IHH cause acrocapitofemoral dysplasia, an autosomal recessive disorder with cone-shaped epiphyses in hands and hips. *Am. J. Hum. Genet.* **72,** 1040–1046.

Henske, E. P. (2005). Tuberous sclerosis and the kidney: From mesenchyme to epithelium, and beyond. *Pediatr. Nephrol.* **20,** 854–857.

Hildebrandt, F., and Zhou, W. (2007). Nephronophthisis-associated ciliopathies. *J. Am. Soc. Nephrol.* **18,** 1855–1871.

Hildebrandt, F., Otto, E., Rensing, C., Nothwang, H. G., Vollmer, M., Adolphs, J., Hanusch, H., and Brandis, M. (1997). A novel gene encoding an SH3 domain protein is mutated in nephronophthisis type 1. *Nat. Genet.* **17,** 149–153.

Hou, X., Mrug, M., Yoder, B. K., Lefkowitz, E. J., Kremmidiotis, G., D'Eustachio, P., Beier, D. R., and Guay-Woodford, L. M. (2002). Cystin, a novel cilia-associated protein, is disrupted in the cpk mouse model of polycystic kidney disease. *J. Clin. Invest.* **109,** 533–540.

Hu, D., and Helms, J. A. (1999). The role of sonic hedgehog in normal and abnormal craniofacial morphogenesis. *Development* **126,** 4873–4884.

Huangfu, D., and Anderson, K. V. (2005). Cilia and Hedgehog responsiveness in the mouse. *Proc. Natl Acad. Sci. USA* **102,** 11325–11330.

Huangfu, D., Liu, A., Rakeman, A. S., Murcia, N. S., Niswander, L., and Anderson, K. V. (2003). Hedgehog signalling in the mouse requires intraflagellar transport proteins. *Nature* **426,** 83–87.

Hughes, J., Ward, C. J., Peral, B., Aspinwall, R., Clark, K., San Millan, J. L., Gamble, V., and Harris, P. C. (1995). The polycystic kidney disease 1 (PKD1) gene encodes a novel protein with multiple cell recognition domains. *Nat. Genet.* **10,** 151–160.

Hummel, C. (1959). *J. Hered.*

Iannaccone, A., Mykytyn, K., Persico, A. M., Searby, C. C., Baldi, A., Jablonski, M. M., and Sheffield, V. C. (2005). Clinical evidence of decreased olfaction in Bardet-Biedl syndrome caused by a deletion in the BBS4 gene. *Am. J. Med. Genet. A* **132,** 343–346.

Ibraghimov-Beskrovnaya, O., and Bukanov, N. (2008). Polycystic kidney diseases: From molecular discoveries to targeted therapeutic strategies. *Cell. Mol. Life Sci.* **65,** 605–619.

Ingham, P. W. (2008). Hedgehog signalling. *Curr. Biol.* **18,** R238–R241.

Janaswami, P. M., Birkenmeier, E. H., Cook, S. A., Rowe, L. B., Bronson, R. T., and Davisson, M. T. (1997). Identification and genetic mapping of a new polycystic kidney disease on mouse chromosome 8. *Genomics* **40,** 101–107.

Jenkins, D., Seelow, D., Jehee, F. S., Perlyn, C. A., Alonso, L. G., Bueno, D. F., Donnai, D., Josifova, D., Mathijssen, I. M., Morton, J. E., *et al.* (2007). RAB23 mutations in Carpenter syndrome imply an unexpected role for hedgehog signaling in cranial-suture development and obesity. *Am. J. Hum. Genet.* **80,** 1162–1170.

Jessen, J. R., Topczewski, J., Bingham, S., Sepich, D. S., Marlow, F., Chandrasekhar, A., and Solnica-Krezel, L. (2002). Zebrafish trilobite identifies new roles for Strabismus in gastrulation and neuronal movements. *Nat. Cell Biol.* **4,** 610–615.

Jiang, S. T., Chiou, Y. Y., Wang, E., Lin, H. K., Lee, S. P., Lu, H. Y., Wang, C. K., Tang, M. J., and Li, H. (2008). Targeted disruption of Nphp1 causes male infertility due to defects in the later steps of sperm morphogenesis in mice. *Hum. Mol. Genet.*

Johnson, K. A., and Gilbert, S. P. (1995). Pathway of the microtubule-kinesin ATPase. *Biophys. J.* **68,** 173S–176Sdiscussion 176S–179S.

Jones, C., and Chen, P. (2007). Planar cell polarity signaling in vertebrates. *BioEssays* **29,** 120–132.

Jones, C., Roper, V. C., Foucher, I., Qian, D., Banizs, B., Petit, C., Yoder, B. K., and Chen, P. (2008). Ciliary proteins link basal body polarization to planar cell polarity regulation. *Nat. Genet.* **40,** 69–77.

Joy, T., Cao, H., Black, G., Malik, R., Charlton-Menys, V., Hegele, R. A., and Durrington, P. N. (2007). Alstrom syndrome (OMIM 203800): A case report and literature review. *Orphanet J. Rare Dis.* **2,** 49.

Kaspareit-Rittinghausen, J., Deerberg, F., Rapp, K. G., and Wcislo, A. (1990). A new rat model for polycystic kidney disease of humans. *Transplant. Proc.* **22,** 2582–2583.

Katsanis, N., Beales, P. L., Woods, M. O., Lewis, R. A., Green, J. S., Parfrey, P. S., Ansley, S. J., Davidson, W. S., and Lupski, J. R. (2000). Mutations in MKKS cause obesity, retinal dystrophy and renal malformations associated with Bardet-Biedl syndrome. *Nat. Genet.* **26,** 67–70.

Keeler, C. E. (1931). "The Laboratory Mouse. Its Origin, Heredity and Culture." Harvard University Press, Cambridge.

Keeler, L. C., Marsh, S. E., Leeflang, E. P., Woods, C. G., Sztriha, L., Al-Gazali, L., Gururaj, A., and Gleeson, J. G. (2003). Linkage analysis in families with Joubert syndrome plus oculo-renal involvement identifies the CORS2 locus on chromosome 11p12–q13.3. *Am. J. Hum. Genet.* **73,** 656–662.

Kelly, M., and Chen, P. (2007). Shaping the mammalian auditory sensory organ by the planar cell polarity pathway. *Int. J. Dev. Biol.* **51,** 535–547.

Kibar, Z., Torban, E., McDearmid, J. R., Reynolds, A., Berghout, J., Mathieu, M., Kirillova, I., De Marco, P., Merello, E., Hayes, J. M., *et al.* (2007). Mutations in VANGL1 associated with neural-tube defects. *N. Engl. J. Med.* **356,** 1432–1437.

Kim, J. C., Badano, J. L., Sibold, S., Esmail, M. A., Hill, J., Hoskins, B. E., Leitch, C. C., Venner, K., Ansley, S. J., Ross, A. J., *et al.* (2004). The Bardet-Biedl protein BBS4 targets cargo to the pericentriolar region and is required for microtubule anchoring and cell cycle progression. *Nat. Genet.* **36,** 462–470.

Kim, J. C., Ou, Y. Y., Badano, J. L., Esmail, M. A., Leitch, C. C., Fiedrich, E., Beales, P. L., Archibald, J. M., Katsanis, N., Rattner, J. B., *et al.* (2005). MKKS/BBS6, a divergent chaperonin-like protein linked to the obesity disorder Bardet-Biedl syndrome, is a novel centrosomal component required for cytokinesis. *J. Cell Sci.* **118,** 1007–1020.

Kimberling, W. J., Kumar, S., Gabow, P. A., Kenyon, J. B., Connolly, C. J., and Somlo, S. (1993). Autosomal dominant polycystic kidney disease: Localization of the second gene to chromosome 4q13–q23. *Genomics* **18,** 467–472.

Kishimoto, N., Cao, Y., Park, A., and Sun, Z. (2008). Cystic kidney gene seahorse regulates cilia-mediated processes and Wnt pathways. *Dev. Cell* **14,** 954–961.

Kolpakova-Hart, E., Jinnin, M., Hou, B., Fukai, N., and Olsen, B. R. (2007). Kinesin-2 controls development and patterning of the vertebrate skeleton by Hedgehog- and Gli3-dependent mechanisms. *Dev. Biol.* **309,** 273–284.

Kolpakova-Hart, E., Nicolae, C., Zhou, J., and Olsen, B. R. (2008). Col2-Cre recombinase is co-expressed with endogenous type II collagen in embryonic renal epithelium and drives development of polycystic kidney disease following inactivation of ciliary genes. *Matrix Biol.* **27,** 505–512.

Kondo, S., Sato-Yoshitake, R., Noda, Y., Aizawa, H., Nakata, T., Matsuura, Y., and Hirokawa, N. (1994). KIF3A is a new microtubule-based anterograde motor in the nerve axon. *J. Cell Biol.* **125,** 1095–1107.

Konig, G., and Hausen, P. (1993). Planar polarity in the ciliated epidermis of Xenopus embryos. *Dev. Biol.* **160,** 355–368.

Koyama, E., Young, B., Nagayama, M., Shibukawa, Y., Enomoto-Iwamoto, M., Iwamoto, M., Maeda, Y., Lanske, B., Song, B., Serra, R., *et al.* (2007). Conditional Kif3a ablation causes abnormal hedgehog signaling topography, growth plate dysfunction, and excessive bone and cartilage formation during mouse skeletogenesis. *Development* **134,** 2159–2169.

Kozminski, K. G., Johnson, K. A., Forscher, P., and Rosenbaum, J. L. (1993). A motility in the eukaryotic flagellum unrelated to flagellar beating. *Proc. Natl Acad. Sci. USA* **90,** 5519–5523.

Kozminski, K. G., Beech, P. L., and Rosenbaum, J. L. (1995). The Chlamydomonas kinesin-like protein FLA10 is involved in motility associated with the flagellar membrane. *J. Cell Biol.* **131,** 1517–1527.

Kramer-Zucker, A. G., Olale, F., Haycraft, C. J., Yoder, B. K., Schier, A. F., and Drummond, I. A. (2005a). Cilia-driven fluid flow in the zebrafish pronephros, brain and Kupffer's vesicle is required for normal organogenesis. *Development* **132,** 1907–1921.

Kramer-Zucker, A. G., Wiessner, S., Jensen, A. M., and Drummond, I. A. (2005b). Organization of the pronephric filtration apparatus in zebrafish requires Nephrin, Podocin and the FERM domain protein Mosaic eyes. *Dev. Biol.* **285,** 316–329.

Kulaga, H. M., Leitch, C. C., Eichers, E. R., Badano, J. L., Lesemann, A., Hoskins, B. E., Lupski, J. R., Beales, P. L., Reed, R. R., and Katsanis, N. (2004). Loss of BBS proteins causes anosmia in humans and defects in olfactory cilia structure and function in the mouse. *Nat. Genet.* **36,** 994–998.

Kyttala, M., Tallila, J., Salonen, R., Kopra, O., Kohlschmidt, N., Paavola-Sakki, P., Peltonen, L., and Kestila, M. (2006). MKS1, encoding a component of the flagellar apparatus basal body proteome, is mutated in Meckel syndrome. *Nat. Genet.* **38,** 155–157.

Lager, D. J., Qian, Q., Bengal, R. J., Ishibashi, M., and Torres, V. E. (2001). The pck rat: A new model that resembles human autosomal dominant polycystic kidney and liver disease. *Kidney Int.* **59,** 126–136.

Lagier-Tourenne, C., Boltshauser, E., Breivik, N., Gribaa, M., Betard, C., Barbot, C., and Koenig, M. (2004). Homozygosity mapping of a third Joubert syndrome locus to 6q23. *J. Med. Genet.* **41,** 273–277.

Lai, C. H., Chou, C. Y., Ch'ang, L. Y., Liu, C. S., and Lin, W. (2000). Identification of novel human genes evolutionarily conserved in *Caenorhabditis elegans* by comparative proteomics. *Genome Res.* **10,** 703–713.

Lantinga-van Leeuwen, I. S., Dauwerse, J. G., Baelde, H. J., Leonhard, W. N., van de Wal, A., Ward, C. J., Verbeek, S., Deruiter, M. C., Breuning, M. H., de Heer, E., *et al.* (2004). Lowering of Pkd1 expression is sufficient to cause polycystic kidney disease. *Hum. Mol. Genet.* **13,** 3069–3077.

Lehman, J. M., Michaud, E. J., Schoeb, T. R., Aydin-Son, Y., Miller, M., and Yoder, B. K. (2008). The Oak Ridge Polycystic Kidney mouse: Modeling ciliopathies of mice and men. *Dev. Dyn.* **237,** 1960–1971.

Li, J. B., Gerdes, J. M., Haycraft, C. J., Fan, Y., Teslovich, T. M., May-Simera, H., Li, H., Blacque, O. E., Li, L., Leitch, C. C., *et al.* (2004). Comparative genomics identifies a flagellar and basal body proteome that includes the BBS5 human disease gene. *Cell* **117,** 541–552.

Li, G., *et al.* (2007). A role for Alstrom syndrome protein, alms1, in kidney ciliogenesis and cellular quiescence. *PLoS Genet.* **3**(1), e8.

Lin, F., Hiesberger, T., Cordes, K., Sinclair, A. M., Goldstein, L. S., Somlo, S., and Igarashi, P. (2003). Kidney-specific inactivation of the KIF3A subunit of kinesin-II inhibits renal ciliogenesis and produces polycystic kidney disease. *Proc. Natl Acad. Sci. USA* **100,** 5286–5291.

Lina, F., and Satlinb, L. M. (2004). Polycystic kidney disease: The cilium as a common pathway in cystogenesis. *Curr. Opin. Pediatr.* **16,** 171–176.

Liu, S., Lu, W., Obara, T., Kuida, S., Lehoczky, J., Dewar, K., Drummond, I. A., and Beier, D. R. (2002). A defect in a novel Nek-family kinase causes cystic kidney disease in the mouse and in zebrafish. *Development* **129,** 5839–5846.

Liu, A., Wang, B., and Niswander, L. A. (2005). Mouse intraflagellar transport proteins regulate both the activator and repressor functions of Gli transcription factors. *Development* **132,** 3103–3111.

Louie, C. M., and Gleeson, J. G. (2005). Genetic basis of Joubert syndrome and related disorders of cerebellar development. *Hum. Mol. Genet.* **14**(Spec No. 2), R235–R242.

Lu, W., Peissel, B., Babakhanlou, H., Pavlova, A., Geng, L., Fan, X., Larson, C., Brent, G., and Zhou, J. (1997). Perinatal lethality with kidney and pancreas defects in mice with a targetted Pkd1 mutation. *Nat. Genet.* **17,** 179–181.

Mandell, J., Koch, W. K., Nidess, R., Preminger, G. M., and McFarland, E. (1983). Congenital polycystic kidney disease. Genetically transmitted infantile polycystic kidney disease in C57BL/6J mice. *Am. J. Pathol.* **113,** 112–114.

Mans, D. A., Lolkema, M. P., van Beest, M., Daenen, L. G., Voest, E. E., and Giles, R. H. (2008). Mobility of the von Hippel-Lindau tumour suppressor protein is regulated by kinesin-2. *Exp. Cell Res.* **314,** 1229–1236.

Marshall, J. D., Bronson, R. T., Collin, G. B., Nordstrom, A. D., Maffei, P., Paisey, R. B., Carey, C., Macdermott, S., Russell-Eggitt, I., Shea, S. E., *et al.* (2005). New Alstrom syndrome phenotypes based on the evaluation of 182 cases. *Arch. Intern. Med.* **165,** 675–683.

Marszalek, J. R., Ruiz-Lozano, P., Roberts, E., Chien, K. R., and Goldstein, L. S. (1999). Situs inversus and embryonic ciliary morphogenesis defects in mouse mutants lacking the KIF3A subunit of kinesin-II. *Proc. Natl Acad. Sci. USA* **96,** 5043–5048.

Marszalek, J. R., Liu, X., Roberts, E. A., Chui, D., Marth, J. D., Williams, D. S., and Goldstein, L. S. (2000). Genetic evidence for selective transport of opsin and arrestin by kinesin-II in mammalian photoreceptors. *Cell* **102,** 175–187.

May, S. R., Ashique, A. M., Karlen, M., Wang, B., Shen, Y., Zarbalis, K., Reiter, J., Ericson, J., and Peterson, A. S. (2005). Loss of the retrograde motor for IFT disrupts localization of Smo to cilia and prevents the expression of both activator and repressor functions of Gli. *Dev. Biol.* **287,** 378–389.

McEwen, D. P., Koenekoop, R. K., Khanna, H., Jenkins, P. M., Lopez, I., Swaroop, A., and Martens, J. R. (2007). Hypomorphic CEP290/NPHP6 mutations result in anosmia caused by the selective loss of G proteins in cilia of olfactory sensory neurons. *Proc. Natl Acad. Sci. USA* **104,** 15917–15922.

McGlinn, E., and Tabin, C. J. (2006). Mechanistic insight into how Shh patterns the vertebrate limb. *Curr. Opin. Genet. Dev.* **16,** 426–432.

McGrath, J., Somlo, S., Makova, S., Tian, X., and Brueckner, M. (2003). Two populations of node monocilia initiate left–right asymmetry in the mouse. *Cell* **114,** 61–73.

Mochizuki, T., Wu, G., Hayashi, T., Xenophontos, S. L., Veldhuisen, B., Saris, J. J., Reynolds, D. M., Cai, Y., Gabow, P. A., Pierides, A., *et al.* (1996). PKD2, a gene for polycystic kidney disease that encodes an integral membrane protein. *Science* **272,** 1339–1342.

Mochizuki, T., Saijoh, Y., Tsuchiya, K., Shirayoshi, Y., Takai, S., Taya, C., Yonekawa, H., Yamada, K., Nihei, H., Nakatsuji, N., *et al.* (1998). Cloning of inv, a gene that controls left/right asymmetry and kidney development. *Nature* **395,** 177–181.

Mollet, G., Salomon, R., Gribouval, O., Silbermann, F., Bacq, D., Landthaler, G., Milford, D., Nayir, A., Rizzoni, G., Antignac, C., *et al.* (2002). The gene mutated in juvenile nephronophthisis type 4 encodes a novel protein that interacts with nephrocystin. *Nat. Genet.* **32,** 300–305.

Mollet, G., Silbermann, F., Delous, M., Salomon, R., Antignac, C., and Saunier, S. (2005). Characterization of the nephrocystin/nephrocystin-4 complex and subcellular localization of nephrocystin-4 to primary cilia and centrosomes. *Hum. Mol. Genet.* **14,** 645–656.

Morgan, D., Eley, L., Sayer, J., Strachan, T., Yates, L. M., Craighead, A. S., and Goodship, J. A. (2002a). Expression analyses and interaction with the anaphase promoting complex protein Apc2 suggest a role for inversin in primary cilia and involvement in the cell cycle. *Hum. Mol. Genet.* **11,** 3345–3350.

Morgan, N. V., Gissen, P., Sharif, S. M., Baumber, L., Sutherland, J., Kelly, D. A., Aminu, K., Bennett, C. P., Woods, C. G., Mueller, R. F., *et al.* (2002b). A novel locus for Meckel-Gruber syndrome, MKS3, maps to chromosome 8q24. *Hum. Genet.* **111,** 456–461.

Morgan, N. V., Bacchelli, C., Gissen, P., Morton, J., Ferrero, G. B., Silengo, M., Labrune, P., Casteels, I., Hall, C., Cox, P., *et al.* (2003). A locus for asphyxiating thoracic dystrophy, ATD, maps to chromosome 15q13. *J. Med. Genet.* **40,** 431–435.

Morris, R. L., and Scholey, J. M. (1997). Heterotrimeric kinesin-II is required for the assembly of motile 9 + 2 ciliary axonemes on sea urchin embryos. *J. Cell Biol.* **138,** 1009–1022.

Moyer, J. H., Lee-Tischler, M. J., Kwon, H. Y., Schrick, J. J., Avner, E. D., Sweeney, W. E., Godfrey, V. L., Cacheiro, N. L., Wilkinson, J. E., and Woychik, R. P. (1994). Candidate gene associated with a mutation causing recessive polycystic kidney disease in mice. *Science* **264,** 1329–1333.

Murcia, N. S., Richards, W. G., Yoder, B. K., Mucenski, M. L., Dunlap, J. R., and Woychik, R. P. (2000). The Oak Ridge Polycystic Kidney (orpk) disease gene is required for left–right axis determination. *Development* **127,** 2347–2355.

Mykytyn, K., Braun, T., Carmi, R., Haider, N. B., Searby, C. C., Shastri, M., Beck, G., Wright, A. F., Iannaccone, A., Elbedour, K., *et al.* (2001). Identification of the gene that, when mutated, causes the human obesity syndrome BBS4. *Nat. Genet.* **28,** 188–191.

Mykytyn, K., Nishimura, D. Y., Searby, C. C., Shastri, M., Yen, H. J., Beck, J. S., Braun, T., Streb, L. M., Cornier, A. S., Cox, G. F., *et al.* (2002). Identification of the gene (BBS1) most commonly involved in Bardet-Biedl syndrome, a complex human obesity syndrome. *Nat. Genet.* **31,** 435–438.

Mykytyn, K., Mullins, R. F., Andrews, M., Chiang, A. P., Swiderski, R. E., Yang, B., Braun, T., Casavant, T., Stone, E. M., and Sheffield, V. C. (2004). Bardet-Biedl syndrome type 4 (BBS4)-null mice implicate Bbs4 in flagella formation but not global cilia assembly. *Proc. Natl Acad. Sci. USA* **101,** 8664–8669.

Nachury, M. V., Loktev, A. V., Zhang, Q., Westlake, C. J., Peranen, J., Merdes, A., Slusarski, D. C., Scheller, R. H., Bazan, J. F., Sheffield, V. C., *et al.* (2007). A core complex of BBS proteins cooperates with the GTPase Rab8 to promote ciliary membrane biogenesis. *Cell* **129,** 1201–1213.

Nagao, S., Hibino, T., Koyama, Y., Marunouchi, T., Konishi, H., and Takahashi, H. (1991). Strain difference in expression of the adult-type polycystic kidney disease gene, pcy, in the mouse. *Jikken Dobutsu* **40,** 45–53.

Nagase, T., Kikuno, R., Ishikawa, K. I., Hirosawa, M., and Ohara, O. (2000). Prediction of the coding sequences of unidentified human genes. XVI. The complete sequences of 150 new cDNA clones from brain which code for large proteins *in vitro*. *DNA Res.* **7,** 65–73.

Nauli, S. M., Alenghat, F. J., Luo, Y., Williams, E., Vassilev, P., Li, X., Elia, A. E., Lu, W., Brown, E. M., Quinn, S. J., *et al.* (2003). Polycystins 1 and 2 mediate mechanosensation in the primary cilium of kidney cells. *Nat. Genet.* **33,** 129–137.

Nauta, J., Ozawa, Y., Sweeney, W. E., Jr., Rutledge, J. C., and Avner, E. D. (1993). Renal and biliary abnormalities in a new murine model of autosomal recessive polycystic kidney disease. *Pediatr. Nephrol.* **7,** 163–172.

Nauta, J., Goedbloed, M. A., Luider, T. M., Hoogeveen, A. T., van den Ouweland, A. M., and Halley, D. J. (1997). The Han:SPRD rat is not a genetic model of human autosomal dominant polycystic kidney disease type 1. *Lab. Anim.* **31,** 241–247.

Nauta, J., Goedbloed, M. A., Herck, H. V., Hesselink, D. A., Visser, P., Willemsen, R., Dokkum, R. P., Wright, C. J., and Guay-Woodford, L. M. (2000). New rat model that phenotypically resembles autosomal recessive polycystic kidney disease. *J. Am. Soc. Nephrol.* **11,** 2272–2284.

Nishimura, D. Y., Searby, C. C., Carmi, R., Elbedour, K., Van Maldergem, L., Fulton, A. B., Lam, B. L., Powell, B. R., Swiderski, R. E., Bugge, K. E., *et al.* (2001). Positional cloning of a novel gene on chromosome 16q causing Bardet-Biedl syndrome (BBS2). *Hum. Mol. Genet.* **10,** 865–874.

Nishimura, D. Y., Fath, M., Mullins, R. F., Searby, C., Andrews, M., Davis, R., Andorf, J. L., Mykytyn, K., Swiderski, R. E., Yang, B., *et al.* (2004). Bbs2-null mice have neurosensory deficits, a defect in social dominance, and retinopathy associated with mislocalization of rhodopsin. *Proc. Natl Acad. Sci. USA* **101,** 16588–16593.

Nishimura, D. Y., Swiderski, R. E., Searby, C. C., Berg, E. M., Ferguson, A. L., Hennekam, R., Merin, S., Weleber, R. G., Biesecker, L. G., Stone, E. M., *et al.*

(2005). Comparative genomics and gene expression analysis identifies BBS9, a new Bardet-Biedl syndrome gene. *Am. J. Hum. Genet.* **77,** 1021–1033.

Noben-Trauth, K., Naggert, J. K., North, M. A., and Nishina, P. M. (1996). A candidate gene for the mouse mutation tubby. *Nature* **380,** 534–538.

Nonaka, S., Tanaka, Y., Okada, Y., Takeda, S., Harada, A., Kanai, Y., Kido, M., and Hirokawa, N. (1998). Randomization of left–right asymmetry due to loss of nodal cilia generating leftward flow of extraembryonic fluid in mice lacking KIF3B motor protein. *Cell* **95,** 829–837.

Nonaka, S., Shiratori, H., Saijoh, Y., and Hamada, H. (2002). Determination of left–right patterning of the mouse embryo by artificial nodal flow. *Nature* **418,** 96–99.

Noor, A., Windpassinger, C., Patel, M., Stachowiak, B., Mikhailov, A., Azam, M., Irfan, M., Siddiqui, Z. K., Naeem, F., Paterson, A. D., *et al.* (2008). CC2D2A, encoding a coiled-coil and C2 domain protein, causes autosomal-recessive mental retardation with retinitis pigmentosa. *Am. J. Hum. Genet.* **82,** 1011–1018.

Nurnberger, J., Bacallao, R. L., and Phillips, C. L. (2002). Inversin forms a complex with catenins and N-cadherin in polarized epithelial cells. *Mol. Biol. Cell* **13,** 3096–3106.

Ocbina, P. J., and Anderson, K. V. (2008). Intraflagellar transport, cilia, and mammalian Hedgehog signaling: Analysis in mouse embryonic fibroblasts. *Dev. Dyn.* **237,** 2030–2038.

Okada, Y., Nonaka, S., Tanaka, Y., Saijoh, Y., Hamada, H., and Hirokawa, N. (1999). Abnormal nodal flow precedes situs inversus in iv and inv mice. *Mol. Cell* **4,** 459–468.

Olbrich, H., Fliegauf, M., Hoefele, J., Kispert, A., Otto, E., Volz, A., Wolf, M. T., Sasmaz, G., Trauer, U., Reinhardt, R., *et al.* (2003). Mutations in a novel gene, NPHP3, cause adolescent nephronophthisis, tapeto-retinal degeneration and hepatic fibrosis. *Nat. Genet.* **34,** 455–459.

Omran, H., Haffner, K., Burth, S., Fernandez, C., Fargier, B., Villaquiran, A., Nothwang, H. G., Schnittger, S., Lehrach, H., Woo, D., *et al.* (2001). Human adolescent nephronophthisis: Gene locus synteny with polycystic kidney disease in pcy mice. *J. Am. Soc. Nephrol.* **12,** 107–113.

Omran, H., *et al.* (2002). Identification of a gene locus for Senior-Loken syndrome in the region of the nephronophthisis type 3 gene. *J. Am. Soc. Nephrol.* **13**(1), 75–79.

Ong, A. C., and Harris, P. C. (2005). Molecular pathogenesis of ADPKD: The polycystin complex gets complex. *Kidney Int.* **67,** 1234–1247.

Onuchic, L. F., Schrick, J. J., Ma, J., Hudson, T., Guay-Woodford, L. M., Zerres, K., Woychik, R. P., and Reeders, S. T. (1995). Sequence analysis of the human hTg737 gene and its polymorphic sites in patients with autosomal recessive polycystic kidney disease. *Mamm. Genome.* **6,** 805–808.

Onuchic, L. F., Furu, L., Nagasawa, Y., Hou, X., Eggermann, T., Ren, Z., Bergmann, C., Senderek, J., Esquivel, E., Zeltner, R., *et al.* (2002). PKHD1, the polycystic kidney and hepatic disease 1 gene, encodes a novel large protein containing multiple immunoglobulin-like plexin-transcription-factor domains and parallel beta-helix 1 repeats. *Am. J. Hum. Genet.* **70,** 1305–1317.

Oti, M., Huynen, M. A., and Brunner, H. G. (2008). Phenome connections. *Trends Genet.* **24,** 103–106.

Otto, E., Hoefele, J., Ruf, R., Mueller, A. M., Hiller, K. S., Wolf, M. T., Schuermann, M. J., Becker, A., Birkenhager, R., Sudbrak, R., *et al.* (2002). A gene mutated in nephronophthisis and retinitis pigmentosa encodes a novel protein, nephroretinin, conserved in evolution. *Am. J. Hum. Genet.* **71,** 1161–1167.

Otto, E. A., Schermer, B., Obara, T., O'Toole, J. F., Hiller, K. S., Mueller, A. M., Ruf, R. G., Hoefele, J., Beekmann, F., Landau, D., *et al.* (2003). Mutations in INVS encoding inversin cause nephronophthisis type 2, linking renal cystic disease to the function of primary cilia and left–right axis determination. *Nat. Genet.* **34,** 413–420.

Otto, E. A., Loeys, B., Khanna, H., Hellemans, J., Sudbrak, R., Fan, S., Muerb, U., O'Toole, J. F., Helou, J., Attanasio, M., *et al.* (2005). Nephrocystin-5, a ciliary IQ domain protein, is mutated in Senior-Loken syndrome and interacts with RPGR and calmodulin. *Nat. Genet.* **37,** 282–288.

Otto, E. A., Trapp, M. L., Schultheiss, U. T., Helou, J., Quarmby, L. M., and Hildebrandt, F. (2008). *NEK8* mutations affect ciliary and centrosomal localization and may cause nephronophthisis. *J. Am. Soc. Nephrol.* **19,** 587–592.

Paavola, P., Salonen, R., Baumer, A., Schinzel, A., Boyd, P. A., Gould, S., Meusburger, H., Tenconi, R., Barnicoat, A., Winter, R., *et al.* (1997). Clinical and genetic heterogeneity in Meckel syndrome. *Hum. Genet.* **101,** 88–92.

Parisi, M. A., Bennett, C. L., Eckert, M. L., Dobyns, W. B., Gleeson, J. G., Shaw, D. W., McDonald, R., Eddy, A., Chance, P. F., and Glass, I. A. (2004). The NPHP1 gene deletion associated with juvenile nephronophthisis is present in a subset of individuals with Joubert syndrome. *Am. J. Hum. Genet.* **75,** 82–91.

Parisi, M. A., Doherty, D., Eckert, M. L., Shaw, D. W., Ozyurek, H., Aysun, S., Giray, O., Al Swaid, A., Al Shahwan, S., Dohayan, N., *et al.* (2006). AHI1 mutations cause both retinal dystrophy and renal cystic disease in Joubert syndrome. *J. Med. Genet.* **43,** 334–339.

Park, T. J., Haigo, S. L., and Wallingford, J. B. (2006). Ciliogenesis defects in embryos lacking inturned or fuzzy function are associated with failure of planar cell polarity and Hedgehog signaling. *Nat. Genet.* **38,** 303–311.

Patel, V., Li, L., Cobo-Stark, P., Shao, X., Somlo, S., Lin, F., and Igarashi, P. (2008). Acute kidney injury and aberrant planar cell polarity induce cyst formation in mice lacking renal cilia. *Hum. Mol. Genet.* **17,** 1578–1590.

Pazour, G. J., Dickert, B. L., and Witman, G. B. (1999). The DHC1b (DHC2) isoform of cytoplasmic dynein is required for flagellar assembly. *J. Cell Biol.* **144,** 473–481.

Pazour, G. J., Dickert, B. L., Vucica, Y., Seeley, E. S., Rosenbaum, J. L., Witman, G. B., and Cole, D. G. (2000). Chlamydomonas IFT88 and its mouse homologue, polycystic kidney disease gene tg737, are required for assembly of cilia and flagella. *J. Cell Biol.* **151,** 709–718.

Pazour, G. J., Baker, S. A., Deane, J. A., Cole, D. G., Dickert, B. L., Rosenbaum, J. L., Witman, G. B., and Besharse, J. C. (2002a). The intraflagellar transport protein, IFT88, is essential for vertebrate photoreceptor assembly and maintenance. *J. Cell Biol.* **157,** 103–113.

Pazour, G. J., San Agustin, J. T., Follit, J. A., Rosenbaum, J. L., and Witman, G. B. (2002b). Polycystin-2 localizes to kidney cilia and the ciliary level is elevated in orpk mice with polycystic kidney disease. *Curr. Biol.* **12,** R378–R380.

Pennekamp, P., Karcher, C., Fischer, A., Schweickert, A., Skryabin, B., Horst, J., Blum, M., and Dworniczak, B. (2002). The ion channel polycystin-2 is required for left–right axis determination in mice. *Curr. Biol.* **12,** 938–943.

Perkins, L. A., Hedgecock, E. M., Thomson, J. N., and Culotti, J. G. (1986). Mutant sensory cilia in the nematode *Caenorhabditis elegans*. *Dev. Biol.* **117,** 456–487.

Perriton, C. L., Powles, N., Chiang, C., Maconochie, M. K., and Cohn, M. J. (2002). Sonic hedgehog signaling from the urethral epithelium controls external genital development. *Dev. Biol.* **247,** 26–46.

Peters, T., Ausmeier, K., and Ruther, U. (1999). Cloning of Fatso (Fto), a novel gene deleted by the Fused toes (Ft) mouse mutation. *Mamm. Genome.* **10,** 983–986.

Peters, T., Ausmeier, K., Dildrop, R., and Ruther, U. (2002). The mouse Fused toes (Ft) mutation is the result of a 1.6-Mb deletion including the entire Iroquois B gene cluster. *Mamm. Genome.* **13,** 186–188.

Piperno, G., and Mead, K. (1997). Transport of a novel complex in the cytoplasmic matrix of Chlamydomonas flagella. *Proc. Natl Acad. Sci. USA* **94,** 4457–4462.

Porter, M. E., and Sale, W. S. (2000). The 9 + 2 axoneme anchors multiple inner arm dyneins and a network of kinases and phosphatases that control motility. *J. Cell Biol.* **151,** F37–F42.

Porter, M. E., Bower, R., Knott, J. A., Byrd, P., and Dentler, W. (1999). Cytoplasmic dynein heavy chain 1b is required for flagellar assembly in Chlamydomonas. *Mol. Biol. Cell* **10,** 693–712.

Qian, F., Germino, F. J., Cai, Y., Zhang, X., Somlo, S., and Germino, G. G. (1997). PKD1 interacts with PKD2 through a probable coiled-coil domain. *Nat. Genet.* **16,** 179–183.

Qin, H., Rosenbaum, J. L., and Barr, M. M. (2001). An autosomal recessive polycystic kidney disease gene homolog is involved in intraflagellar transport in *C. elegans* ciliated sensory neurons. *Curr. Biol.* **11,** 457–461.

Rahmouni, K., Fath, M. A., Seo, S., Thedens, D. R., Berry, C. J., Weiss, R., Nishimura, D. Y., and Sheffield, V. C. (2008). Leptin resistance contributes to obesity and hypertension in mouse models of Bardet-Biedl syndrome. *J. Clin. Invest.* **118,** 1458–1467.

Rana, A. A., Barbera, J. P., Rodriguez, T. A., Lynch, D., Hirst, E., Smith, J. C., and Beddington, R. S. (2004). Targeted deletion of the novel cytoplasmic dynein mD2LIC disrupts the embryonic organiser, formation of the body axes and specification of ventral cell fates. *Development* **131,** 4999–5007.

Robert, A., Margall-Ducos, G., Guidotti, J. E., Bregerie, O., Celati, C., Brechot, C., and Desdouets, C. (2007). The intraflagellar transport component IFT88/polaris is a centrosomal protein regulating G1-S transition in non-ciliated cells. *J. Cell Sci.* **120,** 628–637.

Roepman, R., Letteboer, S. J., Arts, H. H., van Beersum, S. E., Lu, X., Krieger, E., Ferreira, P. A., and Cremers, F. P. (2005). Interaction of nephrocystin-4 and RPGRIP1 is disrupted by nephronophthisis or Leber congenital amaurosis-associated mutations. *Proc. Natl Acad. Sci. USA* **102,** 18520–18525.

Rohatgi, R., Milenkovic, L., and Scott, M. P. (2007). Patched1 regulates hedgehog signaling at the primary cilium. *Science* **317,** 372–376.

Ross, A. J., May-Simera, H., Eichers, E. R., Kai, M., Hill, J., Jagger, D. J., Leitch, C. C., Chapple, J. P., Munro, P. M., Fisher, S., *et al.* (2005). Disruption of Bardet-Biedl syndrome ciliary proteins perturbs planar cell polarity in vertebrates. *Nat. Genet.* **37,** 1135–1140.

Ruiz-Perez, V. L., Ide, S. E., Strom, T. M., Lorenz, B., Wilson, D., Woods, K., King, L., Francomano, C., Freisinger, P., Spranger, S., *et al.* (2000). Mutations in a new gene in Ellis-van Creveld syndrome and Weyers acrodental dysostosis. *Nat. Genet.* **24,** 283–286.

Ruiz-Perez, V. L., Blair, H. J., Rodriguez-Andres, M. E., Blanco, M. J., Wilson, A., Liu, Y. N., Miles, C., Peters, H., and Goodship, J. A. (2007). Evc is a positive mediator of Ihh-regulated bone growth that localises at the base of chondrocyte cilia. *Development* **134,** 2903–2912.

Russell-Eggitt, I. M., Clayton, P. T., Coffey, R., Kriss, A., Taylor, D. S., and Taylor, J. F. (1998). Alstrom syndrome. Report of 22 cases and literature review. *Ophthalmology* **105,** 1274–1280.

Saar, K., Al-Gazali, L., Sztriha, L., Rueschendorf, F., Nur, E. K. M., Reis, A., and Bayoumi, R. (1999). Homozygosity mapping in families with Joubert syndrome identifies a locus on chromosome 9q34.3 and evidence for genetic heterogeneity. *Am. J. Hum. Genet.* **65,** 1666–1671.

Saburi, S., Hester, I., Fischer, E., Pontoglio, M., Eremina, V., Gessler, M., Quaggin, S. E., Harrison, R., Mount, R., and McNeill, H. (2008). Loss of Fat4 disrupts PCP signaling and oriented cell division and leads to cystic kidney disease. *Nat. Genet.* **40,** 1010–1015.

Sage, R. D. (1981). "The Mouse in Biomedical Research," Vol. 1, pp. 40–90. Academic Press, New York.

Salonen, R. (1984). The Meckel syndrome: Clinicopathological findings in 67 patients. *Am. J. Med. Genet.* **18,** 671–689.

Salonen, R., and Paavola, P. (1998). Meckel syndrome. *J. Med. Genet.* **35,** 497–501.
Sanzen, T., Harada, K., Yasoshima, M., Kawamura, Y., Ishibashi, M., and Nakanuma, Y. (2001). Polycystic kidney rat is a novel animal model of Caroli's disease associated with congenital hepatic fibrosis. *Am. J. Pathol.* **158,** 1605–1612.
Sayer, J. A., Otto, E. A., O'Toole, J. F., Nurnberg, G., Kennedy, M. A., Becker, C., Hennies, H. C., Helou, J., Attanasio, M., Fausett, B. V., *et al.* (2006). The centrosomal protein nephrocystin-6 is mutated in Joubert syndrome and activates transcription factor ATF4. *Nat. Genet.* **38,** 674–681.
Schafer, K., Gretz, N., Bader, M., Oberbaumer, I., Eckardt, K. U., Kriz, W., and Bachmann, S. (1994). Characterization of the Han:SPRD rat model for hereditary polycystic kidney disease. *Kidney Int.* **46,** 134–152.
Schneider, M. C., Rodriguez, A. M., Nomura, H., Zhou, J., Morton, C. C., Reeders, S. T., and Weremowicz, S. (1996). A gene similar to PKD1 maps to chromosome 4q22: A candidate gene for PKD2. *Genomics* **38,** 1–4.
Schrick, J. J., Onuchic, L. F., Reeders, S. T., Korenberg, J., Chen, X. N., Moyer, J. H., Wilkinson, J. E., and Woychik, R. P. (1995). Characterization of the human homologue of the mouse Tg737 candidate polycystic kidney disease gene. *Hum. Mol. Genet.* **4,** 559–567.
Schuermann, M. J., Otto, E., Becker, A., Saar, K., Ruschendorf, F., Polak, B. C., Ala-Mello, S., Hoefele, J., Wiedensohler, A., Haller, M., *et al.* (2002). Mapping of gene loci for nephronophthisis type 4 and Senior-Loken syndrome, to chromosome 1p36. *Am. J. Hum. Genet.* **70,** 1240–1246.
Shah, H., *et al.* (2008). Loss of Bardet-Biedl syndrome proteins alters the morphology and function of motile cilia in airway epithelia. *Proc. Natl. Acad. Sci. USA* **105**(9), 3380–3385.
Sheng, G., Xu, X., Lin, Y. F., Wang, C. E., Rong, J., Cheng, D., Peng, J., Jiang, X., Li, S. H., and Li, X. J. (2008). Huntingtin-associated protein 1 interacts with Ahi1 to regulate cerebellar and brainstem development in mice. *J. Clin. Invest.* **118,** 2785–2795.
Shillingford, J. M., Murcia, N. S., Larson, C. H., Low, S. H., Hedgepeth, R., Brown, N., Flask, C. A., Novick, A. C., Goldfarb, D. A., Kramer-Zucker, A., *et al.* (2006). The mTOR pathway is regulated by polycystin-1, and its inhibition reverses renal cystogenesis in polycystic kidney disease. *Proc. Natl Acad. Sci. USA* **103,** 5466–5471.
Signor, D., Wedaman, K. P., Orozco, J. T., Dwyer, N. D., Bargmann, C. I., Rose, L. S., and Scholey, J. M. (1999). Role of a class DHC1b dynein in retrograde transport of IFT motors and IFT raft particles along cilia, but not dendrites, in chemosensory neurons of living *Caenorhabditis elegans*. *J. Cell Biol.* **147,** 519–530.
Simons, M., Gloy, J., Ganner, A., Bullerkotte, A., Bashkurov, M., Kronig, C., Schermer, B., Benzing, T., Cabello, O. A., Jenny, A., *et al.* (2005). Inversin, the gene product mutated in nephronophthisis type II, functions as a molecular switch between Wnt signaling pathways. *Nat. Genet.* **37,** 537–543.
Slavotinek, A. M., Stone, E. M., Mykytyn, K., Heckenlively, J. R., Green, J. S., Heon, E., Musarella, M. A., Parfrey, P. S., Sheffield, V. C., and Biesecker, L. G. (2000). Mutations in MKKS cause Bardet-Biedl syndrome. *Nat. Genet.* **26,** 15–16.
Smith, U. M., Consugar, M., Tee, L. J., McKee, B. M., Maina, E. N., Whelan, S., Morgan, N. V., Goranson, E., Gissen, P., Lilliquist, S., *et al.* (2006). The transmembrane protein meckelin (MKS3) is mutated in Meckel-Gruber syndrome and the wpk rat. *Nat. Genet.* **38,** 191–196.
Sohara, E., Luo, Y., Zhang, J., Manning, D. K., Beier, D. R., and Zhou, J. (2008). *Nek8* regulates the expression and localization of polycystin-1 and polycystin-2. *J. Am. Soc. Nephrol.* **19,** 469–476.
Somlo, S., and Ehrlich, B. (2001). Human disease: Calcium signaling in polycystic kidney disease. *Curr. Biol.* **11,** R356–R360.

Spassky, N., Han, Y. G., Aguilar, A., Strehl, L., Besse, L., Laclef, C., Ros, M. R., Garcia-Verdugo, J. M., and Alvarez-Buylla, A. (2008). Primary cilia are required for cerebellar development and Shh-dependent expansion of progenitor pool. *Dev. Biol.* **317,** 246–259.

Stamataki, D., Ulloa, F., Tsoni, S. V., Mynett, A., and Briscoe, J. (2005). A gradient of Gli activity mediates graded Sonic Hedgehog signaling in the neural tube. *Genes Dev.* **19,** 626–641.

Starich, T. A., Herman, R. K., Kari, C. K., Yeh, W. H., Schackwitz, W. S., Schuyler, M. W., Collet, J., Thomas, J. H., and Riddle, D. L. (1995). Mutations affecting the chemosensory neurons of *Caenorhabditis elegans*. *Genetics* **139,** 171–188.

Stoetzel, C., Laurier, V., Davis, E. E., Muller, J., Rix, S., Badano, J. L., Leitch, C. C., Salem, N., Chouery, E., Corbani, S., *et al.* (2006). BBS10 encodes a vertebrate-specific chaperonin-like protein and is a major BBS locus. *Nat. Genet.* **38,** 521–524.

Stoetzel, C., Muller, J., Laurier, V., Davis, E. E., Zaghloul, N. A., Vicaire, S., Jacquelin, C., Plewniak, F., Leitch, C. C., Sarda, P., *et al.* (2007). Identification of a novel BBS gene (BBS12) highlights the major role of a vertebrate-specific branch of chaperonin-related proteins in Bardet-Biedl syndrome. *Am. J. Hum. Genet.* **80,** 1–11.

Sulik, K., Dehart, D. B., Iangaki, T., Carson, J. L., Vrablic, T., Gesteland, K., and Schoenwolf, G. C. (1994). Morphogenesis of the murine node and notochordal plate. *Dev. Dyn.* **201,** 260–278.

Sullivan-Brown, J., Schottenfeld, J., Okabe, N., Hostetter, C. L., Serluca, F. C., Thiberge, S. Y., and Burdine, R. D. (2008). Zebrafish mutations affecting cilia motility share similar cystic phenotypes and suggest a mechanism of cyst formation that differs from pkd2 morphants. *Dev. Biol.* **314,** 261–275.

Sun, Z., Amsterdam, A., Pazour, G. J., Cole, D. G., Miller, M. S., and Hopkins, N. (2004). A genetic screen in zebrafish identifies cilia genes as a principal cause of cystic kidney. *Development* **131,** 4085–4093.

Supp, D. M., Witte, D. P., Potter, S. S., and Brueckner, M. (1997). Mutation of an axonemal dynein affects left–right asymmetry in inversus viscerum mice. *Nature* **389,** 963–966.

Supp, D. M., Brueckner, M., Kuehn, M. R., Witte, D. P., Lowe, L. A., McGrath, J., Corrales, J., and Potter, S. S. (1999). Targeted deletion of the ATP binding domain of left–right dynein confirms its role in specifying development of left–right asymmetries. *Development* **126,** 5495–5504.

Takahashi, H., Ueyama, Y., Hibino, T., Kuwahara, Y., Suzuki, S., Hioki, K., and Tamaoki, N. (1986). A new mouse model of genetically transmitted polycystic kidney disease. *J. Urol.* **135,** 1280–1283.

Takahashi, H., Calvet, J. P., Dittemore-Hoover, D., Yoshida, K., Grantham, J. J., and Gattone, V. H., II. (1991). A hereditary model of slowly progressive polycystic kidney disease in the mouse. *J. Am. Soc. Nephrol.* **1,** 980–989.

Takeda, S., Yonekawa, Y., Tanaka, Y., Okada, Y., Nonaka, S., and Hirokawa, N. (1999). Left–right asymmetry and kinesin superfamily protein KIF3A: New insights in determination of laterality and mesoderm induction by kif3A−/− mice analysis. *J. Cell Biol.* **145,** 825–836.

Takeda, H., Takami, M., Oguni, T., Tsuji, T., Yoneda, K., Sato, H., Ihara, N., Itoh, T., Kata, S. R., Mishina, Y., *et al.* (2002). Positional cloning of the gene LIMBIN responsible for bovine chondrodysplastic dwarfism. *Proc. Natl Acad. Sci. USA* **99,** 10549–10554.

Tallila, J., Jakkula, E., Peltonen, L., Salonen, R., and Kestila, M. (2008). Identification of CC2D2A as a Meckel syndrome gene adds an important piece to the ciliopathy puzzle. *Am. J. Hum. Genet.* **82,** 1361–1367.

Taulman, P. D., Haycraft, C. J., Balkovetz, D. F., and Yoder, B. K. (2001). Polaris, a protein involved in left–right axis patterning, localizes to basal bodies and cilia. *Mol. Biol. Cell* **12,** 589–599.

Tobin, J. L., and Beales, P. L. (2007). Bardet-Biedl syndrome: Beyond the cilium. *Pediatr. Nephrol.* **22,** 926–936.

Tobin, J. L., and Beales, P. L. (2008). Restoration of renal function in zebrafish models of ciliopathies. *Pediatr. Nephrol.* **23,** 2095–2099.

Tobin, J. L., Di Franco, M., Eichers, E., May-Simera, H., Garcia, M., Yan, J., Quinlan, R., Justice, M. J., Hennekam, R. C., Briscoe, J., *et al.* (2008). Inhibition of neural crest migration underlies craniofacial dysmorphology and Hirschsprung's disease in Bardet-Biedl syndrome. *Proc. Natl Acad. Sci. USA* **105,** 6714–6719.

Torres, V. E. (2004). Therapies to slow polycystic kidney disease. *Nephron Exp. Nephrol.* **98,** e1–e7.

Tran, P. V., Haycraft, C. J., Besschetnova, T. Y., Turbe-Doan, A., Stottmann, R. W., Herron, B. J., Chesebro, A. L., Qiu, H., Scherz, P. J., Shah, J. V., *et al.* (2008). THM1 negatively modulates mouse sonic hedgehog signal transduction and affects retrograde intraflagellar transport in cilia. *Nat. Genet.* **40,** 403–410.

Tsiokas, L., Kim, E., Arnould, T., Sukhatme, V. P., and Walz, G. (1997). Homo- and heterodimeric interactions between the gene products of PKD1 and PKD2. *Proc. Natl Acad. Sci. USA* **94,** 6965–6970.

Upadhya, P., Churchill, G., Birkenmeier, E. H., Barker, J. E., and Frankel, W. N. (1999). Genetic modifiers of polycystic kidney disease in intersubspecific KAT2J mutants. *Genomics* **58,** 129–137.

Upadhya, P., Birkenmeier, E. H., Birkenmeier, C. S., and Barker, J. E. (2000). Mutations in a NIMA-related kinase gene, Nek1, cause pleiotropic effects including a progressive polycystic kidney disease in mice. *Proc. Natl Acad. Sci. USA* **97,** 217–221.

Utsch, B., Sayer, J. A., Attanasio, M., Pereira, R. R., Eccles, M., Hennies, H. C., Otto, E. A., and Hildebrandt, F. (2006). Identification of the first AHI1 gene mutations in nephronophthisis-associated Joubert syndrome. *Pediatr. Nephrol.* **21,** 32–35.

Valente, E. M., Salpietro, D. C., Brancati, F., Bertini, E., Galluccio, T., Tortorella, G., Briuglia, S., and Dallapiccola, B. (2003). Description, nomenclature, and mapping of a novel cerebello-renal syndrome with the molar tooth malformation. *Am. J. Hum. Genet.* **73,** 663–670.

Valente, E. M., Marsh, S. E., Castori, M., Dixon-Salazar, T., Bertini, E., Al-Gazali, L., Messer, J., Barbot, C., Woods, C. G., Boltshauser, E., *et al.* (2005). Distinguishing the four genetic causes of Jouberts syndrome-related disorders. *Ann. Neurol.* **57,** 513–519.

Valente, E. M., Brancati, F., Silhavy, J. L., Castori, M., Marsh, S. E., Barrano, G., Bertini, E., Boltshauser, E., Zaki, M. S., Abdel-Aleem, A., *et al.* (2006a). AHI1 gene mutations cause specific forms of Joubert syndrome-related disorders. *Ann. Neurol.* **59,** 527–534.

Valente, E. M., Silhavy, J. L., Brancati, F., Barrano, G., Krishnaswami, S. R., Castori, M., Lancaster, M. A., Boltshauser, E., Boccone, L., Al-Gazali, L., *et al.* (2006b). Mutations in CEP290, which encodes a centrosomal protein, cause pleiotropic forms of Joubert syndrome. *Nat. Genet.* **38,** 623–625.

van der Hoeven, F., Schimmang, T., Volkmann, A., Mattei, M. G., Kyewski, B., and Ruther, U. (1994). Programmed cell death is affected in the novel mouse mutant Fused toes (Ft). *Development* **120,** 2601–2607.

Vierkotten, J., Dildrop, R., Peters, T., Wang, B., and Ruther, U. (2007). Ftm is a novel basal body protein of cilia involved in Shh signalling. *Development* **134,** 2569–2577.

Vogler, C., Homan, S., Pung, A., Thorpe, C., Barker, J., Birkenmeier, E. H., and Upadhya, P. (1999). Clinical and pathologic findings in two new allelic murine models of polycystic kidney disease. *J. Am. Soc. Nephrol.* **10,** 2534–2539.

Ward, C. J., Hogan, M. C., Rossetti, S., Walker, D., Sneddon, T., Wang, X., Kubly, V., Cunningham, J. M., Bacallao, R., Ishibashi, M., *et al.* (2002). The gene mutated in autosomal recessive polycystic kidney disease encodes a large, receptor-like protein. *Nat. Genet.* **30,** 259–269.

Ward, C. J., Yuan, D., Masyuk, T. V., Wang, X., Punyashthiti, R., Whelan, S., Bacallao, R., Torra, R., LaRusso, N. F., Torres, V. E., *et al.* (2003). Cellular and subcellular localization of the ARPKD protein; fibrocystin is expressed on primary cilia. *Hum. Mol. Genet.* **12,** 2703–2710.

Watanabe, D., Saijoh, Y., Nonaka, S., Sasaki, G., Ikawa, Y., Yokoyama, T., and Hamada, H. (2003). The left–right determinant Inversin is a component of node monocilia and other 9 + 0 cilia. *Development* **130,** 1725–1734.

Watnick, T., and Germino, G. (2003). From cilia to cyst. *Nat. Genet.* **34,** 355–356.

White, D. R., *et al.* (2007). Autozygosity mapping of Bardet-Biedl syndrome to 12q21.2 and confirmation of FLJ23560 as BBS10. *Eur. J. Hum. Genet.* **15**(2), 173–178.

White, M. C., and Quarmby, L. M. (2008). The NIMA-family kinase, Nek1 affects the stability of centrosomes and ciliogenesis. *BMC Cell Biol.* **9,** 29.

Whitehead, J. L., Wang, S. Y., Bost-Usinger, L., Hoang, E., Frazer, K. A., and Burnside, B. (1999). Photoreceptor localization of the KIF3A and KIF3B subunits of the heterotrimeric microtubule motor kinesin II in vertebrate retina. *Exp. Eye Res.* **69,** 491–503.

Wilson, C., Bonnet, C., Guy, C., Idziaszczyk, S., Colley, J., Humphreys, V., Maynard, J., Sampson, J. R., and Cheadle, J. P. (2006). Tsc1 haploinsufficiency without mammalian target of rapamycin activation is sufficient for renal cyst formation in Tsc1+/− mice. *Cancer Res.* **66,** 7934–7938.

Wolf, M. T., Saunier, S., O'Toole, J. F., Wanner, N., Groshong, T., Attanasio, M., Salomon, R., Stallmach, T., Sayer, J. A., Waldherr, R., *et al.* (2007). Mutational analysis of the RPGRIP1L gene in patients with Joubert syndrome and nephronophthisis. *Kidney Int.* **72,** 1520–1526.

Wong, H. C., Mao, J., Nguyen, J. T., Srinivas, S., Zhang, W., Liu, B., Li, L., Wu, D., and Zheng, J. (2000). Structural basis of the recognition of the dishevelled DEP domain in the Wnt signaling pathway. *Nat. Struct. Biol.* **7,** 1178–1184.

Wong, H. C., Bourdelas, A., Krauss, A., Lee, H. J., Shao, Y., Wu, D., Mlodzik, M., Shi, D. L., and Zheng, J. (2003). Direct binding of the PDZ domain of Dishevelled to a conserved internal sequence in the C-terminal region of Frizzled. *Mol. Cell* **12,** 1251–1260.

Wu, G., Mochizuki, T., Le, T. C., Cai, Y., Hayashi, T., Reynolds, D. M., and Somlo, S. (1997). Molecular cloning, cDNA sequence analysis, and chromosomal localization of mouse Pkd2. *Genomics* **45,** 220–223.

Wu, G., D'Agati, V., Cai, Y., Markowitz, G., Park, J. H., Reynolds, D. M., Maeda, Y., Le, T. C., Hou, H., Jr., Kucherlapati, R., *et al.* (1998). Somatic inactivation of Pkd2 results in polycystic kidney disease. *Cell* **93,** 177–188.

Xiong, H., Chen, Y., Yi, Y., Tsuchiya, K., Moeckel, G., Cheung, J., Liang, D., Tham, K., Xu, X., Chen, X. Z., *et al.* (2002). A novel gene encoding a TIG multiple domain protein is a positional candidate for autosomal recessive polycystic kidney disease. *Genomics* **80,** 96–104.

Yoder, B. K. (2007). Role of primary cilia in the pathogenesis of polycystic kidney disease. *J. Am. Soc. Nephrol.* **18,** 1381–1388.

Yoder, B. K., Richards, W. G., Sweeney, W. E., Wilkinson, J. E., Avener, E. D., and Woychik, R. P. (1995). Insertional mutagenesis and molecular analysis of a new gene associated with polycystic kidney disease. *Proc. Assoc. Am. Physicians* **107,** 314–323.

Yoder, B. K., Hou, X., and Guay-Woodford, L. M. (2002). The polycystic kidney disease proteins, polycystin-1, polycystin-2, polaris, and cystin, are co-localized in renal cilia. *J. Am. Soc. Nephrol.* **13,** 2508–2516.

Zhang, F., and Jetten, A. M. (2001). Genomic structure of the gene encoding the human GLI-related, Kruppel-like zinc finger protein GLIS2. *Gene* **280,** 49–57.

Zhang, F., Nakanishi, G., Kurebayashi, S., Yoshino, K., Perantoni, A., Kim, Y. S., and Jetten, A. M. (2002). Characterization of Glis2, a novel gene encoding a Gli-related, Kruppel-like transcription factor with transactivation and repressor functions. Roles in kidney development and neurogenesis. *J. Biol. Chem.* **277,** 10139–10149.

Zhang, Q., Murcia, N. S., Chittenden, L. R., Richards, W. G., Michaud, E. J., Woychik, R. P., and Yoder, B. K. (2003). Loss of the Tg737 protein results in skeletal patterning defects. *Dev. Dyn.* **227,** 78–90.

CHAPTER SIX

Mouse Models of Polycystic Kidney Disease

Patricia D. Wilson

Contents

Abstract

Polycystic kidney disease (PKD) is a diverse group of human monogenic lethal conditions inherited as autosomal dominant (AD) or recessive (AR) traits. Recent development of genetically engineered mouse models of ADPKD, ARPKD, and nephronophthisis/medullary cystic disease (NPHP) are providing additional insights into the molecular mechanisms governing of these disease

Mount Sinai School of Medicine, New York

Current Topics in Developmental Biology, Volume 84
ISSN 0070-2153, DOI: 10.1016/S0070-2153(08)00606-6

processes as well as the developmental differentiation of the normal kidney. Genotypic and phenotypic mouse models are discussed and provide evidence for the fundamental involvement of cell–matrix, cell–cell, and primary cilia–lumen interactions, as well as epithelial proliferation, apoptosis, and polarization. Structure/function relationships between the PKD1, PKD2, PKHD1, and NPHP genes and proteins support the notion of a regulatory multiprotein cystic complex with a mechanosensory function that integrates signals from the extracellular environment. The plethora of intracellular signaling cascades that can impact renal cystic development suggest an exquisitely sensitive requirement for integrated downstream transduction and provide potential targets for therapeutic intervention. Appropriate genocopy models that faithfully recapitulate the phenotypic characteristics of the disease will be invaluable tools to analyze the effects of modifier genes and small molecule inhibitor therapies.

1. Introduction

Since the mouse shares 99% of their genes with humans and most of their physiological and pathological characteristics, they provide an ideal model for the study of human diseases, such as polycystic kidney disease (PKD). With the advent of targeted mutation and knockout technology, this utility has increased dramatically, as this allows for genetic experimentation and experimentation that could not, of course, be carried out in humans. In this way, genetically engineered mouse models of PKD are shedding new light on the mechanisms of these diseases, their modulation, and allow for the preclinical testing of new therapies.

Although hereditary PKD is genetically heterogeneous, it is caused by single mutations in single genes, thus rendering it susceptible to analysis by traditional targeted mutation and knockout analysis. In addition, cell-type-specific knockouts, haploinsufficient, transgenic, and Cre–LoxP conditional strategies have been applied. In addition to genocopy models of *Pkd1, Pkd2, Pkhd1, Nphp2*, and *Nphp3*, several chemical mutagenesis screens and spontaneously occurring mouse models have provided phenocopy models for the study of renal cystic disease. Recent definition of the genes and proteins responsible for the development of renal cysts in these phenocopy models have also shed light onto the mechanisms of cyst formation in the kidney and other organs.

Cell biological and pathophysiological studies of PKD in humans have identified major roles for faulty renal tubule epithelial cell proliferation, ion and fluid secretion, cell–matrix and cell–cell interactions, and differentiation in PKD (Wilson, 2004). The application of manipulated mouse models has allowed genetic confirmation that these pathways are critical and have led to new avenues of discovery and definition.

2. The PKD Genes

In the human population, the most clinically significant type of PKD is inherited as an autosomal dominant trait (ADPKD). Since this has an incidence of 1:750 and affects 500,000 individuals in USA and an estimated 12 million worldwide, this is classified as the most common human lethal monogenetic disease inherited as a dominant trait. Eighty-five percent of ADPKD is caused by a mutation in the *PKD1* gene and 15% by mutation in the *PKD2* gene. Other rarer, but clinically significant hereditary renal cystic diseases include those inherited as autosomal recessive (AR) traits such as ARPKD, the product of mutation in the polycystic kidney and hepatic disease-1 (*PKHD1*) gene and the genetically heterogeneous "nephronophthisis" group of juvenile renal medullary cystic disease that are caused by mutations in *NPHP1–9*. Adult onset AD medullary cystic kidney disease (MCKD) can also be a hereditary condition, and to date two genes have been identified.

The *PKD1* gene is located on human chromosome 16p13.3, and is a very large gene that occupies 52 kb of genomic DNA (Consortium, 1995a,b). The gene is arranged into 46 exons and encodes a large transcript of 14.5 kb. More than 300 different mutations have been described which occur along the entire sequence of the gene and most of which are unique to single families (Rossetti *et al.*, 2007). The types of mutation include frameshifts, nonsense, splicing, rearrangements, deletions, and missense changes and the majority (over 70%) leads to truncations of the encoded protein product. In addition, *PKD1* is a highly polymorphic gene with more than 200 documented nondisease-causing variants. Interestingly, the *PKD1* promoter contains β-catenin-binding sites (Rodova *et al.*, 2002). The *PKD2* gene is located on human chromosome 4q21–23 and is a smaller gene that occupies 70 kb of genomic DNA. It is arranged into 15 exons and encodes a 5.6 kb transcript (Mochizuki *et al.*, 1996). Many different mutations have been detected that occur all along the length of the sequence, the majority of which would lead to truncations (Deltas, 2001; Somlo, 1999).

The *PKHD1* gene is located on human chromosome 6p12.2 (Zerres *et al.*, 1998) and, like *PKD1* is another very large gene that occupies 47 kb of genomic DNA. It is arranged into 66 coding exons and encodes a transcript of 16.2 kb (Onuchic *et al.*, 2002; Wang *et al.*, 2005). More than 300 different mutations have been identified, the majority of which are missense, although frameshift, deletions, insertions nonsense, and splicing mutations have also been detected. As in *PKD1* and *PKD2*, the majority of disease-associated mutations are predicted to lead to truncated transcripts, but unlike for *PKD1* and *PKD2*, some common *PKHD1* mutations have been found in unrelated ARPKD-affected families (Bergmann *et al.*, 2003, 2006; Rossetti *et al.*, 2003).

Interestingly, the *PKHD1* promoter region contains hepatocyte nuclear factor (HNF)-1β-binding sites (Hiesberger *et al.*, 2005).

Several *NPHP* genes have been identified although the vast majority of juvenile nephronophthisis cases are caused by mutation in *NPHP1* that is localized on human chromosome 2q13 (Hildebrandt *et al.*, 1997). The *NPHP2/INVS* gene is located on human chromosome 9q22–q31 and mutations cause an infantile form of nephronophthisis. Mutations in *NPHP3* located on human chromosome 3q22 are responsible for a rare adolescent form of nephronophthisis. Additional rare variants of juvenile nephronophthisis with differing extrarenal manifestations, including retinitis pigmentosa, oculomotor apraxia, Leber congenital amaurosis, cerebellar vermis aplasia, liver fibrosis, cone-shaped epiphyses, and rarely, situs inversus are caused by mutations in *NPHP4/nephroretinin* located on human chromosome 1p36; *NPHP5* located on human chromosome 3q13.33–q21.1 (Senior-Loken syndrome), or *NPHP6/CEP290* (Joubert syndrome) (Hildebrandt and Otto, 2000; Mollet *et al.*, 2005; Olbrich *et al.*, 2003; Omran *et al.*, 2000; Otto *et al.*, 2002).

Two *ADMCKD* genes have been identified to date. *ADMCKD1* is located on human chromosome 1q21 and *ADMCKD2* that is located on human chromosome 16p12 (Hateboer *et al.*, 2001; Wolf *et al.*, 2004).

3. Genocopy Mouse Models of PKD

Given the relative prevalence of the human diseases caused by the *PKD, PKHD, NPHP*, and *MCKD* genes and the depth of knowledge associated with the respective gene structures, it is not surprising that the major efforts have been directed toward *PKD1* and *PKD2* and more recently, *PKHD1* gene manipulation in mice. These have included targeted mutations, some of which were modeled on stop mutations detected in human ADPKD patients. Since complete homozygous knockout of *PKD1* or *PKD2* results in embryonic lethality, recent efforts have concentrated on conditional and inducible strategies, as well as haploinsufficient, hypomorphic models, and the introduction of unstable alleles (Table 6.1).

The ultimate goal of a mouse model of disease is to genetically manipulate the disease gene in question and to derive a phenotype that closely mimics the human disease. The first genetically engineered mouse mutant of PKD1 was a targeted deletion of exon 34 in mouse *Pkd1* that resulted in perinatal death due to renal cystic disease in homozygotes (Lu *et al.*, 1997). Subsequent studies showed that heterozygotes developed multiple bilateral cortical and medullary renal cysts by 16 months of age as well as liver cysts derived from the biliary epithelium (Lu *et al.*, 1999). Limited functional and biological characterization, including identification of apically mislocalized

Table 6.1 Genocopy mouse models of PKD

Pkd1	Exon 34	Targeted deletion	−/− Perinatal lethal, renal cysts No cardiac defects +/− viable, renal, and liver cysts, apical EGFR
	Exon 17–21	Targeted deletion	−/− Embryonic lethal, E13.5–E14.5 renal and liver cysts, cardiovascular and skeletal defects +/− Viable, renal, and liver cysts
	Exon 43	Insertion stop codon	−/− Embryonic lethal, renal cysts vascular fragility
	Null	Insertion pgk-neo exon 4	−/− Embryonic lethal, E16.5, renal and pancreatic cysts, cartilage defects +/− viable, renal, liver, and pancreatic cysts
	Haploinsufficiency 20%	Insertion hypomorphic allele	Viable, renal, and liver cysts
	Haploinsufficiency 15–25%	LoxP	Viable, renal, and liver cysts
	Overexpression	Transgenic, 30 copies	viable, renal cysts
	Kidney-specific conditional null	Tamoxifen-CreLox deletion exon 2–11	Renal cysts
	Conditional null	Tamoxifen-Cre	Renal cysts
Pkd2	WS25	Unstable allele	Viable, renal cysts
	WS183	True null	−/− Embryonic lethal, E13.5–E18, renal and

(continued)

Table 6.1 (*continued*)

			pancreatic cysts, cardiac septation
Pkhd1	Exon 40	Targeted	Liver cysts and fibrosis
	Conditional null	Cre–LoxP exon 2	Liver cysts and fibrosis
Nphp2	inv mouse	Insertional deletion invs	PN lethal d7, situs inversus, renal and liver cysts, cardiac malformations
Nphp3	Null		Embryonic lethal, E 15, situs inversus, cardiac defects
	pcy/ko	Compound heterozygote	Viable, renal cysts
	pcy mouse	Spontaneous hypomorph	Viable, renal cysts
Mckd2/ umod	Null		No phenotype

epidermal growth factor receptor (EGFR) suggested that this model resembled the human disease phenotype with regard to predominant renal cystic disease and hepatic cystic involvement. However, no cardiac manifestations were noted. By contrast, other *Pkd1*-targeted deletion mouse models show different phenotypes and extent of cystic kidney involvement. For instance, targeted deletion of exon 17–21 of the mouse *Pkd1* resulted in embryonic lethality at embryonic day (E) 13.5–14.5 in homozygotes, with predominant cardiovascular abnormalities. Heterozygotes, however, exhibited renal and hepatic cysts (Boulter *et al.*, 2001). It does not seem that the location of the introduced deletion was the cause of this discrepancy, since the introduction of a premature stop codon at exon 43 resulted in embryonic lethality at E15.5 in the homozygotes as a result of vascular fragility (Kim *et al.*, 2000), although renal and hepatic cysts were also noted.

Subsequent approaches focused on manipulation of the levels of *Pkd1*. True null mutants, derived by insertion of a phosphoglycerate kinase (pgk)-neomycin (neo) cassette into exon 4 by homologous recombination, resulted in embryonic lethality of homozygotes at E16.5 renal cystic expansion was noted at E15.5 and of pancreatic cysts from E13.5 as well as polyhydramnios, hydrops fetalis, spina bifida, and osteochondrodysplasia. These studies also showed significant effects of the genetic background (C57/BL6 or BALB/c) on disease severity Heterozygotes were viable and exhibited renal liver and pancreatic cysts from 2 to 24 months of age (Lu *et al.*, 2001). Haploinsufficient models of *Pkd1* in mice show that simple a reduction of the normal levels of *Pkd1* are sufficient to induce a renal cystic phenotype as well as liver and bile duct dilitations similar to those seen in ADPKD. This was seen when levels were reduced to 13–20% by the introduction of a hypomorphic *Pkd1* allele harboring an intronic neo-selectable marker and also after introduction of a loxP site into intron 30 and of a loxP-flanked mc1-neo cassette into intro 24 resulting in a 20–25% level of normal *Pkd1* and protein expression (Jiang *et al.*, 2006; Lantinga-van Leeuwen *et al.*, 2004). Interestingly, transgenic lines with approximately 30 copies of transgene also developed renal cystic disease, hepatic cysts and bile duct proliferation showing that increased expression of *Pkd1* also result in PKD (Pritchard *et al.*, 2000). Taken together these studies strongly argue for the conclusion that *Pkd1* is a developmentally regulated gene whose level of expression must be tightly maintained to lead to normal kidney development and differentiation since reductions or increases lead to renal cystic malformation. These latter studies are of significance, since early studies in human cystic epithelia extracted from ADPKD kidneys had led to a suggestion that renal cyst formation required a "second hit" in the *PKD1* gene in the somatic allele (Qian *et al.*, 1996). These haploinsufficient mouse models with renal cystic development would argue against a requirement for a second hit although a role in disease severity and rapidity of progression can not be ruled out.

The most recent studies have been designed to address the question of developmental versus postnatal effects on *Pkd1*-mediated disease initiation, severity, and progression, since these are critical issues for ADPKD patient management about which there is little current understanding. Using conditional (Tamoxifen-*Cre*) deletion strategies two groups show that deletion of *Pkd1* early in life led to more rapid and severe renal cystic disease (Lantinga-van Leeuwen *et al.*, 2007; Piontek *et al.*, 2007), but only one study noted an apparent developmental switch occurring at postnatal day 13/14, before which cystic disease was rapid and after which cystic disease was slow (Piontek *et al.*, 2007). The reason for this discrepancy is unclear at present, although it should be noted that Lantinga-van Leeuwen *et al.* used a kidney epithelium-specific (KspCad) targeting strategy.

Genetically engineered mouse models have also shown that *Pkd2* gene dosage is of critical importance with regard to disease progression. True null Pkd2/Ws138 homozygotes die in embryonic stages from E13.5 to parturition and develop renal and pancreatic cysts as well as cardiac septation defects. By contrast, the haploinsufficient model Pkd2/WS25, which contains an unstable allele, is viable and dies of renal failure at 20 months (Wu *et al.*, 1998, 2000). In this case, somatic mutation appears to accelerate disease.

The ARPKD gene, *PKHD1*, was identified in 2002 (Ward *et al.*, 2002). Although the *Pck* rat was shown to be a spontaneous model harboring a haploinsufficient mutation in the *Pkhd1* gene, it is recognized that the development of a genetically engineered mouse model would present some significant advantages for the study of mechanisms of this disease.

However, to date both a targeted mutation of exon 40 and Cre–LoxP null deletion at exon 2 have resulted in mouse models with grossly fibrotic and cystic livers but nor renal cystic phenotype has been achieved (Moser *et al.*, 2005; Woollard *et al.*, 2007).

The only *Nphp* gene that has been deliberately genetically targeted in the mouse, to date, is *Nphp3* that is associated with the rare adolescent form of nephronophthisis. *Nphp3* homozygous null mice, generated by insertion into exon 3, are embryonic lethal and show situs inversus and cardiac defects. A compound heterozygous cross between the *Nphp3* null allele and the *pcy* allele found in the spontaneously occurring *pcy* mouse that is hypomorphic for *Nphp3* resulted in renal cystic disease (Bergmann *et al.*, 2008). Interestingly, and somewhat surprisingly, there was no evidence of ciliary loss in this model.

The *inv* (inversion of embryo turning) mouse that results from an insertional mutation in the *invs* gene, causing deletion of exons 3–11, has also been identified as a genocopy model of *Nphp2*. Homozygotes exhibit situs inversus (altered left–right laterality), severe cystic changes of the kidney, and hepatobiliary malformations leading to renal and liver failure and death by postnatal day 7 (Morgan *et al.*, 1998; Phillips *et al.*, 2004; Watanabe *et al.*, 2003).

To date, attempts to model MCKD2 in mouse models by homozygous deletion of UMOD/Tamm–Horsfall protein in mice have failed to generate renal structural abnormalities, although creatinine clearance was decreased by 37% and some distal tubule transporters increased (Bachmann *et al.*, 2005).

4. PKD: The Human Diseases

The hallmark of PKD is the presence of multiple cysts in each kidney, the chronic expansion of which lead to loss of renal function and premature death. A cyst comprises a fluid filled expanded tubule lined by a layer of epithelium resting on a basement membrane. The most common and clinically significant forms of PKD are hereditary and caused by mutation in a single gene. However, multiple renal cyst formation is also a common secondary feature of other hereditary and sporadic clinical syndromes and developmental defects in differentiation (dysplasias).

4.1. Monogenic PKD

The most common and clinically significant form of monogenic PKD is ADPKD which occurs with a frequency of 1 in 800 live births and affects approximately 10 million individuals worldwide. This disease accounts for 7–10% of hemodialysis and renal transplant patients. It comprises two phenotypically similar, but genetically distinct entities caused by mutations, in *PKD1* in 85% of cases and by *PKD2* in 15% of cases (Wilson, 2004). Although the genetic penetrance of ADPKD is 100%, since all affected heterozygous individuals develop multiple renal cysts, there is a wide range in the rate of cystic expansion and this is reflected by a wide distribution of the age of onset of renal failure ranging from the first to the eighth decade of life. Macroscopically, there is massive and progressive bilateral enlargement of the kidneys due to the presence of innumerable, variously sized cysts throughout the cortex and medulla that begin to arise *in utero*. ADPKD cysts develop as epithelial outpouchings of every segment of the renal tubule and rapidly close off from the nephron of origin. After birth, ADPKD cysts continue to expand and destroy the intervening normal renal parenchyma progressively throughout life. In addition to renal cysts, extrarenal manifestations are common and include multiple biliary epithelial cysts in the liver, cystic pancreas and intestine, as well as cardiovascular defects including aortic and intracerebral aneurysms and valvular abnormalities (Wilson and Goilav, 2007).

Although ARPKD is less common than ADPKD, occurring at a frequency of 1:20,000 live births, it is clinically highly significant since it often causes fetal or neonatal death, owing to massive, bilateral renal enlargement associated with impaired lung and liver formation. Those ARPKD patients, who survive the neonatal period, often develop arterial hypertension as infants or children and 20–45% progress to end-stage renal failure by age 15. Liver fibrosis is also a common cause of morbidity and mortality in these children (McDonald and Avner, 1991; Rossetti *et al.*, 2003). Macroscopically, ARPKD kidneys are symmetrically enlarged, but individual ARPKD cysts are smaller than ADPKD cysts and are derived from ectatic expansions of the collecting tubule segment of the nephron that, unlike in ADPKD, remain in longitudinal contact with the nephron of origin (Fig. 6.1).

By contrast to ADPKD and ARPKD, where cysts occur throughout the enlarged kidneys, cortex, and medulla, some hereditary renal cystic diseases present with small kidneys and multiple cysts confined to the renal medulla. Autosomal recessive juvenile nephronophthisis (NPHP) is a frequent cause of chronic renal failure in children and young adults (Hildebrandt and Otto, 2000), who usually present with polyuria, polydipsia, tubular concentration defects, and Fanconi syndrome. Unlike ADPKD and ARPKD, extrarenal manifestations often include growth retardation, tapetoretinal degeneration, skeletal abnormalities, central nervous system malformations, and oculomotor apraxia. In the rare variant, (Joubert syndrome) NPHP is also combined with retinal degeneration, cerebellar vermis aplasia, and mental retardation. The small shrunken NPHP kidneys contain numerous small cysts, usually confined to the corticomedullary junction. Basement membrane thickening and tubulointerstitial damage are also characteristic of this disease. Autosomal dominant forms of medullary cystic kidney disease (AD-MCKD) also exist that are characterized by salt wasting, decreased urinary concentrating ability, and varying rates of progression to end-stage renal failure.

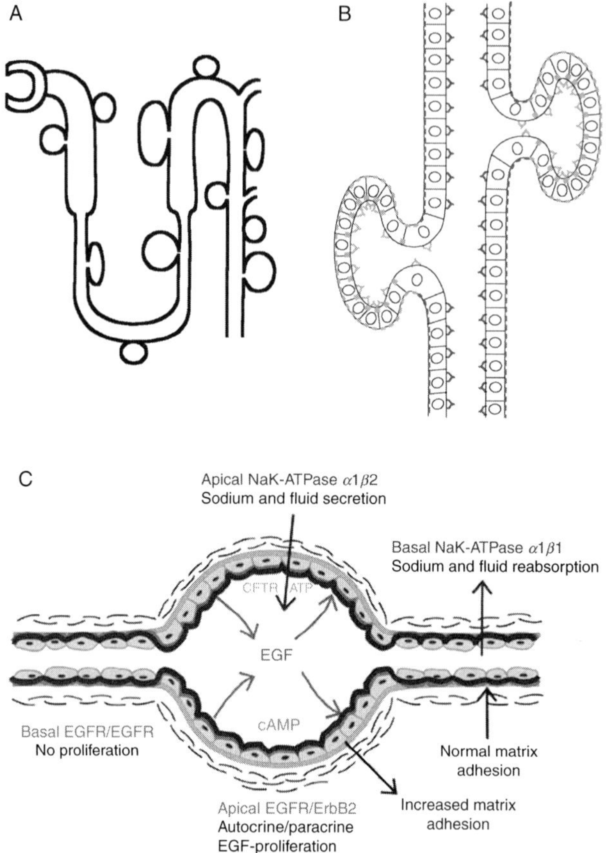

Figure 6.1 Diagram of renal cystic development in ADPKD. (A) Cysts arise from every segment of the nephron during embryonic development and close off from the nephron of origin. (B) Tubule epithelia undergo alterations in planar cell polarity leading to asymmetric angles of cell division, facilitating cystic expansion. (C) Apical mispolarization of membrane proteins such as NaK-ATPase and EGFR lead to increased proliferation and fluid secretion in cystic areas of the nephron.

Macroscopically, kidneys are normal or slightly reduced in size (Waldherr *et al.*, 1982). Hyperuricemia and precocious gout can also be associated with these progressive tubulointerstitial nephropathies.

4.2. PKD, secondary to other hereditary cystic kidney diseases

Renal cysts are often seen as secondary manifestations of genetic proliferative syndromes such as tuberous sclerosis (TSC), Von Hippel-Lindau disease (VHL), Bardet-Biedl syndrome (BBS), and oral-facial-digital syndrome type 1 (OFD-1). Like *VHL* and *TSC1, TSC2* is a tumor suppressor gene and interestingly is located immediately adjacent to *PKD1* on chromosome 16. BBS and OFD-1 are rare hereditary disorders leading to renal impairment secondary to many other extrarenal manifestations including obesity, retinopathy, polydactyly, learning disabilities, hypogenitalism, cleft lip, cleft palate, lingual hamartomas, bifid or lobulated tongue, digit and nasal abnormalities, and facial asymmetry (Katsanis *et al.*, 2000). Hemodialysis-induced acquired renal cystic disease also leads to cyst formation and is associated with an increased incidence in renal carcinoma (Hughson *et al.*, 1986).

Nonhereditary, sporadic disorders of renal development and differentiation also often lead to irregular renal cyst formation including multicystic renal dysplasias in the setting of abnormal metanephric mesenchymal differentiation or ureteral obstruction and can be lethal *in utero*.

5. Phenotypic Mouse Models of PKD

PKDs are complex and several molecular mechanisms have been proposed to play important roles in the formation and expansion of renal cysts, the hallmark of the disease. The occurrence of a renal cystic phenotype across a broad spectrum of genetically engineered, chemically induced, and spontaneously arisen mouse models has also shed light on potential mechanisms underlying PKD.

Several transgenic mouse models producing overexpression of proliferation-related genes and proteins result in the generation of cystic kidneys (see Table 6.2). These include overexpression of oncogenes such as c-myc (SBM mouse); SV40 T-antigen, and T24 ras as well as the sometimes oncogenic epidermal growth factor receptors (EGFR/Erb-B1) and Erb-B2

Table 6.2 Phenotypic overexpression mouse models of PKD

Proliferation	c-myc	Oncogene
	SV40 T-antigen	Oncogene
	T24 ras	Oncogene
	Activated EGFR	Growth factor receptor, oncogene
	Erb-B2	Growth factor receptor, oncogene
	TGFα	Growth factor
	HGF	Growth factor
Developmental regulation	Pax-2	Transcription factor
	p75Cux-1	Transcription factor
Cell adhesion	β-catenin	Adhesion protein, canonical wnt pathway transcription factor

(Richards *et al.*, 1998; Stocklin *et al.*, 1993; Trudel *et al.*, 1998), and the growth factor ligands transforming growth factor-alpha (TGFα) and hepatocyte growth factor (HGF) (Ogborn and Sareen, 1996; Wilson, 1996). This strongly implicates overproliferation as an underlying mechanism involved in renal cystic epithelial overgrowth that is characteristic in PKD. Mice that overexpress the transcription factors Pax-2 and p75Cux-1 are known to play important roles in the specification, induction, and differentiation of renal tubules also develop cystic kidneys (Cadieux *et al.*, 2008; Dressler *et al.*, 1993). Interestingly, overexpression of β-catenin, a mediator of the wnt signaling pathway that can function both as a transcription factor as well as a cell–cell adhesion protein also develop PKD (Saadi-Kheddouci *et al.*, 2001). The latter finding was of interest in the context that overexpression of *Pkd1*, that is also localized to cell–cell adhesion junctions, also caused renal cystic disease (Pritchard *et al.*, 2000). Taken together, the spectrum of transgenic mouse models that can lead to renal cystic disease suggest that a tight control of transcriptional activity and cell adhesive functions during development are critical for the differentiation of a kidney with normal tubular architecture and function.

Inactivation of several genes and proteins by targeted mutation or knockout, chemical, or spontaneous mutagenesis can also lead to the development of PKD and thus help to shed light on the underlying mechanisms associated with renal cystic development (Table 6.3). Interestingly, they fall into some of the same general categories. For instance, inactivation or knockout of the p53 and renal-specific *Vhl* (Von Hippel-Lindau) tumor suppressors lead to renal cyst formation, again suggesting overproliferation is a fundamental feature of renal cystic development (Rankin *et al.*, 2006; Saifudeen *et al.*, 2002). In addition, the inactivation of *Nek* family genes *nek1* and *nek8* which encode never in mitosis (NIM)A-related kinases and are the targets in the spontaneous *kat* (kidney anemia testis) and *jck* (juvenile cystic kidney) mouse models, respectively, include cystic kidneys in their complex phenotypes (Janaswami *et al.*, 1997; Otto *et al.*, 2008).

Disruption of developmental patterning and differentiation by inactivation of several genes also leads to renal cyst formation including the transcription factors HNF-1β, p75Cux, and TAZ (transcriptional activator with PDZ-binding motif) which is thought to link events at the cell plasma membrane and cytoskeleton to nuclear transcription in a 14–3–3-dependent manner (Cadieux *et al.*, 2008; Hiesberger *et al.*, 2005; Makita *et al.*, 2008). Knockout of another PDZ protein, MALS-3 (LIN-7c) also results in cystic kidneys although in this case, the mechanism of action is a failure to interact with Crumbs-3 and discs large complexes causing disruption of apicobasolateral polarization of differentiating renal tubule epithelia (Olsen *et al.*, 2007). The absence of the transcription factor AP-2β or bcl-2, both of

Table 6.3 Phenotypic mouse models of PKD: Gene inactivation

Proliferation	p53		Tumor suppressor
	VHL		Tumor suppressor
Cell cycle	NEK1	*kat* mouse	NIMA-related kinase
	NEK8	*jck* mouse	NIMA-related kinase
	SOCS-1		
Apoptosis inhibition	bcl-2		
	AP-2β		Transcription factor
Developmental regulation	HNF-1β		Transcription factor
	p75Cux-1		Transcription factor
	BMP-7		Secreted glycoprotein
	MALS-3		Apicobasal polarity
	Bicaudal-C	*bpk* mouse	
		jcpk mouse	
Cell–matrix adhesion	Laminin-α_5	Hypomorph	Matrix protein
	$\alpha_3\beta_1$-integrin		Matrix receptor
	Tensin		Focal adhesion protein
Cell–cell adhesion	APC		β-catenin complex, tumor suppressor
	TAZ		14–3–3 homolog
Actin cytoskeleton	Rho-GDIα		Small GTPase
Cilia	Tg737		Polaris
	Cystin	*cpk* mouse	
	Kif3α		Subunit of kinesin-II, microtubule motor

which inhibit apoptosis, also lead to the development of cystic kidneys (Moser *et al.*, 1997; Veis *et al.*, 1993).

Knockout or inactivation of several genes and proteins involved in cell–matrix and cell–cell adhesive interactions also lead to renal cystic disease including the matrix protein laminin-α_5; the matrix receptor $\alpha_3\beta_1$-integrin; the focal adhesion protein tensin; and the β-catenin complex protein APC (Kreidberg *et al.*, 1996; Lo *et al.*, 1997; Qian *et al.*, 2005; Shannon *et al.*, 2006). Taken in context with the known localization of *PKD1*-, *PKHD1*-, and *NPHP1–4*-encoded proteins at cell–matrix focal adhesion contacts and/or cell–cell adherens junctions of renal epithelia, these models suggest that cell–cell and cell–matrix adhesion-related functions are critical for normal renal tubule epithelial development. This notion is further supported by the renal cystic phenotypes in mice lacking TAZ or the small GTPase Rho-GDIα (Makita *et al.*, 2008; Togawa *et al.*, 1999).

Cystic proteins encoded by *PKD1, PKD2, PKHD1*, and *NPHP1–4* have also been localized to the apical primary nonmotile cilium of renal epithelial cells that are thought to play a role in sensory recognition in the tubule lumen environment and subsequent mechanotransduction of signals into the cell. This conclusion is further supported by the development of renal cystic phenotypes in mice that lack ciliary or cilia-associated genes including *Tg737/Polaris* in the *orpk* (Oak Ridge Polycystic Kidney) mouse; *cystin* in the *cpk* (congenital polycystic kidney) mouse; or the *Kif3A* subunit of the kinesin-II microtubule motor (Hou *et al.*, 2002; Lin *et al.*, 2003; Murcia *et al.*, 2000; Yoder *et al.*, 2002).

6. Normal Mammalian Kidney Development

Cell and developmental biology together with genetic approaches have helped to define many of the molecular pathways involved in mammalian metanephric development. The mammalian kidney originates from the intermediate mesoderm. After transient formation of the pronephros and mesonephros, the final metanephros develops and differentiates into the mature kidney (Sorokin and Ekblom, 1992). At murine E9 and human E22–24, the ureteric bud develops as an outpushing of the caudal portion of the mesonephric Wolffian duct, which grows toward and invades the undifferentiated metanephric mesenchyme at murine E11 and human E35–37. Reciprocal signaling between the ureteric bud epithelium and the undifferentiated metanephric mesenchyme leads to elongation and iterative branching of the ureteric bud and induction and differentiation of nephrons at the ureteric bud tips (Ekblom, 1996) (Fig. 6.2). Mesenchymal cells begin to condense around the tip at murine E11.5 and undergo mesenchymal to epithelial conversion forming the vesicle, comma, and S-body intermediates that by E13.5 differentiate into the glomerular and tubular epithelia of the immature nephron that then proliferate, elongate, and differentiate into defined epithelia of the glomerulus (podocytes), proximal tubule, descending and ascending limbs of Henle's loop, and the distal convoluted tubule. The ureteric bud gives rise to the collecting tubules of the cortex and medulla.

The complex reciprocal interactions between the metanephric mesenchyme and ureteric bud involved in metanephric development and differentiation are orchestrated by many differentially expressed transcription factors that drive growth factor secretion, extracellular matrix factor regulation, and receptor expression. These interactions takes place iteratively, forming arcades of nephrons connected to the collecting system (Osathanondh and Potter, 1963), resulting in a complete complement of 1,000,000 nephrons in each human kidney, at term. By contrast, mouse

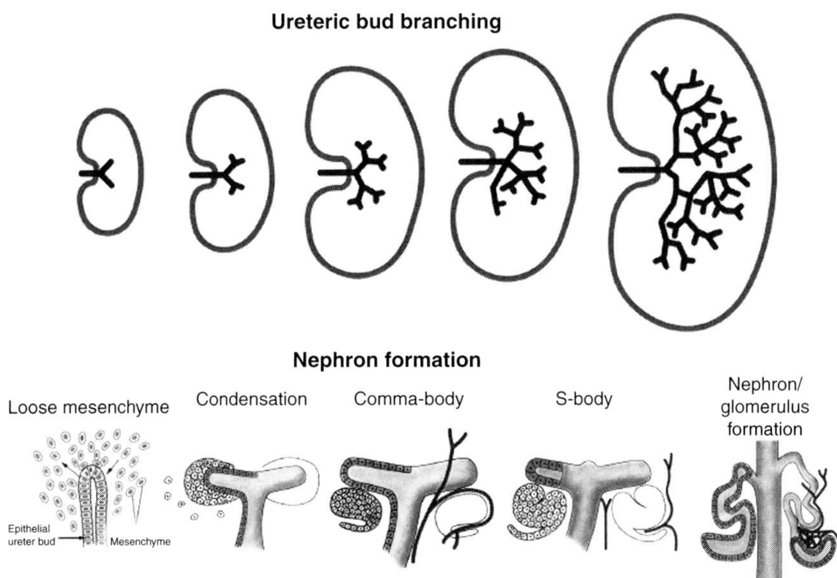

Figure 6.2 Diagram of renal development. Upper panel: ureteric bud branching. Lower panel: induction of metanephric mesenchyme to differentiate renal tubules. (See Color Insert.)

metanephric development continues for approximately 14 days after birth and results in the formation of 10,000–20,000 nephrons in each kidney.

Early patterning of the kidney mesoderm depends on incompletely understood complex interactions between the transcription factor genes *Pax2, Pax8, Eya*, and *Six* aided by *Lim1* and *Odd1* (reviewed in Dressler, 2006). Ureteric bud outgrowth and branching is a complex process that is controlled in its earliest stages by the tyrosine kinase receptor c-Ret expressed in the ureteric bud and its ligand Glial-derived neurotrophic factor (GDNF) expressed in the metanephric mesenchyme (Costantini and Shakya, 2006). Several gene products function within the ureteric bud to tightly regulate its outgrowth, elongation, insertion, and branching including the negative regulators: Hox-D11, Six-2, BMP-4, and Sprouty-1 (Basson *et al.*, 2005; Miyazaki *et al.*, 2000) and the positive regulators: Hox-B7, Gata-3, and Wnt-9b (Carroll *et al.*, 2005). Similarly, opposing regulators are expressed within the metanephric mesenchyme including the positive regulators: WT-1, Sal-1, and Hox-11 which complexes with Eya-1 and Pax-2 to upregulate GDNF (Gong *et al.*, 2007; Kreidberg *et al.*, 1993; Nishinakamura *et al.*, 2001) and the negative regulators: FoxC1, FoxC2 as well as the transmembrane protein Slit-2 and its receptor Robo-2 (Grieshammer *et al.*, 2004; Kume *et al.*, 2000).

Early induction of the metanephric mesenchyme to form condensates is associated with upregulation of α_8-integrin mediated by Hox-11 (Valerius *et al.*, 2002), while vesicle differentiation involves Wnt-4 (Stark *et al.*, 1994). Several growth factors play important roles, notably the transforming growth factor-beta (TGFβ) superfamily, FGF-2 and FGF-8 as well as leukemia inhibitory factor (LIF) (Barasch *et al.*, 1997, 1999; Dressler, 2006). In addition, recent studies have implicated Notch-2 signaling in proximal tubule specification; WT-1 and LIM-1 in the definition of the glomerular podocyte lineage; and Brn1 in differentiation of the distal tubule (Cheng and Kopan, 2005; Schedl, 2007).

7. The PKD Proteins

Since many of the majority of human cystic genes have been cloned, the structures of their encoded protein products have been deduced, antibodies raised, and their tissue and cellular distributions analyzed (Fig. 6.3).

PKD1-encoded Polycystin-1 (PC-1) is a large (>460 kDa), modular protein with a long extracellular N-terminal portion, 11 transmembrane domains, and a short (200 amino acid) intracellular C-terminal domain. The extracellular portion of PC-1 contains two cysteine-flanked leucine-rich repeats (LRR) capable of binding collagen, fibronectin, and laminin in cell-free assays; a cell wall integrity and stress response component (WSC) homology domain and C-type lectin domain capable ofbinding carbohydrate, a low-density lipoprotein (LDL)-A domain and 16 immunoglobulin-like *PKD1* repeats. Overall, the modular makeup of the long extracellular portion of PCI suggests that it participates in protein-protein and protein–carbohydrate interactions, consistent with cell–cell or cell–matrix attachment function. The additional presence of a latrophilin-like (GPS) and a receptor for egg jelly (REJ) domain just proximal to the first transmembrane domain suggests that the N-terminal may be proteolytically cleaved and possibly secreted into the extracellular space (Hughes *et al.*, 1995; Lohning *et al.*, 1996; Qian *et al.*, 2002; Yu *et al.*, 2007). Eleven transmembrane domains ensure that the PC-1 protein is firmly embedded in the cell membrane, as has been confirmed by immunolocalization studies. The intracellular C-terminal tail of PC-1 contains several tyrosine and serine residues that have been identified as targets for phosphorylation and activation by c-Src (at tyrosine (Y)4237); focal adhesion kinase (FAK at Y4127); protein kinase A (PKA at S4252); protein kinase X (PRKX at S4161), facilitating activation of downstream intracellular signaling cascades (Li *et al.*, 1999, 2002). In addition, the PC-1 C-terminal domain contains a src homology (SH)3-binding site, a G protein activating site and a coiled-coil domain consistent with protein–protein binding and activation functions (Parnell *et al.*, 1998). PC-1 has also been shown to interact with β-integrin, and other members of the focal adhesion complex linked to

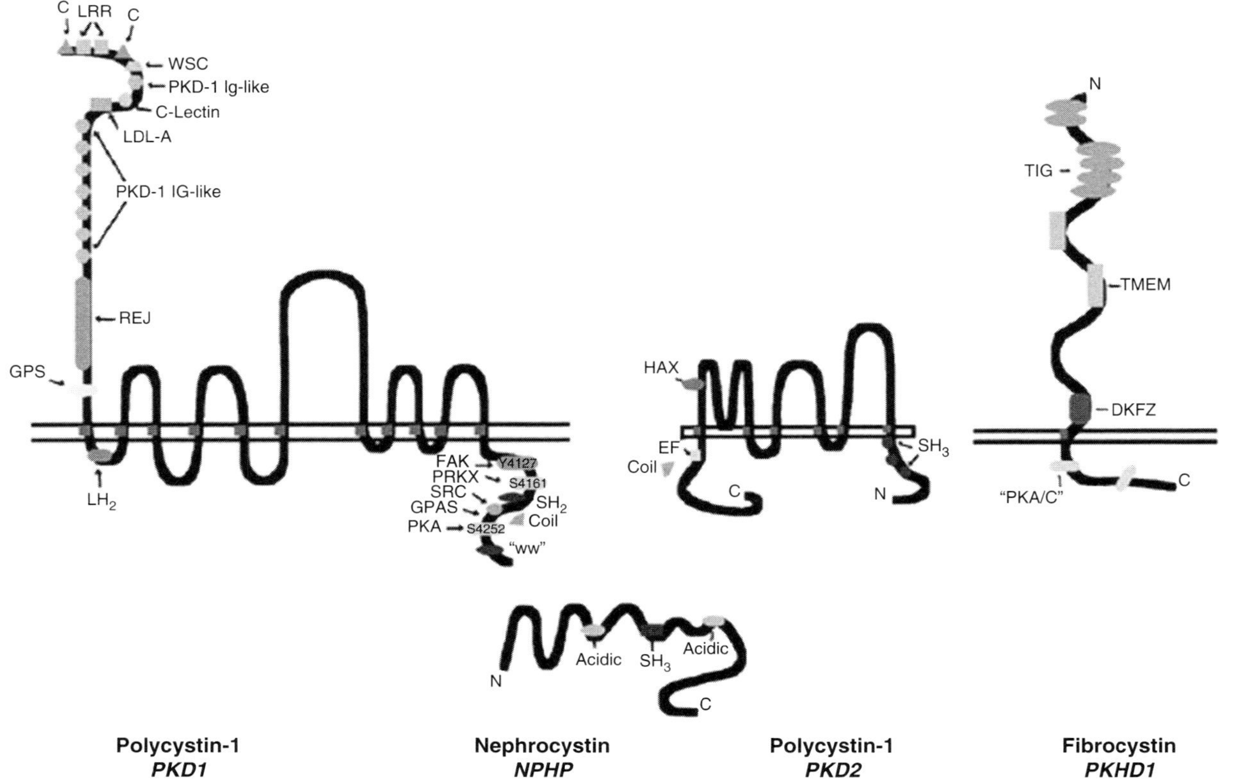

Figure 6.3 Domain structure of cystic proteins: Polycystin-1, Polycystin-2, Fibrocystin-1, and Nephrocystin-1. (See Color Insert.)

the actin cytoskeleton, as well with E-cadherin and β-catenin (Geng *et al.*, 2000; Wilson *et al.*, 1999). Overall, structure/function and localization analysis strongly suggests that PC-1 functions as a plasma membrane receptor that receives and transduces signals from the extracellular environment (Wilson, 2004).

PKD2-encoded Polycystin-2 (PC-2) is a smaller 110 kDa protein with intracellular NH_2- and COOH-terminals and six transmembrane domains with overall ion channel signature. Its NH2-terminal contains an SH3 domain and its COOH-terminal contains a coiled-coil domain as well as a calcium-binding (EF hand) domain, again suggesting protein–protein interactions. Structure/function and localization analysis strongly suggests that PC-2 acts as a nonselective cation channel that is capable of conducting calcium in cell and endoplasmic reticulum membranes (Hanaoka *et al.*, 2000; Koulen *et al.*, 2002; Nauli *et al.*, 2003). PC-2 may also be phosphorylated and interacts with the actin-binding protein α-actinin as well as PC-1 (Li *et al.*, 2005a,d; Qian *et al.*, 1997; Tsiokas *et al.*, 1997).

PKHD1-encoded Fibrocystin-1 (FC-1) is a large (>450 kDa), modular protein with a long extracellular NH2-terminal, one membrane-spanning domain and a short intracellular C-terminal (Ward *et al.*, 2002). In addition to this, overall similarity in structure to PC-1, the extracellular N-terminal of Fibrocystin-1 also contains domains predicted to facilitate protein–protein interactions and two putative serine consensus sequence sites for phosphorylation by PKA or PKC. Although fewer structure/function and localization studies have yet been conducted, initial studies are consistent with a plasma membrane localization and a receptor function.

NPHP1-encoded Nephrocystin-1 is a relatively small (733 amino acid) intracellular protein without transmembrane domains. An N-terminal coiled-coil domain adjacent to an SH3 domain predicted protein–protein interaction capability. The SH3 domain is also flanked by acidic E-rich domains and a "nephrocystin homology" domain of unknown function. Structure/function analysis has shown that Nephrocystin-1 binds to the focal adhesion complex and actin-binding proteins p130-cas, pyk-2 (proline-rich tyrosine kinase), tensin, and filamins, as well as to PC-1 and tubulins (Benzing *et al.*, 2001; Donaldson *et al.*, 2000; Fliegauf *et al.*, 2006). Interactions of Nephrocystin-2/Inversin, a 1062 amino acid nephrocystin with ankyrin repeats, have also been documented with β-catenin, N-cadherin, calmodulin, and the anaphase-promoting complex (apc)-2 (Morgan *et al.*, 2002).

The *MCKD2*-encoded Tamm–Horsfall protein is a glycosyl phosphatidyl inositol (GPI)-anchored membrane glycoprotein that is shed into the urine in very large amounts but whose function remains obscure.

Depending on cell type and developmental context, the Polycystins, Fibrocystin, and Nephrocystins have been shown to be localized to three major specialized areas of the cell that acts as signaling centers to collect and respond to signals from the extracellular environment by sensing force: the cell–matrix focal adhesion complex attachment sites at the basal membrane; the lateral cell–cell adhesion junctions; and the primary cilium present on the apical surface of renal epithelial cells (Fig. 6.4).

PC-1, PC-2, and Fibrocystin-1 are developmentally regulated showing high levels of expression in the fetal kidneys and low levels in adult kidneys (Polgar *et al.*, 2005; Van Adelsberg *et al.*, 1997; Wilson *et al.*, 1999). The majority of protein expression is seen associated with the ureteric

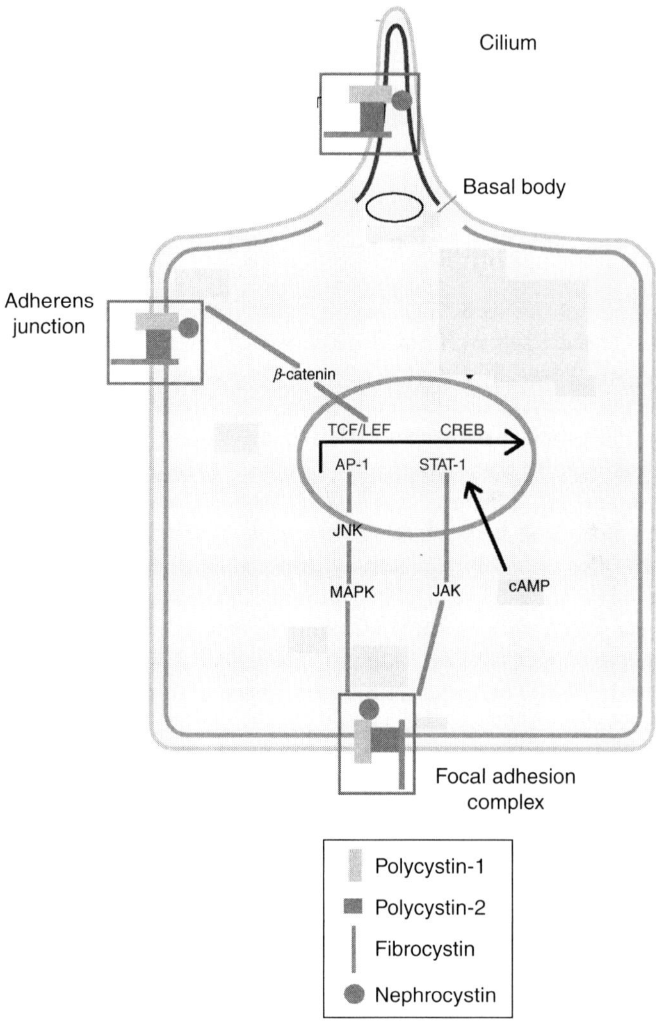

Figure 6.4 Diagram of a renal epithelial cell showing cystic protein localization sites and interacting proteins. (See Color Insert.)

Proper morphogenesis of the metanephric kidney involves the exquisitely coordinated regulation of epithelial proliferation, adhesion, migration, differentiation, cell and tubule geometry, maturation, and polarized distribution of membrane proteins. Mouse models in which manipulation of these classes of genes have resulted in renal cystic phenotypes have not only provided invaluable genetic confirmation of several formerly implicated pathways but have also shed important new light onto the understanding of the complexities underlying mechanisms of cyst formation in the kidney.

8.2. Proliferation/apoptosis

Multiple genetically engineered phenotypic mouse models confirm the important role for enhance proliferation as a means to develop multiple renal cysts. These include overexpression of oncogenes, growth factor receptors, growth factor ligands, and inactivation of tumor suppressors or apoptosis. These models also confirm the importance of EGFR family members (ErbB1 and ErbB2) and their ligands TGFα in renal cystic expansion. More surprisingly, perhaps, a role for the NIMA-related kinase family of cell cycle regulators (*Nek1* and *Nek8*) has been identified suggesting that coordination of centrosome structure and function in mitotic progression may be compromised in PKD. Support of this notion has been provided by studies showing that Nek8 interacts with PC-2 (Sohara *et al.*, 2008). Centrosomes are highly structured organelles at the poles of the mitotic spindle and at the base of cilia where they are thought to act as microtubule-organizing centers. The majority of the cystic proteins have been localized to the collecting tubule cilium and basal body. Furthermore PC-1, PC-2, NPHP2, and NPHP4 show a dynamic pattern of expression and localization during cell division (C. Iomini, personal communication).

Given the large number of mutations that can cause cystic kidneys, it is likely that integration of several biological processes and pathways are critical. In this context, important evidence exists linking a focal adhesion scaffolding protein, HEF-1 that has a well-defined role in integrin-dependent attachment the control of cell division. Not only does HEF-1 relocalize to the spindle asters during mitosis but it activates Aurora A and Nek2 kinases at the centrosome. (Fielding *et al.*, 2008; Pugacheva and Golemis, 2005). Furthermore, possibilities for interactions between the focal adhesion and cilium are suggested by the discovery of the localization of β_1-integrin in the cilium (Praetorius *et al.*, 2004).

8.3. Role of cilia

The biogenesis of cilia is important for normal development. Most cilia in the body are motile and mutations causing their functional deficiencies lead to multiorgan diseases including retinal degeneration, brain malformations, obesity, polydactyly, and randomization of left–right asymmetry (Ansley *et al.*, 2003). Mutations in nonmotile ciliary proteins lead to renal cystic disease. Nonmotile, primary cilia are single, microtubule-containing hair-like structures of 2–5 μm in length inserted into and extending from a basal body derived from the mother centriole of most vertebrate cells including the principal cells of renal collecting tubule epithelia. Microtubule bundles forming the core of the cilium are arranged circumferentially in nine doublets surrounding a microtubule free center (9+0) configuration. Protein cargo is delivered to and removed from the cilia via a complex network of intraflagellar transport (IFT) proteins under the control of kinesin and dynein molecular motors, respectively. Cilia have been implicated in the pathogenesis of PKD, since most cystic proteins PC-1, PC-2, FC-1, and NPHP1–4 are localized to basal bodies and some to the ciliary membrane, often in a punctate pattern of unknown significance. These proteins are also localized to focal adhesions and cell–cell adherens junctions. Mutation analysis suggests that cystic proteins are not essential for ciliogenesis and no abnormalities have been detected in ciliary structure in human PKD. However, this does not rule out the possibility of functional abnormalities in human PKD epithelia. By contrast, IFT proteins are localized solely to cilia and basal bodies and can be essential for ciliogenesis and mutations, resulting in severe and diverse defects including situs inversus, retinitis pigmentosa, ocular motor apraxia, cerebellar abnormalities, and obesity in addition to cystic kidneys. The extreme phenotype of the *Nphp2/inv* mouse presents an anomaly since it shows left–right reversal while no situs inversus has been seen in patients with juvenile NPHP.

Little is known of the molecular functions of nonmotile renal epithelial cilia, except that their deformation results in intracellular influx of calcium leading to the hypothesis that they can act as sensors of fluid flow or mechanical stress (Nauli *et al.*, 2003). Mutational studies suggest a more important role during development than in the adult and morphological evidence shows that they are elaborated during the later stages of renal development after initial apicobasal polarization of epithelia (Davenport *et al.*, 2007). It has been suggested that they may play a role in noncanonical wnt signaling and the establishment of planar cell polarity and that they may be important in establishing and/or maintaining appropriate tubule lumen diameters.

8.4. Matrix adhesion/migration

Phenotypic mouse models show that overexpression of β-catenin or knockout of its cytoplasmic protein complex partner, APC or knockout of a variety of cell–matrix interaction and actin-binding proteins including laminin-α_5, $\alpha_3\beta_1$-integrin, tensin, TAZ, and Rho-GDIα result in cystic kidneys (Table 6.3). Focal adhesion complexes are multimolecular assemblies linking the extracellular matrix to the actin cytoskeleton via integrin receptors that acts as signaling interfaces through which cells sense and respond to changes in their external environment. Focal adhesions regulate cell adhesion, spreading, and migration and act as traction points that enable specific cell shape changes and facilitate cell migration by means of site-specific turnover. Cell migration is an integrated process that requires coordination of signaling molecules such as kinases, ubiquitin ligases, proteases, and phosphatases. Focal adhesion complex number, size, regulation, and function, is abnormal in ADPKD and ARPKD leading to increased matrix adhesion and decreased migration of epithelial cells (Israeli *et al.*, 2007; Polgar *et al.*, 2005).

Both the PC-1-containing focal adhesion complex and the PC-1-containing cell–cell adherens complex show abnormalities in composition, turnover, regulation, and function in mouse and human PKD cyst-lining epithelial cells. This identifies these cellular sites, interacting protein partners and downstream signaling pathways as key players in PKD. Knockout of many focal adhesion-associated proteins in mice, including FAK, src family kinases, paxillin, and p130cas, although lethal prior to kidney formation, like PKD1 mutations, lead to impaired cell migration. It is clear that integration between cell–matrix and cell–cell adhesion functions is critical for coordinated differentiation of renal epithelial tubular cells during development and much crosstalk between these pathways together with growth factor receptors is known to occur. It seems feasible that the cystic proteins could function as organizers of these complexes at the focal adhesions, and cell–cell adherens junctions and together with their ciliary complexes, act as signaling receptors, integrators, and responders to signals from the extracellular environment either in the base of the cell from the matrix; at the lateral aspect of the cell as it responds to interactions with adjacent cells and to the apical lumen fluid and diametrically opposed cells in the tubule.

8.5. Polarity

Normal kidney development and physiology require polarization of epithelia that line renal tubules. Distinct sets of transporters, channels, and growth factor receptors are asymmetrically distributed on the apical versus basolateral membranes of each of the distinct nephron segments and allows for vectorial transport of solutes along the nephron. Several pathophysiologically significant abnormalities in epithelial cell apicobasal polarity have been identified in human ADPKD cyst-lining epithelia including apically mispolarized NaK-ATPase and EGFR. Apicobasal polarity is established early during the embryonic developmental patterning and differentiation of the metanephric kidney. As in other organs, the basolateral membrane is defined and separated from the apical membrane by the establishment lateral cell–cell adhesions and occluding tight junctions and by basal interactions with the extracellular matrix. Crumbs, a PDZ domain-containing molecule first identified in *Drosophila melanogaster* as an apical membrane protein has three isoforms in mammals: RP12 that is mutated in patients with retinitis pigmentosa and Leber congenital amaurosis: Crumbs-2 found in brain, eye, and kidney and Crumbs-3 that has a broad expression pattern and is know to play an important role in epithelial polarity and tight junction formation and to complex with the scaffold proteins PALS 1 (protein associated with Lin-7) and partition-defective (Par)-6. Importantly, mice lacking the small PDZ protein mammalian LIN-7c (MALS-3) have hypomorphic, cystic, and fibrotic kidneys MALS-3 mediates stable assembly of the Crumbs tight junction complex with the discs large basolateral complex and these complexes are disrupted in renal epithelia from MALS-3 knockout mice (Olsen *et al.*, 2007). PAR-3 also forms a complex with atypical (a)PKC and PAR-6 that excludes basolateral proteins from the apical domain. PAR-3 is not only important for the formation of junctional complexes but is also required for the growth and elongation of the primary cilium which it does by interacting with kinesin-II, the microtubule motor responsible for anterograde transport of IFT particles to the tip of the growing cilium (Sfakianos *et al.*, 2007). aPKC/PAR-3 deletion in mice is embryonically lethal causing aberrant development of epithelia which suggests integration of ciliary growth and establishment of apicobasal polarity. This is further supported by the observation that the PAR–aPKC complex can associate with the tumor suppressor VHL, which is required for growth of microtubules during ciliogenesis in the kidney and mutations in which are associated with renal cystic formation. Tumor suppressors and other factors linked to signaling pathways implicated in cancer are frequently able to regulate apicobasal polarity and cell growth simultaneously which they may achieve by means of independent effector molecules. This, therefore, suggests that a loss of normal apicobasal polarity can be combined with increased epithelial growth as is the case in PKD.

Loss of E-cadherin, leads to the loss of integrity of cell–cell junctions and has been noted in ADPKD where N-cadherin substitutes but impairs PC-1–multiprotein complex integrity. Such impairments can affect signaling pathways that cluster at sites of cell–cell contact including EGFR and wnt, both of which are abnormal in ADPKD. Recently, it has been shown that ligand-activated ErbB2 receptor binds directly to Par-6 leading to recruitment of aPKC and disruption of apicobasal polarity (Aranda *et al.*, 2006). Erb-B2-mediated inhibition of apoptosis has also been shown to require binding of PAR-6 to the receptor thus linking Erb-B2, polarity and apoptosis, all of which are abnormal in ADPKD. Furthermore, β_4-integrin binds directly to ErbB2 and promotes ErbB2-induced proliferation and disrupts polarity (Guo *et al.*, 2006).

In addition to apicobasal polarity, planar cell polarity is involved in cell differentiation and the shaping of tissues and organs. During the lengthening of renal tubules during embryonic development, mitotic orientation of cells along the axis of normal tubules and becomes distorted in cystic models of PKD caused by HNF-1β inactivation. Specifically, 23% of mitoses showed distortion of their mitotic angles. These studies suggested that appropriately oriented cell division is necessary to maintain a constant tubule diameter in the elongating kidney tubule (Fischer *et al.*, 2006). Future studies will determine whether the noncanonical wnt pathway, Dishevelled and/or Vangl2 are abnormal in PKD. Interestingly, basal bodies have been shown to direct planar polarization of sensory hair cells (Jones and Chen, 2007).

8.6. Differentiation

The ureteric bud undergoes branching morphogenesis, followed by a growth phase, maturation, and differentiation into the collecting duct system of the kidney. At the tips of the developing ureteric bud, the adjacent induced metanephric mesenchyme undergoes condensation followed by differentiation of a polarized epithelium. Epithelial differentiation is a complex, multistep process that first involves the formation of polarized cell scaffold which is initiated by cell–matrix inductive interactions followed by definition and elaboration of lateral cell–cell adherens junctions and apicolateral occluding tight junctions that allow for the separation of the apical from the basolateral membrane domains of the polarized cell. Once this framework is formed, this then provides template on which to differentiate and specify the renal epithelium with regard to the specific apical or basolateral insertion of specific proteins such as ion channels and transporters as well as growth factor and ligand-responsive receptors. Finally, cell proliferation ceases and terminal differentiation takes place during which nephron segment-specific epithelia complete their maturation with their distinct structural and functional properties associated with the postnatal expression of a specific complement of apicobasolateral membrane proteins and alterations in cell shape including definition of epithelial cell height and elaboration of basolateral membrane infoldings and apical membrane microvilli and cilia. Cell culture studies suggest that the initial sites of localization of the polycystins after attachment to extracellular matrix are the focal adhesion contacts (Mollet *et al.*, 2005; Wilson, 2004). This is also the predominant localization pattern seen *in vivo* in the epithelium of the invading ureteric bud epithelium in E11–E13 embryonic mouse kidneys (Polgar *et al.*, 2005; Wilson, 1997). As cell density increases *in vitro*, and during the later stages of embryonic kidney development (E17) *in vivo*, PC-1, Fibrocystin-1, and NPHPs are also found at sites of cell–cell attachment and in the developing cilium (Silberberg *et al.*, 2005; Ward *et al.*, 2003; Zhang *et al.*, 2004).

This circumstantial evidence suggests that the cystic protein complex is situated at key points in the epithelial cell during different stage of differentiation and is consistent with playing an important role in the control of the definition and maintenance of renal epithelia cell differentiation, an important component of which is the appropriate stage-specific morphogenetic formation and maintenance of tubule lumen diameter. The apparently dynamic nature of relative expression levels and subcellular sites of distribution relative to developmental stage and differentiation status of the epithelial cell further suggests that there is likely to be associated dynamic intracellular signaling cascades stimulated in response to mechanical stimuli from the extracellular matrix/basement membrane, cell–cell and apical cilia/lumen interfaces. Although a plethora of intracellular signaling cascades have been implicated in cystic protein signal transduction, it remains to be determined whether there is cell interaction site-dependent specificity, how they are integrated or whether they too are developmentally regulated.

9. Future Perspectives

The ultimate goal of genetic engineering of mouse models of PKD as for any disease is to shed light on the underlying molecular mechanisms of the human disease. Some headway has been made in recent years with the development of true genocopy models of ADPKD, some of which truly recapitulate the phenotypic characteristics of the human disease. It is to be expected that similar models will be developed for ARPKD and NPHP in the foreseeable future. The current difficulties in generating true genocopy/phenotypic models may relate to the strong influence of genetic background in the mixed strains used to generate the mice and more emphasis is likely to be placed on renal-specific knockdown strategies as well as the use of congenic mouse strains.

The wide spectrum of disease severity and progression to renal failure in ADPKD patients is of critical prognostic impact and mouse models will be important tools to dissect out the effects of potential modifier genes. *PKD1, PKD2*, and *PKHD1* have all been shown to be highly polymorphic and the effects of these polymorphisms could be tested on homogeneous backgrounds. Since it has been reported that large N-terminal deletion mutations of *PKD1* may favor rapid progression while certain substitution mutations in *PKHD1* may be related to a milder course, animal models may have a role for the future study of genotype–phenotype correlations (Furu *et al.*, 2003; Rossetti *et al.*, 2007).

Finally, an important goal is to develop therapies that retard PKD and thus improve the quality of life of patients with this disease. At present, their only "treatment" is renal functional replacement by hemodialysis or transplantation. Since cell and molecular biological studies have led to the identification of proliferation and secretion as major pathophysiological features of continuous cystic enlargement, several putative "therapies" have been proposed (Gattone *et al.*, 2003; Torres *et al.*, 2004). With the accumulation of more knowledge and understanding of intracellular mechanisms, more potential targets are likely to be identified. Together with human cell lines of PKD and age-matched normal renal epithelia, the advent of genocopy and phenotypically faithful mouse models of ADPKD, ARPKD, and NPHP will afford the ideal preclinical testing ground for the new and innovative drug therapies of the future.

ACKNOWLEDGMENTS

I am indebted to Drs. Carlo Iomini, Beatrice Goilav, and Irina Barash for stimulating discussions, and to Jaime Pei for graphic design.

REFERENCES

Ansley, S. J., Badano, J. L., Blacque, O. E., Hill, J., Hoskins, B. E., Leitch, C. C., Kim, J. C., Ross, A. J., Eichers, E. R., Teslovich, T. M., Mah, A. K., Johnsen, R. C., *et al.* (2003). Basal body dysfunction is a likely cause of pleiotropic Bardet-Biedl syndrome. *Nature* **425,** 628–633.

Aranda, V., Haire, T., Nolan, M. E., Calarco, J. P., Rosenberg, A. Z., Fawcett, J. P., Pawson, T., and Muthuswamy, S. K. (2006). Par6–aPKC uncouples ErbB2 induced disruption of polarized epithelial organization from proliferation control. *Nat. Cell Biol.* **8,** 1235–1245.

Bachmann, S., Mutig, K., Bates, J., Welker, P., Geist, B., Gross, V., Luft, F. C., Alenina, N., Bader, M., Thiele, B. J., Prasadan, K., Raffi, H. S., *et al.* (2005). Renal effects of Tamm–Horsfall protein (uromodulin) deficiency in mice. *Am. J. Physiol. Renal Physiol.* **288,** F559–F567.

Barasch, J., Qiao, J., McWilliams, G., Chen, D., Oliver, J. A., and Herzlinger, D. (1997). Ureteric bud cells secrete multiple factors, including bFGF, which rescue renal progenitors from apoptosis. *Am. J. Physiol.* **273,** F757–F767.

Barasch, J., Yang, J., Ware, C. B., Taga, T., Yoshida, K., Erdjument-Bromage, H., Tempst, P., Parravicini, E., Malach, S., Aranoff, T., and Oliver, J. A. (1999). Mesenchymal to epithelial conversion in rat metanephros is induced by LIF. *Cell* **99,** 377–386.

Basson, M. A., Akbulut, S., Watson-Johnson, J., Simon, R., Carroll, T. J., Shakya, R., Gross, I., Martin, G. R., Lufkin, T., McMahon, A. P., Wilson, P. D., Costantini, F. D., *et al.* (2005). Sprouty1 is a critical regulator of GDNF/RET-mediated kidney induction. *Dev. Cell* **8,** 229–239.

Battini, L., Fedorova, E., Macip, S., Li, X., Wilson, P. D., and Gusella, G. L. (2006). Stable knockdown of polycystin-1 confers integrin-alpha2beta1-mediated anoikis resistance. *J. Am. Soc. Nephrol.* **17,** 3049–3058.

Benzing, T., Gerke, P., Hopker, K., Hildebrandt, F., Kim, E., and Walz, G. (2001). Nephrocystin interacts with Pyk2, p130(Cas), and tensin and triggers phosphorylation of Pyk2. *Proc. Natl Acad. Sci. USA* **98,** 9784–9789.

Bergmann, C., Senderek, J., Sedlacek, B., Pegiazoglou, I., Puglia, P., Eggermann, T., Rudnik-Schoneborn, S., Furu, L., Onuchic, L. F., De Baca, M., Germino, G. G., Guay-Woodford, L., *et al.* (2003). Spectrum of mutations in the gene for autosomal recessive polycystic kidney disease (ARPKD/PKHD1). *J. Am. Soc. Nephrol.* **14,** 76–89.

Bergmann, C., Frank, V., Kupper, F., Schmidt, C., Senderek, J., and Zerres, K. (2006). Functional analysis of PKHD1 splicing in autosomal recessive polycystic kidney disease. *J. Hum. Genet.* **51,** 788–793.

Bergmann, C., Fliegauf, M., Bruchle, N. O., Frank, V., Olbrich, H., Kirschner, J., Schermer, B., Schmedding, I., Kispert, A., Kranzlin, B., Nurnberg, G., Becker, C., *et al.* (2008). Loss of nephrocystin-3 function can cause embryonic lethality, Meckel-Gruber-like syndrome, situs inversus, and renal–hepatic–pancreatic dysplasia. *Am. J. Hum. Genet.* **82,** 959–970.

Bhunia, A. K., Piontek, K., Boletta, A., Liu, L., Qian, F., Xu, P. N., Germino, F. J., and Germino, G. G. (2002). PKD1 induces p21(waf1) and regulation of the cell cycle via direct activation of the JAK–STAT signaling pathway in a process requiring PKD2. *Cell* **109,** 157–168.

Boca, M., D'Amato, L., Distefano, G., Polishchuk, R. S., Germino, G. G., and Boletta, A. (2007). Polycystin-1 induces cell migration by regulating phosphatidylinositol 3-kinase-dependent cytoskeletal rearrangements and GSK3beta-dependent cell–cell mechanical adhesion. *Mol. Biol. Cell* **18,** 4050–4061.

Boulter, C., Mulroy, S., Webb, S., Fleming, S., Brindle, K., and Sandford, R. (2001). Cardiovascular, skeletal, and renal defects in mice with a targeted disruption of the Pkd1 gene. *Proc. Natl Acad. Sci. USA* **98,** 12174–12179.

Boucher, C., Case, R., Thurston, K., Li, X., Needham, A., Romero, E., Ward, H., Hyink, D., Qamar, S., Roitbak, T., Powell, S., Ward, C., *et al.* (2008). Receptor protein tyrosine phosphatases are novel components of the polycystin complex. *J. Am. Soc. Nephrol.*

Cadieux, C., Harada, R., Paquet, M., Cote, O., Trudel, M., Nepveu, A., and Bouchard, M. (2008). Polycystic kidneys caused by sustained expression of Cux1 isoform p75. *J. Biol. Chem.* **283,** 13817–13824.

Cai, Y., Anyatonwu, G., Okuhara, D., Lee, K. B., Yu, Z., Onoe, T., Mei, C. L., Qian, Q., Geng, L., Wiztgall, R., Ehrlich, B. E., and Somlo, S. (2004). Calcium dependence of polycystin-2 channel activity is modulated by phosphorylation at Ser812. *J. Biol. Chem.* **279,** 19987–19995.

Carroll, T. J., Park, J. S., Hayashi, S., Majumdar, A., and McMahon, A. P. (2005). Wnt9b plays a central role in the regulation of mesenchymal to epithelial transitions underlying organogenesis of the mammalian urogenital system. *Dev. Cell* **9,** 283–292.

Chauvet, V., Tian, X., Husson, H., Grimm, D. H., Wang, T., Hiesberger, T., Igarashi, P., Bennett, A. M., Ibraghimov-Beskrovnaya, O., Somlo, S., and Caplan, M. J. (2004). Mechanical stimuli induce cleavage and nuclear translocation of the polycystin-1 C terminus. *J. Clin. Invest.* **114,** 1433–1443.

Cheng, H. T., and Kopan, R. (2005). The role of Notch signaling in specification of podocyte and proximal tubules within the developing mouse kidney. *Kidney Int.* **68,** 1951–1952.

Consortium, The American PKD. (1995a). Analysis of the genomic sequence for the autosomal dominant polycystic kidney disease (PKD1) gene predicts the presence of a leucine-rich repeat. *Hum. Mol. Genet.* **4,** 575–582.

Consortium, The International PK D. (1995b). Polycystic kidney disease: The complete structure of the PKD1 gene and its protein. *Cell* **81,** 289–298.

Costantini, F., and Shakya, R. (2006). GDNF/Ret signaling and the development of the kidney. *BioEssays* **28,** 117–127.

Daikha-Dahmane, F., Narcy, F., Dommergues, M., Lacoste, M., Beziau, A., and Gubler, M. C. (1997). Distribution of alpha-integrin subunits in fetal polycystic kidney diseases. *Pediatr. Nephrol.* **11,** 267–273.

Dalkilic, I., and Kunkel, L. M. (2003). Muscular dystrophies: Genes to pathogenesis. *Curr. Opin. Genet. Dev.* **13,** 231–238.

Davenport, J. R., Watts, A. J., Roper, V. C., Croyle, M. J., van Groen, T., Wyss, J. M., Nagy, T. R., Kesterson, R. A., and Yoder, B. K. (2007). Disruption of intraflagellar

transport in adult mice leads to obesity and slow-onset cystic kidney disease. *Curr. Biol.* **17,** 1586–1594.
Davidow, C. J., Maser, R. L., Rome, L. A., Calvet, J. P., and Grantham, J. J. (1996). The cystic fibrosis transmembrane conductance regulator mediates transepithelial fluid secretion by human autosomal dominant polycystic kidney disease epithelium *in vitro*. *Kidney Int.* **50,** 208–218.
Deltas, C. C. (2001). Mutations of the human polycystic kidney disease 2 (PKD2) gene. *Hum. Mutat.* **18,** 13–24.
Donaldson, J. C., Dempsey, P. J., Reddy, S., Bouton, A. H., Coffey, R. J., and Hanks, S. K. (2000). Crk-associated substrate p130(Cas) interacts with nephrocystin and both proteins localize to cell–cell contacts of polarized epithelial cells. *Exp. Cell Res.* **256,** 168–178.
Dressler, G. R. (2006). The cellular basis of kidney development. *Annu. Rev. Cell Dev. Biol.* **22,** 509–529.
Dressler, G. R., Wilkinson, J. E., Rothenpieler, U. W., Patterson, L. T., Williams-Simons, L., and Westphal, H. (1993). Deregulation of Pax-2 expression in transgenic mice generates severe kidney abnormalities. *Nature* **362,** 65–67.
Du, J., and Wilson, P. D. (1995). Abnormal polarization of EGF receptors and autocrine stimulation of cyst epithelial growth in human ADPKD. *Am. J. Physiol.* **269,** C487–C495.
Ekblom, P. (1996). Genetics of kidney development. *Curr. Opin. Nephrol. Hypertens.* **5,** 282–287.
Fielding, A. B., Dobreva, I., McDonald, P. C., Foster, L. J., and Dedhar, S. (2008). Integrin-linked kinase localizes to the centrosome and regulates mitotic spindle organization. *J. Cell Biol.* **180,** 681–689.
Fischer, E., Legue, E., Doyen, A., Nato, F., Nicolas, J. F., Torres, V., Yaniv, M., and Pontoglio, M. (2006). Defective planar cell polarity in polycystic kidney disease. *Nat. Genet.* **38,** 21–23.
Fliegauf, M., Horvath, J., von Schnakenburg, C., Olbrich, H., Muller, D., Thumfart, J., Schermer, B., Pazour, G. J., Neumann, H. P., Zentgraf, H., Benzing, T., and Omran, H. (2006). Nephrocystin specifically localizes to the transition zone of renal and respiratory cilia and photoreceptor connecting cilia. *J. Am. Soc. Nephrol.* **17,** 2424–2433.
Furu, L., Onuchic, L. F., Gharavi, A., Hou, X., Esquivel, E. L., Nagasawa, Y., Bergmann, C., Senderek, J., Avner, E., Zerres, K., Germino, G. G., Guay-Woodford, L. M., *et al.* (2003). Milder presentation of recessive polycystic kidney disease requires presence of amino acid substitution mutations. *J. Am. Soc. Nephrol.* **14,** 2004–2014.
Gattone, V. H., II, Wang, X., Harris, P. C., and Torres, V. E. (2003). Inhibition of renal cystic disease development and progression by a vasopressin V2 receptor antagonist. *Nat. Med.* **9,** 1323–1326.
Geng, L., Burrow, C. R., Li, H. P., and Wilson, P. D. (2000). Modification of the composition of polycystin-1 multiprotein complexes by calcium and tyrosine phosphorylation. *Biochim. Biophys. Acta* **1535,** 21–35.
Gong, K. Q., Yallowitz, A. R., Sun, H., Dressler, G. R., and Wellik, D. M. (2007). A Hox–Eya–Pax complex regulates early kidney developmental gene expression. *Mol. Cell. Biol.* **27,** 7661–7668.
Grieshammer, U., Le, M., Plump, A. S., Wang, F., Tessier-Lavigne, M., and Martin, G. R. (2004). SLIT2-mediated ROBO2 signaling restricts kidney induction to a single site. *Dev. Cell* **6,** 709–717.
Guo, W., Pylayeva, Y., Pepe, A., Yoshioka, T., Muller, W. J., Inghirami, G., and Giancotti, F. G. (2006). Beta 4 integrin amplifies ErbB2 signaling to promote mammary tumorigenesis. *Cell* **126,** 489–502.

Hanaoka, K., Devuyst, O., Schwiebert, E. M., Wilson, P. D., and Guggino, W. B. (1996). A role for CFTR in human autosomal dominant polycystic kidney disease. *Am. J. Physiol.* **270,** C389–C399.

Hanaoka, K., Qian, F., Boletta, A., Bhunia, A. K., Piontek, K., Tsiokas, L., Sukhatme, V. P., Guggino, W. B., and Germino, G. G. (2000). Co-assembly of polycystin-1 and -2 produces unique cation-permeable currents. *Nature* **408,** 990–994.

Hateboer, N., Gumbs, C., Teare, M. D., Coles, G. A., Griffiths, D., Ravine, D., Futreal, P. A., and Rahman, N. (2001). Confirmation of a gene locus for medullary cystic kidney disease (MCKD2) on chromosome 16p12. *Kidney Int.* **60,** 1233–1239.

Hiesberger, T., Shao, X., Gourley, E., Reimann, A., Pontoglio, M., and Igarashi, P. (2005). Role of the hepatocyte nuclear factor-1beta (HNF-1beta) C-terminal domain in Pkhd1 (ARPKD) gene transcription and renal cystogenesis. *J. Biol. Chem.* **280,** 10578–10586.

Hildebrandt, F., and Otto, E. (2000). Molecular genetics of nephronophthisis and medullary cystic kidney disease. *J. Am. Soc. Nephrol.* **11,** 1753–1761.

Hildebrandt, F., Otto, E., Rensing, C., Nothwang, H. G., Vollmer, M., Adolphs, J., Hanusch, H., and Brandis, M. (1997). A novel gene encoding an SH3 domain protein is mutated in nephronophthisis type 1. *Nat. Genet.* **17,** 149–153.

Hou, X., Mrug, M., Yoder, B. K., Lefkowitz, E. J., Kremmidiotis, G., D'Eustachio, P., Beier, D. R., and Guay-Woodford, L. M. (2002). Cystin, a novel cilia-associated protein, is disrupted in the cpk mouse model of polycystic kidney disease. *J. Clin. Invest.* **109,** 533–540.

Hughes, J., Ward, C. J., Peral, B., Aspinwall, R., Clark, K., San Millan, J. L., Gamble, V., and Harris, P. C. (1995). The polycystic kidney disease 1 (PKD1) gene encodes a novel protein with multiple cell recognition domains. *Nat. Genet.* **10,** 151–160.

Hughson, M. D., Buchwald, D., and Fox, M. (1986). Renal neoplasia and acquired cystic kidney disease in patients receiving long-term dialysis. *Arch. Pathol. Lab. Med.* **110,** 592–601.

Israeli, S., Amsler, K., and Wilson, P. (2007). Loss of fibrocystin leads to focal adhesion complex protein phosphorylation and functional abnormalities in autosomal recessive polycystic kidney disease (ARPKD) epithelia. *Am. Soc. Cell Biol.* abst 1786.

Janaswami, P. M., Birkenmeier, E. H., Cook, S. A., Rowe, L. B., Bronson, R. T., and Davisson, M. T. (1997). Identification and genetic mapping of a new polycystic kidney disease on mouse chromosome 8. *Genomics* **40,** 101–107.

Jiang, S. T., Chiou, Y. Y., Wang, E., Lin, H. K., Lin, Y. T., Chi, Y. C., Wang, C. K., Tang, M. J., and Li, H. (2006). Defining a link with autosomal-dominant polycystic kidney disease in mice with congenitally low expression of Pkd1. *Am. J. Pathol.* **168,** 205–220.

Joly, D., Morel, V., Hummel, A., Ruello, A., Nusbaum, P., Patey, N., Noel, L. H., Rousselle, P., and Knebelmann, B. (2003). Beta4 integrin and laminin 5 are aberrantly expressed in polycystic kidney disease: Role in increased cell adhesion and migration. *Am. J. Pathol.* **163,** 1791–1800.

Joly, D., Ishibe, S., Nickel, C., Yu, Z., Somlo, S., and Cantley, L. G. (2006). The polycystin 1-C-terminal fragment stimulates ERK-dependent spreading of renal epithelial cells. *J. Biol. Chem.* **281,** 26329–26339.

Jones, C., and Chen, P. (2007). Planar cell polarity signaling in vertebrates. *Bioessays* **29,** 120–132.

Katsanis, N., Beales, P. L., Woods, M. O., Lewis, R. A., Green, J. S., Parfrey, P. S., Ansley, S. J., Davidson, W. S., and Lupski, J. R. (2000). Mutations in MKKS cause obesity, retinal dystrophy and renal malformations associated with Bardet-Biedl syndrome. *Nat. Genet.* **26,** 67–70.

Kim, E., Arnould, T., Sellin, L. K., Benzing, T., Fan, M. J., Gruning, W., Sokol, S. Y., Drummond, I., and Walz, G. (1999). The polycystic kidney disease 1 gene product modulates Wnt signaling. *J. Biol. Chem.* **274,** 4947–4953.

Kim, K., Drummond, I., Ibraghimov-Beskrovnaya, O., Klinger, K., and Arnaout, M. A. (2000). Polycystin 1 is required for the structural integrity of blood vessels. *Proc. Natl Acad. Sci. USA* **97,** 1731–1736.

Kim, H., Jeong, W., Ahn, K., Ahn, C., and Kang, S. (2004). Siah-1 interacts with the intracellular region of polycystin-1 and affects its stability via the ubiquitin–proteasome pathway. *J. Am. Soc. Nephrol.* **15,** 2042–2049.

Kim, I., Fu, Y., Hui, K., Moeckel, G., Mai, W., Li, C., Liang, D., Zhao, P., Ma, J., Chen, X. Z., George, A. L., Jr., Coffey, R. J., *et al.* (2008). Fibrocystin/polyductin modulates renal tubular formation by regulating polycystin-2 expression and function. *J. Am. Soc. Nephrol.* **19,** 455–468.

Koulen, P., Cai, Y., Geng, L., Maeda, Y., Nishimura, S., Witzgall, R., Ehrlich, B. E., and Somlo, S. (2002). Polycystin-2 is an intracellular calcium release channel. *Nat. Cell Biol.* **4,** 191–197.

Kreidberg, J. A., Sariola, H., Loring, J. M., Maeda, M., Pelletier, J., Housman, D., and Jaenisch, R. (1993). WT-1 is required for early kidney development. *Cell* **74,** 679–691.

Kreidberg, J. A., Donovan, M. J., Goldstein, S. L., Rennke, H., Shepherd, K., Jones, R. C., and Jaenisch, R. (1996). Alpha 3 beta 1 integrin has a crucial role in kidney and lung organogenesis. *Development* **122,** 3537–3547.

Kume, T., Deng, K., and Hogan, B. L. (2000). Murine forkhead/winged helix genes Foxc1 (Mf1) and Foxc2 (Mfh1) are required for the early organogenesis of the kidney and urinary tract. *Development* **127,** 1387–1395.

Lantinga-van Leeuwen, I. S., Dauwerse, J. G., Baelde, H. J., Leonhard, W. N., van de Wal, A., Ward, C. J., Verbeek, S., Deruiter, M. C., Breuning, M. H., de Heer, E., and Peters, D. J. (2004). Lowering of Pkd1 expression is sufficient to cause polycystic kidney disease. *Hum. Mol. Genet.* **13,** 3069–3077.

Lantinga-van Leeuwen, I. S., Leonhard, W. N., van der Wal, A., Breuning, M. H., de Heer, E., and Peters, D. J. (2007). Kidney-specific inactivation of the Pkd1 gene induces rapid cyst formation in developing kidneys and a slow onset of disease in adult mice. *Hum. Mol. Genet.* **16,** 3188–3196.

Le, N. H., van der Bent, P., Huls, G., van de Wetering, M., Loghman-Adham, M., Ong, A. C., Calvet, J. P., Clevers, H., Breuning, M. H., van Dam, H., and Peters, D. J. (2004). Aberrant polycystin-1 expression results in modification of activator protein-1 activity, whereas Wnt signaling remains unaffected. *J. Biol. Chem.* **279,** 27472–27481.

Li, H. P., Geng, L., Burrow, C. R., and Wilson, P. D. (1999). Identification of phosphorylation sites in the PKD1-encoded protein C-terminal domain. *Biochem. Biophys. Res. Commun.* **259,** 356–363.

Li, X., Li, H. P., Amsler, K., Hyink, D., Wilson, P. D., and Burrow, C. R. (2002). PRKX, a phylogenetically and functionally distinct cAMP-dependent protein kinase, activates renal epithelial cell migration and morphogenesis. *Proc. Natl Acad. Sci. USA* **99,** 9260–9265.

Li, Q., Montalbetti, N., Shen, P. Y., Dai, X. Q., Cheeseman, C. I., Karpinski, E., Wu, G., Cantiello, H. F., and Chen, X. Z. (2005a). Alpha-actinin associates with polycystin-2 and regulates its channel activity. *Hum. Mol. Genet.* **14,** 1587–1603.

Li, X., Hyink, D. P., Polgar, K., Gusella, G. L., Wilson, P. D., and Burrow, C. R. (2005b). Protein kinase X activates ureteric bud branching morphogenesis in developing mouse metanephric kidney. *J. Am. Soc. Nephrol.* **16,** 3543–3552.

Li, X., Luo, Y., Starremans, P. G., McNamara, C. A., Pei, Y., and Zhou, J. (2005c). Polycystin-1 and polycystin-2 regulate the cell cycle through the helix–loop–helix inhibitor Id2. *Nat. Cell Biol.* **7,** 1202–1212.

Li, Y., Wright, J. M., Qian, F., Germino, G. G., and Guggino, W. B. (2005d). Polycystin 2 interacts with type-I IP3 receptor to modulate intracellular Ca2+ signaling. *J. Biol. Chem.* **280,** 41298–41306.

Lin, F., Hiesberger, T., Cordes, K., Sinclair, A. M., Goldstein, L. S., Somlo, S., and Igarashi, P. (2003). Kidney-specific inactivation of the KIF3A subunit of kinesin-II inhibits renal ciliogenesis and produces polycystic kidney disease. *Proc. Natl Acad. Sci. USA* **100,** 5286–5291.

Lo, S. H., Yu, Q. C., Degenstein, L., Chen, L. B., and Fuchs, E. (1997). Progressive kidney degeneration in mice lacking tensin. *J. Cell Biol.* **136,** 1349–1361.

Lohning, C., Pohlschmidt, M., Glucksmann-Kuis, M. A., Duyk, G., Bork, P., Schneider, M. O., Reeders, S. T., and Frischauf, A. M. (1996). Structural motifs of the PKD1 protein. *Nephrol. Dial. Transplant.* **11**(Suppl. 6), 2–4.

Lu, W., Peissel, B., Babakhanlou, H., Pavlova, A., Geng, L., Fan, X., Larson, C., Brent, G., and Zhou, J. (1997). Perinatal lethality with kidney and pancreas defects in mice with a targeted Pkd1 mutation. *Nat. Genet.* **17,** 179–181.

Lu, W., Fan, X., Basora, N., Babakhanlou, H., Law, T., Rifai, N., Harris, P. C., Perez-Atayde, A. R., Rennke, H. G., and Zhou, J. (1999). Late onset of renal and hepatic cysts in Pkd1-targeted heterozygotes. *Nat. Genet.* **21,** 160–161.

Lu, W., Shen, X., Pavlova, A., Lakkis, M., Ward, C. J., Pritchard, L., Harris, P. C., Genest, D. R., Perez-Atayde, A. R., and Zhou, J. (2001). Comparison of Pkd1-targeted mutants reveals that loss of polycystin-1 causes cystogenesis and bone defects. *Hum. Mol. Genet.* **10,** 2385–2396.

Mai, W., Chen, D., Ding, T., Kim, I., Park, S., Cho, S. Y., Chu, J. S., Liang, D., Wang, N., Wu, D., Li, S., Zhao, P., *et al.* (2005). Inhibition of Pkhd1 impairs tubulomorphogenesis of cultured IMCD cells. *Mol. Biol. Cell* **16,** 4398–4409.

Makita, R., Uchijima, Y., Nishiyama, K., Amano, T., Chen, Q., Takeuchi, T., Mitani, A., Nagase, T., Yatomi, Y., Aburatani, H., Nakagawa, O., Small, E. V., *et al.* (2008). Multiple renal cysts, urinary concentration defects, and pulmonary emphysematous changes in mice lacking TAZ. *Am. J. Physiol. Renal Physiol.* **294,** F542-F553.

McDonald, R. A., and Avner, E. D. (1991). Inherited polycystic kidney disease in children. *Semin. Nephrol.* **11,** 632–642.

Menezes, L. F., Cai, Y., Nagasawa, Y., Silva, A. M., Watkins, M. L., Da Silva, A. M., Somlo, S., Guay-Woodford, L. M., Germino, G. G., and Onuchic, L. F. (2004). Polyductin, the PKHD1 gene product, comprises isoforms expressed in plasma membrane, primary cilium, and cytoplasm. *Kidney Int.* **66,** 1345–1355.

Miyazaki, Y., Oshima, K., Fogo, A., Hogan, B. L., and Ichikawa, I. (2000). Bone morphogenetic protein 4 regulates the budding site and elongation of the mouse ureter. *J. Clin. Invest.* **105,** 863–873.

Mochizuki, T., Wu, G., Hayashi, T., Xenophontos, S. L., Veldhuisen, B., Saris, J. J., Reynolds, D. M., Cai, Y., Gabow, P. A., Pierides, A., Kimberling, W. J., Breuning, M. H., *et al.* (1996). PKD2, a gene for polycystic kidney disease that encodes an integral membrane protein. *Science* **272,** 1339–1342.

Mollet, G., Silbermann, F., Delous, M., Salomon, R., Antignac, C., and Saunier, S. (2005). Characterization of the nephrocystin/nephrocystin-4 complex and subcellular localization of nephrocystin-4 to primary cilia and centrosomes. *Hum. Mol. Genet.* **14,** 645–656.

Morgan, D., Turnpenny, L., Goodship, J., Dai, W., Majumder, K., Matthews, L., Gardner, A., Schuster, G., Vien, L., Harrison, W., Elder, F. F., Penman-Splitt, M., *et al.* (1998). Inversin, a novel gene in the vertebrate left–right axis pathway, is partially deleted in the inv mouse. *Nat. Genet.* **20,** 149–156.

Morgan, D., Eley, L., Sayer, J., Strachan, T., Yates, L. M., Craighead, A. S., and Goodship, J. A. (2002). Expression analyses and interaction with the anaphase promoting complex protein Apc2 suggest a role for Inversin in primary cilia and involvement in the cell cycle. *Hum. Mol. Genet.* **11,** 3345–3350.

Moser, M., Pscherer, A., Roth, C., Becker, J., Mucher, G., Zerres, K., Dixkens, C., Weis, J., Guay-Woodford, L., Buettner, R., and Fassler, R. (1997). Enhanced apoptotic cell death of renal epithelial cells in mice lacking transcription factor AP-2beta. *Genes Dev.* **11,** 1938–1948.

Moser, M., Matthiesen, S., Kirfel, J., Schorle, H., Bergmann, C., Senderek, J., Rudnik-Schoneborn, S., Zerres, K., and Buettner, R. (2005). A mouse model for cystic biliary dysgenesis in autosomal recessive polycystic kidney disease (ARPKD). *Hepatology* **41,** 1113–1121.

Murcia, N. S., Richards, W. G., Yoder, B. K., Mucenski, M. L., Dunlap, J. R., and Woychik, R. P. (2000). The Oak Ridge Polycystic Kidney (orpk) disease gene is required for left–right axis determination. *Development* **127,** 2347–2355.

Nauli, S. M., Alenghat, F. J., Luo, Y., Williams, E., Vassilev, P., Li, X., Elia, A. E., Lu, W., Brown, E. M., Quinn, S. J., Ingber, D. E., and Zhou, J. (2003). Polycystins 1 and 2 mediate mechanosensation in the primary cilium of kidney cells. *Nat. Genet.* **33,** 129–137.

Nickel, C., Benzing, T., Sellin, L., Gerke, P., Karihaloo, A., Liu, Z. X., Cantley, L. G., and Walz, G. (2002). The polycystin-1 C-terminal fragment triggers branching morphogenesis and migration of tubular kidney epithelial cells. *J. Clin. Invest.* **109,** 481–489.

Nishinakamura, R., Matsumoto, Y., Nakao, K., Nakamura, K., Sato, A., Copeland, N. G., Gilbert, D. J., Jenkins, N. A., Scully, S., Lacey, D. L., Katsuki, M., Asashima, M., and Yokota, T. (2001). Murine homolog of SALL1 is essential for ureteric bud invasion in kidney development. *Development* **128,** 3105–3115.

Norman, J. T., Gatti, L., and Wilson, P. D. (1993). Abnormal matrix metalloproteinase (MMP) regulation in human autosomal dominant polycystic kidney disease (ADPKD) (abstract). *J. Am. Soc. Nephrol.* **4,** 819.

Nurnberger, J., Bacallao, R. L., and Phillips, C. L. (2002). Inversin forms a complex with catenins and N-cadherin in polarized epithelial cells. *Mol. Biol. Cell* **13,** 3096–3106.

Ogborn, M. R., and Sareen, S. (1996). Transforming growth factor alpha and epidermal growth factor expression in experimental murine polycystic kidney disease. *Pediatr. Nephrol.* **10,** 181–184.

Olbrich, H., Fliegauf, M., Hoefele, J., Kispert, A., Otto, E., Volz, A., Wolf, M. T., Sasmaz, G., Trauer, U., Reinhardt, R., Sudbrak, R., Antignac, C., *et al.* (2003). Mutations in a novel gene, NPHP3, cause adolescent nephronophthisis, tapeto-retinal degeneration and hepatic fibrosis. *Nat. Genet.* **34,** 455–459.

Olsen, O., Funke, L., Long, J. F., Fukata, M., Kazuta, T., Trinidad, J. C., Moore, K. A., Misawa, H., Welling, P. A., Burlingame, A. L., Zhang, M., and Bredt, D. S. (2007). Renal defects associated with improper polarization of the CRB and DLG polarity complexes in MALS-3 knockout mice. *J. Cell Biol.* **179,** 151–164.

Omran, H., Fernandez, C., Jung, M., Haffner, K., Fargier, B., Villaquiran, A., Waldherr, R., Gretz, N., Brandis, M., Ruschendorf, F., Reis, A., and Hildebrandt, F. (2000). Identification of a new gene locus for adolescent nephronophthisis, on chromosome 3q22 in a large Venezuelan pedigree. *Am. J. Hum. Genet.* **66,** 118–127.

Onuchic, L. F., Furu, L., Nagasawa, Y., Hou, X., Eggermann, T., Ren, Z., Bergmann, C., Senderek, J., Esquivel, E., Zeltner, R., Rudnik-Schoneborn, S., Mrug, M., *et al.* (2002). PKHD1, the polycystic kidney and hepatic disease 1 gene, encodes a novel large protein containing multiple immunoglobulin-like plexin-transcription-factor domains and parallel beta-helix 1 repeats. *Am. J. Hum. Genet.* **70,** 1305–1317.

Osathanondh, V., and Potter, E. L. (1963). Development of human kidney as shown by microdissection. I. Preparation of tissue with reasons for possible misinterpretation of observations. *Arch. Pathol.* **76,** 271–276.

Otto, E., Hoefele, J., Ruf, R., Mueller, A. M., Hiller, K. S., Wolf, M. T., Schuermann, M. J., Becker, A., Birkenhager, R., Sudbrak, R., Hennies, H. C., Nurnberg, P., *et al.* (2002). A gene mutated in nephronophthisis and retinitis pigmentosa encodes a novel protein, nephroretinin, conserved in evolution. *Am. J. Hum. Genet.* **71,** 1161–1167.

Otto, E. A., Trapp, M. L., Schultheiss, U. T., Helou, J., Quarmby, L. M., and Hildebrandt, F. (2008). NEK8 mutations affect ciliary and centrosomal localization and may cause nephronophthisis. *J. Am. Soc. Nephrol.* **19,** 587–592.

Parnell, S. C., Magenheimer, B. S., Maser, R. L., Rankin, C. A., Smine, A., Okamoto, T., and Calvet, J. P. (1998). The polycystic kidney disease-1 protein, polycystin-1, binds and activates heterotrimeric G-proteins *in vitro*. *Biochem. Biophys. Res. Commun.* **251,** 625–631.

Parnell, S. C., Magenheimer, B. S., Maser, R. L., Zien, C. A., Frischauf, A. M., and Calvet, J. P. (2002). Polycystin-1 Activation of c-Jun N-terminal kinase and AP-1 is mediated by heterotrimeric G proteins. *J. Biol. Chem.* **277,** 19566–19572.

Phillips, C. L., Miller, K. J., Filson, A. J., Nurnberger, J., Clendenon, J. L., Cook, G. W., Dunn, K. W., Overbeek, P. A., Gattone, V. H., II, and Bacallao, R. L. (2004). Renal cysts of inv/inv mice resemble early infantile nephronophthisis. *J. Am. Soc. Nephrol.* **15,** 1744–1755.

Piontek, K., Menezes, L. F., Garcia-Gonzalez, M. A., Huso, D. L., and Germino, G. G. (2007). A critical developmental switch defines the kinetics of kidney cyst formation after loss of Pkd1. *Nat. Med.* **13,** 1490–1495.

Polgar, K., Burrow, C. R., Hyink, D. P., Fernandez, H., Thornton, K., Li, X., Gusella, G. L., and Wilson, P. D. (2005). Disruption of polycystin-1 function interferes with branching morphogenesis of the ureteric bud in developing mouse kidneys. *Dev. Biol.* **286,** 16–30.

Praetorius, H. A., Praetorius, J., Nielsen, S., Frokiaer, J., and Spring, K. R. (2004). Beta1-integrins in the primary cilium of MDCK cells potentiate fibronectin-induced Ca2+ signaling. *Am. J. Physiol. Renal Physiol.* **287,** F969–F978.

Pritchard, L., Sloane-Stanley, J. A., Sharpe, J. A., Aspinwall, R., Lu, W., Buckle, V., Strmecki, L., Walker, D., Ward, C. J., Alpers, C. E., Zhou, J., Wood, W. G., *et al.* (2000). A human PKD1 transgene generates functional polycystin-1 in mice and is associated with a cystic phenotype. *Hum. Mol. Genet.* **9,** 2617–2627.

Pugacheva, E. N., and Golemis, E. A. (2005). The focal adhesion scaffolding protein HEF1 regulates activation of the Aurora-A and Nek2 kinases at the centrosome. *Nat. Cell Biol.* **7,** 937–946.

Qian, F., Watnick, T. J., Onuchic, L. F., and Germino, G. G. (1996). The molecular basis of focal cyst formation in human autosomal dominant polycystic kidney disease type I. *Cell* **87,** 979–987.

Qian, F., Germino, F. J., Cai, Y., Zhang, X., Somlo, S., and Germino, G. G. (1997). PKD1 interacts with PKD2 through a probable coiled-coil domain. *Nat. Genet.* **16,** 179–183.

Qian, F., Boletta, A., Bhunia, A. K., Xu, H., Liu, L., Ahrabi, A. K., Watnick, T. J., Zhou, F., and Germino, G. G. (2002). Cleavage of polycystin-1 requires the receptor for egg jelly domain and is disrupted by human autosomal-dominant polycystic kidney disease 1-associated mutations. *Proc. Natl Acad. Sci. USA* **99,** 16981–16986.

Qian, C. N., Knol, J., Igarashi, P., Lin, F., Zylstra, U., Teh, B. T., and Williams, B. O. (2005). Cystic renal neoplasia following conditional inactivation of apc in mouse renal tubular epithelium. *J. Biol. Chem.* **280,** 3938–3945.

Rankin, E. B., Tomaszewski, J. E., and Haase, V. H. (2006). Renal cyst development in mice with conditional inactivation of the von Hippel-Lindau tumor suppressor. *Cancer Res.* **66,** 2576–2583.

Richards, W. G., Sweeney, W. E., Yoder, B. K., Wilkinson, J. E., Woychik, R. P., and Avner, E. D. (1998). Epidermal growth factor receptor activity mediates renal cyst formation in polycystic kidney disease. *J. Clin. Invest.* **101,** 935–939.

Rodova, M., Islam, M. R., Maser, R. L., and Calvet, J. P. (2002). The polycystic kidney disease-1 promoter is a target of the beta-catenin/T-cell factor pathway. *J. Biol. Chem.* **277,** 29577–29583.

Roitbak, T., Ward, C. J., Harris, P. C., Bacallao, R., Ness, S. A., and Wandinger-Ness, A. (2004). A polycystin-1 multiprotein complex is disrupted in polycystic kidney disease cells. *Mol. Biol. Cell* **15,** 1334–1346.

Roitbak, T., Surviladze, Z., Tikkanen, R., and Wandinger-Ness, A. (2005). A polycystin multiprotein complex constitutes a cholesterol-containing signalling microdomain in human kidney epithelia. *Biochem. J.* **392,** 29–38.

Rossetti, S., Torra, R., Coto, E., Consugar, M., Kubly, V., Malaga, S., Navarro, M., El-Youssef, M., Torres, V. E., and Harris, P. C. (2003). A complete mutation screen of PKHD1 in autosomal-recessive polycystic kidney disease (ARPKD) pedigrees. *Kidney Int.* **64,** 391–403.

Rossetti, S., Consugar, M. B., Chapman, A. B., Torres, V. E., Guay-Woodford, L. M., Grantham, J. J., Bennett, W. M., Meyers, C. M., Walker, D. L., Bae, K., Zhang, Q. J., Thompson, P. A., *et al.* (2007). Comprehensive molecular diagnostics in autosomal dominant polycystic kidney disease. *J. Am. Soc. Nephrol.* **18,** 2143–2160.

Saadi-Kheddouci, S., Berrebi, D., Romagnolo, B., Cluzeaud, F., Peuchmaur, M., Kahn, A., Vandewalle, A., and Perret, C. (2001). Early development of polycystic kidney disease in transgenic mice expressing an activated mutant of the beta-catenin gene. *Oncogene* **20,** 5972–5981.

Saifudeen, Z., Dipp, S., and El-Dahr, S. S. (2002). A role for p53 in terminal epithelial cell differentiation. *J. Clin. Invest.* **109,** 1021–1030.

Schedl, A. (2007). Renal abnormalities and their developmental origin. *Nat. Rev. Genet.* **8,** 791–802.

Sfakianos, J., Togawa, A., Maday, S., Hull, M., Pypaert, M., Cantley, L., Toomre, D., and Mellman, I. (2007). Par3 functions in the biogenesis of the primary cilium in polarized epithelial cells. *J. Cell Biol.* **179,** 1133–1140.

Shannon, M. B., Patton, B. L., Harvey, S. J., and Miner, J. H. (2006). A hypomorphic mutation in the mouse laminin alpha5 gene causes polycystic kidney disease. *J. Am. Soc. Nephrol.* **17,** 1913–1922.

Silberberg, M., Charron, A. J., Bacallao, R., and Wandinger-Ness, A. (2005). Mispolarization of desmosomal proteins and altered intercellular adhesion in autosomal dominant polycystic kidney disease. *Am. J. Physiol. Renal Physiol.* **288,** F1153–F1163.

Sohara, E., Luo, Y., Zhang, J., Manning, D. K., Beier, D. R., and Zhou, J. (2008). Nek8 regulates the expression and localization of polycystin-1 and polycystin-2. *J. Am. Soc. Nephrol.* **19,** 469–476.

Somlo, S. (1999). The PKD2 gene: Structure, interactions, mutations, and inactivation. *Adv. Nephrol. Necker Hosp.* **29,** 257–275.

Sorokin, L., and Ekblom, P. (1992). Development of tubular and glomerular cells of the kidney. *Kidney Int.* **41,** 657–664.

Stark, K., Vainio, S., Vassileva, G., and McMahon, A. P. (1994). Epithelial transformation of metanephric mesenchyme in the developing kidney regulated by Wnt-4. *Nature* **372,** 679–683.

Stocklin, E., Botteri, F., and Groner, B. (1993). An activated allele of the c-erbB-2 oncogene impairs kidney and lung function and causes early death of transgenic mice. *J. Cell Biol.* **122,** 199–208.

Streets, A. J., Moon, D. J., Kane, M. E., Obara, T., and Ong, A. C. (2006). Identification of an N-terminal glycogen synthase kinase 3 phosphorylation site which regulates the functional localization of polycystin-2 *in vivo* and *in vitro*. *Hum. Mol. Genet.* **15,** 1465–1473.

Tani, T., von Koskull, H., and Virtanen, I. (1996). Focal adhesion kinase pp125FAK is associated with both intercellular junctions and matrix adhesion sites in vivo. *Histochem. Cell Biol.* **105,** 17–25.

Thi, M. M., Tarbell, J. M., Weinbaum, S., and Spray, D. C. (2004). The role of the glycocalyx in reorganization of the actin cytoskeleton under fluid shear stress: A "bumper-car" model. *Proc. Natl Acad. Sci. USA* **101,** 16483–16488.

Togawa, A., Miyoshi, J., Ishizaki, H., Tanaka, M., Takakura, A., Nishioka, H., Yoshida, H., Doi, T., Mizoguchi, A., Matsuura, N., Niho, Y., Nishimune, Y., *et al.* (1999). Progressive impairment of kidneys and reproductive organs in mice lacking Rho GDIalpha. *Oncogene* **18,** 5373–5380.

Torres, V. E., Wang, X., Qian, Q., Somlo, S., Harris, P. C., and Gattone, V. H. (2004). Effective treatment of an orthologous model of autosomal dominant polycystic kidney disease. *Nat. Med.* **10,** 363–364.

Trudel, M., Barisoni, L., Lanoix, J., and D'Agati, V. (1998). Polycystic kidney disease in SBM transgenic mice: Role of c-myc in disease induction and progression. *Am. J. Pathol.* **152,** 219–229.

Tsiokas, L., Kim, E., Arnould, T., Sukhatme, V. P., and Walz, G. (1997). Homo- and heterodimeric interactions between the gene products of PKD1 and PKD2. *Proc. Natl Acad. Sci. USA* **94,** 6965–6970.

Valerius, M. T., Patterson, L. T., Feng, Y., and Potter, S. S. (2002). Hoxa 11 is upstream of Integrin alpha8 expression in the developing kidney. *Proc. Natl Acad. Sci. USA* **99,** 8090–8095.

Van Adelsberg, J., Chamberlain, S., and D'Agati, V. (1997). Polycystin expression is temporally and spatially regulated during renal development. *Am. J. Physiol.* **272,** F602–F609.

Veis, D. J., Sorenson, C. M., Shutter, J. R., and Korsmeyer, S. J. (1993). Bcl-2-deficient mice demonstrate fulminant lymphoid apoptosis, polycystic kidneys, and hypopigmented hair. *Cell* **75,** 229–240.

Waldherr, R., Lennert, T., Weber, H. P., Fodisch, H. J., and Scharer, K. (1982). The nephronophthisis complex. A clinicopathologic study in children. *Virchows Arch. A Pathol. Anat. Histopathol.* **394,** 235–254.

Wang, S., Luo, Y., Wilson, P. D., Witman, G. B., and Zhou, J. (2004). The autosomal recessive polycystic kidney disease protein is localized to primary cilia, with concentration in the basal body area. *J. Am. Soc. Nephrol.* **15,** 592–602.

Wang, X., Gattone, V., II, Harris, P. C., and Torres, V. E. (2005). Effectiveness of vasopressin V2 receptor antagonists OPC-31260 and OPC-41061 on polycystic kidney disease development in the PCK rat. *J. Am. Soc. Nephrol.* **16,** 846–851.

Ward, C. J., Hogan, M. C., Rossetti, S., Walker, D., Sneddon, T., Wang, X., Kubly, V., Cunningham, J. M., Bacallao, R., Ishibashi, M., Milliner, D. S., *et al.* (2002). The gene mutated in autosomal recessive polycystic kidney disease encodes a large, receptor-like protein. *Nat. Genet.* **30,** 259–269.

Ward, C. J., Yuan, D., Masyuk, T. V., Wang, X., Punyashthiti, R., Whelan, S., Bacallao, R., Torra, R., LaRusso, N. F., Torres, V. E., and Harris, P. C. (2003). Cellular and subcellular localization of the ARPKD protein; fibrocystin is expressed on primary cilia. *Hum. Mol. Genet.* **12,** 2703–2710.

Watanabe, D., Saijoh, Y., Nonaka, S., Sasaki, G., Ikawa, Y., Yokoyama, T., and Hamada, H. (2003). The left–right determinant Inversin is a component of node monocilia and other 9+0 cilia. *Development* **130,** 1725–1734.

Weimbs, T. (2006). Regulation of mTOR by polycystin-1: Is polycystic kidney disease a case of futile repair? *Cell Cycle* **5,** 2425–2429.

Wilson, P. (1996). Pathogenesis of polycystic kidney disease: Altered cellular function. *In* "Polycystic Kidney Disease" (M. Watson and V. Torres, Eds.), pp. 125–163. Oxford Medical, Oxford.

Wilson, P. D. (1997). Epithelial cell polarity and disease. *Am. J. Physiol.* **272,** F434–F442.

Wilson, P. D. (2004). Polycystic kidney disease. *N. Engl. J. Med.* **350,** 151–164.

Wilson, P. D., and Goilav, B. (2007). Cystic disease of the kidney. *Annu. Rev. Pathol.* **2,** 341–368.

Wilson, P. D., Hreniuk, D., and Gabow, P. A. (1992). Abnormal extracellular matrix and excessive growth of human adult polycystic kidney disease epithelia. *J. Cell. Physiol.* **150,** 360–369.

Wilson, P. D., Geng, L., Li, X., and Burrow, C. R. (1999). The PKD1 gene product, "polycystin-1," is a tyrosine-phosphorylated protein that colocalizes with alpha2beta1-integrin in focal clusters in adherent renal epithelia. *Lab. Invest.* **79,** 1311–1323.

Wilson, P. D., Devuyst, O., Li, X., Gatti, L., Falkenstein, D., Robinson, S., Fambrough, D., and Burrow, C. R. (2000). Apical plasma membrane mispolarization of NaK-ATPase in polycystic kidney disease epithelia is associated with aberrant expression of the beta2 isoform. *Am. J. Pathol.* **156,** 253–268.

Wilson, S. J., Amsler, K., Hyink, D., Li, X., Lu, W., Zhou, J., Burrow, C. R., and Wilson, P. (2006). Inhibition of HER-2(neu/ErbB2) restores normal function and structure to polycystic kidney disease (PKD) epithelia. *Biochim. Biophys. Acta Mol. Basis Dis.* **1762,** 647–655.

Wolf, M. T., van Vlem, B., Hennies, H. C., Zalewski, I., Karle, S. M., Puetz, M., Panther, F., Otto, E., Fuchshuber, A., Lameire, N., Loeys, B., and Hildebrandt, F. (2004). Telomeric refinement of the MCKD1 locus on chromosome 1q21. *Kidney Int.* **66,** 580–585.

Woollard, J. R., Punyashtiti, R., Richardson, S., Masyuk, T. V., Whelan, S., Huang, B. Q., Lager, D. J., vanDeursen, J., Torres, V. E., Gattone, V. H., LaRusso, N. F., Harris, P. C., *et al.* (2007). A mouse model of autosomal recessive polycystic kidney disease with biliary duct and proximal tubule dilatation. *Kidney Int.* **72,** 328–336.

Wu, G., D'Agati, V., Cai, Y., Markowitz, G., Park, J. H., Reynolds, D. M., Maeda, Y., Le, T. C., Hou, H., Jr., Kucherlapati, R., Edelmann, W., and Somlo, S. (1998). Somatic inactivation of Pkd2 results in polycystic kidney disease. *Cell* **93,** 177–188.

Wu, G., Markowitz, G. S., Li, L., D'Agati, V. D., Factor, S. M., Geng, L., Tibara, S., Tuchman, J., Cai, Y., Park, J. H., van Adelsberg, J., Hou, H., Jr., *et al.* (2000). Cardiac defects and renal failure in mice with targeted mutations in Pkd2. *Nat. Genet.* **24,** 75–78.

Wu, Y., Dai, X. Q., Li, Q., Chen, C. X., Mai, W., Hussain, Z., Long, W., Montalbetti, N., Li, G., Glynne, R., Wang, S., Cantiello, H. F., *et al.* (2006). Kinesin-2 mediates physical and functional interactions between polycystin-2 and fibrocystin. *Hum. Mol. Genet.* **15,** 3280–3292.

Xie, Z., and Tsai, L. H. (2004). Cdk5 phosphorylation of FAK regulates centrosome-associated microtubules and neuronal migration. *Cell Cycle* **3,** 108–110.

Yoder, B. K., Tousson, A., Millican, L., Wu, J. H., Bugg, C. E., Jr., Schafer, J. A., and Balkovetz, D. F. (2002). Polaris, a protein disrupted in orpk mutant mice, is required for assembly of renal cilium. *Am. J. Physiol. Renal Physiol.* **282,** F541–F552.

Yu, S., Hackmann, K., Gao, J., He, X., Piontek, K., Garcia-Gonzalez, M. A., Menezes, L. F., Xu, H., Germino, G. G., Zuo, J., and Qian, F. (2007). Essential role of cleavage of Polycystin-1 at G protein-coupled receptor proteolytic site for kidney tubular structure. *Proc. Natl Acad. Sci. USA* **104,** 18688–18693.

Zatti, A., Chauvet, V., Rajendran, V., Kimura, T., Pagel, P., and Caplan, M. J. (2005). The C-terminal tail of the polycystin-1 protein interacts with the Na,K-ATPase alpha-subunit. *Mol. Biol. Cell* **16,** 5087–5093.

Zerres, K., Rudnik-Schoneborn, S., Steinkamm, C., Becker, J., and Mucher, G. (1998). Autosomal recessive polycystic kidney disease. *J. Mol. Med.* **76,** 303–309.

Zhang, M. Z., Mai, W., Li, C., Cho, S. Y., Hao, C., Moeckel, G., Zhao, R., Kim, I., Wang, J., Xiong, H., Wang, H., Sato, Y., *et al.* (2004). PKHD1 protein encoded by the gene for autosomal recessive polycystic kidney disease associates with basal bodies and primary cilia in renal epithelial cells. *Proc. Natl Acad. Sci. USA* **101,** 2311–2316.

CHAPTER SEVEN

Fraying at the Edge: Mouse Models of Diseases Resulting from Defects at the Nuclear Periphery

Tatiana V. Cohen* *and* Colin L. Stewart†

Contents

* Center for Genetic Medicine, Children's National Medical Center, 111 Michigan Avenue, N.W. Washington, DC 20010
† Institute of Medical Biology, 8A Biomedical Grove, Immunos, Singapore 138668

Current Topics in Developmental Biology, Volume 84
ISSN 0070-2153, DOI: 10.1016/S0070-2153(08)00607-8

Abstract

Eukaryotic cells compartmentalize their genetic material within the nucleus. The boundary separating the genetic material from the cytoplasm is the nuclear envelope (NE) and lamina. Historically, the NE was perceived as functioning primarily as a barrier regulating the entry and exit of macromolecules between the nucleus and cytoplasm via the nuclear pore complexes (NPCs) that traverse the nuclear membranes. However, recent findings have caused a fundamental reassessment with regard to NE and lamina functions. Evidence now points to the NE and lamina functioning as a "hub" in regulating and perhaps integrating critical cellular functions that include chromatin organization, transcriptional regulation, mechanical integrity of the cell, signaling pathways, as well as acting as a key component of the cytoskeleton. Such an integral role for the nuclear boundary has emerged from increased interest into the functions of the NE/lamina, which has been largely stimulated by the discovery that some 24 different diseases and anomalies are caused by defects in proteins of the NE and lamina.

1. Introduction

The nuclear envelope (NE) consists of the inner and outer nuclear membranes (INM and ONM, respectively) that are separated by the perinuclear space (PNS). These membranes are connected at the point they are traversed by the nuclear pore complexes (NPCs). The ONM is also contiguous with the cytoplasmic endoplasmic reticulum (ER), making the ER, INM, and ONM one continuous membrane system with lumen of the ER extending into the PNS. The other component of the NE, and underlying the INM, is the nuclear lamina—a thin proteinaceous meshwork of some 10–20 nm, whose thickness varies between different cell types (Hoger *et al.*, 1991). The principal components of the lamina are intermediate filament proteins—the nuclear lamins. Most adult mammalian somatic cells contain four major lamin proteins, A, B1, B2, and C. The lamins are grouped into two classes, A-type (A, A10, and C), and B-type (B1, B2, and B3). Separate genes encode lamins B1 and B2 with lamin B3 being produced as a minor spliced variant of Lamin B2 (Burke and Stewart, 2006; Furukawa and Hotta, 1993). A single gene, *LMNA*, encodes the A-type lamins, which arise through alternative splicing of a common pre-mRNA. A minor spiced *LMNA* variant, Lamin C2, is also produced, and as with lamin B3 is found in the testes. The nuclear lamina has important roles in regulating DNA synthesis, RNA transcription, chromatin organization and, in the selective retention of INM proteins (Goldman *et al.*, 2002). Transcriptional cofactors also associate with the lamins, suggesting that the lamina and NE are important in transcriptional regulation (Heessen and Fornerod, 2007;

Fig. 7.1). In mammals, lamins are developmentally regulated, with all cells expressing at least one lamin B, whereas A-type lamins are absent in early embryonic development and in certain stem cell populations in adults (Rober *et al.*, 1989; Stewart and Burke, 1987).

Although the INM, ONM, and ER comprise one continuous membrane system, each membrane is characterized by its association with a unique set of proteins, with for instance, the ER containing the reticulon and DP1/Yop1p families of proteins that are required for the ER's assembly and maintenance as a tubular structure (Shnyrova *et al.*, 2008). Both the ONM and ER, but not the INM, are associated with ribosomes. Some of the more intriguing NE associated proteins are the Nesprins, a recently described family of proteins that predominantly localize to the ONM (Crisp and Burke, 2008). In contrast to the ONM and ER, proteomic studies have identified some 70 transmembrane proteins that are associated with the rat liver INM, some of which have already been extensively studied (see below) (Schirmer *et al.*, 2003).

2. The Laminopathies

The principal stimulus to reinvigorating interest into the functions of the NE and lamina has been the recent discoveries that at least 24 inherited diseases and anomalies, ranging from muscular dystrophies, premature aging-like syndromes to anomalies affecting skeletal and fat homeostasis, are caused by mutations in the *LMNA* gene and in genes encoding some of the NE associated proteins (Worman and Bonne, 2007). The largest of this group of these diseases, the laminopathies, is associated with defects in the

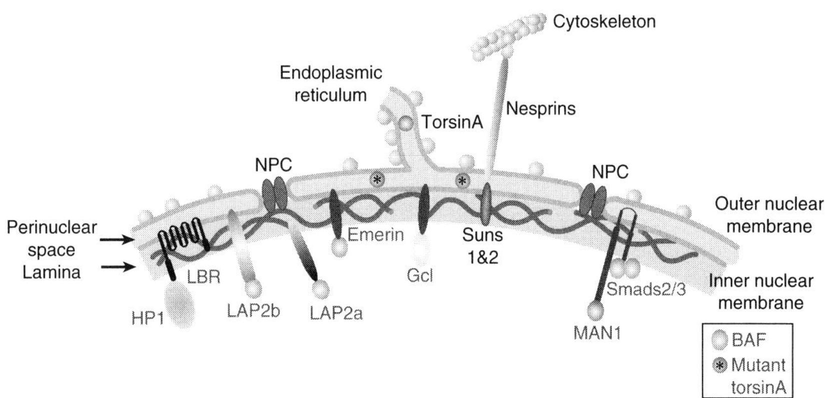

Figure 7.1 A diagram outlining the principal components of the NE and lamina to which 24 diseases and anomalies have been linked. (NPC - Nuclear Pore comples, LAP- Lamin associated protein, HP- heterochromatin protein, Gcl- germ cell lethal, LBR- Lamin B receptor.) (See Color Insert.)

A-type lamins and are classified as the primary and secondary laminopathies. The primary laminopathies are caused by mutations in the *LMNA*. The secondary laminopathies are caused by mutations in the gene encoding the enzyme ZMPSTE24, an endoprotease essential for the posttranslational maturation of prelamin A to mature lamin A.

The primary laminopathies can be classified into three groups:

1. The first and largest consists of diseases affecting striated muscle. These include the autosomal dominant form of Emery–Dreifuss muscular dystrophy (AD-EDMD), dilated cardiomyopathy (DCM), and limb-girdle muscular dystrophy 1B (LMG1B). Within this group is a peripheral neuropathy, Charcot–Marie–Tooth neuropathy type 2B (CMT2B), which arises due to demyelination of motor nerves (Worman and Courvalin, 2005).
2. The second group of laminopathies has minimal, if any effect on muscle, but influences white fat distribution and skeletal development. The two diseases are Dunnigan-type familial partial lipodystrophy (FPLD) and mandibuloacral dysplasia (MAD).
3. The third group of diseases is the premature aging or progeroid syndromes, Hutchinson–Gilford progeria syndrome (HGPS) and some cases of atypical Werner's syndrome.

   To date, some 200 mutations have been identified in the *LMNA* gene; a database on the "nuclear envelopathies" can be found at http://www.umd.be: The *LMNA* gene is thus unique in that no other gene has been described that when mutated produces so many different diseases. What is all the more remarkable is that the A-type lamins are almost ubiquitously expressed in most adult tissues. This poses the question as to how do these different mutations within the same ubiquitously expressed protein result in many diseases affecting specific tissues?

3. Mouse Models for the A-Type Laminopathies

3.1. Laminopathies affecting striated muscle

The first mutations linked to the *LMNA* gene were those causing the autosomal dominant form of Emery-Dreifuss muscular dystrophy (AD-EMD2; Bonne *et al.*, 1999). As with the X-linked form, which results from mutations in the INM associated protein emerin (*EMD*—see below) (Bione *et al.*, 1994), AD-EMD is characterized by progressive wasting of specific muscle groups, as well as the defining cardiac conduction defects that distinguish it from other muscular dystrophies. *LMNA* mutations causing AD-EDMD display wide variability in their severity with the conduction defects leading to death being more acute than observed in the X-linked form of the disease (Morris, 2001). Other *LMNA* mutations are also associated with dilated cardiomyopathy with conduction system disease (DCM-CD1) with no apparent skeletal muscle dystrophy (Fatkin *et al.*, 1999). The conduction defects observed in DCM-CD1 are similar to those in patients with EDMD, suggesting cardiomyopathy represents one extreme in a phenotypic continuum in which skeletal muscle involvement is not observed. Dilation of heart chambers, hypertrophy, arrhythmic conduction defects, and cardiac arrest are all found with *LMNA*-associated cardiomyopathy. At the other end of the spectrum, limb-girdle muscular dystrophy 1B (LGMD1B), also caused by mutations in *LMNA*, is associated with tendon contractures and fewer cardiac complications (Muchir *et al.*, 2000). Muscle wasting in the proximal limbs is the main clinical feature of LGMD1B. The heterogeneity in disease phenotypes, even among members of a single family carrying the same *LMNA* mutation, suggests the pathological outcome maybe strongly influenced by genetic or environmental modifiers (Brodsky *et al.*, 2000; Muntoni *et al.*, 2006). To date, diseases affecting striated muscle comprise about 60% of the laminopathies. However only some 50% of patients diagnosed with AD-EDMD or EDMD have *EMD* or *LMNA* mutations, indicating that mutations in other genes, perhaps those encoding proteins that interact with the lamins (e.g., LAP2 and Nesprins 1 and 2), may account for the remaining 50% of patients (Brodsky *et al.*, 2000; Taylor *et al.*, 2005; Zhang *et al.*, 2007a).

Four mouse lines have been established with mutations in *Lmna*, each of which, to varying extent model of the striated muscle laminopathies (Table 7.1). Mice with no lamin A and C develop normally to birth, but then show post-natal growth retardation starting at 2 weeks and by 6–7 weeks of age are dead with death being associated with muscular dystrophy and cardiomyopathy (Nikolova *et al.*, 2004; Sullivan *et al.*, 1999). There is however no report of a human patient completely lacking lamins A/C, apart from an individual that may have been haploinsufficient for *LMNA*, (Bonne *et al.*, 1999) and a fetus that died late in gestation that was homozygous for a premature stop codon in *LMNA* (van Engelen *et al.*, 2005) perhaps questioning the relevance of this mouse model. However, 1-year-old mice heterozygous for *Lmna* develop atrioventricular (AV) conduction defects with atrial and ventricular arrhythmias analogous to those in humans with *LMNA* mutations. *Lmna*+/− cardiomyocytes have impaired cell and sarcomere contractility, with the AV node cells having abnormal nuclei, which undergo apoptosis and are replaced by fibroblasts (Wolf *et al.*, 2008).

Two lines of mice, each carrying a missense mutation (histidine-to-proline substitution at amino acid 222 (H222P) (Arimura *et al.*, 2005) and asparagine-to-lysine substitution at amino acid 195 (N195K) (Mounkes *et al.*, 2005), were also established. Both mutations were originally identified in families with AD-EDMD. Adult male mice homozygous for the H222P

Table 7.1 Summary of the mouse lines generated to model the nuclear envelopathies and laminopathies

Disease	Gene and representative mutations	Human pathology	Mutations introduced into mouse genes	Mouse Phenotype	Refs
Autosomal-Dominant Emery-Dreifuss Muscular Dystrophy (AD-EDMD)	*LMNA* (H222P, R527P, R249Q, R453W)	Progressive atrophy of skeletal muscles, stiffening of tendons, cardiac conduction defects	*Lmna*−/− Lamin A and C null, $Lmna^{H222P/H222P}$ mis-sense mutation	Postnatal lethality associated with muscular dystrophy and cardiomyopathy	Arimura *et al.* (2005); Sullivan *et al.* (1999)
X-linked Emery-Dreifuss Muscular Dystrophy (X-EDMD)	*EMD* (Emerin)	Progressive atrophy of skeletal muscles, stiffening of tendons, cardiac conduction defects	*Emd y/−*	Minor conduction defects no overt muscle pathology, *in vitro* myogenesis impaired	Melcon *et al.* (2006); Ozawa *et al.* (2006)
Emery-Dreifuss Muscular Dystrophy	*SYNE1* (R257H, V572L, and E646K) and *SYNE2* (T89M)	EDMD or EDMD like syndromes	Deletion of KASH domain in *Syne1* or *Syne2*	Loss of either *Syne1* alone disrupts post-synaptic nuclear clustering *Syne2* no overt phenotype. Combined deficiency results in perinatal lethality	Zhang *et al.* (2007)
Autosomal recessive cerebellar ataxia	*SYNE1* (R2906X, Q7640X, and Premature stop at position 5244)	Impaired walking with lack of limb co-ordination	Deletion of KASH domain in *Syne1*	Disrupts post-synaptic nuclear clustering No overt phenotype	Zhang *et al.* (2007)
Dilated-cardiomyopathy DCM-CD1	*LMNA* (N195K, E161K, R571S, M371K)	Enlargement of the ventricles, conduction defects with reduced systolic function	$Lmna^{N195K/N195K}$ mis-sense mutation Lmna M371K cDNA with mis-sense mutation	Mice die at ~3 months from cardiac conduction defects minimal muscular dystrophy Expressed in heart resulting in cardiomyopathy and early postnatal lethality	Mounkes *et al.* (2005); Wang *et al.* (2006)

Dilated-cardiomyopathy	*LAP2α*	Cardiomyopathy	*Lap2α−/−*	Proliferative defects in paw epidermis and erythroid compartment	Naetar *et al.* (2008)
Hutchinson-Gilford Progeria Syndrome (HGPS)	*LMNA* (G608G, E145K)	Alopecia, growth retardation, reduced bone density, craniofacial abnormalities, premature death	*Lmna*$^{\Delta 9/\Delta 9}$ Splicing mutation and in-frame deletion of exon 9 *Lmna*$^{HG/+}$replacement of LaminA with Progerin hBAC with G606G base change Keratin 14 and Keratin 5 promoter driven expression of Progerin	*Lmna*$^{\Delta 9/\Delta 9}$ homzygotes die at 4 weeks, growth retardation, hyperkeratosis, skeletal, and craniofacial defects *Lmna*$^{HG/+}$ die at 6 months with osteoporosis, alopecia. *Lmna*$^{HG/HG}$ severely retarded in postnatal growth, death at 3 weeks At 1 year progressive loss of smooth muscle cells in medial layer of large arteries. Skin and craniofacial defects, abnormal nuclei	Mounkes *et al.* (2003); Sagelius *et al.* (2008); Varga *et al.* (2006); Wang *et al.* (2008); Yang *et al.* (2005)
Autosomal dominant leukodystrophy	*LMNB1* duplication	Progressive demyelination, similar to multiple sclerosis	*Lmbr−/−* Gene trap insertion into *Lmnb1*	Perinatal lethal possibly due to respiratory failure	Vergnes *et al.* (2004)
Restrictive Dermopathy/ Tight-Skin	*ZMPSTE24* (frameshift with premature stop codon)	Intrauterine growth retardation, hyperkeratosis, dysplastic clavicles, mineralization defects peri-natal death	*Zmpste24*$^{-/-}$	Mice retain farnesylated pre-lamin A. Death at 6–7 months, with rib fractures, osteoporosis, muscle weakness	Bergo *et al.* (2002); Liu *et al.* (2006); Pendas *et al.* (2002)

(continued)

Table 7.1 (*continued*)

Disease	Gene and representative mutations	Human pathology	Mutations introduced into mouse genes	Mouse Phenotype	Refs
Buschke–Ollendorff syndrome, familial Melorheostosis, and melorheostosis with osteopoikilosis	*LEMD-3 (MAN1)*		Gene trap insertion $Man1^{Gt/Gt}$	$Man1^{Gt/+}$ overtly normal, $Man1^{Gt/Gt}$ embryonic lethal vascular defects	Cohen *et al.* (2007); Ishimura *et al.* (2006)
Pelger–Huët anomaly, Hem/Greenberg skeletal	*LBR*	As heterozygotes hypolobulation of granulocyte nuclei, as homozygotes Hem/Greenberg skeletal dysplasia	Ichthyosis ic^{j}, gene trap insertion $LBR^{Gt/Gt}$	Homozygotes alopecia, syndactyly, nuclear hypolobulation, hydrocephalus, variable lethality	Cohen *et al.* (2008); Shultz *et al.* (2003)
Early onset torsion dystonia (DYT1)	torsinA	Deficits in movement control, muscle contractions, abnormal posturing	Transgenic, knock-out and knock-in lines established	Motor deficits, hyperactivity, and perinatal lethality	Dang *et al.* (2005); Goodchild and Dauer (2004, 2005); Goodchild *et al.* (2005); Yokoi *et al.* (2008)

mutation, develop a stiff walking posture, with cardiac fibrosis, chamber dilation, with conduction defects, and die by nine months of age. Female homozygotes also exhibit these pathologies but at a later age and survive for longer. The *Lmna*$^{H222P/H222P}$ mice may represent a good model for studying laminopathies affecting striated muscles as they develop a dystrophic condition in both skeletal and cardiac musculature that is similar to the human disease.

The missense mutation N195K in *LMNA* acts in an autosomal-dominant manner and causes DCM in humans. A mouse line (*Lmna*$^{N195K/N195K}$) homozygous for the same mutation, develop cardiac conduction defects consistent with DCM-CD1, with the mice dying at three months due to cardiac arrhythmia. The mice showed minimal or were free of muscular dystrophy. The transcription factor Hf1b/Sp4 and the gap junction proteins connexin 40 and connexin 43 were misexpressed and/or mislocalized in the mutant hearts. Desmin staining revealed a loss of organization at sarcomeres and intercalated disks suggesting that *LMNA* mutations may cause cardiomyopathy by disrupting the internal organization of the cardiomyocyte and/or altering the expression of transcription factors essential to normal cardiac development, aging, and function.

Transgenic mice have been produced, in which a mutant form of lamin A, M371K (which causes EDMD), was specifically expressed in the heart using a heart-specific a-myosin heavy chain promoter. Mice expressing the mutant transgene are born at a lower frequency than would be expected, and these died by 2–7 weeks of age. The hearts showed extensive disruption to the cardiomyocytes with many having abnormal nuclei. Their early death made it difficult to establish a line of mice from which offspring carrying the transgene could be routinely derived. However, these results demonstrate that expression of a *Lmna* mutant that induces alterations in nuclear morphology can cause tissue and organ damage in mice that express the normal complement of endogenous lamins (Wang *et al.*, 2006).

3.2. Laminopathies affecting adipose and skeletal tissues

The second group of laminopathies, Dunnigan-type familial partial lipodystrophy (FPLD) and mandibuloacral dysplasia (MAD) do not affect muscle tissue (Cao and Hegele, 2000; Novelli *et al.*, 2002; Shackleton *et al.*, 2000). FPLD is inherited as an autosomal dominant trait, with about 85% of the cases (from some 200 affected individuals) being associated with a missense mutation at some Arg482. FPLD is characterized by the loss of subcutaneous white adipose tissue from the limbs, gluteal region, and areas of the trunk, with a concomitant accumulation of white adipose tissue in the neck, face, and abdominal regions. These changes begin at puberty, whereas children are overtly normal prior to adolescence, suggesting a possible hormonal influence on the initiation of disease phenotypes (Vigouroux *et al.*, 2000). The remodeling and redistribution of adipose tissue is accompanied by progressive insulin resistance, which often results in type II diabetes mellitus. These patients are also hyperlipidemic and exhibit an increased susceptibility to atherosclerotic heart disease (Vigouroux and Capeau, 2005).

The aberrant adipose tissue redistribution in lipodystrophy may arise due to an autonomous defect in subsets of mesenchymal or adipocyte precursors, as white fat is a heterogeneous tissue (Fruhbeck, 2008). In lipodystrophy, the body may be attempting to compensate for this loss of fat, in some areas, by an accumulation of fat in others. Neither the levels of lamin A and C expression nor the ratio of these two A-type lamins to each other vary significantly in the subcutaneous, omental, and neck fat depots of normal individuals, suggesting that an intrinsic fat depot-specific pattern of A-type lamin expression does not underlie the fat depot abnormalities associated with FPLD (Cutler *et al.*, 2002; Lelliott *et al.*, 2002). Although *Lmna−/−* mice exhibit reduced stores of white fat, they are cachexic and do not exhibit the hallmark insulin resistance and plasmid hypertriglycideridemia found in humans with FPLD (Cutler *et al.*, 2002).

A rare autosomal recessive mutation in the carboxy-terminal globular domain of A-type lamins is responsible for mandibuloacral dysplasia (MAD; Novelli *et al.*, 2002; Simha *et al.*, 2003), with 94% cases having a missense mutation at residue 527 (R527H). MAD is a disease with many of the metabolic and fat depot redistribution phenotypes of lipodystrophy, but with an expanded set of skeletal abnormalities, including osteolytic lesions in the bones. The primary sites of skeletal malformations in MAD are the craniofacial region, termini of the digits, and clavicles.

3.3. Laminopathies affecting axonal myelination

A single autosomal recessive mutation in the rod domain of A-type lamins results in the peripheral neuropathy Charcot–Marie–Tooth syndrome Type 2b (CMT2B1) (Chaouch *et al.*, 2003; De Sandre-Giovannoli *et al.*, 2002). Families homozygous for the R298C lamin variant exhibit absent deep-tendon reflexes, distal amyotrophy, motor deficits, and loss of large myelinated nerve fibers. The muscular weakening associated with this disease is, maybe, a secondary result of loss of innervations and subsequent muscle atrophy. Lamin-associated CMT2B illustrates that nuclear lamina defects utilize more than one mechanism to cause musculoskeletal defects. Intriguingly, neurons in the sciatic nerve of the *Lmna* null mice showed extensive demyelination (De Sandre-Giovannoli *et al.*, 2002). A mouse line homozygous for the R298C mutation has been derived, although, to date, no overt effect on myelination or locomotor activity has been observed (Kozlov *et al.*, unpublished observations).

4. Progeroid Syndromes

Hutchinson-Gilford progeria syndrome (HGPS) is a rare dominantly inherited disease in which patients show some aspects of premature aging, including growth retardation, loss of subcutaneous fat, hair loss, a reduction in bone density, and weakened muscle function (Gordon *et al.*, 2007; Merideth *et al.*, 2008). The average age of death in HGPS is twelve to fifteen years, usually due to myocardial infarction or stroke caused by artherosclerosis. Artherosclerosis in the HGPS patients is not linked to abnormal systemic lipids levels (Gordon *et al.*, 2005), but may be associated with smooth muscle depletion in the sclerotic vessels and disruption to the microvasculature in various tissues (Gordon *et al.*, 2007; Stehbens *et al.*, 2001). Individuals with HGPS do not show any increase in tumor susceptibility, cataract formation, or cognitive degeneration, features often associated with normal aging. This classifies HGPS as a segmental progeroid syndrome, as it may only partially reproduce some aspects of the normal aging processes (Ershler *et al.*, 2008; Martin, 1989).

A second premature aging condition, Werner's syndrome, is inherited as an autosomal recessive trait due to mutations in *WRN*, a 3′–5′ RecQ DNA helicase-exonuclease. The disease exhibits a high incidence of cancers, early-onset cataracts, arthrosclerosis, diabetes, premature graying of hair, and early death, usually in the late 1940s, from myocardial infarction (Fry, 2002; Hickson, 2003; Oshima, 2000). Approximately 15% or Werner's patients have mutations in *LMNA* instead of *WRN* (Chen *et al.*, 2003). These atypical Werner's patients have short stature, alopecia, osteoporosis, lipodystrophy, diabetes, and muscle atrophy and tend to live longer than patients with the most common HGPS mutation. Additional recessive missense mutations have also been identified in *LMNA* and ZMPSTE24 that result in syndromes resembling both progeria and MAD (Csoka *et al.*, 2004; Plasilova *et al.*, 2004).

Compound heterozygous and homozygous missense mutations in the endoprotease ZMPSTE24 result in a few cases of mandibuloacral disease (MAD) and progeroid-like phenotypes (Agarwal *et al.*, 2003; Shackleton *et al.*, 2005). Recessive mutations resulting in complete absence of ZMPSTE24 cause restrictive dermopathy, which is characterized by intrauterine growth retardation, rigid, or tight skin with prominent superficial vessels, defects in bone mineralization, dysplastic clavicles, and early postnatal death (Agarwal *et al.*, 2003; Moulson *et al.*, 2005; Navarro *et al.*, 2005; Shackleton *et al.*, 2005). Hypomorphic ZMPSTE24 alleles can lead to the accumulation of unprocessed prelamin A in addition to mature lamin A, indicating residual activity of the mutated ZMPSTE24 protein. In ZMPSTE24-null cells, lamin C and prelamin A are present, but no mature lamin A is produced (Bergo *et al.*, 2002; Pendas *et al.*, 2002).

4.1. Progeric mouse models

The most common mutation resulting in HGPS is a splicing defect in exon 11 of the *LMNA* gene, due to a *de novo* single-base substitution; a C-to-T transition at nucleotide 1824 of the coding sequence (G608G). The G608G mutation introduces a cryptic donor splice site resulting in a 150-bp deletion and consequently a 50-amino acid in frame deletion in prelamin A, with lamin C being unaffected. The truncated prelamin A in HGPS is often called "Progerin" or Δ50 (De Sandre-Giovannoli *et al.*, 2003; Eriksson *et al.*, 2003).

The first gene-targeted *Lmna* mutant with progeria was created by the fortuitous introduction of a splicing defect, resulting in the deletion of exon 9 of *Lmna*, ($Lmna^{\Delta/\Delta}$) with the consequent in frame removal of 40 amino acids from the carboxyl-terminal globular domain (Table 7.1). This mutation produced a truncated LaminA protein that remains farnesylated. Mice homozygous for the $Lmna^{\Delta/\Delta}$ mutation develop to term and are seemingly normal at birth. However, postnatal development is associated with multiple pathologies resembling HGPS. Loss of subcutaneous fat, decreased bone density, osteoporosis, abnormal dentition, thin hyperkeratotic skin, growth retardation and death by 4 weeks of age were some of the most striking features of these mice (Mounkes *et al.*, 2003). Fibroblasts cultured from various tissues of post-natal $Lmna^{\Delta/\Delta}$ mice, have misshapen nuclei, undergo a rapid onset of senescence, and death, a characteristic similar to that described for fibroblasts isolated from HGPS patients (Bridger and Kill, 2004; Mounkes *et al.*, 2003).

A second mouse line was engineered to only produce Progerin ($Lmna^{HG}$) (Yang *et al.*, 2005). Heterozygous fibroblasts from these mice express large amounts of progerin and have misshapen nuclei. $Lmna^{HG/+}$ mice are normal at birth, but their post-weaning growth rate is retarded. The $Lmna^{HG/+}$ mice develop many phenotypes common to HGPS, including osteoporosis, alopecia, micrognathia, reduced subcutaneous fat, and osteolysis of the clavicle. These phenotypes are progressive, with the mice dying by 6–7 months of age. Homozygous mice ($Lmna^{HG/HG}$) have severe osteoporosis, spontaneous bone fractures and die before weaning. In none of these mouse lines ($Lmna^{HG/+}$ $Lmna^{HG/HG}$ or $Lmna^{\Delta 9/\Delta 9}$) have the arterial lesions characteristic of HGPS been found (Stewart and Hernandez, unpublished data; Yang *et al.*, 2006).

Three other transgenic Progeric mouse lines have also been produced (Table 7.1). The first carries a 164-kb human bacterial artificial chromosome with the *LMNA* gene containing the (G608G) mutation causing HGPS (Varga *et al.*, 2006). The amount of Progerin expressed in the tissues of the BAC transgenic mice, relative to human and mouse lamin A/C, was not established. The transgenic mice do not manifest any of the early onset pathologies of progeria, such as retarded growth or bone disease, nor do they exhibit atherosclerotic lesions in the large vessels. However, by 1 year they did show a loss of smooth muscle cells in the aorta, a feature reported in the autopsies of some HGPS individuals. The other transgenic lines were engineered to express Progerin in a tissue specific manner. Expression of Progerin under the Keratin-5-promoter tissues resulted in abnormalities in the skin and teeth, loss of dermal fat, hair follicles, sweat glands, and abnormal incisors (Sagelius *et al.*, 2008). However, expression of Progerin under control of the Keratin-14 promoter only resulted in misshapen nuclei, but did not result in any tissue abnormalities (Wang *et al.*, 2008), for reasons that are unclear, as both Keratins-5 and 14 are expressed in the same cells in the basal epidermis (Lersch and Fuchs, 1988).

4.1.1. *ZMPSTE24*-deficient mice

Human patients with complete loss of *ZMPSTE24* develop restrictive dermopathy, or tight skin and die birth. In contrast, mice lacking *Zmpste24* do not (Bergo *et al.*, 2002; Liu *et al.*, 2005; Navarro *et al.*, 2005; Pendas *et al.*, 2002). *Zmpste24*$^{-/-}$ mice are seemingly normal at birth, however postnatal growth is retarded. They eventually develop incisor abnormalities, kyphosis, hair loss, and an arthritic gait with muscle weakness. Bone abnormalities with osteolytic lesions in the ribs at the costovertebral junction, which lead to fractured ribs by ~20 weeks of age, are also present. Bone density is reduced in 3-month-old *Zmpste24*$^{-/-}$ mice (Bergo *et al.*, 2002), although their bones have similar numbers of osteoclasts compared to wild-type siblings. They have vacuolated osteoblasts lacking typical stacks of rough endoplasmic reticulum, suggesting that defects in the osteoblast lineage may be the cause of the *Zmpste24*$^{-/-}$ skeletal pathology (Young *et al.*, 2005). Death usually occurs at ~6 months. The mice, therefore, have a longer equivalent lifespan than humans lacking *ZMPSTE24* or children with progeria. Furthermore, the *Zmpste24*$^{-/-}$ mice also show an increase in autophagy and turnover of cellular constituents, rather than a decline that is associated with normal ageing (Marino *et al.*, 2008).

Zmpste24$^{-/-}$ cells show enhanced rates of senescence in culture, which is associated with changes in the expression of genes controlled by the tumor suppressor p53 (Liu *et al.*, 2005; Pendas *et al.*, 2002). Loss of ZMPSTE24 results in increased DNA damage and activation of a cell senescence. These defects can be partly ameliorated by deleting p53, resulting in mice that appear healthier (Varela *et al.*, 2005), and probably completely ameliorated by reducing the levels of lamin A by removing one allele of *Lmna* by intercrossing the *Zmpste*$^{+/-}$ mice with *Lmna* KO mice (Fong *et al.*, 2004; Varga *et al.*, 2006).

4.2. Defective posttranslational processing of lamin A as the basis for progeria

During normal translational processing, lamins A, B1, and B2 undergo a series of sequential posttranslational modifications. Each of these proteins contains a carboxyl-terminal *CaaX* motif, ("*C*" is cysteine, "*a*" is an aliphatic amino acid, and "*X*" can be many different amino acids). The *CaaX* motif sequence for prelamin A is CSIM and CAIM for lamin B1; lamin C does not have a *CaaX* motif. Posttranslational modification of the lamins begins with the addition of a farnesyl lipid to the cysteine, by the cytosolic protein farnesyltransferase (Holtz *et al.*, 1989; Sinensky *et al.*, 1994). Subsequently, the –*aaX* amino acids are removed in the ER by specific endoproteases, with RCE1 cleaving lamin B1. The cleavage of the –*aaX* from prelamin A cleavage appears to be a redundant function between the endoproteases ZMPSTE24 and RCE1. The isoprenylated cysteine is then methylated in the ER by a membrane methyltransferase, ICMT. These modifications result in the carboxyl-terminal domains of the lamins becoming hydrophobic, facilitating their association with the INM, and are required for the assembly of lamins A, B1, and B2 into the lamina. Prelamin A, but not the B-type lamins, undergoes a final endoproteolytic cleavage, by ZMPSTE24, that removes an additional 15 amino acids, including the farnesylated and methylated cysteine, to produce the mature lamin A (Bergo *et al.*, 2002; Gerace *et al.*, 1984; Pendas *et al.*, 2002) (Fig. 7.1).

In HGPS patients with Progerin, the Progerin variant remains farnesylated, as the 50 amino acid deletion removes the second ZMSTE24 endoproteolytic site used to cleave the farnesylated and methylated cysteine. Much evidence has pointed to the retention of the farnesyl group by Progerin as perhaps being the major factor underlying the pathology of HGPS. The presence of Progerin results in the nuclei becoming grossly misshapen, and this is accompanied by increased thickening of the lamina, alterations in chromatin organization and delays in mitotic progression, with a significant percentage of cells becoming binucleate (Dechat *et al.*, 2007; Goldman *et al.*, 2004; Liu *et al.*, 2006; Shumaker *et al.*, 2006). It is not clear if all these pathological effects are solely due to the retention of the farnesyl moiety or are compounded by the 50 amino acid deletion that potentially may affect protein turnover or interactions with other nuclear factors (Liu *et al.*, 2006). In addition, significant numbers of cases of diagnosed with progeria are caused by other, frequently missense mutations, in *LMNA*. Whether these other mutations resulting in Progeria are also due to post-translational processing defects and/or retention of the farnesyl group remains to be determined (Csoka *et al.*, 2004; Plasilova *et al.*, 2004).

Reduction in the levels of the farnesylated progerin has become a therapeutic target for Progeria. Short-term *in vitro* treatment of HGPS or ZMPSTE24 null cells with farnesyl transferase inhibitors (FTIs) rectifies the alterations in nuclear morphology (Glynn and Glover, 2005; Toth *et al.*, 2005; Yang *et al.*, 2005). FTI treatment of *Zmpste24* null mice or mice heterozygous for Progerin significantly inhibited their rate of weight loss, although the FTI treatments also reduced the body weights of wild-type mice (Fong *et al.*, 2006) and did not prevent the prelamin A or Progerin being geranylgeranylated (Varela *et al.*, 2008). Despite this, FTI treatment significantly improved muscle strength, longevity, and reduced the incidence of rib fractures that are a hallmark of the loss of *Zmpste24* in mice (Fong *et al.*, 2006). Similar improvements to the viability and pathology of the Progerin expressing mice (*Lmna*^{HG}) were reported following their treatment with FTIs (Yang *et al.*, 2006, 2008). Together, these findings have initiated a clinical trial to determine whether FTIs will be useful for treating children with Progeria (http://www.progeriaresearch.org/). Recent results have, however, indicated that treatment of *Zmpste24* null mice with statins and aminobisphosphonates, which when combined, inhibit the prenylation of Lamin A maybe even more effective at alleviating the progeroid pathologies (Varela *et al.*, 2008).

The mouse studies provide some basis for optimism, in that FTI/Statin treatment may retard disease progression—but not reverse the disease, as lifetime treatment with the FTI/Statins may be necessary to maintain Progerin in a nonfarnesylated state. It is also uncertain what effect chronic expression of unfarnesylated Progerin may have on individuals.

5. Laminopathies Associated with Mutations in the B-Type Lamins

To date, no missense mutations in *LMNB1* or *LMNB2* have been linked to any disease. This may be because the B-type lamins are expressed at lower levels than the A-type lamins, and that only a few cell types express only the B-type lamins alone, with either lamin A or the other lamin B being able to compensate for the mutated lamin B gene. However, adult onset autosomal-dominant Leukodystrophy, a neurodegenerative disease caused by progressive myelin loss in the central nervous system, is associated with a duplication of the lamin B1 gene. It is not clear how duplication of the lamin B1 gene results in such a tissue-specific disease, although it may lead to the generation of autoantibodies resulting in demyelination (Padiath *et al.*, 2006).

5.1. Lamin B1–deficient mice

Lamin B1 deficient with a gene trap insertion mutation in *Lmnb1* mice have also been established. The mutation results in a truncated lamin B1 protein containing the amino-terminal head domain and a α-helical central rod domain fused to *βgeo*, but lacking 273 amino acids at the carboxyl-terminus (Vergnes *et al.*, 2004). Heterozygous *Lmnb1* gene-trap mice develop to term but homozygotes die a few minutes after birth. The *Lmnb1*$^{-/-}$ embryos are smaller than wild-type siblings, exhibiting craniofacial and skeletal abnormalities. The perinatal mortality in *Lmnb1*$^{-/-}$ mice may be due to respiratory failure, as many of the *Lmnb1−/−* mice neonates have abnormal lung histology with fewer inflated alveoli than in wild-type mice. Fibroblasts from *Lmnb1−/−* embryos exhibit grossly misshapen nuclei with frequent blebbing, reduced replication rates, increased polyploidy, and prematurely senesce. Interphase chromosomal position is also affected which may disrupt gene expression in specific chromosomal regions (Malhas *et al.*, 2007).

6. Molecular Mechanisms Underlying the Laminopathies

6.1. Mechanical stress hypothesis

With these seemingly diverse diseases having tissue specific defects, a major question that emerges, is how do so many different diseases arise from mutations in the same protein that is expressed in the majority of adult cell types? Two principal theories have been posited to account for the molecular basis to this spectrum of degenerative diseases. The first hypothesis is that disruption of the lamina results in a physical weakening of the nucleus and its ability to withstand mechanical stresses. In mechanically stressed tissues, such as contracting striated muscle, this weakening results in increased cellular apoptosis and necrosis with the consequent loss of the cells and tissue.

Structural studies on the lamin A protein have provided evidence to support this hypothesis. The mutations causing the muscular dystrophies and DCM are distributed throughout the A-type lamins. Many of these mutations disrupt assembly of the lamins and their incorporation into the lamina (Ostlund *et al.*, 2001; Raharjo *et al.*, 2001). These mutations result in abnormal nuclear morphologies, separation of the ONM from the INM and the relocalization of proteins in the NE, such as emerin and nesprin-3 to the ER. A comparison between lamin A–deficient cells with lamin B1 gene trap mouse cells revealed the A-type lamins are the principal contributors to the biophysical properties of the lamina, providing mechanical strength and stiffness to the nuclei, with loss of lamin A resulting in increased levels in apoptosis and necrosis following mechanical induced stress (Lammerding *et al.*, 2004).

Although these observations are consistent with the notion that mutations in *LMNA* physically weaken the nucleus, a consensus is emerging that such a weakening does not provide a complete explanation. In patients, mutation, or loss of emerin results in muscle wasting and cardiac conduction defects similar to those caused by *LMNA* mutations. In mice, *Emd*-deficient nuclei do not exhibit a physical weakening. However, *Emd*-null fibroblasts, which have morphologically and physically normal nuclei (Melcon *et al.*, 2006), are more prone to apoptosis following mechanically induced strain and show reduced levels of *iex-1* (immediate early response gene-1) induced by mechanical strain, compared to normal cells (Lammerding *et al.*, 2005). In addition, *Lmna*$^{N195K/N195K}$ mice have either no or minimal dystrophy in their muscles. Yet *Lmna*$^{N195K/N195K}$ nuclei are severely morphological abnormal and are just as weak as the *Lmna*$^{-/-}$ nuclei (J. Lammerding and C. L. Stewart, unpublished observations). Together, these findings suggest that it might be too simplistic to view physical weakening of the nuclei as the prime cause of striated muscle defects in the laminopathies.

The discovery of the LINC complex, comprised of the NE associated Synes/Nesprins, SUN domain proteins and lamina, that tether the nucleus to the cytoskeleton has added a new dimension in providing insights into the molecular basis of some of the diseases caused by mutations in the *LMNA* gene (Crisp *et al.*, 2006; Tzur *et al.*, 2006). The nesprins and their localization to the NE were first described in the worm *Caenorhabditis elegans*. In *C. elegans*, ANC-1, a large rod-like protein with multiple spectrin repeats, anchors nuclei in the syncytial hypoderm and gut to the cytoskeleton via an N-terminal actin binding domain (Starr and Han, 2002). At its C-terminus, ANC-1 traverses the ONM, with a ~40 amino acid motif, the highly conserved KASH domain, extending into the PNS. Molecular and genetic studies revealed that ANC-1 is retained at the ONM by the KASH domain interacting with another INM associated protein UNC-84. In turn, UNC-84's location depends on the single *C. elegans* lamin protein. ANC-1, UNC-84, and the single lamin therefore form a complex at the NE, so tethering the nucleus to the cytoplasmic cytoskeleton (Tzur *et al.*, 2006). In mammals, at least four Nesprins (Nesprins 1–4) have been identified (Crisp and Burke, 2008). All possess a KASH domain at their C-termini. However, their N termini differ in that Nesprins 1 and 2 have a calponin homology domain resulting in their interaction with the actin cytoskeleton. Nesprin 3 has a plectin homology domain allowing it to interact with intermediate filaments (Wilhelmsen *et al.*, 2005), and Nesprin-4 interacts with kinesin light chains (Roux *et al.* submitted for publication). The Nesprins also bind to UNC-84 orthologues, termed the SUN-domain proteins, of which 5 have been identified in mammals (Crisp and Burke, 2008). SUNs 1 and 2 are widely distributed among most tissues, with SUN1 and 2 interacting with the KASH domains of all 4 Nesprins (Crisp and Burke, 2008; Stewart-Hutchinson *et al.*, 2008). The nucleoplasmic domains of the SUN proteins also interact with the lamins, with this interaction being important for the localization of SUN2 to the INM, at least in some cells (Haque *et al.*, 2006; Padmakumar *et al.*, 2005). The nucleoplasmic domains of the SUN proteins potentially bind to a large number of nuclear proteins involved in transcription and chromatin organization (C. L. Stewart, unpublished data). The remaining three SUN domain proteins appear to be only found in the testes (Bray *et al.*, 2002; C. L. Stewart, unpublished observations).

6.2. Gene expression hypothesis

As the mechanical stress hypothesis does not provide a complete explanation for the muscle envelopathies, a second hypothesis–the gene expression hypothesis–has been invoked. The basis for this hypothesis is that the both A and B type lamins interact either directly or indirectly with many chromatin organization proteins, transcription factors, and transcriptional regulatory factors. Furthermore, signaling pathways that culminate in transcription factor activity may also be regulated at NE and lamina. The gene expression hypothesis should not however be seen at excluding effects that mechanical stress may have on cells. Indeed, much preliminary evidence from studying both mouse and human cells carrying laminopathy mutations indicate that gene expression, signaling pathways, and mechanical integrity are all perturbed.

In human fibroblasts, Lamin B1 binds to some ~1300 chromosomal domains with these domains tending to contain genes that are either not expressed or are expressed at low levels (Guelen *et al.*, 2008). Inducing the tethering of genes or chromosomal regions to the NE, also often suppresses their expression supporting the notion that gene silencing is somehow associated with localization to the nuclear periphery (Finlan *et al.*, 2008; Kumaran and Spector, 2008; Reddy *et al.*, 2008). Loss of *Lmnb1*, however, does not appear to compromise the mechanical properties of the nucleus (Lammerding *et al.*, 2006).

Lamin A-null cells exhibit weakening of the nucleus, a reduction in cytoplasmic elasticity, and viscosity, together with a reduction in cell motility (Lee *et al.*, 2007). These alterations are also associated with an inability of the *Lmna* null cells to fully activate the stress induced Nf-κB signaling pathway that correlates with a reduction in the levels of stress induced genes, such as *egr*-1 and *iex-1*, the latter which functions as an anti-apoptotic factor in cardiomyocytes (Lammerding *et al.*, 2004).

The lamina also functions as a scaffold for the localization of factors to the NE that are important in the regeneration of muscle and other tissues affected by the lamin mutations. Loss of the A-type lamins results in a redistribution of emerin from the nucleus to the endoplasmic reticulum, suggesting that it is the reduction in emerin levels at the NE which contributes to the etiology of AD- and X-linked EDMD (Sullivan *et al.*, 1999). Loss of emerin from the NE also results in the centrosome detaching from the NE (Lee *et al.*, 2007; Salpingidou *et al.*, 2007).

Emerin was the first NE associated protein identified, that when mutated, resulted in a disease, the X-linked form of EDMD (Bione *et al.*, 1994). Emerin binds to several transcription factors including the muscle-specific transcription factor Lmo7 (Holaska and Wilson, 2007; Holaska *et al.*, 2006). Emerin interacts with the transcriptional repressors Btf and Gcl and the RNA splicing factor YT521-B (Holaska and Wilson, 2006).

Emerin may also regulate β-catenin accumulation in the nucleus, as there is an increase in the nuclear accumulation of β-catenin in fibroblasts from patients with *EMD* mutations (Markiewicz *et al.*, 2006). Such effects on β-catenin localization or activity were not, however, observed in emerin deficient mouse fibroblasts (C. L. Stewart and L. Hernandez, unpublished observations).

In mice, emerin deficiency does not phenocopy human X-linked EDMD. *Emd*-deficient mice appear overtly normal and do not show any muscle or cardiac abnormalities, perhaps delineating a key difference that age and stress presents to human muscle (Melcon *et al.*, 2006; Ozawa *et al.*, 2006). However, evidence suggests that satellite cell-mediated muscle regeneration depends, to some extent, on emerin (Melcon *et al.*, 2006).

Following injury or disease, muscle satellite cells are activated to regenerate the muscle fiber with which they are associated. The satellite cells proliferate then exit the cell cycle and fuse with the underlying myotubes, a process mediated by the retinoblastoma gene product (Rb) (Mancini *et al.*, 1994; Takahashi *et al.*, 2003). Phosphorylation of Rb is regulated throughout the cell cycle with Rb being dephosphorylated at the end of M phase and hyperphosphorylated during progression from G1 to S (Mancini *et al.*, 1994). When hypophosphorylated, Rb complexes with the E2F family of transcription factors. In these complexes, E2F is inactive and unable to transactivate cell cycle inhibitor genes, such as p21, resulting in continued cell proliferation. When Rb is hyperphosphorylated, its activity is inhibited and can no longer bind to E2F, cell cycle arrest genes are induced, resulting in cells exiting the cycle.

Lamin A/C is part of the pRb-E2F complex required for cell cycle exit (Johnson *et al.*, 2004; Mancini *et al.*, 1994; Ozaki *et al.*, 1994). Regenerating muscle of *Emd*-deficient mice shows upregulation of the Rb/MyoD transcriptional pathway components including *Crebbp*, a MyoD acetylase, *Nap1L1*, *Cri-1*, and *Cdk1* and a delay in the induction of myogenic genes (Melcon *et al.*, 2006). A similar upregulation of Rb/MyoD pathway components was also observed in human EDMD muscle. Accordingly, disruption of lamin A/C or emerin may result in destabilization of Rb complexes, resulting in a compensatory upregulation of Rb and MyoD target genes and defects in muscle regeneration (Bakay *et al.*, 2006). These observations implicate inappropriate regulation of Rb in the pathogenesis of NE muscular dystrophies (Bakay *et al.*, 2006). This may also explain why immortalized $Lmna^{-/-}$ myoblasts, but not primary $Lmna^{-/-}$myoblasts, are impaired in their *in vitro* differentiation to myotubes (Frock *et al.*, 2006), as the immortalized myoblasts may be defective in Rb regulation and/or expression.

While the Rb/MyoD pathway regulates skeletal muscle differentiation, the MAPK pathway appears to be important in the cardiomyopathies caused by mutations in *LMNA* (Bueno *et al.*, 2001). A second line of *Emd*-null mice shows minimal motor coordination defects and vacuolization within cardiomyocytes (Ozawa *et al.*, 2006). Pathway analysis of heart tissue showed perturbation of the MAPK pathway in this mouse line (Muchir *et al.*, 2007a). Similar findings were obtained with the examination of hearts from $Lmna^{H222P/H222P}$ mice (Muchir *et al.*, 2007b). In the $Lmna^{H222P/H222P}$ and *Emd*-null hearts, MAPK pathway components including *c-Jun*, *Atf2*, *Atf4*, *Nfat2*, and *Nfat4* are upregulated and this upregulation and these changes to the MAPK pathway are implicated in inducing cardiomyopathy (Bueno *et al.*, 2001). Although it is unclear whether the pathological differences between the two lines of *Emd* null mice are due to background strain differences, it is plausible that two different pathways, MyoD/Rb and MAPK, respectively, contribute to the pathogenesis in skeletal and cardiac muscle.

The FPLD and MAD mutations are both predicted to have minimal effects on lamina structure, assembly, or emerin localization (Ostlund *et al.*, 2001; Raharjo *et al.*, 2001). The molecular basis as to how these specific mutations result in these diseases remains obscure. As the affected amino acids are both located at the outside of the Ig-fold domain at the C-terminus of the A-type lamins, their substitution may perturb interactions between the A-type lamins and other nuclear proteins (Burke and Stewart, 2002), such as the adipogenic factor SREBP1 (Lloyd *et al.*, 2002), rather than disrupting the structure and/or assembly of the lamins as is predicted for the dystrophic lamin mutations. Human cells, either with an FPLD, MAD, or one of the mutations responsible for atypical Werner's syndrome may also show increased levels of prelamin A (Capanni *et al.*, 2005; Caron *et al.*, 2007), suggesting that these mutations affect the processing of lamin A resulting in an accumulation of this immature form of lamin A protein.

6.3. Laminopathies associated with the LINC complex

The most prevalent forms of muscular dystrophy are caused by mutations in components of the dystrophin–sarcoglycan–dystroglycan complex that links the extracellular matrix (ECM) of the muscle cell to the muscle cytoskeleton at the cell membrane. Duchenne, the most common form of muscular dystrophy, results from mutations in the large cytoplasmic spectrin repeat-rich protein dystrophin (Ervasti and Sonnemann, 2008). Dystrophin connects the membrane sarcoglycan/dystroglycan complex to the actin cytoskeleton, creating a direct physical link running from the ECM via the sarcoglycan/dystrophin complex, and cytoskeleton to the LINC complex and finally to the nucleoplasm. Disruption of this physical link, by mutations in many of its components, results in different forms of myofibre degeneration and muscular dystrophy. This includes recently reported mutations in the components of the human LINC complex in which mutations in Syne/Nesprins-1 and 2, result in nuclear abnormalities, emerin redistribution with EDMD or EDMD-like pathologies (Wheeler *et al.*,

2007). In addition, a group of Canadian families diagnosed with late onset cerebellar ataxia carry mutations in the *SYNE1* gene (Dupre *et al.*, 2007; Gros-Louis *et al.*, 2007). In mice, disruption of Syne1/Nesprin-1, but not Syne2/Nesprin2 expression or deletion of the KASH domain causes an increase in centrally localized nuclei in muscle fibres, resulting in synaptic nuclei clustering. Mice lacking either of the Syne/Nesprin KASH domains exhibited no overt pathology. However mice doubly deficient for both KASH domains die perinataly due to respiratory problems that may be associated with abnormal phrenic nerve enervation of the diaphragm (Zhang *et al.*, 2007b).

As far as other LINC components are concerned, loss of SUN-1 results in male and female meiotic defects (Ding *et al.*, 2007), a somewhat surprising result given that SUN1 is widely expressed in many somatic tissues, although its loss may be compensated by the similarly widely expressed SUN2. Together, these findings raise the tantalizing possibility that many of the muscular dystrophies, cardiomyopathies, and some ataxias may have a common molecular basis, with disruption to different components in the nucleo-cytoplasmic link ultimately affecting cytoskeletal organization and function, as well as possibly directly affecting nucleoplasmic/chromatin organization and gene expression.

7. Diseases and Anomalies Caused by Mutations in Other NE Associated Proteins

A growing number of NE integral and associated proteins have been identified, with some 70 proteins alone being described in a proteomic study of rat liver nuclei (Schirmer *et al.*, 2003). Among the best studied integral NE proteins containing at least one transmembrane domain are emerin, MAN1/LEMD3, LAP1, LAP2, and LBR (reviewed in (Stewart *et al.*, 2007). Some of these proteins (LAP2, emerin, and MAN1/LEMD3) share a common 40 amino acid motif (LEM domain), which interacts with the small DNA binding protein BAF (barrier-to-autointegration factor). By binding with BAF the LEM domain proteins indirectly are linked to DNA (reviewed in Dorner *et al.*, 2007). Additional LEM domain proteins have also been recently identified (Brachner *et al.*, 2005; Lee and Wilson, 2004; Ulbert *et al.*, 2006)

7.1. Nuclear envelopathies associated with MAN1/LEMD3

MAN1/LEMD3 is an integral protein of the INM (Bengtsson, 2007; Paulin-Levasseur *et al.*, 1996). The N-terminus of MAN1 contains a LEM domain, which binds to both lamins A/C and chromatin via BAF.

Following a double pass transmembrane region, the C-terminus contains an RRM domain, which for interacts with SMAD transcription factors (Bengtsson, 2007).

Heterozygous autosomal dominant loss-of-function mutations in *LEMD3* result in Osteopoikilosis, Buschke–Ollendorff syndrome (BOS) and melorheostosis, disorders marked by increased bone density (Ben-Asher *et al.*, 2005; Hellemans *et al.*, 2006; Hellemans *et al.*, 2004). More recently, a novel mutation in MAN1/LEMD3 was identified in Familial Cutaneous Collagenomas, marked by cardiomyopathy and osteopoikilosis (Hershkovitz *et al.*, 2007).

TGFβ/BMP/activin signaling pathways activate SMAD transcription factors and are essential in regulating mouse embryogenesis (Goumans and Mummery, 2000; Itoh *et al.*, 2000). Due to the widespread cellular processes governed by the SMADs, their transcriptional activity is tightly regulated. MAN1/LEMD3 physically interacts with SMADs (Hellemans *et al.*, 2004; Lin *et al.*, 2005; Osada *et al.*, 2003) using the nucleoplasmic RRM domain at the C-terminus of MAN1 that binds to and seemingly sequesters SMADs1–3 at the NE. Such sequestration regulates SMAD nuclear translocation, phosphorylation, and transcriptional activity, by a yet undefined mechanism (Lin *et al.*, 2005; Pan *et al.*, 2005).

Embryos homozygous for the gene-trap mutation into MAN1/LEMD3 (*Man1*$^{GT/GT}$) die around E8.5. Lethality is associated with ectopic expression of the anterior marker, *Nodal*, and abnormal nuclear morphologies (Ishimura *et al.*, 2006). Angiogenesis is also defective in the *Man1*$^{GT/GT}$ embryos, with yolk-sac vasculogenesis and vessel formation being disorganized (Cohen *et al.*, 2007; Ishimura *et al.*, 2006). TGFβ signaling pathway plays a central role in the development of the vascular system in mouse embryos (Mummery, 2001; Rossant and Howard, 2002). The *Man1*$^{GT/GT}$ nuclei contain increased levels of phosphorylated SMADs2/3, components of the Alk-5 pathway, resulting in upregulation of the targets of SMAD transcription including fibronectin and the cell cycle inhibitors p15ink4b, p21waf1, and p27kip1 (Cohen *et al.*, 2007; Ishimura *et al.*, 2006). Overexpression of these factors in the mutant cells results in increased ECM deposition, low rates of endothelial cell proliferation, inhibition of EC migration and angiogenic branching culminating in a reduced capacity of the ECs for forming mature blood vessels. Interestingly, X-ray scans of *Man1*$^{GT/+}$ mice did not reveal any skeletal abnormalities, suggesting that mutations in MAN1/LEMD3 may not result in osteopoikilosis in mice (Cohen *et al.*, 2007).

These studies suggested that MAN1/LEMD3 regulates the export of SMADs from the nucleus, perhaps by acting as a scaffold for a yet unidentified SMAD specific phosphatase (Chen *et al.*, 2006; Cohen *et al.*, 2007; Knockaert *et al.*, 2006). A-type lamins may also influence TGFβ1-mediated

SMAD phosphorylation by regulating protein phosphatase 2A activity, although it remains unclear whether MAN1/LEMD3 is involved in this regulation, despite the mobility of MAN1/LEMD3 in the NE being influenced by the A-type lamins (Ostlund *et al.*, 2006; Van Berlo *et al.*, 2005).

7.2. Nuclear envelopathies and LAP2 mutations

Another LEM-domain containing protein is the Lamina-Associated Polypeptide 2 (LAP2, or Thymopoietin), that comprise a family of alternatively spliced polypeptides, α, β, γ, δ, ϵ, and ζ (Dorner *et al.*, 2007; Wagner and Krohne, 2007). Most of the LAP2 polypeptides are integral NE proteins with the exception of α and ζ. LAP2α is localized exclusively to the nucleoplasm where it interacts with nucleoplasmic lamin A/C. Complexes between LAP2α, lamin A/C, and Rb/E2F are necessary for cell cycle exit in differentiating cells, probably by preventing the proteolytic degradation of Rb (Dorner *et al.*, 2007). Proliferating fibroblasts have high levels of LAP2α, but these levels decline once the fibroblasts reach confluence (Pekovic *et al.*, 2007). Together, these studies suggest that abnormal lamin A/C–LAP2α–Rb complexes may exist in some laminopathies and as a consequence affect cell proliferation.

A mutation in LAP2α is linked to dilated cardiomyopathy (Taylor *et al.*, 2005). Mice deficient in LAP2α are viable and fertile, with young mice developing impaired systolic function that culminates in extensive fibrosis in old individuals (Naetar *et al.*, 2008 comm). The LAP2α mice also show epidermal hyperproliferation in paw epidermis and in the erythroid compartment that is associated with defective exit from the cell cycle and persistent Rb hyperphosphorylation (Naetar *et al.*, 2008). Further examination of these mice may shed light on the regulation of Rb in differentiating cells and the role of the lamina in regulating tissue regeneration.

The lack of overt pathologies in *Emd*-deficient mice suggested the possibility of redundancy between the various LEM domain proteins. To date, such redundancy or epistasis has not been identified in intercrosses between the MAN1/LEMD3, LAP2α, and emerin-deficient mice (T. V. Cohen and C. L. Stewart, unpublished observations). However, recent findings suggested that LAP2α and emerin maybe required in the early replication of murine and HIV retroviruses following their infection of cells (Jacque and Stevenson, 2006). This requirement was examined in both single and compound LAP2α and emerin-deficient mice. Both macrophages and fibroblasts derived from these mice were easily infected with retroviruses, indicating that in mice, and perhaps in contrast to human cells, both emerin and LAP2α are dispensable for retroviral infectivity (Mulky *et al.*, 2008; Shun *et al.*, 2007).

7.3. Nuclear envelopathies associated with the Lamin B receptor

The Lamin B receptor (LBR) is a 58kDa NE protein that binds to the B-type lamins, and with a lower affinity to lamin A (Worman *et al.*, 1988). LBR is an evolutionary conserved multifunctional integral membrane protein of the INM. The N-terminal third of the protein is hydrophilic, containing a nucleoplasmic domain. This domain interacts with chromatin in a cell cycle dependent manner and may be important for NE reassembly (Pyrpasopoulou *et al.*, 1996). The N-terminus also interacts with p34/p32–a putative splicing complex (Nikolakaki *et al.*, 1997; Simos and Georgatos, 1994), the chromatin binding protein HP1, HA95–an RNA helicase A binding protein, and p18, another integral membrane protein (Martins *et al.*, 2000; Nikolakaki *et al.*, 1996; Ye and Worman, 1996). During nuclear reassembly at mitosis, the LBR is important in targeting the NE to chromatin in an importin β mediated process (Braunagel *et al.*, 2007; Ma *et al.*, 2007). The C-terminal region of LBR contains 8 transmembrane domains and exhibits sterol Δ^{14}-reductase activity *in vitro*, similar to another factor DHCR14 (TM7SF2, SR-1) (Holmer *et al.*, 1998; Silve *et al.*, 1998).

Heterozygous mutations in the *LBR* gene result in the autosomal dominant Pelger-Huët anomaly, a disorder marked by hyposegmentation in the granulocyte/neutrophil nuclei (Hoffmann *et al.*, 2002). Homozygotes for a splicing anomaly in the *LBR* gene exhibit a more severe blood phenotype and sometimes exhibit impaired cognitive development, heart defects, and bone deformities (Hoffmann *et al.*, 2007; Waterham *et al.*, 2003). Homozygous nonsense mutations in *LBR* result in the autosomal recessive chondrodystrophy, HEM/Greenberg Skeletal Dysplasia, a severe inborn metabolic disorder characterized by hydrops, skeletal abnormalities including shortening, a "moth-eaten appearance", ectopic calcifications of bones, intestinal, lung, and kidney malformations and prenatal lethality (Konstantinidou *et al.*, 2008; Waterham *et al.*, 2003). Embryonic fibroblasts, homozygous for the LBR mutation, express elevated levels of cholesta-8, 14-dien-3-β-ol, a key metabolite of the sterol-D^{14}-reductase activity of LBR, suggesting that LBR may be essential to proper cholesterol synthesis during development. This is, however, now disputed due to possible redundancy with another cholesterol metabolizing enzyme *Dhcr14* (Wassif *et al.*, 2007).

In mice, missense or nonsense mutations in the *Lbr* gene result in ichthyosis (ic^J) of which the most overt phenotypes include alopecia, postnatal growth retardation, early lethality, and syndactyly (Shultz *et al.*, 2003). Within the hematopoietic compartment of ic^J mice, splenic lymphocytes show clumping of heterochromatin. In the peripheral blood, neutrophils and eosinophils are immature and the nuclei are hypolobulated.

A gene-trap insertion into the *Lbr* locus (*Lbr*$^{GT/GT}$) that truncates the LBR, results in Heterochromatin protein-1α (HP1α) accumulation at nuclear foci and increased frequency of micronuclei, indicative of possible defects in cell division at mitosis (Lindberg *et al.*, 2007; Ma *et al.*, 2007). *Lbr*$^{GT/GT}$ mice are phenotypically similar to *ic*^{J} mice, with white blood cells containing condensed nuclei. Granulocyte/neutrophil numbers were increased, with an immature appearance and absence of mature ring forms (Cohen *et al.*, in press). Accompanying this was a decrease in immature B cell numbers whereas mature B cells, T cells and erythrocytes remained unchanged. Chromatin clumping and an increase in granulocyte numbers is also observed in Specific Granulocyte Deficiency in humans (Lekstrom-Himes and Xanthopoulos, 1999) and mice deficient in CCAAT/enhancer-binding protein epsilon (C/EBPε) (Lekstrom-Himes, 2001), suggesting that C/EBPε and LBR are functionally interrelated. The *Lbr* gene is indeed transcriptionally regulated by C/EBPε. However, some of the activities of granulocytes that are absent in C/EBPε granulocytes, such as myeloperoxidase immunoreactivity, and bacterial killing were still functional in the *Lbr*$^{GT/GT}$ granulocytes revealing that, despite having an abnormal nuclear morphology, some granulocyte functions are not compromised by the loss of LBR (Cohen *et al.*, 2008).

7.4. Nuclear envelopathies involving torsinA

A single deletion of a glutamate residue in the TorsinA protein results in early onset autosomal dominant torsion dystonia (DYT1), a disorder characterized by involuntary movements due to hyperactive dopaminergic systems in the striatum (Cookson and Clarimon, 2005). TorsinA, a member of the AAA-family of ATPases, acts as a chaperone regulating the processing of proteins through the secretory pathway (Hewett *et al.*, 2007; Ozelius *et al.*, 1997). Although TorsinA is localized predominantly to the ER where it interacts with a luminal protein, LULL1 (TOR1AIP/NET8), some torsinA is also localized to the NE where it interacts with the luminal domain of lamina-associated polypeptide 1 (LAP1; Goodchild and Dauer, 2005; Naismith *et al.*, 2004). Although TorsinA is widely expressed in many different cells types, the mutant form appears to only affect post-migratory neurons indicating these cells have some unique requirement for TorsinA, such as exocytosis, and the uptake of neurotransmitters, which are defective because of the relocation of the mutant TorsinA protein to the NE (Granata *et al.*, 2008). Alternatively, it has been suggested that the relocation of the mutant TorsinA to the NE may perturb NE-cytoskeletal interactions, perhaps by affecting the Nesprin-SUN LINC complex (Gerace, 2004).

Several mouse lines carrying a mutated torsin have been generated. A transgenic line was developed carrying primary dystonia-associated GAG deletion of exon 5 of TORIA (Goodchild and Dauer, 2004). Examination of brains from the transgenic mice determined that the mutated form of torsinA accumulates and forms membranous inclusion bodies in the nuclei of neurons. These mice showed delayed performance in some motor tasks and higher concentrations of dopamine metabolites DOPAC and homovanillic acid (Zhao *et al.*, 2008). In another mouse model, a ''knock-in'' approach was used, resulting in a similar cellular pathology of abnormal nuclear membranes in embryonic neurons. A knockout mouse line was also developed (Goodchild *et al.*, 2005). Importantly, comparison of the knock-in and the knockout models yielded many similarities. Homozygotes from both lines die shortly after birth due to a failure to feed and show abnormal nuclear membrane morphology. Although the heterozygotes are viable and appear normal, subtle behavioral changes were detected (Dang *et al.*, 2005). Additional mouse lines of dystonia include a knockdown model (Dang *et al.*, 2006) and a conditional model with the mutated torsinA expressed in the cortex (Yokoi *et al.*, 2008). Both of these lines displayed phenotypes of motor deficits and hyperactivity. Insights gained from these mouse lines have led to a model for the treatment of dystonia using siRNA to knock-down mutated torsinA (Gonzalez-Alegre *et al.*, 2003).

Intriguingly, one of the severe forms of lipodystrophy, Berardinelli-Seip Congenital Lipodystrophy, may also be caused by aberrant relocalization of an ER protein to the NE. Berardinelli-Seip Congenital Lipodystrophy is caused by mutations in *BSCL2* or Seipin, a protein normally localized to the ER, where it maybe required for the formation of lipid vesicles (Szymanski *et al.*, 2007). A mutation *BSCL2* resulting in an amino acid substitution causes the predominately ER localized protein to relocate to the NE where it is apparently non functional (Payne *et al.*, 2008).

8. Conclusions

In recent years, our understanding of the function of the NE and lamina has undergone a renaissance. We, now, view the NE/Lamina as having a key role in integrating many cellular processes, with particular regard to linking the cytoskeleton to the inside of the nucleus and chromatin. Mouse lines carrying mutations in many of the NE proteins and lamins have been instrumental in deepening our understanding as to the functions of the NE and lamina, by revealing the A-type lamins to have multiple functions. In doing this, they have also provided us with models for many of the diseases that are caused by defects in the lamins and NE proteins. These models are providing a valuable resource in the search for therapies that may help treat the different diseases. The fact that proteomic studies on the NE have suggested some 70 proteins are located at the NE and that the composition of the NE varies between different cell types (Kavanagh *et al.*, 2007), suggests that many other conditions and anomalies may be found to be linked to the NE. Study of these proteins may well reveal novel functions for the NE and lamina in regulating cellular physiology.

REFERENCES

Agarwal, A. K., Fryns, J. P., Auchus, R. J., and Garg, A. (2003). *Hum. Mol. Genet.* **12,** 1995–2001.

Arimura, T., Helbling-Leclerc, A., Massart, C., Varnous, S., Niel, F., Lacene, E., Fromes, Y., Toussaint, M., Mura, A. M., Keller, D. I., Amthor, H., Isnard, R., *et al.* (2005). *Hum. Mol. Genet.* **14,** 155–169.

Bakay, M., Wang, Z., Melcon, G., Schiltz, L., Xuan, J., Zhao, P., Sartorelli, V., Seo, J., Pegoraro, E., Angelini, C., Shneiderman, B., Escolar, D., *et al.* (2006). *Brain* **129,** 996–1013.

Ben-Asher, E., Zelzer, E., and Lancet, D. (2005). *Isr. Med. Assoc. J.* **7,** 273–274.

Bengtsson, L. (2007). *FEBS J.* **274,** 1374–1382.

Bergo, M. O., Gavino, B., Ross, J., Schmidt, W. K., Hong, C., Kendall, L. V., Mohr, A., Meta, M., Genant, H., Jiang, Y., Wisner, E. R., Van Bruggen, N., Carano, R. A., Michaelis, S., Griffey, S. M., and Young, S. G. (2002). *Proc. Natl. Acad. Sci. USA* **99,** 13049–13054.

Bione, S., Maestrini, E., Rivella, S., Mancini, M., Regis, S., Romeo, G., and Toniolo, D. (1994). *Nat. Genet.* **8,** 323–327.

Bonne, G., Di Barletta, M. R., Varnous, S., Becane, H. M., Hammouda, E. H., Merlini, L., Muntoni, F., Greenberg, C. R., Gary, F., Urtizberea, J. A., Duboc, D., Fardeau, M., Toniolo, D., and Schwartz, K. (1999). *Nat. Genet.* **21,** 285–288.

Brachner, A., Reipert, S., Foisner, R., and Gotzmann, J. (2005). *J. Cell Sci.* **118,** 5797–5810.

Braunagel, S. C., Williamson, S. T., Ding, Q., Wu, X., and Summers, M. D. (2007). *Proc. Natl. Acad. Sci. USA* **104,** 9307–9312.

Bray, J. D., Chennathukuzhi, V. M., and Hecht, N. B. (2002). *Genomics* **79,** 799–808.

Bridger, J. M., and Kill, I. R. (2004). *Exp. Gerontol.* **39,** 717–724.

Brodsky, G. L., Muntoni, F., Miocic, S., Sinagra, G., Sewry, C., and Mestroni, L. (2000). *Circulation* **101,** 473–476.

Bueno, O. F., De Windt, L. J., Lim, H. W., Tymitz, K. M., Witt, S. A., Kimball, T. R., and Molkentin, J. D. (2001). *Circ. Res.* **88,** 88–96.

Burke, B., and Stewart, C. L. (2002). *Nat. Rev. Mol. Cell Biol.* **3,** 575–585.

Burke, B., and Stewart, C. L. (2006). *Annu. Rev. Genomics Hum. Genet.* **7,** 369–405.

Cao, H., and Hegele, R. A. (2000). *Hum. Mol. Genet.* **9,** 109–112.

Capanni, C., Mattioli, E., Columbaro, M., Lucarelli, E., Parnaik, V. K., Novelli, G., Wehnert, M., Cenni, V., Maraldi, N. M., Squarzoni, S., and Lattanzi, G. (2005). *Hum. Mol. Genet.* **14,** 1489–1502.

Caron, M., Auclair, M., Donadille, B., Bereziat, V., Guerci, B., Laville, M., Narbonne, H., Bodemer, C., Lascols, O., Capeau, J., and Vigouroux, C. (2007). *Cell Death Differ.* **14,** 1759–1767.

Chaouch, M., Allal, Y., De Sandre-Giovannoli, A., Vallat, J. M., Amer-el-Khedoud, A., Kassouri, N., Chaouch, A., Sindou, P., Hammadouche, T., Tazir, M., Levy, N., and Grid, D. (2003). *Neuromuscul. Disord.* **13,** 60–67.

Chen, H. B., Shen, J., Ip, Y. T., and Xu, L. (2006). *Genes Dev.* **20,** 648–653.

Chen, L., Lee, L., Kudlow, B. A., Dos Santos, H. G., Sletvold, O., Shafeghati, Y., Botha, E. G., Garg, A., Hanson, N. B., Martin, G. M., Mian, I. S., Kennedy, B. K., *et al.* (2003). *Lancet* **362,** 440–445.

Cohen, T. V., Klarmann, K. D., Sakchaisri, K., Cooper, J. P., Kuhns, D., Anver, M., Johnson, P. F., Williams, S. C., Keller, J. R., and Stewart, C. L. (2008). *Hum. Mol. Genet.*
Cohen, T. V., Kosti, O., and Stewart, C. L. (2007). *Development* **134,** 1385–1395.
Cookson, M. R., and Clarimon, J. (2005). *Neuron* **48,** 875–877.
Crisp, M., and Burke, B. (2008). *FEBS Lett.* **582,** 2023–2032.
Crisp, M., Liu, Q., Roux, K., Rattner, J. B., Shanahan, C., Burke, B., Stahl, P. D., and Hodzic, D. (2006). *J. Cell Biol.* **172,** 41–53.
Csoka, A. B., Cao, H., Sammak, P. J., Constantinescu, D., Schatten, G. P., and Hegele, R. A. (2004). *J. Med. Genet.* **41,** 304–308.
Cutler, D. A., Sullivan, T., Marcus-Samuels, B., Stewart, C. L., and Reitman, M. L. (2002). *Biochem. Biophys. Res. Commun.* **291,** 522–527.
Dang, M. T., Yokoi, F., McNaught, K. S., Jengelley, T. A., Jackson, T., Li, J., and Li, Y. (2005). *Exp. Neurol.* **196,** 452–463.
Dang, M. T., Yokoi, F., Pence, M. A., and Li, Y. (2006). *Neurosci. Res.* **56,** 470–474.
De Sandre-Giovannoli, A., Bernard, R., Cau, P., Navarro, C., Amiel, J., Boccaccio, I., Lyonnet, S., Stewart, C. L., Munnich, A., Le Merrer, M., and Levy, N. (2003). *Science* **300,** 2055.
De Sandre-Giovannoli, A., Chaouch, M., Kozlov, S., Vallat, J. M., Tazir, M., Kassouri, N., Szepetowski, P., Hammadouche, T., Vandenberghe, A., Stewart, C. L., Grid, D., and Levy, N. (2002). *Am. J. Hum. Genet.* **70,** 726–736.
Dechat, T., Shimi, T., Adam, S. A., Rusinol, A. E., Andres, D. A., Spielmann, H. P., Sinensky, M. S., and Goldman, R. D. (2007). *Proc. Natl. Acad. Sci. USA* **104,** 4955–4960.
Ding, X., Xu, R., Yu, J., Xu, T., Zhuang, Y., and Han, M. (2007). *Dev. Cell* **12,** 863–872.
Dorner, D., Gotzmann, J., and Foisner, R. (2007). *FEBS J.* **274,** 1362–1373.
Dupre, N., Gros-Louis, F., Chrestian, N., Verreault, S., Brunet, D., de Verteuil, D., Brais, B., Bouchard, J. P., and Rouleau, G. A. (2007). *Ann. Neurol.* **62,** 93–98.
Eriksson, M., Brown, W. T., Gordon, L. B., Glynn, M. W., Singer, J., Scott, L., Erdos, M. R., Robbins, C. M., Moses, T. Y., Berglund, P., Dutra, A., Pak, E., *et al.* (2003). *Nature* **423,** 293–298.
Ershler, W. B., Ferrucci, L., and Longo, D. L. (2008). *N. Engl. J. Med.* **358,** 2409–2410; author reply 2410–2411.
Ervasti, J. M., and Sonnemann, K. J. (2008). *Int. Rev. Cytol.* **265,** 191–225.
Fatkin, D., MacRae, C., Sasaki, T., Wolff, M. R., Porcu, M., Frenneaux, M., Atherton, J., Vidaillet, H. J., Jr., Spudich, S., De Girolami, U., Seidman, J. G., Seidman, C., *et al.* (1999). *N. Engl. J. Med.* **341,** 1715–1724.
Finlan, L. E., Sproul, D., Thomson, I., Boyle, S., Kerr, E., Perry, P., Ylstra, B., Chubb, J. R., and Bickmore, W. A. (2008). *PLoS Genet.* **4,** e1000039.
Fong, L. G., Frost, D., Meta, M., Qiao, X., Yang, S. H., Coffinier, C., and Young, S. G. (2006). *Science* **311,** 1621–1623.
Fong, L. G., Ng, J. K., Meta, M., Cote, N., Yang, S. H., Stewart, C. L., Sullivan, T., Burghardt, A., Majumdar, S., Reue, K., Bergo, M. O., and Young, S. G. (2004). *Proc. Natl. Acad. Sci. USA* **101,** 18111–18116.
Frock, R. L., Kudlow, B. A., Evans, A. M., Jameson, S. A., Hauschka, S. D., and Kennedy, B. K. (2006). *Genes Dev.* **20,** 486–500.
Fruhbeck, G. (2008). *Methods Mol. Biol.* **456,** 1–22.
Fry, M. (2002). *Sci. Aging. Knowledge Environ.* **2002,** re2.
Furukawa, K., and Hotta, Y. (1993). *Embo. J.* **12,** 97–106.
Gerace, L. (2004). *Proc. Natl. Acad. Sci. USA* **101,** 8839–8840.
Gerace, L., Comeau, C., and Benson, M. (1984). *J. Cell Sci. Suppl.* **1,** 137–160.
Glynn, M. W., and Glover, T. W. (2005). *Hum. Mol. Genet.* **14,** 2959–2969.

Goldman, R. D., Gruenbaum, Y., Moir, R. D., Shumaker, D. K., and Spann, T. P. (2002). *Genes Dev.* **16,** 533–547.
Goldman, R. D., Shumaker, D. K., Erdos, M. R., Eriksson, M., Goldman, A. E., Gordon, L. B., Gruenbaum, Y., Khuon, S., Mendez, M., Varga, R., and Collins, F. S. (2004). *Proc. Natl. Acad. Sci. USA* **101,** 8963–8968.
Gonzalez-Alegre, P., Miller, V. M., Davidson, B. L., and Paulson, H. L. (2003). *Ann. Neurol.* **53,** 781–787.
Goodchild, R. E., and Dauer, W. T. (2004). *Proc. Natl. Acad. Sci. USA* **101,** 847–852.
Goodchild, R. E., and Dauer, W. T. (2005). *J. Cell Biol.* **168,** 855–862.
Goodchild, R. E., Kim, C. E., and Dauer, W. T. (2005). *Neuron* **48,** 923–932.
Gordon, L. B., Harten, I. A., Patti, M. E., and Lichtenstein, A. H. (2005). *J. Pediatr.* **146,** 336–341.
Gordon, L. B., McCarten, K. M., Giobbie-Hurder, A., Machan, J. T., Campbell, S. E., Berns, S. D., and Kieran, M. W. (2007). *Pediatrics* **120,** 824–833.
Goumans, M. J., and Mummery, C. (2000). *Int. J. Dev. Biol.* **44,** 253–265.
Granata, A., Watson, R., Collinson, L. M., Schiavo, G., and Warner, T. T. (2008). *J. Biol. Chem.* **283,** 7568–7579.
Gros-Louis, F., Dupre, N., Dion, P., Fox, M. A., Laurent, S., Verreault, S., Sanes, J. R., Bouchard, J. P., and Rouleau, G. A. (2007). *Nat. Genet.* **39,** 80–85.
Guelen, L., Pagie, L., Brasset, E., Meuleman, W., Faza, M. B., Talhout, W., Eussen, B. H., de Klein, A., Wessels, L., de Laat, W., and van Steensel, B. (2008). *Nature* **453,** 948–951.
Haque, F., Lloyd, D. J., Smallwood, D. T., Dent, C. L., Shanahan, C. M., Fry, A. M., Trembath, R. C., and Shackleton, S. (2006). *Mol. Cell Biol.* **26,** 3738–3751.
Heessen, S., and Fornerod, M. (2007). *EMBO Rep.* **8,** 914–919.
Hellemans, J., Debeer, P., Wright, M., Janecke, A., Kjaer, K. W., Verdonk, P. C., Savarirayan, R., Basel, L., Moss, C., Roth, J., David, A., De Paepe, A., Coucke, P., and Mortier, G. R. (2006). *Hum. Mutat.* **27,** 290.
Hellemans, J., Preobrazhenska, O., Willaert, A., Debeer, P., Verdonk, P. C., Costa, T., Janssens, K., Menten, B., Van Roy, N., Vermeulen, S. J., Savarirayan, R., Van Hul, W., *et al.* (2004). *Nat. Genet.* **36,** 1213–1218.
Hershkovitz, D., Amitai, B., and Sprecher, E. (2007). *Br. J. Dermatol.* **156,** 375–377.
Hewett, J. W., Tannous, B., Niland, B. P., Nery, F. C., Zeng, J., Li, Y., and Breakefield, X. O. (2007). *Proc. Natl. Acad. Sci. USA* **104,** 7271–7276.
Hickson, I. D. (2003). *Nat. Rev. Cancer* **3,** 169–178.
Hoffmann, K., Dreger, C. K., Olins, A. L., Olins, D. E., Shultz, L. D., Lucke, B., Karl, H., Kaps, R., Muller, D., Vaya, A., Aznar, J., Ware, R. E., *et al.* (2002). *Nat. Genet.* **31,** 410–414.
Hoffmann, K., Sperling, K., Olins, A. L., and Olins, D. E. (2007). *Chromosoma.*
Hoger, T. H., Grund, C., Franke, W. W., and Krohne, G. (1991). *Eur. J. Cell Biol.* **54,** 150–156.
Holaska, J. M., Rais-Bahrami, S., and Wilson, K. L. (2006). *Hum. Mol. Genet.* **15,** 3459–3472.
Holaska, J. M., and Wilson, K. L. (2006). *Anat. Rec. A Discov. Mol. Cell Evol. Biol.* **288,** 676–680.
Holaska, J. M., and Wilson, K. L. (2007). *Biochemistry* **46,** 8897–8908.
Holmer, L., Pezhman, A., and Worman, H. J. (1998). *Genomics* **54,** 469–476.
Holtz, D., Tanaka, R. A., Hartwig, J., and McKeon, F. (1989). *Cell* **59,** 969–977.
Ishimura, A., Ng, J. K., Taira, M., Young, S. G., and Osada, S. (2006). *Development* **133,** 3919–3928.
Itoh, S., Itoh, F., Goumans, M. J., and Ten Dijke, P. (2000). *Eur. J. Biochem.* **267,** 6954–6967.
Jacque, J. M., and Stevenson, M. (2006). *Nature* **441,** 641–645.

Johnson, B. R., Nitta, R. T., Frock, R. L., Mounkes, L., Barbie, D. A., Stewart, C. L., Harlow, E., and Kennedy, B. K. (2004). *Proc. Natl. Acad. Sci. USA* **101,** 9677–9682.

Kavanagh, D. M., Powell, W. E., Malik, P., Lazou, V., and Schirmer, E. C. (2007). *Subcell. Biochem.* **43,** 51–76.

Knockaert, M., Sapkota, G., Alarcon, C., Massague, J., and Brivanlou, A. H. (2006). *Proc. Natl. Acad. Sci. USA* **103,** 11940–11945.

Konstantinidou, A., Karadimas, C., Waterham, H. R., Superti-Furga, A., Kaminopetros, P., Grigoriadou, M., Kokotas, H., Agrogiannis, G., Giannoulia-Karantana, A., Patsouris, E., and Petersen, M. B. (2008). *Prenat. Diagn.* **28,** 309–312.

Kumaran, R. I., and Spector, D. L. (2008). *J. Cell Biol.* **180,** 51–65.

Lammerding, J., Fong, L. G., Ji, J. Y., Reue, K., Stewart, C. L., Young, S. G., and Lee, R. T. (2006). *J. Biol. Chem.* **281,** 25768–25780.

Lammerding, J., Hsiao, J., Schulze, P. C., Kozlov, S., Stewart, C. L., and Lee, R. T. (2005). *J. Cell Biol.* **170,** 781–791.

Lammerding, J., Schulze, P. C., Takahashi, T., Kozlov, S., Sullivan, T., Kamm, R. D., Stewart, C. L., and Lee, R. T. (2004). *J. Clin. Invest.* **113,** 370–378.

Lee, J. S., Hale, C. M., Panorchan, P., Khatau, S. B., George, J. P., Tseng, Y., Stewart, C. L., Hodzic, D., and Wirtz, D. (2007). *Biophys. J.* **93,** 2542–2552.

Lee, K. K., and Wilson, K. L. (2004). *Symp. Soc. Exp. Biol.* 329–339.

Lekstrom-Himes, J., and Xanthopoulos, K. G. (1999). *Blood* **93,** 3096–3105.

Lekstrom-Himes, J. A. (2001). *Stem Cells* **19,** 125–133.

Lelliott, C. J., Logie, L., Sewter, C. P., Berger, D., Jani, P., Blows, F., O'Rahilly, S., and Vidal-Puig, A. (2002). *J. Clin. Endocrinol. Metab.* **87,** 728–734.

Lersch, R., and Fuchs, E. (1988). *Mol. Cell Biol.* **8,** 486–493.

Lin, F., Morrison, J. M., Wu, W., and Worman, H. J. (2005). *Hum. Mol. Genet.* **14,** 437–445.

Lindberg, H. K., Wang, X., Jarventaus, H., Falck, G. C., Norppa, H., and Fenech, M. (2007). *Mutat. Res.* **617,** 33–45.

Liu, B., Wang, J., Chan, K. M., Tjia, W. M., Deng, W., Guan, X., Huang, J. D., Li, K. M., Chau, P. Y., Chen, D. J., Pei, D., Pendas, A. M., *et al.* (2005). *Nat. Med.* **11,** 780–785.

Liu, Y., Rusinol, A., Sinensky, M., Wang, Y., and Zou, Y. (2006). *J. Cell Sci.* **119,** 4644–4649.

Lloyd, D. J., Trembath, R. C., and Shackleton, S. (2002). *Hum. Mol. Genet.* **11,** 769–777.

Ma, Y., Cai, S., Lv, Q., Jiang, Q., Zhang, Q., Sodmergen, Z., Zhai, Z., and Zhang, C. (2007). *J. Cell Sci.* **120,** 520–530.

Malhas, A., Lee, C. F., Sanders, R., Saunders, N. J., and Vaux, D. J. (2007). *J. Cell Biol.* **176,** 593–603.

Mancini, M. A., Shan, B., Nickerson, J. A., Penman, S., and Lee, W. H. (1994). *Proc. Natl. Acad. Sci. USA* **91,** 418–422.

Marino, G., Ugalde, A. P., Salvador-Montoliu, N., Varela, I., Quiros, P. M., Cadinanos, J., van der Pluijm, I., Freije, J. M., and Lopez-Otin, C. (2008). *Hum. Mol. Genet.*

Markiewicz, E., Tilgner, K., Barker, N., van de Wetering, M., Clevers, H., Dorobek, M., Hausmanowa-Petrusewicz, I., Ramaekers, F. C., Broers, J. L., Blankesteijn, W. M., Salpingidou, G., Wilson, R. G., *et al.* (2006). *Embo. J.* **25,** 3275–3285.

Martin, G. M. (1989). *Genome* **31,** 390–397.

Martins, S. B., Eide, T., Steen, R. L., Jahnsen, T., Skalhegg, B. S., and Collas, P. (2000). *J. Cell Sci.* **113**(Pt 21), 3703–3713.

Melcon, G., Kozlov, S., Cutler, D. A., Sullivan, T., Hernandez, L., Zhao, P., Mitchell, S., Nader, G., Bakay, M., Rottman, J. N., Hoffman, E. P., and Stewart, C. L. (2006). *Hum. Mol. Genet.* **15,** 637–651.

Merideth, M. A., Gordon, L. B., Clauss, S., Sachdev, V., Smith, A. C., Perry, M. B., Brewer, C. C., Zalewski, C., Kim, H. J., Solomon, B., Brooks, B. P., Gerber, L. H., *et al.* (2008). *N. Engl. J. Med.* **358,** 592–604.

Morris, G. E. (2001). *Trends Mol. Med.* **7,** 572–577.

Moulson, C. L., Go, G., Gardner, J. M., van der Wal, A. C., Smitt, J. H., van Hagen, J. M., and Miner, J. H. (2005). *J. Invest. Dermatol.* **125,** 913–919.

Mounkes, L. C., Kozlov, S., Hernandez, L., Sullivan, T., and Stewart, C. L. (2003). *Nature* **423,** 298–301.

Mounkes, L. C., Kozlov, S. V., Rottman, J. N., and Stewart, C. L. (2005). *Hum. Mol. Genet.* **14,** 2167–2180.

Muchir, A., Bonne, G., van der Kooi, A. J., van Meegen, M., Baas, F., Bolhuis, P. A., de Visser, M., and Schwartz, K. (2000). *Hum. Mol. Genet.* **9,** 1453–1459.

Muchir, A., Pavlidis, P., Bonne, G., Hayashi, Y. K., and Worman, H. J. (2007a). *Hum. Mol. Genet.* **16,** 1884–1895.

Muchir, A., Pavlidis, P., Decostre, V., Herron, A. J., Arimura, T., Bonne, G., and Worman, H. J. (2007b). *J. Clin. Invest.* **117,** 1282–1293.

Mulky, A., Cohen, T. V., Kozlov, S. V., Korbei, B., Foisner, R., Stewart, C. L., and KewalRamani, V. N. (2008). *J. Virol.* **82,** 5860–5868.

Mummery, C. L. (2001). *Microsc. Res. Tech.* **52,** 374–386.

Muntoni, F., Bonne, G., Goldfarb, L. G., Mercuri, E., Piercy, R. J., Burke, M., Yaou, R. B., Richard, P., Recan, D., Shatunov, A., Sewry, C. A., and Brown, S. C. (2006). *Brain* **129,** 1260–1268.

Naetar, N., Korbell, B., Kozlov, S., Kerenyi, M. A., Dorner, D., Kral, R., Gotic., Fuchs, P., Cohen, T., Bittner, R. Stewart, C. L., and Foisner, R. (2008). Loss of nucleoplasmic LAP2alpha-lamin A complexes causes erythroid and epidermal progenitor hyperproliferation. *Nat. Cell Biol.* Oct 12. [Epub ahead of print]

Naismith, T. V., Heuser, J. E., Breakefield, X. O., and Hanson, P. I. (2004). *Proc. Natl. Acad. Sci. USA* **101,** 7612–7617.

Navarro, C. L., Cadinanos, J., De Sandre-Giovannoli, A., Bernard, R., Courrier, S., Boccaccio, I., Boyer, A., Kleijer, W. J., Wagner, A., Giuliano, F., Beemer, F. A., Freije, J. M., *et al.* (2005). *Hum. Mol. Genet.* **14,** 1503–1513.

Nikolakaki, E., Meier, J., Simos, G., Georgatos, S. D., and Giannakouros, T. (1997). *J. Biol. Chem.* **272,** 6208–6213.

Nikolakaki, E., Simos, G., Georgatos, S. D., and Giannakouros, T. (1996). *J. Biol. Chem.* **271,** 8365–8372.

Nikolova, V., Leimena, C., McMahon, A. C., Tan, J. C., Chandar, S., Jogia, D., Kesteven, S. H., Michalicek, J., Otway, R., Verheyen, F., Rainer, S., Stewart, C. L., *et al.* (2004). *J. Clin. Invest.* **113,** 357–369.

Novelli, G., Muchir, A., Sangiuolo, F., Helbling-Leclerc, A., D'Apice, M. R., Massart, C., Capon, F., Sbraccia, P., Federici, M., Lauro, R., Tudisco, C., Pallotta, R., *et al.* (2002). *Am. J. Hum. Genet.* **71,** 426–431.

Osada, S., Ohmori, S. Y., and Taira, M. (2003). *Development* **130,** 1783–1794.

Oshima, J. (2000). *Bioessays* **22,** 894–901.

Ostlund, C., Bonne, G., Schwartz, K., and Worman, H. J. (2001). *J. Cell Sci.* **114,** 4435–4445.

Ostlund, C., Sullivan, T., Stewart, C. L., and Worman, H. J. (2006). *Biochemistry* **45,** 1374–1382.

Ozaki, T., Saijo, M., Murakami, K., Enomoto, H., Taya, Y., and Sakiyama, S. (1994). *Oncogene* **9,** 2649–2653.

Ozawa, R., Hayashi, Y. K., Ogawa, M., Kurokawa, R., Matsumoto, H., Noguchi, S., Nonaka, I., and Nishino, I. (2006). *Am. J. Pathol.* **168,** 907–917.

Waterham, H. R., Koster, J., Mooyer, P., Noort Gv, G., Kelley, R. I., Wilcox, W. R., Wanders, R. J., Hennekam, R. C., and Oosterwijk, J. C. (2003). *Am. J. Hum. Genet.* **72,** 1013–1017.

Wheeler, M. A., Davies, J. D., Zhang, Q., Emerson, L. J., Hunt, J., Shanahan, C. M., and Ellis, J. A. (2007). *Exp. Cell Res.* **313,** 2845–2857.

Wilhelmsen, K., Litjens, S. H., Kuikman, I., Tshimbalanga, N., Janssen, H., van den Bout, I., Raymond, K., and Sonnenberg, A. (2005). *J. Cell Biol.* **171,** 799–810.

Wolf, C. M., Wang, L., Alcalai, R., Pizard, A., Burgon, P. G., Ahmad, F., Sherwood, M., Branco, D. M., Wakimoto, H., Fishman, G. I., See, V., Stewart, C. L., *et al.* (2008). *J. Mol. Cell Cardiol.* **44,** 293–303.

Worman, H. J., and Bonne, G. (2007). *Exp. Cell Res.* **313,** 2121–2133.

Worman, H. J., and Courvalin, J. C. (2005). *Int. Rev. Cytol.* **246,** 231–279.

Worman, H. J., Yuan, J., Blobel, G., and Georgatos, S. D. (1988). *Proc. Natl. Acad. Sci. USA* **85,** 8531–8534.

Yang, S. H., Bergo, M. O., Toth, J. I., Qiao, X., Hu, Y., Sandoval, S., Meta, M., Bendale, P., Gelb, M. H., Young, S. G., and Fong, L. G. (2005). *Proc. Natl. Acad. Sci. USA* **102,** 10291–10296.

Yang, S. H., Meta, M., Qiao, X., Frost, D., Bauch, J., Coffinier, C., Majumdar, S., Bergo, M. O., Young, S. G., and Fong, L. G. (2006). *J. Clin. Invest.* **116,** 2115–2121.

Yang, S. H., Qiao, X., Fong, L. G., and Young, S. G. (2008). *Biochim. Biophys. Acta* **1781,** 36–39.

Ye, Q., and Worman, H. J. (1996). *J. Biol. Chem.* **271,** 14653–14656.

Yokoi, F., Dang, M. T., Mitsui, S., Li, J., and Li, Y. (2008). *J. Biochem.* **143,** 39–47.

Young, S. G., Fong, L. G., and Michaelis, S. (2005). *J. Lipid. Res.* **46,** 2531–2558.

Zhang, Q., Bethmann, C., Worth, N. F., Davies, J. D., Wasner, C., Feuer, A., Ragnauth, C. D., Yi, Q., Mellad, J. A., Warren, D. T., Wheeler, M. A., Ellis, J. A., *et al.* (2007a). *Hum. Mol. Genet.* **16,** 2816–2833.

Zhang, X., Xu, R., Zhu, B., Yang, X., Ding, X., Duan, S., Xu, T., Zhuang, Y., and Han, M. (2007b). *Development* **134,** 901–908.

Zhao, Y., DeCuypere, M., and LeDoux, M. S. (2008). *Exp. Neurol.* **210,** 719–730.

CHAPTER EIGHT

Mouse Models for Human Hereditary Deafness

Michel Leibovici, Saaid Safieddine, *and* Christine Petit

Contents

Abstract

Hearing impairment is a frequent condition in humans. Identification of the causative genes for the early onset forms of isolated deafness began 15 years ago and has been very fruitful. To date, approximately 50 causative genes have been identified. Yet, limited information regarding the underlying pathogenic mechanisms can be derived from hearing tests in deaf patients. This chapter describes the success of mouse models in the elucidation of some pathophysiological processes in the auditory sensory organ, the cochlea. These models have revealed a variety of defective structures and functions at the origin of

Institut Pasteur, Unité de Génétique et Physiologie de l'Audition, 25 rue du Dr. Roux, F75015 Paris, France; Inserm UMRS587, Collège de France, UPMC University, Paris, France

Current Topics in Developmental Biology, Volume 84
ISSN 0070-2153, DOI: 10.1016/S0070-2153(08)00608-X

deafness genetic forms. This is illustrated by three different examples: (1) the DFNB9 deafness form, a synaptopathy of the cochlear sensory cells where otoferlin is defective; (2) the Usher syndrome, in which deafness is related to abnormal development of the hair bundle, the mechanoreceptive structure of the sensory cells to sound; (3) the DFNB1 deafness form, which is the most common form of inherited deafness in Caucasian populations, mainly caused by connexin-26 defects that alter gap junction communication between nonsensory cochlear cells.

Abbreviations

Cx, CX	connexin
dB	decibels
ENU	*N*-ethyl-*N*-nitrosourea
HL	hearing level
IHC	inner hair cell
OAE	otoacoustic emissions
OHC	outer hair cell
SPL	sound pressure level

1. Introduction

Deafness is the most frequent sensory defect. Its overall impact is determined by the severity of the hearing impairment and its onset. Children affected by severe to profound congenital deafness will encounter major obstacles for oral language acquisition, whereas ageing people becoming severely deaf suffer from loss of social communication. Congenital deafness affects 1 out of 700–1000 newborns, while about 30% of the population over 60 years of age suffer from hearing loss impeding conversational exchange, mainly due to presbycusis.

Efforts to identify deafness genes (Petit, 1996) have established that gene defects are the major cause of childhood and young adult severe to profound hearing impairment in developed countries (Denoyelle *et al.*, 1999).

In this chapter, we discuss the irreplaceable contribution of mouse models to the deciphering of the pathogenic processes underlying human hereditary deafness forms. Following an overview on sound processing in the peripheral auditory system, and a brief presentation of inherited deafness and hearing tests in humans and mouse, we then discuss how mouse models have enlightened the pathophysiology of three different forms of congenital deafness.

2. The Mammalian Peripheral Auditory System

The three components shaping the mammalian ear are the air-filled external and middle ears, and the fluid-filled inner ear. The inner ear consists of two sensory organs: the vestibule, the balance organ, and the cochlea, the hearing organ (Fig. 8.1A). The cochlea and its afferent innervation both derive from the otic placode during embryogenesis and together make up the peripheral auditory system. The cochlea is composed of three fluid-filled canals (Fig. 8.1B) that spiral up together in approximately two and a half turns in humans: the scala vestibuli, the scala tympani, and in between the scala media. The scala vestibuli and scala tympani are filled with perilymph, a liquid of ionic composition similar to cerebrospinal fluid (Na^+ rich, K^+ poor); they communicate at the cochlea's apex by the helicotrema, and are separated from the middle ear by the oval and round windows, respectively. The scala media is limited by the sensory epithelium at the bottom, the Reissner membrane at the top, and the stria vascularis on the side. The scala media is closed at both extremities and contains the endolymph, a liquid of unusual ionic composition, close to that of intracellular fluid (K^+ rich, Na^+ poor). The auditory sensory epithelium (organ of Corti) is housed in the membranous labyrinth (cochlear duct). (Fig. 8.1C).

The organ of Corti amplifies the mechanical input signal evoked by sound and transduces it in an output electrical signal transferred to the brain. Sound processing in the peripheral auditory system relies on various functions: (1) the frequency analysis of sound waves, based on the systematic variation of mechanical and electrical properties of the organ of Corti along the cochlea; (2) the mechanoelectrical transduction operated by the stereociliary bundles of cochlear sensory cells (hair cells) (Fig. 8.1D); (3) the amplification of sound-evoked vibrations operated by one category of hair cells, the outer hair cells (OHCs); (4) the neurotransmitter release by the second category of hair cells, the inner hair cells (IHCs), which results in the production of action potentials by the primary auditory neurons; (5) the cochlear homeostasis (specifically the ionic homeostasis) that relies on various cochlear structures, including the stria vascularis. Affecting any one of these processes may result in hearing impairment.

2.1. Frequency analysis of sound waves in the cochlea

The organ of Corti is delineated by the basilar membrane on which it sits, and by the tectorial membrane, which overlies it. The organ itself is composed of two sensory cell types, IHCs and OHCs, and many different types of supporting cells (Fig. 8.1D). IHCs are organized in one row, whereas the OHCs form three rows, each row running the length of the

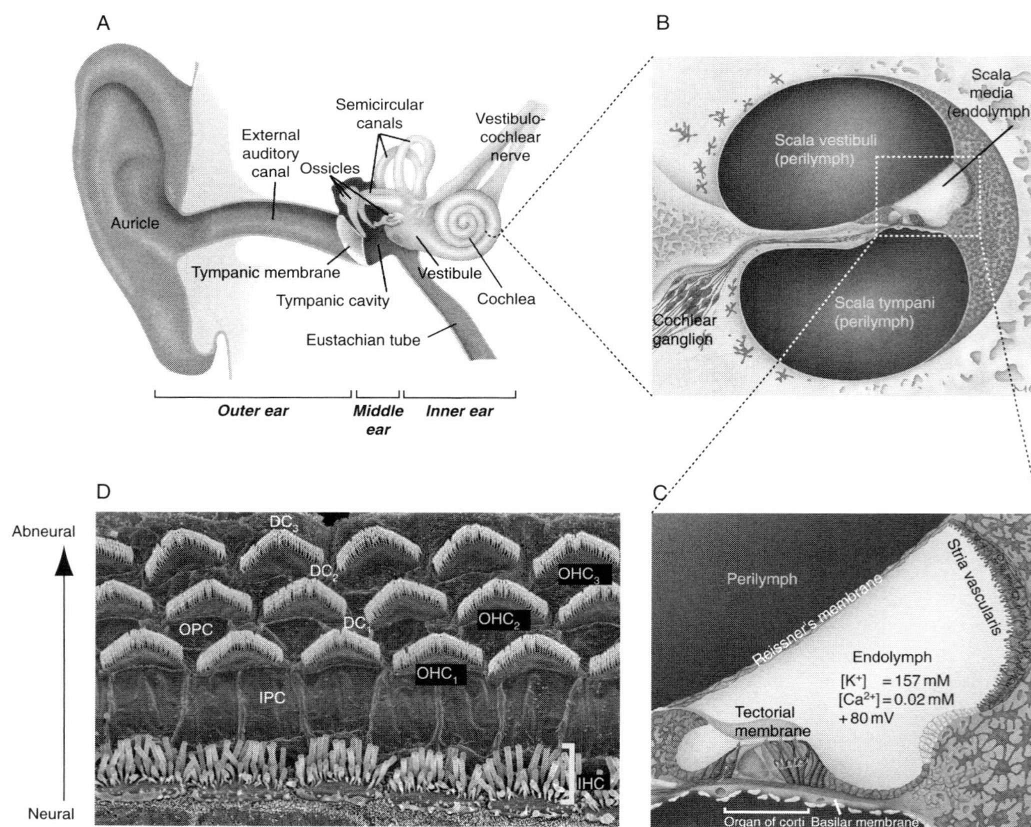
A
Auricle
External auditory canal
Ossicles
Semicircular canals
Vestibulo-cochlear nerve
Tympanic membrane
Tympanic cavity
Vestibule
Cochlea
Eustachian tube
Outer ear
Middle ear
Inner ear
B
Scala media (endolymph)
Scala vestibuli (perilymph)
Scala tympani (perilymph)
Cochlear ganglion
C
Perilymph
Reissner's membrane
Stria vascularis
Endolymph
[K+] = 157 mM
[Ca2+] = 0.02 mM
+80 mV
Tectorial membrane
Organ of corti
Basilar membrane
D
Abneural
Neural
DC3
DC2
DC1
OPC
IPC
OHC1
OHC2
OHC3
IHC

cochlea. Because of the nonpermeant tight junctions between epithelial cells, the apical surfaces of the hair cells are bathed by endolymph, while their basolateral surfaces are bathed by perilymph. All along the longitudinal cochlear axis from the base to the apex, the thickness and stiffness of the basilar and tectorial membranes vary gradually (for a review, see Raphael and Altschuler, 2003). In parallel, the length and stiffness of the OHC lateral walls and hair bundles change. These variations constrain the cochlear vibrations in response to a sound of given frequency to develop at a particular emplacement along the cochlear axis. The basilar membrane plays an essential role in this process, by acting as a frequency analyzer that decomposes complex sounds into their pure tone constituents and targets each frequency to a characteristic place along the cochlea (Robles and Ruggero, 2001; von Békésy, 1960). A tonotopic map is thus established, making the organ of Corti respond to high-frequency sounds at the cochlea's base and to low-frequency sounds at its apex.

2.2. Mechanoelectrical transduction by the hair bundle

Each IHC or OHC is crowned by a unique array of thick and stiff microvilli named stereocilia (or stereovilli) that project a few microns from their apical surface and form the so-called hair bundle (Figs. 8.1D and 8.2). The hair

---

Figure 8.1 The mammalian ear is composed of three compartments: outer, middle, and inner ear. (A) The outer ear collects incoming sound vibration, while the ossicles of the middle ear transmit the vibration of the tympanic membrane to the inner ear, a fluid-filled organ. The auditory organ of the inner ear, a coiled-shape duct named cochlea, transduces sound waves into nerve impulses. (B) A cross section through the cochlear duct shows the three fluid chambers filled with perilymph (scala vestibuli and scala tympani), or endolymph, an extracellular fluid of unusual ion composition, with high K^+ (~157 mM) and low Ca^{2+} (~0.02 mM) concentrations (scala media). Tight junctions between cells of the scala media prevent ion leaking between the perilymphatic and endolymphatic compartments. (C) An enlargement of the scala media displays the main structures responsible for sound perception. The stria vascularis is a secretory epithelium that accounts for the high K^+ ion concentration of the endolymph and the positive potential of the scala media (+80 mV to +100 mV) compared to scalae tympani and vestibuli. The organ of Corti is the sensory epithelium of the cochlea, which contains outer (green) and inner (red) sensory cells as well as supporting cells. The organ of Corti is sandwiched between the tectorial membrane at the top and the basilar membrane (blue) at the bottom. (D) A scanning electron micrograph of the organ of Corti (without the tectorial membrane) shows the three rows of outer hair cells (OHC) and the single row of inner hair cells (IHC) with their apical hair bundles. OHCs and their supporting cells (Deiters cells, DC) are denoted from 1 to 3 from the neural to the abneural edge of the organ of Corti. Inner pillar cells (IPC) and outer pillar cells (OPC) separate the IHC row from the three OHC rows. Note that all hair bundles are orientated in the same direction (vertices of the "V"-shaped hair bundles pointing to the abneural edge), so that the sound pressure wave produces a coherent mechanical stimulation of the hair bundles. (See Color Insert.)

bundle is an asymmetric structure, both morphologically and functionally polarized: the stereocilia are arranged in rows of increasing heights, the bundle responding optimally to deflections directed to and away from the tallest row of stereocilia. Two types of fibrous links bridge stereocilia in the mature hair bundle: the tip link and a set of lateral and apical links (top connectors) (Fig. 8.2B). The tip links connect adjacent stereocilia of different rows and are required for the opening and closure of the mechanoelectrical transduction channels. According to the "gating spring" model, the currently most widely accepted model of mechanoelectrical transduction (Corey and Hudspeth, 1983; Furness and Hackney, 1985; Howard and Hudspeth, 1988; Pickles *et al.*, 1984), each tip link is connected to a mechanoelectrical transduction channel at one or both of its ends. Deflection of the hair bundle toward the tallest row of stereocilia (excitatory direction) leads to tip link tensing and opening of mechanoelectrical transduction channels (Figs. 8.2B and 8.4A), resulting in a cationic influx (carried mainly by K^+, but also by Ca^{2+} ions) that depolarizes the hair cell. The lateral apical links join neighboring stereocilia within and between rows (Furness and Hackney, 1985). The hair bundle works as a single functional unit. A force applied anywhere in the hair bundle results in a coherent

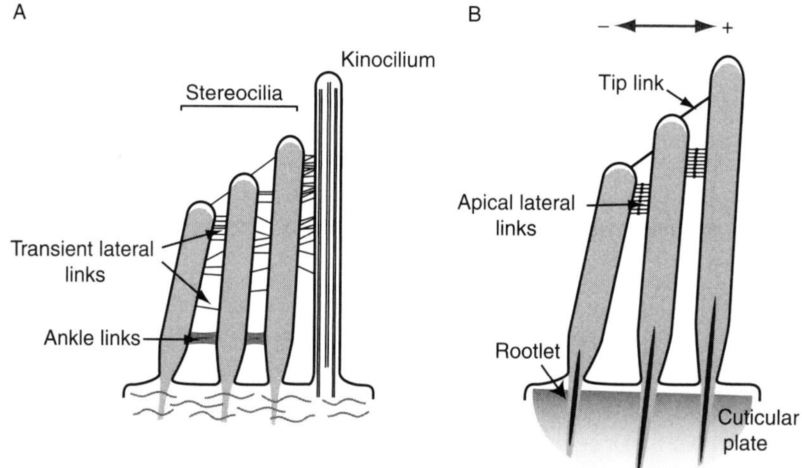

Figure 8.2 The hair bundle links. (A) Immature hair bundles, between postnatal days 2 and 5 in the mouse, have three types of links that are, transient lateral links connecting stereocilia together and with the kinocilium (the transient primary cilium), ankle links connecting stereocilia at their base, and tip links that connect the tip of each stereocilium with the side of the adjacent taller one. Of these, only the tip links persist in the mature hair bundle. (B) In the mature hair bundle, the kinocilium has disappeared and two different types of links are observed. The tip link is believed to gate the mechanotransduction channel. The role of the apical lateral links (also called horizontal top connectors) is still poorly understood. Double headed arrow indicates the direction of the stimulus.

displacement of all stereocilia in the bundle, down to the level of thermal noise (Kozlov *et al.*, 2007). The hair bundle is a mechanical receptor structure, capable of converting nanometer stereocilia displacements induced by sound waves into measurable membrane potential changes. This conversion process occurs in less than a few microseconds, and thus does not involve any enzymatic reaction. Hair cell membrane depolarization leads to different outcomes in OHCs and IHCs.

2.3. Sound amplification by outer hair cells

OHCs amplify the incoming sound stimulation and enhancing frequency selectivity of the cochlear response. When OHCs are affected, the hearing thresholds are elevated and the tuning abilities of the cochlea are reduced. In response to sound waves, the basilar membrane moves up and down leading to an oscillatory motion of the organ of Corti. During the upward phase of a stimulation cycle, the OHC hair bundle, whose tallest stereocilia are anchored in the tectorial membrane, is deflected toward the abneural edge of the organ of Corti (excitatory direction), making the mechanoelectrical transduction channels to open. In the downward phase, the stereocilia are pushed back beyond their equilibrium state, which results in the closure of mechanoelectrical transduction channels. While the IHCs are the genuine auditory sensory cells, stimulation of their hair bundles resulting in neurotransmitter release, OHC hair bundle stimulation triggers a unique property of OHCs to locally enhance the sound-induced vibrations of the basilar membrane, in a positive feedback process referred to as the cochlear amplifier (Dallos, 1992; Davis, 1983; Gold, 1948). The amplification ensured by OHCs is a dynamic mechanism. Whether it relies on somatic electromotility, that is, the contraction–decontraction of the OHC body lateral wall driven by the sound-induced depolarization–hyperpolarization cycle (Ashmore, 1987; Brownell *et al.*, 1985; Dallos *et al.*, 2008; He *et al.*, 2004; Liberman *et al.*, 2002; Mellado Lagarde *et al.*, 2008 (Fig. 8.3)), or an "active" movement of the OHC hair bundles (Chan and Hudspeth, 2005; Kennedy *et al.*, 2005; Liberman *et al.*, 2004), or both, is still under debate (Furness *et al.*, 2008; Kennedy *et al.*, 2006); for reviews see Ashmore (2008), Fettiplace and Hackney (2006), Ren and Gillespie (2007). This amplification contributes to the high sensitivity of the cochlea in response to very low intensity sounds, that is, close to the thermal noise. It operates only for sounds with intensity inferior to 60–70 dB sound pressure level (SPL) and works nonlinearly (Robles and Ruggero, 2001), such as the lower the sound intensity, the higher the amplification gain. In addition, the OHC amplification response is highly tuned, developing only within a narrow region of the cochlea around the characteristic place corresponding to a given frequency. It thus sharpens the frequency discrimination operated by the basilar membrane and the response of the IHCs. Alteration of

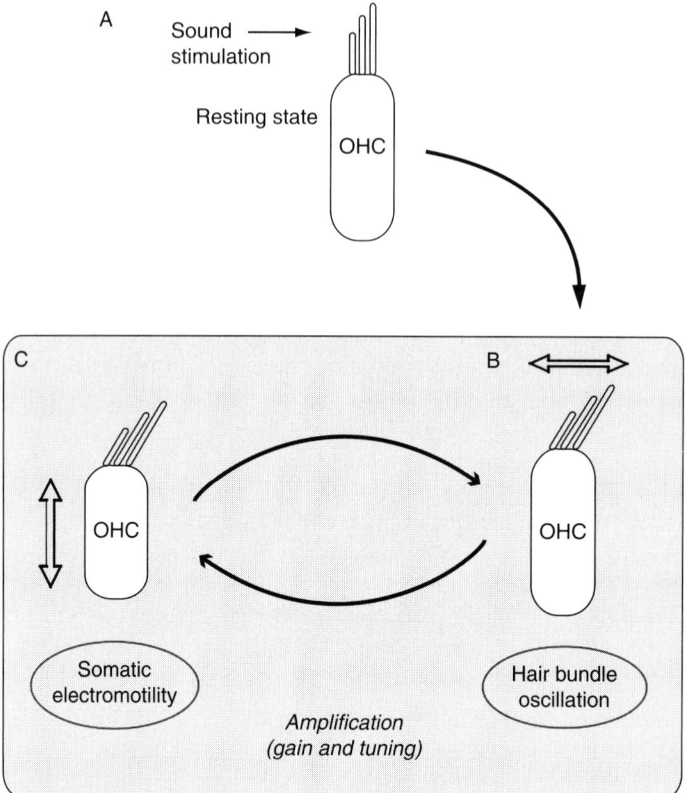

Figure 8.3 Schematic representation of the OHC amplification process. (A) Upon sound stimulation, the vibration of the basilar membrane underlying the organ of Corti produces a shearing motion between OHC hair bundles and the tectorial membrane, leading to hair bundle periodic deflection (B) that in turn periodically opens mechanotransduction channels, causing K^{+} ion influx, and therefore cell depolarization. Cell depolarization produces conformational changes of an abundant protein of the OHC lateral wall (prestin), resulting in periodic shortening and elongation of the cell, a process known as electromotility (C). OHC electromotility contributes to a positive feedback loop that locally enhances the motion of the basilar membrane leading to amplification. When sound stimulation stops, OHCs exit the amplification cycle and return to their resting state (A).

structures involved in cochlear amplification result in moderate to severe hearing impairment. In patients carrying mutations exclusively affecting tectorial membrane components (Mustapha *et al.*, 1999; Verhoeven *et al.*, 1998) and in the mutant mice defective for these tectorial membrane proteins or for prestin, an integral membrane protein of the OHC lateral wall that is critically implicated in OHC electromotility (Mustapha *et al.*, 1999; Verhoeven *et al.*, 1998); reviewed by (Friedman *et al.*, 2007), the shift of the hearing threshold indeed usually does not exceed 60–70 dB. In contrast, IHC or auditory nerve defects can result in profound deafness (see Section 4.1).

2.4. Neurotransmission at the inner hair cell synapses

Stimulated IHCs respond to membrane depolarization by releasing neurotransmitter at their basal pole. In contrast to OHCs, the IHCs are almost completely entrenched with their supporting cells, and are located above a region of the basilar membrane that does not vibrate in response to sound. In addition, their hair bundles are freestanding in the subtectorial space, while the tips of the tallest OHC stereocilia are embedded in the tectorial membrane (Kimura, 1966). As a result, IHC hair bundles are thought to be deflected by the radial motion of the endolymphatic fluid trapped between the tectorial membrane and the cell apical surface (Nowotny and Gummer, 2006). The mechanoelectrical transduction current depolarizes the cell, thereby opening basolateral voltage-gated Ca^{2+} channels. The resulting Ca^{2+} influx at the synaptic active zones in turn triggers synaptic vesicle exocytosis, which releases glutamate onto glutamate receptors present on the dendritic end of the primary auditory neurons (Glowatzki and Fuchs, 2002) (Fig. 8.4A).

The IHC synapse has unique structural features and functional characteristics. This synapse is equipped with a peculiar structural specialization, the synaptic ribbon, an electron-dense structure of submicron diameter that marks the center of the synaptic active zone. The ribbon is surrounded by synaptic vesicles attached to it or located in its close vicinity (Liberman, 1980; Sobkowicz *et al.*, 1986) (Fig. 8.4C). The IHC synapse contains many proteins involved in synaptic vesicular trafficking and fusion described in conventional synapses (including syntaxin, SNAP25, and VAMP), but apparently lacks synaptotagmin I (Safieddine and Wenthold, 1999; Wenthold *et al.*, 2002). In contrast to conventional synapses, where neurotransmitter release occurs upon the production of action potentials, mechanoelectrical transduction currents do not trigger firing of action potentials in the IHCs. Instead, the IHC responds to sound stimulation by a finely graded plasma membrane depolarization, which results in a graded neurotransmitter release at a rate varying in proportion to the stimulus intensity (Fuchs, 2005; Juusola *et al.*, 1996). The IHC is also specialized to maintain a high level of tonic neurotransmitter release over long time periods of sustained sound stimulation (Beutner *et al.*, 2001; Griesinger *et al.*, 2005).

Each IHC synapses with 10–30 afferent auditory neurons. These are bipolar neurons with different firing rates and thresholds (Liberman, 1980; Merchan-Perez and Liberman, 1996). Thus, the output of each IHC is sampled by different auditory neurons, which independently encode partial information about the sound frequency and intensity, and collectively provide a faithful representation of sound characteristics. The frequency information is directly represented by the firing rate of the auditory neurons (only up to sound frequencies of about 3 kHz). The dendrites of the auditory neurons synapse on the IHC, their central axons (that make up the cochlear nerve) project to the cochlear nucleus in the brainstem, and

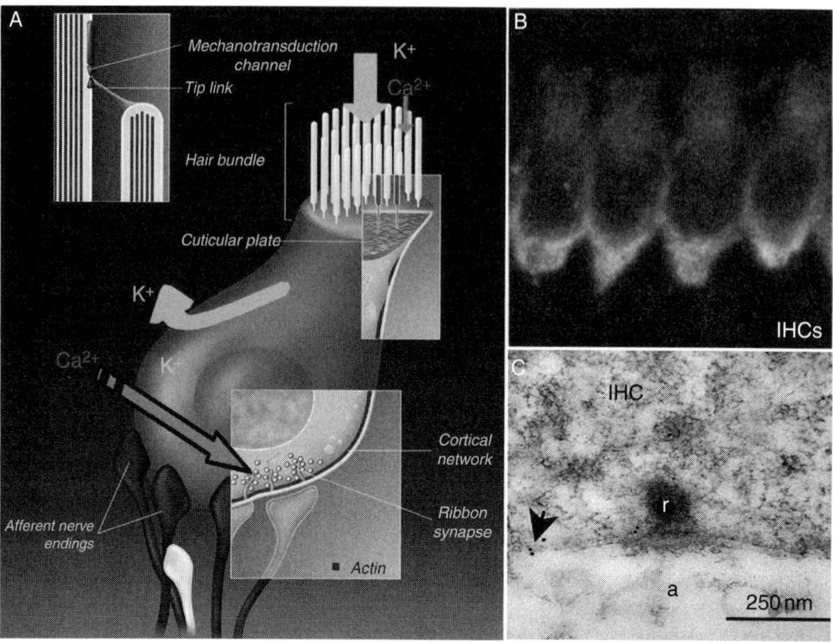

Figure 8.4 The IHC. (A) Schematic representation of the IHC and its innervation. Deflection of the hair bundle opens the mechanotransduction channels. Calcium channels open upon cell depolarization resulting in Ca^{2+} influx at the active zone where otoferlin is enriched, as shown in IHC stained for otoferlin (B and C). The Ca^{2+} ions bind to the calcium sensor, probably otoferlin, thereby triggering the fusion of synaptic vesicles with the presynaptic plasma membrane and neurotransmitter release onto the glutamate receptors located on the auditory nerve fibers. (B) Immunogold electron microscopy localized otoferlin around the ribbon (r) facing and afferent nerve fiber (a) and to the presynaptic plasma membrane (arrowhead). (See Color Insert.)

their cell bodies compose the cochlear ganglion. Auditory information is thereafter processed in the central auditory system that includes the superior olive, the inferior colliculus, the medial geniculate body of the thalamus, and the primary auditory cortex in the temporal lobe of the brain (Rouiller, 1997). Each of the central auditory structures is tonotopically organized, thereby preserving the cochlear tonotopic information all along the auditory pathway.

2.5. K^+ secretion in the endolymph and production of the endocochlear potential by the stria vascularis

A +80 to +100 mV transepithelial electrical potential difference, named the endolymphatic or endocochlear potential, exists between the endolymphatic and perilymphatic cochlear compartments. Both the production of the endocochlear potential and K^+ secretion in the endolymph are achieved by the stria vascularis, a bilayered highly vascularized epithelium enclosing an intrastrial fluid space (Fig. 8.5). The capillary network found in the stria vascularis is the source of oxygen and glucose required for K^+ secretion and endocochlear potential production. The spatial separation between the stria vascularis and the organ of Corti has the benefit of isolating the hair cells from the "noise" associated with blood flow in the capillaries

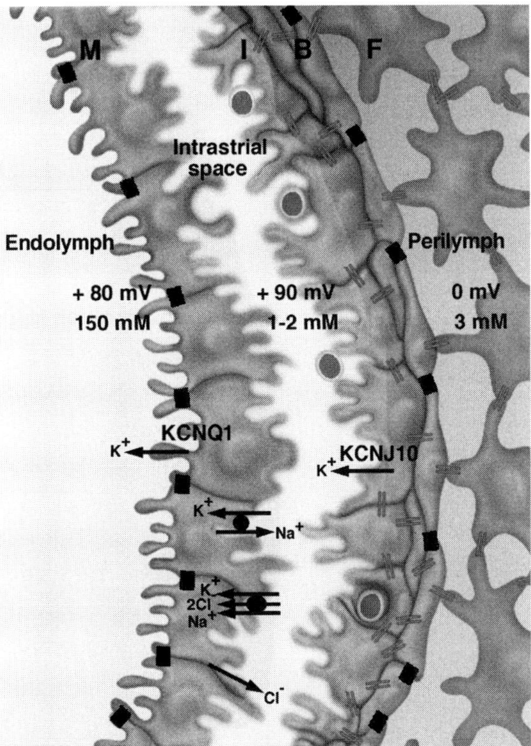

Figure 8.5 The stria vascularis and its electrogenic machinery: the stria vascularis consists of two epithelial cell layers; the basal (B) and intermediate (I) cell layer faces the spiral ligament connective tissue, and the marginal cell layer (M) faces the cochlear duct filled with endolymph. These layers limit the intrastrial space, an extracellular fluid compartment that contains a dense capillary network (red circles). Tight junctions between marginal cells and between basal cells are denoted by black boxes. Gap junctions (green double bars) connect intermediate and basal cells of the stria vascularis, and fibrocytes (F) of the spiral ligament. The endocochlear potential is generated across the basal and intermediate cell layer. The K^+ concentration and the electric potential are indicated in the three extracellular fluid spaces, namely the perilymph, intrastrial fluid, and endolymph. Each electric potential is indicated relative to that of the perilymph, which is taken as reference (0 mV). Adapted from (Cohen-Salmon *et al.*, 2007), Copyright 2007 National Academy of Sciences, U.S.A. (See Color Insert.)

(Wangemann, 2006). The endocochlear potential has a considerable impact on sensory transduction as it accounts for approximately half of the driving force for K^+ and Ca^{2+} entry in hair cells during the mechanoelectrical transduction process. Molecular mechanisms involved in endocochlear potential production, in K^+ secretion and K^+ reabsorption implicate various K^+ channels, including Kcnj10, Kcnq1, and Kcnq4, and cotransporters, including Slc12a2, KCC3, and KCC4 (Fig. 8.5). When the corresponding genes are mutated in humans or in mice, this leads to hearing impairment due to alterations of cochlear ionic homeostasis (reviewed by Lang *et al.*, 2007; Wangemann, 2006). Another important, still under debate, aspect of cochlear homeostasis is the routing of K^+ ions inside the cochlea. It has been suggested that gap junctions play a role in K^+ ions circulation within the cochlea. Two gap junction networks have been described in the cochlea, an epithelial network and a connective network. The epithelial network includes nonsensory cells of the organ of Corti and cells lying laterally (toward the spiral ligament) and medially (toward the inner sulcus), whereas the connective tissue network consists of the spiral ligament fibrocytes and stria vascularis basal and intermediate cells (Ahmad *et al.*, 2003; Forge *et al.*, 2003; Kikuchi *et al.*, 1995; Lautermann *et al.*, 1998). Gap junctions have not been detected in hair cells or in marginal cells of the stria vascularis. Within the two gap junction networks, connexin-26 (Cx26) and connexin-30 (Cx30), the major connexins in the cochlea, are colocalized and make hybrid gap junction channels that probably display different permeation functions according to the Cx26/Cx30 ratio (Forge *et al.*, 2003; Sun *et al.*, 2005; Yum *et al.*, 2007). In view of the localization of gap junctions in the cochlea, the influx of endolymphatic K^+ ions into hair cells that occurs during mechanoelectrical transduction, and their efflux in the perilymph at the base of these cells, together with the constant secretion of K^+ in the endolymph by the stria vascularis, have led to suggest that cochlear gap junction networks provide an intracellular "recycling" pathway for K^+ ions (Kikuchi *et al.*, 1995, 2000; Spicer and Schulte, 1996). It has been postulated that supporting cells uptake and funnel K^+ away from the body of sensory cells, a process referred as "spatial buffering." Direct evidence for spatial buffering of K^+ ions and K^+ intracellular recycling pathway through gap junctions, however, is wanting.

Fine assessment of each of the earlier cited functions of the cochlea is critical to investigate the pathogenesis of hearing impairments. For this purpose, clinical tests have been developed. After a brief overview on human hereditary deafness, we focus on these noninvasive tests and discuss why mouse models for human deafness are so critical to grasp significant knowledge on human deafness pathogenesis.

Table 8.1 Genes underlying isolated deafness as a result of primary defects in hair cells, nonsensory cells, the tectorial membrane, or unknown cell type, and corresponding mouse mutants

Primary defect	Gene	Gene product	Forms of human deafness	Mouse mutants
Hair cells	*MYO7A*	Myosin VIIA (motor protein)	DFNB2 ± retinopathy[a] (Usher 1B); DFNA11	Shaker-1 (*sh1*); Headbanger (*hdb*)
	MYO15	Myosin XV (motor protein)	DFNB3	Shaker-2 (*sh2*)
	MYO6	Myosin VI (motor protein)	DFNA22 ± cardiomyopathy[a]; DFNB37	Snell's waltzer (*sv*); *Twist* (*Twt*)
	MYO3A	Myosin IIIA (motor protein)	DFNB30	
	MYO1A	Myosin IA (motor protein)	DFNA48	
	MYH9	Myosin IIA (motor protein)	DFNA17 ± giant platelets[a] (Fechtner)	*Myh9*$^{+/-b,\ e}$
	ACTG1	γ-Actin (cytoskeletal protein)	DFNA20/26	
	ESPN	Espin (actin-bundling protein)	DFNB36; DFNA$_i$[d]	Jerker (*je*)
	RDX	Radixin (actin-binding protein)	DFNB24	*Rdx*$^{-/-b}$
	TRIOBP	TRIOBP (actin-binding protein)	DFNB28	
	HDIA1	Diaphanous-1 (formin-homologous cytoskeleton regulatory protein)	DFNA1	
	CLDN14	Claudin-14 (tight junction protein)	DFNB29	*Cldn14*$^{-/-b}$
	TRIC	Tricellulin (tight junction protein)	DFNB49	
	USH1C			

(*continued*)

Table 8.1 *(continued)*

Primary defect	Gene	Gene product	Forms of human deafness	Mouse mutants
		Harmonin (PDZ domain-containing protein)	DFNB18 ± retinopathy[a] (Usher 1C)	Deaf circler (*dfcr*); *Ush1c*$^{-/-b}$
	WHRN	Whirlin (PDZ domain-containing protein)	DFNB31 ± retinopathy[a] (Usher 2D)	Whirler (*wi*)
	CDH23	Cadherin-23 (integral membrane adhesion protein)	DFNB12 ± retinopathy[a] (Usher 1D)	Waltzer (*v*)
	PCDH15	Protocadherin-15 (integral membrane adhesion protein)	DFNB23 ± retinopathy[a] (Usher 1F)	Ames waltzer (*av*)
	TMIE	TMIE (transmembrane domain-containing protein)	DFNB6	Spinner (*sr*); Circling (*Cir*)
	STRC	Stereocilin (extracellular protein)	DFNB16	*Strc*$^{-/-b}$
	SLC26A5	Prestin (anion transporter)	DFNB61	*Slc26a5*$^{-/-b}$
	KCNQ4	KCNQ4 (K^+ channel subunit)	DFNA2	*Kcnq4*$^{-/-b}$
	TMC1	TMC1 (transmembrane channel-like protein)	DFNB7/11; DFNA36	Deafness (*dn*); Beethoven (*bth*)
	TMHS	TMHS (transmembrane protein)	DFNB67	Hurry-scurry (*hscy*)
	OTOF	Otoferlin (synaptic exocytosis protein)	DFNB9	*Otof*$^{-/-b}$; *Pachanga* (*Pga*)
		vesicular glutamate transporter-3	DFNA25	*Vglut3*$^{-/-b}$

	VGLUT3/ SLC17A8			
	POU4F3	POU4F3 (transcription factor)	DFNA15	*Brn3*$^{c-/-b}$; Dreidl (*ddl*)
Nonsensory cells	*CX26 / GJB2*	Connexin-26 (gap junction protein)	DFNB1; DFNA3 ± keratodermia[a] (Vohwinkel, palmoplantar keratodermia, KID, Bart-Pumphrey)	*Cx26*$^{Otog\text{-}Cre\,c,\,g}$
	CX30 / GJB6	Connexin-30 (gap junction protein)	DFNB1; DFNA3 ± keratodermia[a] (KID)	*Cx30*$^{-/-b,\,g}$; *Cx26*$^{+/-}$/ *Cx30*$^{+/-b}$
	CX31 / GJB3	Connexin-31 (gap junction protein)	DFNA2$_i$[d]	
	PDS / SLC26A4	Pendrin (I^-/Cl^- transporter)	DFNB4 ± thyroid goitre[a] (Pendred)	*Pds*$^{-/-b}$
	CRYM	μ-Cristallin (thyroid hormone-binding protein)	DFNAi[d]	
	OTOA	Otoancorin (cell surface protein)	DFNB22	
	COCH	Cochlin (extracellular matrix component)	DFNA9	*Coch*$^{-/-b,\,e}$
	TMPRSS3	TMPRSS3 (transmembrane serine protease)	DFNB8/10	
	MYH14	Myosin IIC (motor protein)	DFNA4	
	CCDC50	Ymer (effector of EGF-mediated cell signaling)	DFNA44	
	EYA4	EYA4 (transcriptional coactivator)	DFNA10	*Eya4*$^{-/-b}$

(continued)

Table 8.1 (*continued*)

Primary defect	Gene	Gene product	Forms of human deafness	Mouse mutants
	POU3F4	POU3F4 (transcription factor)	DFN3	Sex-linked fidget (*slf*); *Brn4*$^{-/-b}$
	ESRRB	ESRRB (estrogen-related receptor gene)	DFNB35	
Tectorial membrane	*COL11A2*	Collagen XI (α2-chain) (extracellular matrix component)	DFNA13 ± osteochondrodysplasia[a] (Stickler 2)	*Col11a2*−/−[b]
	TECTA	α-Tectorin (extracellular matrix component)	DFNA8/12 DFNB21	*Tecta*−/−[b]
Unknown target cells	*DFNA5*	(unidentified)	DFNA5	*Dfna5*$^{-/-b}$
	WFS1	Wolframin (endoplasmic reticulum membrane protein)	DFNA6/14/38 ± diabetes & optic atrophy[a] (Wolfram)	Tilted? (*tlt*)
	TFCP2L3	TFCP2L3 (transcription factor)	DFNA28	
	DFNB59	Pejvakin[f]	DFNB59	*Dfnb59*$^{-/-b}$; *Sirtaki* (*Sti*)
	MTRNR1	*Mitochondrial 12S rRNA*		
	MTTS1	*Mitochondrial tRNAser(UCN)*		

[a] Syndromic deafness.
[b] Mutant obtained by gene knockout.
[c] Mutant obtained by targeted gene inactivation in the cochlear and vestibular sensory epithelia.
[d] Subscript i denotes undefined locus number.
[e] Mutant not deaf.
[f] indications for a primary defect in hair cells, peripheral, and brainstem auditory neurons.
[g] See also Table 8.3.

3. Human Congenital Deafness

3.1. Clinical and genetic aspects

Five main clinical criteria are currently used to classify hearing impairment in humans. The emplacement of the primary lesion defines two deafness types: conductive deafness, resulting from dysfunction of the external or middle ear, and perceptive or sensorineural deafness, caused by a defect located anywhere along the auditory pathway, from the cochlea to the auditory cerebral cortex. The onset of hearing loss permits to distinguish between prelingual and postlingual deafness, that is, hearing impairment beginning before and after the onset of spoken language, respectively. Audiometric measurements permit to assess the degree of hearing loss that can be mild (loss of 20–40 dB HL), moderate (loss of 40–70 dB HL), severe (loss of 70–90 dB HL), or profound (loss higher than 90 dB HL) (see definition of dB HL in the later text). The range of sound frequencies whose perception is affected separates low (<500 Hz), middle (500–2000 Hz), and high (>2000 Hz) frequency hearing loss (Kochhar *et al.*, 2007; Petit, 2006). Finally, hearing impairment can be either isolated (no other clinical manifestation) or syndromic that is, associated with other symptoms or anomalies. Genetic studies have shown that sensorineural forms of hereditary isolated deafness accounts for about 60% of congenital hearing impairment cases in developed countries (Denoyelle *et al.*, 1999). Isolated forms of hereditary deafness are classified by their mode of transmission. DFN, DFNA, and DFNB refer to deafness forms inherited on the X chromosome-linked, autosomal dominant, and autosomal recessive modes of transmission, respectively (Petit, 2006). A Y chromosome-linked mode of transmission and a maternal inheritance linked to the mitochondrial genome have also been identified. To date about 140 loci underlying nonsyndromic hereditary hearing loss have been reported and 47 genes have been identified. Approximately half of the known genes (28/47) encode proteins that are present in hair cells (see Table 8.1). Among them some are also responsible for syndromic forms of deafness (see Section 4.2). Although occasionally the nature of the encoded protein and its immunolocalization provide a first insight into the underlying pathogenic process (e.g., when a tectorial membrane component is defective, see Table 8.1), in most instances this is not the case. Various tests devoted to the evaluation of hearing in humans have been developed that provide information regarding the earlier discussed physiological processes (see Section 2). To investigate hearing in mouse models, these tests have been adapted to mouse and new procedures have been developed to a more in-depth functional characterization of the peripheral audition system.

3.2. Human clinical investigations and their limits

In patients, a panel of tests allows the evaluation of the hearing impairment and some assessment of the anatomic site of the lesion involved. Some of them are subjective tests and others are objective ones.

Among the subjective explorations, the standard procedure for assessing hearing impairment is the measurement of the hearing thresholds for pure tones, that is, single frequency tones. The acoustic stimulus is presented through either air conduction (by ear phones, loudspeaker, ...) or bone conduction (by bone vibrator). Air conduction explores the entire auditory system, whereas bone conduction, which circumvents sound propagation through the auditory canal and the three ossicles of the middle ear, only tests the sensorineural auditory system, that is, from cochlea to auditory cortex. A difference between the thresholds obtained by the two techniques (with air conduction threshold being higher than bone conduction threshold) indicates a contribution of a conductive component (outer or middle ear) to the hearing impairment. Hearing thresholds can be evaluated in decibels (dB) SPL. dB SPL are the logarithms of the squared ratio between the pressure amplitudes of the test tone and a reference pressure of 20 μPa. It follows that a 60 dB shift of the hearing threshold corresponds to a 10^3-fold increase in pressure amplitude of the weakest detected sound. For clinical purposes, it is customary to refer a pure-tone auditory threshold to the average one normally obtained at the same frequency (in dB HL, for hearing level). The response to pure tones provides limited information regarding the ability of the individuals to hear spectrally complex sounds, including speech—the primary means of communication between persons. Various behavioral tests allow the exploration of a speech intelligibility deficit. Dissociation between hearing thresholds and speech intelligibility, mostly seen in late onset forms of deafness, is generally attributed to IHC and/or auditory neuron defects.

The main objective test of hearing is the recording of the auditory brainstem responses. Electrodes placed on the scalp record the synchronized action potentials evoked by a broadband frequency sound (click) or short pulses of pure tones. Latency, amplitude, and shape of the various waves (labeled in roman numerals from I to VI), normally observed for sounds louder than 10–20 dB above the audiometric threshold, provide information on the electrical response of the ascending auditory pathways from the peripheral portion of the auditory nerve (wave I) to the medial geniculate body of the thalamus (wave VI). Because auditory brainstem response collection involves synchronous averaging of several hundred electroencephalographic responses to a repeated click, the presence of auditory brainstem response waves requires synchronization of the response in the various relays of the neuronal circuits. Wave II likely comes from the proximal portion of the auditory nerve, and waves III, IV, and V are attributed to the cochlear nucleus, superior olivary complex, and lateral lemniscus, respectively (the superior olivary complex response is bilateral, and a controlateral response dominates in the following neuronal relays). This highly informative test is also the only one that allows an evaluation of hearing thresholds in newborns and infants.

Otoacoustic emissions (OAEs) are sounds recorded by a sensitive miniature microphone placed in the auditory canal. They are classified as spontaneous OAEs and evoked OAEs. Evoked OAEs are either transiently evoked by a click or a tone burst, or are distortion products produced in response to two continuous pure tones of frequencies f_1 and f_2, with a ratio f_2/f_1 of 1.20–1.25. The most robust distortion product is the so-called cubic distortion product that occurs at a frequency of $2f_1 - f_2$. OAEs are assumed to be produced by the OHCs, and propagated backwards to the middle ear and to the ear canal. Recording of evoked OAEs, is extensively used as a screening test for audition in newborns. A caveat to the use of OAEs as a screening tests is that they only explore the activity of the OHCs, thus they are known to persist in some forms of sensorineural deafness (referred to as auditory neuropathies) that affect the IHCs or the downstream auditory pathway (see Section 4.1). Yet, as a diagnostic tool, evoked OAEs should find a broader interest, if only because the presence of normal OAE levels in the absence of auditory brainstem response waves defines the concept of, possibly not so rare, auditory neuropathies (Starr *et al.*, 1996). Another, more evident caveat is that the existence of a conductive hearing loss can hamper OAE detection, even though the cochlea produces them normally.

Tympanometry is an objective measure of acoustic impedance of the middle ear that is used to diagnose middle ear disorders. Acoustic reflex (ipsilateral and controlateral) evaluates the auditory system ability to modify the middle ear impedance by acting on the muscle of the last middle ear ossicle, the stapes. This reflex (also called stapedial reflex arc) is elicited by 90–95 dB SPL pure tones (500–4000 Hz). It is both ipsilateral and controlateral and involves the cochlea, auditory nerve, ventral cochlear nucleus, brainstem nuclei, as well as the ipsilateral and controlateral facial motor nuclei and facial nerves.

Electrocochleography detects components of the electrical cochlear response that, nowadays, can be recorded with noninvasive electrodes. The recorded electrical potentials include (1) the cochlear microphonics, corresponding to alternating currents following the waveform of the input sound and attributed to mechanoelectrical transduction currents generated by hair cells (mainly OHCs) from the basal region of the cochlea, (2) the summating potential, a direct potential of complex origin with a mixed contribution of OHCs and IHCs, and (3) the compound action potential that reflects the synchronous response of the auditory nerve to tone bursts.

Temporal bone imaging, mainly performed by computerized tomography and magnetic resonance imaging, enables detection of anomalies within the bone and in the air cavities, as well as opacities related to any soft tissue anomalies.

Although when combined, the aforementioned tests have the ability to frame the impaired step of sound proccessing, the pathogenesis of hearing impairment cannot be deciphered in humans in the absence of any possible direct observation of the human cochlea.

The progressive confidence in mutant mice as animal models for human deafness is based on the fact that, without exception until now, a hearing impairment has been observed in mice carrying null mutations in the genes orthologous to human deafness genes; the shift of the hearing thresholds is, however, sometimes different. On this basis, we assume that the pathogenic process is the same, a conclusion reinforced by the fact that the same auditory tests give consistent results when used in humans and mice carrying the same mutation (Delmaghani *et al.*, 2006; Kharkovets *et al.*, 2006; Kubisch *et al.*, 1999; Roux *et al.*, 2006; Ruel *et al.*, 2008; Seal *et al.*, 2008; Varga *et al.*, 2006). Nevertheless, for human hearing impairment affecting low frequencies only, the mouse is not expected to be an appropriate model. Indeed, the frequency ranges of mouse and human hearing have only about three octaves in common (from 2.5 to 20 kHz), with nonoverlapping frequencies roughly extending seven octaves lower in humans and two octaves higher in mice.

Not only have all the earlier mentioned objective audiometric tests been developed for the mice (see Fig. 8.6), but also numerous other tests allowing quantitative assessment of biophysical and physiological criteria of the functioning cochlea at deeper levels than possible in humans. These include measurements of the endocochlear potential, K^+ and Ca^{2+} concentrations in the endolymph, basilar and tectorial membrane motion, frequency tuning curves of individual auditory neurons, presynaptic and postsynaptic currents, *ex vivo* recording of the mechanoelectrical transduction currents, and *in vivo* recordings of hair cell receptor potentials. In addition, histological analysis of the cochlea, analysis of the hair bundle structure by scanning electron microscopy (see Figs. 8.1D and 8.7) and immunodetection of proteins of interest can be carried out. As a result, virtually all we have learnt so far regarding the pathogenic mechanisms underlying the genetic forms of human deafness is derived from the study of mouse models.

4. Mouse Models for Human Hereditary Deafness

A variety of pathogenic processes underlying human deafness forms have now been deciphered thanks to mouse models (see Friedman *et al.*, 2007 for review). We here discuss only a subset of them that should illustrate contrasted situations regarding the levels of understanding of these processes that mouse models have enabled to reach so far.

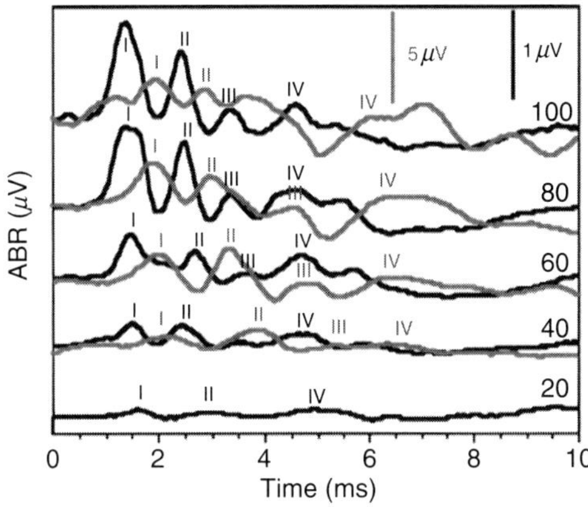

Figure 8.6 ABR waveforms rom wild type (black) and *Dfnb59* knockin (red) mice. Waves I–IV are marked. Labels on the right right indicate stimulus levels (from 20 to 100 dB SPL). Note that the vertical scales are different for the *Dfnb59* knockin mouse and the wild type mouse. In the *Dfnb59* knockin mouse, ABR waves I–II (auditory nerve) and III–IV (brainstem) have reduced amplitudes compared to the wild type mouse and the latencies of all four waves are significantly augmented, indicating a dysfunction in synaptic transmission or neuronal conduction. Reproduced from (Delmaghani *et al.*, 2006). (See Color Insert.)

First, we will discuss a mouse model of the DFNB9 form of isolated deafness. This genetic form belongs to the only subset of sensorineural deafness forms that can be clinically distinguished from all the others, namely the auditory neuropathies (Starr *et al.*, 1996). The DFNB9 mouse model unambiguously points to a failure of IHC synaptic exocytosis. Even though many issues remain to be addressed regarding the biological activity of the encoded protein, otoferlin, this mouse model was informative enough to guide cochlear implant use in DFNB9 patients.

Second, we will consider the multiple mouse models of Usher syndrome (USH) (a dual auditory and visual sensory defect), which revealed abnormal hair bundle morphogenesis as a major cause of the hearing impairment in this disease. Biochemical and genetic evidence based on these mouse models allowed to involve the proteins encoded by USH genes into two molecular networks associated with different sets of hair bundle links, thereby establishing the critical role of these links in hair bundle morphogenesis.

Finally, we will examine the most common congenital form of deafness in Caucasian populations, DFNB1, which is due to connexin-26 deficiency. Because this gap junction protein is normally present in most of the cochlear cells, and the corresponding gene inactivation in the mouse is lethal at embryonic stage, the underlying pathogenic processes tend to be difficult to grasp in the available, still incomplete, mouse models of the disease.

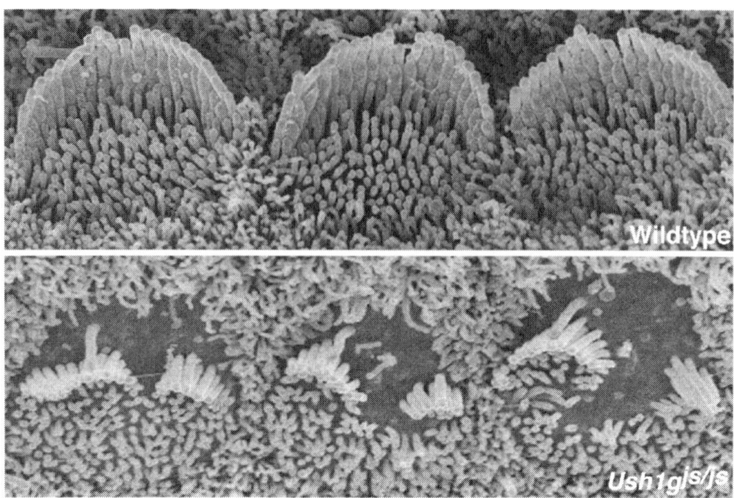

Figure 8.7 Scanning electron micrograph of wild type and mutant hair bundles. In the *Ush1g*$^{js/js}$ mouse mutant at P0, IHC hair bundles are split (bottom panel) compared with the control (top panel). Micrographs were courtesy provided by G.M. Lefèvre.

4.1. DFNB9, an inner hair cell synaptopathy

DFNB9-affected patients suffer from congenital profound deafness. Using linkage analysis combined with a candidate gene approach, *OTOF*, encoding otoferlin has been identified as responsible for this autosomal recessive form of deafness (Yasunaga *et al.*, 1999). *OTOF* encodes several long and short transcripts (Yasunaga *et al.*, 2000), some of which are only expressed in the cochlea, while others are detected both in the brain and the cochlea (Yasunaga *et al.*, 2000). Otoferlin contains six C2 domains, followed by a single C-terminus membrane-spanning domain (Roux *et al.*, 2006). Several of these C2 domains are predicted to bind Ca^{2+} (Jimenez and Bashir, 2007).

The DFNB9 deafness meets the clinical criteria for auditory neuropathy (Loundon *et al.*, 2005; Rodriguez-Ballesteros *et al.*, 2003; Varga *et al.*, 2003). This term was initially used to describe sensorineural hearing impairment characterized by the absence or severe distortion of the auditory brainstem responses whereas OAEs are preserved (Starr *et al.*, 1996). Therefore, although auditory neuropathies are generally believed to be of retrocochlear origin, likely affecting auditory neurons, clinical symptoms are also compatible with an IHC defect.

In the mature cochlea, otoferlin is expressed in the IHCs where it is located at the presynaptic plasma membrane and ribbon-associated synaptic vesicles (Roux *et al.*, 2006) (Fig. 8.4B and 8.4C). Moreover, otoferlin interacts with the synaptic SNARE complex of the IHC. This is reminiscent of synaptotagmin I interactions, the major Ca^{2+} sensor of fast synaptic vesicle exocytosis at conventional synapses (reviewed in Chapman, 2002; Sudhof, 2002). These findings led to the proposal that otoferlin is involved in IHC synaptic vesicle exocytosis, possibly as a Ca^{2+} sensor. It may substitute for synaptotagmin I that is undetectable in IHCs (see Section 2.4).

The otoferlin-null mouse (*Otof*$^{-/-}$) turned out to be highly informative regarding DFNB9 pathogenesis. Audiological tests showed that these mice are profoundly deaf with persisting distortion product AEs, just like DFNB9 patients. Morphological analysis of the organ of Corti in mutant mice did not show gross anomalies. In addition, direct stimulation of the auditory nerve showed normal conduction of the afferent neurons innervating the IHCs. As mentioned earlier, transduction of the sound stimulus into membrane depolarization in a normal IHC results in an increase of Ca^{2+} influx raising Ca^{2+} concentration at the synaptic active zone. Ca^{2+} ions bind to a calcium sensor, thereby triggering the fusion of synaptic vesicles with the presynaptic plasma membrane. Electrophysiological study of IHCs in these mutant mice showed that although the Ca^{2+} currents were normal, these currents failed to trigger synaptic vesicle exocytosis (Roux *et al.*, 2006). This finding narrowed down the implication of otoferlin in the synaptic functioning to the coupling of Ca^{2+} influx to the synaptic vesicle machinery. This is consistent with a role of otoferlin, as a Ca^{2+} sensor triggering synaptic vesicle–plasma membrane fusion at the IHC ribbon synapse. Alternatively, otoferlin, by interacting with the SNARE complex, could increase its affinity for the Ca^{2+} sensor. Several lines of evidence suggest that a multivesicular fusion takes place at the IHC synapse (Glowatzki and Fuchs, 2002). The wide dynamic range that the six C2 domains of otoferlin may provide in Ca^{2+} sensing could be suitable for this type of synaptic vesicle fusion. Whatever the precise mechanism involving otoferlin in IHC synaptic exocytosis, the functional characterization of *Otof*$^{-/-}$ mice identifies a novel nosological entity among auditory neuropathies, namely the auditory hair cell synaptopathy. Ultrastructural analysis of the *Otof*$^{-/-}$ IHC synapse uncovered a secondary degenerative process of the IHC active zone and some afferent neurons, starting at P15, which is likely to take place also in DFNB9 patients. Whether patients affected by auditory neuropathies can benefit from cochlear implants, an electroacoustical device that directly stimulates the auditory nerve, is an unsolved issue that the clarification of the underlying pathogenic processes should contribute to solve. Since the mouse model of DFNB9 has unambiguously shown that IHCs, not the auditory neurons, are the primary target cells of the genetic defect, early cochlear implant is expected to be of major help for DFNB9 patients.

4.2. Usher syndrome

USH is the most frequent cause of monogenic sensory disability associating sensorineural hearing impairment and blindness due to retinitis pigmentosa. It is transmitted on the autosomal recessive mode. Three clinical subtypes

(USHI, USHII, and USHIII) can be distinguished. In the most severe form (USHI), patients suffer from severe to profound congenital hearing impairment and balance defects that manifest in young children, whose ability to maintain a sitting posture is delayed as is their further developing of the ability to walk. Night blindness is the first symptom of retinitis pigmentosa, which is characterized by a juvenile age of onset (during the first decade or early second decade). It is closely followed by a progressive narrowing of the visual field and a decrease of the visual acuity. In patients affected by USHII, deafness, also congenital, is moderate to severe with no sign of balance defect and retinitis pigmentosa usually appears in the second decade. USHIII is not as strictly defined, with progressive hearing impairment, variable balance problems and retinitis pigmentosa. The prevalence of the syndrome has been estimated to 1 out of 25,000 individuals. USHI and II are the most frequent subtypes. However, due to a founder effect, USHIII accounts for 42% of the Usher cases in the Finnish population, as against 2% in most other countries (Pakarinen *et al.*, 1995).

Because photoreceptor cells and hair cells are the primary targets of the deficit in most retinitis pigmentosa and early onset forms of deafness respectively, the common structures and properties of these sensory cells are appealing hypothetical substrates for this dual sensory disorder. The synapses of the two sensory cells are similar and differ from conventional synapses by the presence of an electron-dense body, a ribbon associated to the synaptic active zone and surrounded by tethered synaptic vesicles (see Section 2.4). Moreover, these two types of sensory cells do not produce action potentials, but graded variations of their membrane potential trigger their synaptic exocytosis. An immotile cilium is present in the two sensory cells: the kinocilium in developing hair cells (Fig. 8.2A) and the connecting cilium in photoreceptor cells. The connecting cilium extends between the inner and outer segments of the photoreceptor, the cell compartments of protein synthesis and of phototransduction, respectively. It is a transit site for newly synthesized proteins en route toward the outer discs and for light-dependent translocation of phototransduction proteins (Sokolov *et al.*, 2002).

Ten chromosomal loci for USH (USH1B,C,D,F,G, USH2A,C,D, USH3A,B) have been reported. Nine of the corresponding genes have been identified through positional cloning strategies, namely *USH1B/MYO7a* (myosin VIIa gene), *USH1C* (harmonin gene), *USH1D/CDH23* (cadherin-23 gene), *USH1F/PCDH15* (protocadherin-15 gene), *USH1G/SANS* (sans gene), *USH2A* (usherin gene), *USH2C/VLGR1* (very large G-protein-coupled receptor 1 gene), *USH2D/WHRN*, (whirlin gene) and *USH3A/USH3A* (clarin gene). For several of these genes, the existence of spontaneous and ENU (*N*-ethyl-*N*-nitrosourea)-mutagenized mouse mutants in the orthologous genes has been a decisive help in their identification. Notably, some missense or leaky splice site mutations in the *USH1* genes (*USH1B, USH1C, USH1D*, and *USH1F*) lead to isolated deafness.

The same type of mutations in *USH2* genes are causative for isolated retinis pigmentosa (*USH2A*) (Rivolta *et al.*, 2000) or atypical USH (Liu *et al.*, 1999).

4.2.1. USH type I

The five causative genes for USHI syndrome encode a set of proteins referred to as the USH1 proteins: a molecular motor, myosin VIIa (USH1B) (Weil *et al.*, 1995), two transmembrane Ca^{2+}-dependent adhesion proteins, cadherin-23 (USH1D) (Bolz *et al.*, 2001; Bork *et al.*, 2001), and protocadherin-15 (USH1F) (Ahmed *et al.*, 2001; Alagramam *et al.*, 2001), a PDZ domain-containing protein, harmonin, likely to be a submembranous scaffolding protein (USH1C) (Bitner-Glindzicz *et al.*, 2000; Verpy *et al.*, 2000), and a putative cytoskeletal protein with ankyrin repeats, named sans (USH1G) (Weil *et al.*, 2003). The two cadherin and protocadherin genes encode several transmembrane isoforms characterized by a large number of extracellular repeats and the absence of a consensus binding site to *β*-catenin, a key protein for the anchoring of classical cadherins to the actin filaments (Cavey *et al.*, 2008; Moreno *et al.*, 2003; Perez-Yamada *et al.*, 2005). Analysis of harmonin transcripts predicts more than 10 isoforms belonging to three protein subclasses. The harmonin b subclass binds to the actin filaments.

All of what we have learned so far regarding the USHI pathogenesis of deafness is based on the study of either spontaneous, ENU-mutagenized or genetically engineered (see Table 8.2) mouse mutants carrying mutations in each of the *USH1* orthologous genes. Indeed, they all faithfully reproduce the hearing and balance defects of the human disease. In contrast, no retinitis pigmentosa can be detected in *Ush1* mouse models. Only marginal electrophysiological anomalies of the retina have been reported. Whether this discrepancy is caused by physiological differences between human and mouse photoreceptors, or by genetic differences with functional redundancy existing for Ush1 proteins in the mouse but not in the human retina, or whether it is attributable to the difference in the lifespans of each of the two species is still unknown.

USH1 mouse models have shown that hair cells are the primary target cells of the hearing deficit (see El-Amraoui and Petit, 2005 for a review). A recent comparative study of the mouse models for each of the five USH1 genetic forms, has revealed hair bundle morphogenetic defects common to all of them (Lefèvre *et al.*, 2008). Consistently, Ush1 proteins are present and colocalize in the developing hair bundle as soon as it emerges from the hair cell apical surface. Moreover, these proteins show a number of direct *in vitro* interactions. In particular, harmonin binds to all other Ush1 proteins, and myosin VIIa also binds to all of them except cadherin-23.

Three growing stages can be distinguished during normal development of the murine hair bundle. Apical protrusions from the apical cell surface of the hair cells, first noticeable around E15.5, rapidly gather into a bundle and grow uniformly. The kinocilium, a transient genuine cilium present at the apex of cochlear sensory cells up to P12, initially occupies a central position, and then migrates to the periphery of the forming hair bundle. In a second stage, from P0 to P5, the kinocilium progressively reaches its final position which defines the planar polarity axis of the developing hair bundle (Jones *et al.*, 2008; and see Jones and Chen, 2007; Wang and Nathans, 2007 as reviews). The stereocilia row next to the kinocilium begins to grow, and as a result of a differential elongation of the various stereocilia rows, the hair

Table 8.2 Usher syndrome: human and mouse causative genes and encoded proteins

Genetic form	Human gene	Protein	Mouse models
Usher 1B	*MYO7A*	Myosin VIIa (actin based motor protein)	Shaker-1 (*sh1*) Headbanger (*hdb*)
Usher 1C	*USH1C*	Harmonin (PDZ domain-containing protein)	Deaf circler (*dfcr*) *Ush1c*$^{-/-}$ [a]
Usher 1D	*CDH23*	Cadherin23 (integral membrane adhesion protein)	Waltzer (*v*)
Usher 1F	*PCDH15*	Protocadherin15 (integral membrane adhesion protein)	Ames waltzer (*av*)
Usher 1G	*SANS*	Sans (putative scaffold protein)	Jackson shaker (*js*)
Usher 2A	*USH2A*	Usherin (integral membrane protein)	*Ush2a*$^{-/-}$ [a]
Usher 2C	*VLGR1*	Vlgr1 (G-protein-coupled receptor)	*Vlgr1/del7TM* *Vlgr1*$^{-/-}$ [a]
Usher 2D	*WHRN*	Whirlin (PDZ domain-containing protein)	Whirler (*wi*)
Usher 3A	*CLRN1*	Clarin (integral membrane protein)	–

[a] Mutant obtained by gene knockout.

Usher 1B (Gibson *et al.*, 1995; Weil *et al.*, 1995). Usher 1C (Bitner-Glindzicz *et al.*, 2000; Johnson *et al.*, 2003; Lefèvre *et al.*, 2008; Verpy *et al.*, 2000). Usher 1D (Bolz *et al.*, 2001; Bork *et al.*, 2001; Di Palma *et al.*, 2001). Usher 1F (Ahmed *et al.*, 2001; Alagramam *et al.*, 2001). Usher 1G (Kikkawa *et al.*, 2003; Weil *et al.*, 2003). Usher 2A (Eudy *et al.*, 1998; Liu *et al.*, 2007).Usher 2C (McMillan and White, 2004; Weston *et al.*, 2004; Yagi *et al.*, 2005). Usher 2D (Ebermann *et al.*, 2007; Mburu *et al.*, 2003). Usher 3A (Adato *et al.*, 2002; Joensuu *et al.*, 2001).

bundle staircase pattern forms. The final phase of the hair bundle development, between P5 and P12, is characterized by the simultaneous growth of all stereocilia, followed by the disappearance of the kinocilium.

Among the common hair bundle morphological anomalies observed in USH1 mouse models, two were detected as soon as E17–E18. On these days, almost each IHC and OHC hair bundle of these mutants was fragmented into 2–4 clumps of stereocilia, and even though the kinocilium was located at the periphery of the apical cell surface, it appeared to have significantly deviated from its normal position (ranging from 25° to 52° compared to about 6° to 12° in wild-type mice, at P0). In addition, in cadherin-23 or protocadherin-15-deficient mice, the kinocilium was often dissociated from the stereocilia. The third anomaly common to all USH1 mouse models was an elongation defect of the medium and small stereocilia rows after P0. A series of immunolabeling and biochemical evidence are substantiating that together cadherin-23 and protocadherin-15 make up several hair bundle links, that is some early lateral interstereocilia links, the stereokinociliary links, and the tip link (see Fig. 8.2A). Immunolocalization of harmonin b and its *in vitro* interactions with the cytoplasmic regions of the two cadherins qualify it to anchor them to the stereocilia actin filaments. The fragmentation of the stereocilia bundle and the independence of the kinocilium from the stereocilia bundle observed in Ush1-deficient mice, thus strongly suggest a critical role of early lateral interstereocilia links and stereokinociliary links in the cohesiveness of the growing hair bundle (Fig. 8.7). As the fragmentation of the hair bundle produces stereocilia clumps in which some interstereocilia fibrous links are present and no further fragmentation is noted later on, this suggests that the rupture of the hair bundle is the result of tension forces exerted during its early developmental step. They may be traction forces induced either by the cell movements, the so-called convergent extension process (Keller, 2006; Lefèvre *et al.*, 2008) that seems to occur in the neuroepithelium, concomitantly to the hair bundle growth, or by the reorientation of the kinocilium driven by the movement of its basal body somehow connected to the rootlets of the stereocilia. The hair bundle aroused as a result of defective links is very likely unable to withstand these forces.

The stereocilia growth defect in Usher I mouse models differs from all the others reported so far, namely in myosin XV, whirlin, or espin (Mburu *et al.*, 2003; Naz *et al.*, 2004; Wang *et al.*, 1998) defects (responsible for deafness forms in humans) (see Table 8.1), in which the size of every stereocilia row is reduced. Pulling forces applied to the stereocilia tips by the tip link or apical links composed of cadherin-23 and protocadherin-15, thereby possibly controlling the polymerization of stereocilia actin filaments (Hill and Kirschner, 1982), may account for the selective growth defect of the short and middle stereocilia rows in Ush1-deficient mice.

Finally, the absence of harmonin b in the hair bundle of myosin VIIa- or sans-defective mouse mutants is consistent with the presence of the three depicted hair bundle abnormal features also in these mutants (i.e., hair bundle fragmentation, kinocilium deviation, and stereocilia growth defects).

Even though physiological evidence is still lacking, the subcellular localization of cadherin-23, protocadherin-15 at the level of the tip link (Kazmierczak *et al.*, 2007) and of harmonin b at the tip link upper insertion point (Lefèvre *et al.*, 2008), strongly suggest that they are components of the mechanotransduction machinery that is detected as starting to operate in the wild type mouse at P1, that is, when the hair bundle is already disorganized in *Ush1* mutants.

4.2.2. USH type II

The USH2A gene codes for a transmembrane usherin isoform with a long extracellular domain composed of up to 33 fibronectin type III repeats and four different types of laminin-like domains (van Wijk *et al.*, 2004). The USH2C gene encodes Vlgr1 (very large G protein-coupled receptor-1), also a transmembrane protein that belongs to the B-subclass of G-protein-coupled receptors. Its large extracellular region mostly consists of 35 copies of Calx-β domains. Finally, the USH2D gene encodes whirlin, a PDZ domain containing protein, which is the closest homolog of harmonin.

Just like for Usher type I, only the mouse models enabled to get insight into the pathogenesis of Usher type II hearing impairment. However, in contrast to Usher type I, retinal studies, so far limited to usherin-deficient mice, have shown photoreceptor degeneration (Liu *et al.*, 2007). Although no indication about the underlying retinal pathogenic process is presently available, usherin seems to be mainly expressed at the membrane of the apical inner segment that surrounds the connecting cilium.

Several lines of evidence support the fact that Vlgr1 and usherin are components of the ankle links and form an ankle link molecular complex with whirlin acting as a submembranous scaffolding protein. Indeed, the three Ush2 proteins are colocalized at the stereocilia basis as long as ankle links are present, from P0 up to almost the final maturation step of the hair bundle. Moreover, whirlin directly interacts *in vitro* with the cytoplasmic regions of usherin and Vlgr1. Finally, in mice carrying mutations in *Vlgr1*, no ankle link could be detected (McGee *et al.*, 2006; Michalski *et al.*, 2007).

Mouse mutants defective for Vlgr1 or whirlin have been studied (Mburu *et al.*, 2003; McGee *et al.*, 2006; Michalski *et al.*, 2007). Both display abnormally shaped IHC and OHC hair bundles. Most OHC hair bundles are U-shaped or round-shaped instead of displaying the characteristic V-shape seen in wild type mice. IHCs have an asymmetrical instead of symmetrical arch-shaped hair bundles.

Table 8.3 Gap junction defective mouse models and rescuing model

Mouse models	EP (in mV)	Endolymphatic K^+ (in mM)	Hearing loss	Pathohistological findings	Vestibular function	References
$Cx26^{OtogCre}$ (conditional KO in the epithelial GJ network)	~40 (vs 110)	~85 (vs 150)	Moderate	Apoptosis in the organ of Corti (supporting cells, OHC) Collapse of tunnel of Corti Reticular lamina breaks	Not affected	Cohen-Salmon *et al.* (2002)
Tg *Cx26* R75W (dominant mutation)	Not affected	Not affected ?	Profound	Collapse of tunnel of Corti IHC and OHC death	Not affected	Kudo *et al.* (2003)
$Cx30^{-/-}$ (ubiquitous KO)	Absent	~45 (vs 150)	Severe	Apoptosis in the organ of Corti (mainly IHC and OHC) Endothelial breaks in stria vascularis	Not affected, but saccular hair cell loss	Teubner *et al.* (2003), Cohen-Salmon *et al.* (2007), Qu *et al.* (2007)
$Cx26^{+/-}$ / $Cx30^{+/-}$	~45 (vs 80)	Not affected	Moderate	No anomaly of the cochlear tissues	Not affected	Michel *et al.* (2003)
BAC^{Cx26}-rescue of $Cx30^{-/-}$	Normal	Not determined	No	Prevention of hair cell apoptosis	Not affected	Ahmad *et al.* (2007)

The electrical response of Vlgr1-deficient IHCs to hair bundle stimulation showed that a mechanoelectrical transduction current is evoked not only by a deflection of the hair bundle in the excitatory direction (see Section 2.2) but also in the opposite direction, the inhibitory direction (Michalski *et al.*, 2007). Videomonitoring of the stereocilia displacements upon stimulation in the "inhibitory" direction revealed their uncoordinated movements in the mutants only, which presumably increases tension in the tip links and opens the mechanoelectrical transduction channels. In conjunction with the abnormal shapes of the hair bundle, this suggests that ankle links mediate the distribution of cohesive forces throughout the hair bundle. Both the morphological and physiological anomalies are more prominent in the OHCs than in the IHCs. The relative preservation of the IHCs likely accounts for the residual hearing function in USHII patients.

Interestingly, myosin VIIa is particularly concentrated at the ankle link level (Hasson *et al.*, 1997) and is directly interacting with each of the Ush2 proteins (Michalski *et al.*, 2007). Moreover, in myosin VIIa defective mice, usherin and Vlgr1 are absent from the hair bundle. It ensues that myosin VIIa defect in fact results both in USHI-type and USHII-type hair bundle anomalies. Whether myosin VIIa is a general conveyor of proteins within the hair bundle, and/or maintains proteins within given subregions of the hair bundle remains to be clarified. In this regard, it is worth noting that the kinetic properties of this myosin, that is, high duty ratio and low ATPase activity, qualify it as a generator of local tension forces, as long as the ATP concentration is not very high (El-Amraoui *et al.*, 2008).

Altogether, important insights have been obtained regarding the pathogenesis of the hearing impairment in Usher type I and II, based on mouse models. From these data, it follows that hair bundle anomalies of USHI and USHII are expected to appear in the affected fetus before the twelfth and the twenty-fifth week in IHC and OHC, respectively (i.e., when IHC and OHC hair bundles normally get mature). With the purpose of uniting views regarding the pathogenesis of hearing and retinal defects, fibrous links of the photoreceptors extending between the apical region of the inner segment and the connecting cilium, should also be the primary targets of the retinitis pigmentosa (Petit, 2001). Still, the function of these links is not known. Finally, in spite of the conflictual nature of the data on the immunolocalization of USH1 and USH2 proteins at the photoreceptor synapse (see Williams, 2008), it is not yet possible to exclude the existence of synaptic defects in both hair cells and photoreceptor cells of patients affected by Usher syndrome.

4.3. Mouse models for gap junction defects

In 1994, the first human locus for an autosomal recessive form of deafness, DFNB1, was reported (Guilford *et al.*, 1994). It was then rapidly stated that this genetic form of deafness was the most prevalent among nonsyndromic congenital hearing loss (Maw *et al.*, 1995). The gene encoding Cx26 was soon identified as the causative gene for DFNB1 (Kelsell *et al.*, 1997) but also for DFNA3, a rare dominant form of deafness (Denoyelle *et al.*, 1998). In the last decade, several groups focused their efforts on the identification of other connexin genes potentially involved in hearing impairment. Connexins 29, 30, 31, and 32 genes were reported as responsible for deafness in humans (del Castillo *et al.*, 2002; Grifa *et al.*, 1999; Liu *et al.*, 2000; Xia *et al.*, 1998; Yang *et al.*, 2007) even though mouse models did not always provide convincing evidence of their implication in the disease, except for connexin-30 (Eiberger *et al.*, 2006; Nelles *et al.*, 1996; Plum *et al.*, 2001; Tang *et al.*, 2006; Teubner *et al.*, 2003). Here, we focus on Cx26 and Cx30 encoded by *Gjb2* and *Gjb6*, respectively, for the following reasons: (1) mutations in *CX26* (or *GJB2*) are the major cause of congenital hearing impairment (Denoyelle *et al.*, 1999); for a review see (Cohen-Salmon *et al.*, 2005); (2) A variety of mutations have been found to be responsible for DFNB1 and many of them have been functionally validated in heterologous systems; (3) immunofluorescence and biochemical evidences indicate that Cx26 and Cx30 are the main connexins of the cochlear gap junctions, and make up of the heterotypic/heteromeric gap junction channels (Ahmad *et al.*, 2003; Forge *et al.*, 2003; Lautermann *et al.*, 1998); (4) Mouse models of Cx26 and Cx30 deficiencies have provided an entry point to understand the related disease pathogenesis (Cohen-Salmon *et al.*, 2002, 2007; Kudo *et al.*, 2003; Teubner *et al.*, 2003).

CX26/GJB2 and *CX30/GJB6* are located close together on the same chromosome (13q12). DFNB1 is mainly caused by biallelic point mutations in *CX26*, but some patients carry a *CX26* point mutation on one chromosome and a deletion encompassing *CX30* on the other chromosome. Finally, a few DFNB1 patients have biallelic deletions encompassing *CX30* (del Castillo *et al.*, 2002, 2003, 2005; Estivill *et al.*, 1998; Kelley *et al.*, 1998; Zelante *et al.*, 1997). In addition, rare mutations in *CX26* or *CX30* have been shown to cause the dominant deafness DFNA3 (Denoyelle *et al.*, 1998; Grifa *et al.*, 1999). Most importantly, mutations in *CX26* are highly prevalent especially in Caucasian population, and account for 30–50% of nonsyndromic hearing impairment congenital cases, so that testing *CX26* point mutations (and to a lesser extent *CX30* deletions) is a priority in the molecular diagnosis of nonsyndromic hearing impairment and genetic counseling. *CX26* mutations underlie a prelingual form of deafness and the severity of the hearing impairment ranges from mild to profound. Systematic analyses of audiometric data in large cohorts of patients carrying *CX26* mutations have shown a correlation between the nature of the mutations and the severity of the hearing impairment (Azaiez *et al.*, 2004; Cryns *et al.*, 2004; Snoeckx *et al.*, 2005). Audiometric curves are either flat or slope with the high frequencies being preferentially affected (Denoyelle *et al.*, 1999; Wilcox *et al.*, 2000). The vestibular function is normal and no temporal bone anomalies are found by computerized tomography (Denoyelle *et al.*, 1999). Histological examination of postmortem cochlear specimen has shown a complete degeneration of the hair cells, whereas the cochlear ganglion appeared to be preserved (Jun *et al.*, 2000).

Connexins are integral membrane proteins that assemble into hexamers to form hemichannels or connexons. A channel is composed of two connexons facing each other, from adjacent cells. Channels are grouped into gap junctions, specialized cytoplasmic membrane areas conferring permeability for ions, and small metabolites of relative molecular mass up to 1200 Da. Gap junction channels can be formed by either identical or different connexin subunits (21 different connexins have been identified in man) leading to homotypic or hybrid (heterotypic or heteromeric) channels, respectively (Laird, 2008). Homomeric and hybrid gap junction channels differ in their biochemical and electrical properties according to their connexin compositions (for a review, see Zhao *et al.*, 2006).

Gap junctions are spread within the cochlea and virtually form two main cellular networks (see Section 2.5). Analysis of the phenotype of mouse mutants will mainly provide information on functions carried out by these networks rather than by a particular cell type. Examination of *Cx* mutants has focused on two important physiological parameters: the endocochlear potential and the endolymphatic K^+ concentration, which depend on the proper functioning of the stria vascularis and the integrity of the epithelia lining the endolymphatic compartment (see Fig. 8.5). A first attempt to knockout *Cx26* in the mouse failed to provide any information about auditory impairment as homozygous mutant mice died *in utero* (Gabriel *et al.*, 1998). The absence of Cx26 indeed affects the placental uptake of glucose (and probably others nutrients) by the mouse embryo, whereas Cx26 is not expressed in the human placenta (for a review, see Malassiné and Cronier, 2005). To circumvent mouse embryonic lethality, *Cx26* was then specifically knocked out in the cochlear epithelial gap junction network, by using the conditional cre-loxP strategy (Cohen-Salmon *et al.*, 2002). In another approach, the human mutation R75W was transgenically expressed in both epithelial and connective gap junction networks, resulting in a dominant-negative effect (Kudo *et al.*, 2003). Notably, none of these mouse models completely mimics the human DFNB1 situation, since the conditional knockout of *Cx26* does not affect the cochlear connective gap junction network, and the R75W *Cx26* mutation is associated with a dominant form of deafness and keratodermia (Richard *et al.*, 1998). Important conclusions, however, have been reached from these models. Both mutant mice have a hearing impairment, which is moderate in *Cx26* conditional knockout mice and profound in the transgenic *Cx26* R75W mice. Morphological examination of the cochlea indicated that Cx26 is not required for the development of the organ of Corti, but is necessary for cell survival in this sensory epithelium. Indeed, cell death by apoptosis was observed from P14 onwards in the Cx26 conditional knockout mice. In addition, the endocochlear potential value was reduced by more than a half, and a significant decrease in the endolymphatic K^+ concentration was found in these mice (Cohen-Salmon *et al.*, 2002). These abnormalities were correlated with the disruption of the electrochemical barrier of the sensory epithelium. In the *Cx26* R75W mutant, hair cell death was observed but, surprisingly, the endocochlear potential was normal (the endolymphatic K^+ concentration was not measured) (Kudo *et al.*, 2003). In both studies, it has been proposed that the absence or decrease of Cx26 in the epithelial gap junction network results in a local increase in extracellular K^+ concentration in the basal region of hair cells that might be excitotoxic for these cells. Apparently, Cx30, which colocalizes with Cx26 in cochlear gap junctions, was not able to compensate for the lack of Cx26 in the epithelial gap junction network.

Cx30 has also been knocked-out, and while heterozygous mice do not show any defect, $C30^{-/-}$ animals display a severe to profound hearing impairment (Teubner *et al.*, 2003). Phenotypic analysis of $Cx30^{-/-}$ mice from P13 onwards showed a complete lack of the endocochlear potential before any histological alterations of the cochlea or any decrease in the endolymphatic K^+ concentration. Substantial cell death affecting hair cells and supporting cells, as well as a decrease of endolymphatic K^+ occurred from P18 onwards (Teubner *et al.*, 2003). In line with the absence of the endocochlear potential production by the stria vascularis in these mutant mice, the endothelial barrier of the capillaries running inside the stria vascularis was disrupted. In contrast, capillaries in the spiral ligament were not altered, indicating that the defect was specific to the stria vascularis (Cohen-Salmon *et al.*, 2007). The disruption of the stria vascularis endothelial barrier is likely to result in an electric shunt that might be sufficient to account for the absence of endocochlear potential. The possibility that the KCNJ10 K^+-channel of Stria Vascularis intermediate cells, which plays a major role in endocochlear potential production, was also functionally affected, however, could not be ruled out. Because *Cx30* is not expressed in stria vascularis capillaries of wild-type animals, the rupture of the stria vascularis endothelial barrier in $Cx30^{-/-}$ mice is likely to result from an indirect effect. A local increase in homocysteine concentration, a potential deleterious factor for endothelia, was found in the stria vascularis of $Cx30^{-/-}$ mice (Cohen-Salmon *et al.*, 2007).

The lack of compensation between connexins Cx26 and 30 in the earlier cited mouse models may reflect the requirement of hybrid gap junctions made of Cx26/Cx30 for normal hearing. Alternatively, the expression of *CX26* and *CX30* could be coregulated in such a way that the lack of one connexin decreases the expression of the other one. This issue has been addressed in the cochlea and the vestibule (Ahmad *et al.*, 2007; Qu *et al.*, 2007). A reduced level of Cx26 in the $Cx30^{-/-}$ cochlea, possibly due to an increased degradation of homotypic Cx26 gap junctions that are less stable than hybrid gap junctions, was observed (Ahmad *et al.*, 2007). Interestingly, the hearing function was completely rescued in transgenic mice overexpressing *Cx26* on a *Cx30*$^{-/-}$ background. In these mice, endocochlear potential values and the cochlear morphology were normal, indicating that homotypic Cx26/Cx26 gap junction channels may compensate for hybrid Cx26/Cx30 and possibly homotypic Cx30/Cx30 gap junction channels in the cochlea. Therefore, there seems to be a quantitative rather than qualitative requirement of cochlear gap junctions for normal hearing, even if hybrid gap junction channels are found under physiological conditions (Ahmad *et al.*, 2003, 2007; Forge *et al.*, 2003; Lautermann *et al.*, 1998). This conclusion is consistent with the increased auditory brainstem response thresholds and decreased endocochlear potential values found in *Cx26*$^{+/-}$/*Cx30*$^{+/-}$ double heterozygous mice that are expected to have as diverse cochlear gap junction channels as wild-type mice, though in lower amounts (Michel *et al.*, 2003). It will be of interest to examine whether the hearing impairment in the *Cx26*$^{-/-}$ conditional knockout mice can be rescued by overexpressing *Cx30* in the cochlea. Of note, a similar compensation process has been found in the vestibule where saccular (but not utricular or ampullar) hair cells degenerate in *Cx30*$^{-/-}$ mice. Indeed, overexpression of *Cx26* in the vestibule of these mice was able to rescue the degeneration of saccular hair cells (Qu *et al.*, 2007). Notably, the loss of saccular hair cells in *Cx30*$^{-/-}$ mice does not manifest by an obvious vestibular dysfunction (Qu *et al.*, 2007; Teubner *et al.*, 2003).

In addition to these *in vivo* models developed to understand the function of Cx26 and Cx30 in the inner ear and the pathogenic processes resulting from their absence, *in vitro* approaches have been used to study missense mutations encountered in human deafness that impede gap junction channel permeability (for a review, see Cohen-Salmon *et al.*, 2005). Such studies have permitted to confirm the pathogenicity of the p.M34T aminoacid substitution in Cx26 (Bicego *et al.*, 2006), and to show that some mutations in *Cx26* and *Cx30* may alter or modify the permeability of gap junction channels to certain cytoplasmic signaling molecules and metabolites necessary for cochlear function (Beltramello *et al.*, 2005; Zhang *et al.*, 2005).

In summary, the earlier mentioned *Cx* mouse models (Table 8.3) have been instrumental in elucidating physiological parameters initially affected in the DFNB1 form of deafness (i.e., endocochlear potential and endolymphatic K^+ concentration). Analysis of the partially knockout *Cx26* model has led to the suggestion that K^+ buffering by gap junctions in the perilymphatic space surrounding the base of the sensory cells may be critical to the survival of these cells, while the *Cx30* ubiquitous knockout points to an additional critical role of gap junction channels in the stria vascularis, in the production of the endocochlear potential. Because of the heterogeneity of cell types involved in the two cochlear gap junction networks, further deciphering of the DFNB1 pathogenic processes will require conditional inactivation in specific cell types. This should also permit, for instance, to test the K^+ "spatial buffering" hypothesis, and to get an insight into the functional integration of the epithelial and connective gap junction networks of the cochlea. Still, the main obstacle to such a conditional knockout approach so far is the lack of available gene promoters specific of each cochlear cell type.

5. Concluding Remarks

The early-onset forms of hereditary sensorineural hearing impairment are in their vast majority due to primary defects of the peripheral auditory system, mostly the cochlea. Clinical exams provide interesting but inaccurate information regarding the underlying pathogenic processes. Fortunately, most mouse models of human deafness genetic forms faithfully mimic them. They allow to assess a spectrum of functional characteristics of the peripheral auditory system, to collect detailed morphological observations and to integrate them in a meaningful pathogenic scenario. Moreover, they permit to track cochlear developmental failures, especially in the hair cells, which underlie numerous prelingual deafness forms. Of note, most of the hair cells' differentiation is achieved after birth in the mouse (whereas the cochlea is fully mature in the 7–8-month-old human fetus), which makes this species very much amenable for experimental testing of hair cell developmental defects. Significant progress has thus been achieved in deafness pathophysiology in the past 12 years or so. The development of therapeutic approaches still calls for clarification regarding the roles of proteins encoded by the various deafness genes, beyond the first pathogenic step, which is the only one that can be explored in full knockout models. Indeed, there is evidence that several of these proteins are critically involved in subsequently set up cochlear functions. Only timely-controlled knockout or knockdown strategies in the mouse will enable to address this issue. Moreover, a large proportion of the deafness genes encode multiple protein isoforms, whose functional relevance regarding pathogenesis needs to be elucidated.

ACKNOWLEDGMENTS

The authors wish to thank Jean-Pierre Hardelin for fruitful discussions, Paul Avan, Jacques Boutet de Monvel, and Martine Cohen-Salmon for critical reading of the manuscript, Dominique Weil and Jacqueline Levilliers for help in the preparation of the document.
The work in the authors' laboratory is supported by the European Community (EuroHear FP6 project), French National Research Agency, R. and G. Strittmatter Foundation, A. and M. Suchert Forschung contra Blindheit-Initiative Usher Syndrom and Fondation Louis-Jeantet.

REFERENCES

Adato, A., Vreugde, S., Joensuu, T., Avidan, N., Hamalainen, R., Belenkiy, O., Olender, T., Bonne-Tamir, B., Ben-Asher, E., Espinos, C., Millan, J. M., Lehesjoki, A. E., *et al.* (2002). USH3A transcripts encode clarin-1, a four-transmembrane-domain protein with a possible role in sensory synapses. *Eur. J. Hum. Genet.* **10,** 339–350.

Ahmad, S., Chen, S., Sun, J., and Lin, X. (2003). Connexins 26 and 30 are co-assembled to form gap junctions in the cochlea of mice. *Biochem. Biophys. Res. Commun.* **307,** 362–368.

Ahmad, S., Tang, W., Chang, Q., Qu, Y., Hibshman, J., Li, Y., Sohl, G., Willecke, K., Chen, P., and Lin, X. (2007). Restoration of connexin26 protein level in the cochlea completely rescues hearing in a mouse model of human connexin30-linked deafness. *Proc. Natl. Acad. Sci. USA* **104,** 1337–1341.

Ahmed, Z. M., Riazuddin, S., Bernstein, S. L., Ahmed, Z., Khan, S., Griffith, A. J., Morell, R. J., Friedman, T. B., Riazuddin, S., and Wilcox, E. R. (2001). Mutations of the protocadherin gene *PCDH15* cause Usher syndrome type 1F. *Am. J. Hum. Genet.* **69,** 25–34.

Alagramam, K. N., Yuan, H., Kuehn, M. H., Murcia, C. L., Wayne, S., Srisailpathy, C. R. S., Lowry, R. B., Knaus, R., Van Laer, L., Bernier, F. P., Schwartz, S., Lee, C., *et al.* (2001). Mutations in the novel protocadherin *PCDH15* cause Usher syndrome type 1F. *Hum. Mol. Genet.* **10,** 1709–1718.

Ashmore, J. (2008). Cochlear outer hair cell motility. *Physiol. Rev.* **88,** 173–210.

Ashmore, J. F. (1987). A fast motile response in guinea-pig outer hair cells: The cellular basis of the cochlear amplifier. *J. Physiol.* **388,** 323–347.

Azaiez, H., Chamberlin, G. P., Fischer, S. M., Welp, C. L., Prasad, S. D., Taggart, R. T., del Castillo, I., Van Camp, G., and Smith, R. J. (2004). GJB2: The spectrum of deafness-causing allele variants and their phenotype. *Hum. Mutat.* **24,** 305–311.

Beltramello, M., Piazza, V., Bukauskas, F. F., Pozzan, T., and Mammano, F. (2005). Impaired permeability to Ins(1,4,5)P3 in a mutant connexin underlies recessive hereditary deafness. *Nat. Cell Biol.* **7,** 63–69.

Beutner, D., Voets, T., Neher, E., and Moser, T. (2001). Calcium dependence of exocytosis and endocytosis at the cochlear inner hair cell afferent synapse. *Neuron* **29,** 681–690.

Bicego, M., Beltramello, M., Melchionda, S., Carella, M., Piazza, V., Zelante, L., Bukauskas, F. F., Arslan, E., Cama, E., Pantano, S., Bruzzone, R., D'Andrea, P., *et al.* (2006). Pathogenetic role of the deafness-related M34T mutation of Cx26. *Hum. Mol. Genet.* **15,** 2569–2587.

Bitner-Glindzicz, M., Lindley, K. J., Rutland, P., Blaydon, D., Smith, V. V., Milla, P. J., Hussain, K., Furth-Lavi, J., Cosgrove, K. E., Shepherd, R. M., Barnes, P. D., O'Brien, R. E., *et al.* (2000). A recessive contiguous gene deletion causing infantile hyperinsulinism, enteropathy and deafness identifies the Usher type 1C gene. *Nat. Genet.* **26,** 56–60.

Bolz, H., von Brederlow, B., Ramirez, A., Bryda, E. C., Kutsche, K., Nothwang, H. G., Seeliger, M., del, C. S. C. M., Vila, M. C., Molina, O. P., Gal, A., and Kubisch, C. (2001). Mutation of CDH23, encoding a new member of the cadherin gene family, causes Usher syndrome type 1D. *Nat. Genet.* **27,** 108–112.

Bork, J. M., Peters, L. M., Riazuddin, S., Bernstein, S. L., Ahmed, Z. M., Ness, S. L., Polomeno, R., Ramesh, A., Schloss, M., Srisailpathy, C. R. S., Wayne, S., Bellman, S., *et al.* (2001). Usher syndrome 1D and nonsyndromic autosomal recessive deafness DFNB12 are caused by allelic mutations of the novel cadherin-like gene *CDH23*. *Am. J. Hum. Genet.* **68,** 26–37.

Brownell, W. E., Bader, C. R., Bertrand, D., and de Ribaupierre, Y. (1985). Evoked mechanical responses of isolated cochlear outer hair cells. *Science* **227,** 194–196.
Cavey, M., Rauzi, M., Lenne, P. F., and Lecuit, T. (2008). A two-tiered mechanism for stabilization and immobilization of E-cadherin. *Nature* **453,** 751–756.
Chan, D. K., and Hudspeth, A. J. (2005). Ca2+ current-driven nonlinear amplification by the mammalian cochlea *in vitro.*. *Nat. Neurosci.* **8,** 149–155.
Chapman, E. R. (2002). Synaptotagmin: A Ca(2+) sensor that triggers exocytosis? *Nat. Rev. Mol. Cell Biol.* **3,** 498–508.
Cohen-Salmon, M., del Castillo, F. J., and Petit, C. (2005). Connexins responsible for hereditary deafness. The tale unfolds. *In* "Gap Junctions in Development and Disease" (E. Winterhager, ed.), pp. 111–134. Springer-Verlag, Berlin, Heidelberg.
Cohen-Salmon, M., Ott, T., Michel, V., Hardelin, J. P., Perfettini, I., Eybalin, M., Wu, T., Marcus, D. C., Wangemann, P., Willecke, K., and Petit, C. (2002). Targeted ablation of connexin26 in the inner ear epithelial gap junction network causes hearing impairment and cell death. *Curr. Biol.* **12,** 1106–1111.
Cohen-Salmon, M., Regnault, B., Cayet, N., Caille, D., Demuth, K., Hardelin, J. P., Janel, N., Meda, P., and Petit, C. (2007). Connexin30 deficiency causes instrastrial fluid-blood barrier disruption within the cochlear stria vascularis. *Proc. Natl. Acad. Sci. USA* **104,** 6229–6234.
Corey, D. P., and Hudspeth, A. J. (1983). Analysis of the microphonic potential of the bullfrog's sacculus. *J. Neurosci.* **3,** 942–961.
Cryns, K., Orzan, E., Murgia, A., Huygen, P. L., Moreno, F., del Castillo, I., Chamberlin, G. P., Azaiez, H., Prasad, S., Cucci, R. A., Leonardi, E., Snoeckx, R. L., *et al.* (2004). A genotype-phenotype correlation for GJB2 (connexin 26) deafness. *J. Med. Genet.* **41,** 147–154.
Dallos, P. (1992). The active cochlea. *J. Neurosci.* **12,** 4575–4585.
Dallos, P., Wu, X., Cheatham, M. A., Gao, J., Zheng, J., Anderson, C. T., Jia, S., Wang, X., Cheng, W. H., Sengupta, S., He, D. Z., and Zuo, J. (2008). Prestin-based outer hair cell motility is necessary for mammalian cochlear amplification. *Neuron* **58,** 333–339.
Davis, H. (1983). An active process in cochlear mechanics. *Hear. Res.* **9,** 79–90.
del Castillo, F. J., Rodríguez-Ballesteros, M., Álvarez, A., Hutchin, T., Leonardi, E., de Oliveira, C. A., Azaiez, H., Brownstein, Z., Avenarius, M. R., Marlin, S., Pandya, A., Shahin, H., *et al.* (2005). A novel deletion involving the connexin-30 gene, del(GJB6-D13S1854), found in trans with mutations in the GJB2 gene (connexin-26) in subjects with autosomal recessive nonsyndromic hearing impairment. *J. Med. Genet.*
del Castillo, I., Moreno-Pelayo, M. A., Del Castillo, F. J., Brownstein, Z., Marlin, S., Adina, Q., Cockburn, D. J., Pandya, A., Siemering, K. R., Chamberlin, G. P., Ballana, E., Wuyts, W., *et al.* (2003). Prevalence and evolutionary origins of the del (GJB6-D13S1830) mutation in the DFNB1 locus in hearing-impaired subjects: A multicenter study. *Am. J. Hum. Genet.* **73,** 1452–1458.
del Castillo, I., Villamar, M., Moreno-Pelayo, M. A., del Castillo, F. J., Alvarez, A., Telleria, D., Menendez, I., and Moreno, F. (2002). A deletion involving the connexin 30 gene in nonsyndromic hearing impairment. *N. Engl. J. Med.* **346,** 243–249.
Delmaghani, S., del Castillo, F. J., Michel, V., Leibovici, M., Aghaie, A., Ron, U., Van Laer, L., Ben-Tal, N., Van Camp, G., Weil, D., Langa, F., Lathrop, M., *et al.* (2006). Mutations in the gene encoding pejvakin, a newly identified protein of the afferent auditory pathway, cause DFNB59 auditory neuropathy. *Nat. Genet.* **38,** 770–778. Epub 2006 Jun 25.
Denoyelle, F., Lina-Granade, G., Plauchu, H., Bruzzone, R., Chaib, H., Levi-Acobas, F., Weil, D., and Petit, C. (1998). Connexin 26 gene linked to a dominant deafness. *Nature* **393,** 319–320.

Denoyelle, F., Marlin, S., Weil, D., Moatti, L., Chauvin, P., Garabedian, E. N., and Petit, C. (1999). Clinical features of the prevalent form of childhood deafness, DFNB1, due to a connexin-26 gene defect: Implications for genetic counselling. *Lancet* **353,** 1298–1303.

Di Palma, F., Holme, R. H., Bryda, E. C., Belyantseva, I. A., Pellegrino, R., Kachar, B., Steel, K. P., and Noben-Trauth, K. (2001). Mutations in Cdh23, encoding a new type of cadherin, cause stereocilia disorganization in waltzer, the mouse model for Usher syndrome type 1D. *Nat. Genet.* **27,** 103–107.

Ebermann, I., Scholl, H. P., Charbel Issa, P., Becirovic, E., Lamprecht, J., Jurklies, B., Millan, J. M., Aller, E., Mitter, D., and Bolz, H. (2007). A novel gene for Usher syndrome type 2: Mutations in the long isoform of whirlin are associated with retinitis pigmentosa and sensorineural hearing loss. *Hum. Genet.* **121,** 203–211.

Eiberger, J., Kibschull, M., Strenzke, N., Schober, A., Bussow, H., Wessig, C., Djahed, S., Reucher, H., Koch, D. A., Lautermann, J., Moser, T., Winterhager, E., *et al.* (2006). Expression pattern and functional characterization of connexin29 in transgenic mice. *Glia* **53,** 601–611.

El-Amraoui, A., Bahloul, A., and Petit, C. (2008). Myosin VII. *In* "Myosins: A Superfamily of Molecular Motors" (L. M. Coluccio, ed.), Vol. 7, pp. 353–373. Springer, New York.

El-Amraoui, A., and Petit, C. (2005). Usher I syndrome: Unravelling the mechanisms that underlie the cohesion of the growing hair bundle in inner ear sensory cells. *J. Cell Sci.* **118,** 4593–4603.

Estivill, X., Fortina, P., Surrey, S., Rabionet, R., Melchionda, S., D'Agruma, L., Mansfield, E., Rappaport, E., Govea, N., Mila, M., Zelante, L., and Gasparini, P. (1998). Connexin-26 mutations in sporadic and inherited sensorineural deafness. *Lancet* **351,** 394–398.

Eudy, J. D., Weston, M. D., Yao, S., Hoover, D. M., Rehm, H. L., Ma-Edmonds, M., Yan, D., Ahmad, I., Cheng, J. J., Ayuso, C., Cremers, C., Davenport, S., *et al.* (1998). Mutation of a gene encoding a protein with extracellular matrix motifs in Usher syndrome type IIa. *Science* **280,** 1753–1757.

Fettiplace, R., and Hackney, C. M. (2006). The sensory and motor roles of auditory hair cells. *Nat. Rev. Neurosci.* **7,** 19–29.

Forge, A., Becker, D., Casalotti, S., Edwards, J., Marziano, N., and Nevill, G. (2003). Gap junctions in the inner ear: Comparison of distribution patterns in different vertebrates and assessement of connexin composition in mammals. *J. Comp. Neurol.* **467,** 207–231.

Friedman, L. M., Dror, A. A., and Avraham, K. B. (2007). Mouse models to study inner ear development and hereditary hearing loss. *Int. J. Dev. Biol.* **51,** 609–631.

Fuchs, P. A. (2005). Time and intensity coding at the hair cell's ribbon synapse. *J. Physiol.* **566,** 7–12.

Furness, D. N., and Hackney, C. M. (1985). Cross-links between stereocilia in the guinea pig cochlea. *Hear. Res.* **18,** 177–188.

Furness, D. N., Mahendrasingam, S., Ohashi, M., Fettiplace, R., and Hackney, C. M. (2008). The dimensions and composition of stereociliary rootlets in mammalian cochlear hair cells: Comparison between high- and low-frequency cells and evidence for a connection to the lateral membrane. *J. Neurosci.* **28,** 6342–6353.

Gabriel, H. D., Jung, D., Butzler, C., Temme, A., Traub, O., Winterhager, E., and Willecke, K. (1998). Transplacental uptake of glucose is decreased in embryonic lethal connexin26-deficient mice. *J. Cell Biol.* **140,** 1453–1461.

Gibson, F., Walsh, J., Mburu, P., Varela, A., Brown, K. A., Antonio, M., Beisel, K. W., Steel, K. P., and Brown, S. D. (1995). A type VII myosin encoded by the mouse deafness gene shaker-1. *Nature* **374,** 62–64.

Glowatzki, E., and Fuchs, P. A. (2002). Transmitter release at the hair cell ribbon synapse. *Nat. Neurosci.* **5,** 147–154.

Gold, T. (1948). Hearing. II. The physical basis of the action of the cochlea. *Proc. R. Soc. Lond. B Biol. Sci.* **135,** 492–498.

Griesinger, C. B., Richards, C. D., and Ashmore, J. F. (2005). Fast vesicle replenishment allows indefatigable signalling at the first auditory synapse. *Nature* **435,** 212–215.

Grifa, A., Wagner, C. A., D'Ambrosio, L., Melchionda, S., Bernardi, F., Lopez-Bigas, N., Rabionet, R., Arbones, M., Monica, M. D., Estivill, X., Zelante, L., Lang, F., *et al.* (1999). Mutations in GJB6 cause nonsyndromic autosomal dominant deafness at DFNA3 locus. *Nat. Genet.* **23,** 16–18.

Guilford, P., Ben Arab, S., Blanchard, S., Levilliers, J., Weissenbach, J., Belkahia, A., and Petit, C. (1994). A non-syndrome form of neurosensory, recessive deafness maps to the pericentromeric region of chromosome 13q. *Nat. Genet.* **6,** 24–28.

Hasson, T., Gillespie, P. G., Garcia, J. A., MacDonald, R. B., Zhao, Y., Yee, A. G., Mooseker, M. S., and Corey, D. P. (1997). Unconventional myosins in inner-ear sensory epithelia. *J. Cell Biol.* **137,** 1287–1307.

He, D. Z., Jia, S., and Dallos, P. (2004). Mechanoelectrical transduction of adult outer hair cells studied in a gerbil hemicochlea. *Nature* **429,** 766–770.

Hill, T. L., and Kirschner, M. W. (1982). Subunit treadmilling of microtubules or actin in the presence of cellular barriers: Possible conversion of chemical free energy into mechanical work. *Proc. Natl. Acad. Sci. USA* **79,** 490–494.

Howard, J., and Hudspeth, A. J. (1988). Compliance of the hair bundle associated with gating of mechanoelectrical transduction channels in the bullfrog's saccular hair cell. *Neuron* **1,** 189–199.

Jimenez, J. L., and Bashir, R. (2007). In silico functional and structural characterisation of ferlin proteins by mapping disease-causing mutations and evolutionary information onto three-dimensional models of their C2 domains. *J. Neurol. Sci.* **260,** 114–123.

Joensuu, T., Hamalainen, R., Yuan, B., Johnson, C., Tegelberg, S., Gasparini, P., Zelante, L., Pirvola, U., Pakarinen, L., Lehesjoki, A. E., de la Chapelle, A., and Sankila, E. M. (2001). Mutations in a novel gene with transmembrane domains underlie Usher syndrome type 3. *Am. J. Hum. Genet.* **69,** 673–684.

Johnson, K. R., Gagnon, L. H., Webb, L. S., Peters, L. L., Hawes, N. L., Chang, B., and Zheng, Q. Y. (2003). Mouse models of USH1C and DFNB18: Phenotypic and molecular analyses of two new spontaneous mutations of the Ush1c gene. *Hum. Mol. Genet.* **12,** 3075–3086.

Jones, C., and Chen, P. (2007). Planar cell polarity signaling in vertebrates. *Bioessays* **29,** 120–132.

Jones, C., Roper, V. C., Foucher, I., Qian, D., Banizs, B., Petit, C., Yoder, B. K., and Chen, P. (2008). Ciliary proteins link basal body polarization to planar cell polarity regulation. *Nat. Genet.* **40,** 69–77.

Jun, A. I., McGuirt, W. T., Hinojosa, R., Green, G. E., Fischel-Ghodsian, N., and Smith, R. J. (2000). Temporal bone histopathology in connexin 26-related hearing loss. *Laryngoscope* **110,** 269–275.

Juusola, M., French, A. S., Uusitalo, R. O., and Weckstrom, M. (1996). Information processing by graded-potential transmission through tonically active synapses. *Trends Neurosci.* **19,** 292–297.

Kazmierczak, P., Sakaguchi, H., Tokita, J., Wilson-Kubalek, E. M., Milligan, R. A., Muller, U., and Kachar, B. (2007). Cadherin 23 and protocadherin 15 interact to form tip-link filaments in sensory hair cells. *Nature* **449,** 87–91.

Keller, R. (2006). Mechanisms of elongation in embryogenesis. *Development* **133,** 2291–2302.

Kelley, P. M., Harris, D. J., Comer, B. C., Askew, J. W., Fowler, T., Smith, S. D., and Kimberling, W. J. (1998). Novel mutations in the connexin 26 gene (GJB2) that cause autosomal recessive (DFNB1) hearing loss. *Am. J. Hum. Genet.* **62,** 792–799.

Kelsell, D. P., Dunlop, J., Stevens, H. P., Lench, N. J., Liang, J. N., Parry, G., Mueller, R. F., and Leigh, I. M. (1997). Connexin 26 mutations in hereditary non-syndromic sensorineural deafness. *Nature* **387,** 80–83.

Kennedy, H. J., Crawford, A. C., and Fettiplace, R. (2005). Force generation by mammalian hair bundles supports a role in cochlear amplification. *Nature* **433,** 880–883.

Kennedy, H. J., Evans, M. G., Crawford, A. C., and Fettiplace, R. (2006). Depolarization of cochlear outer hair cells evokes active hair bundle motion by two mechanisms. *J. Neurosci.* **26,** 27572766.

Kharkovets, T., Dedek, K., Maier, H., Schweizer, M., Khimich, D., Nouvian, R., Vardanyan, V., Leuwer, R., Moser, T., and Jentsch, T. J. (2006). Mice with altered KCNQ4 K+ channels implicate sensory outer hair cells in human progressive deafness. *Embo. J.* **25,** 642–652.

Kikuchi, T., Kimura, R. S., Paul, D. L., and Adams, J. C. (1995). Gap junctions in the rat cochlea: Immunohistochemical and ultrastructural analysis. *Anat. Embryol. (Berl).* **191,** 101–118.

Kikuchi, T., Kimura, R. S., Paul, D. L., Takasaka, T., and Adams, J. C. (2000). Gap junction systems in the mammalian cochlea. *Brain Res. Brain Res. Rev.* **32,** 163–166.

Kikkawa, Y., Shitara, H., Wakana, S., Kohara, Y., Takada, T., Okamoto, M., Taya, C., Kamiya, K., Yoshikawa, Y., Tokano, H., Kitamura, K., Shimizu, K., Wakabayashi, Y., Shiroishi, T., Kominami, R., and Yonekawa, H. (2003). Mutations in a new scaffold protein Sans cause deafness in Jackson shaker mice. *Hum. Mol. Genet.* **12,** 453–461.

Kimura, R. S. (1966). Hairs of the cochlear sensory cells and their attachment to the tectorial membrane. *Acta Otolaryngol.* **61,** 55–72.

Kochhar, A., Hildebrand, M. S., and Smith, R. J. (2007). Clinical aspects of hereditary hearing loss. *Genet. Med.* **9,** 393–408.

Kozlov, A. S., Risler, T., and Hudspeth, A. J. (2007). Coherent motion of stereocilia assures the concerted gating of hair-cell transduction channels. *Nat. Neurosci.* **10,** 87–92.

Kubisch, C., Schroeder, B. C., Friedrich, T., Lutjohann, B., El-Amraoui, A., Marlin, S., Petit, C., and Jentsch, T. J. (1999). KCNQ4, a novel potassium channel expressed in sensory outer hair cells, is mutated in dominant deafness. *Cell* **96,** 437–446.

Kudo, T., Kure, S., Ikeda, K., Xia, A. P., Katori, Y., Suzuki, M., Kojima, K., Ichinohe, A., Suzuki, Y., Aoki, Y., Kobayashi, T., and Matsubara, Y. (2003). Transgenic expression of a dominant-negative connexin26 causes degeneration of the organ of Corti and non-syndromic deafness. *Hum. Mol. Genet.* **12,** 995–1004.

Laird, D. W. (2008). Closing the gap on autosomal dominant connexin-26 and connexin-43 mutants linked to human disease. *J. Biol. Chem.* **283,** 2997–3001.

Lang, F., Vallon, V., Knipper, M., and Wangemann, P. (2007). Functional significance of channels and transporters expressed in the inner ear and kidney. *Am. J. Physiol. Cell Physiol.* **293,** C1187–1208.

Lautermann, J., ten Cate, W. J., Altenhoff, P., Grummer, R., Traub, O., Frank, H., Jahnke, K., and Winterhager, E. (1998). Expression of the gap-junction connexins 26 and 30 in the rat cochlea. *Cell Tissue Res.* **294,** 415–420.

Lefèvre, G., Michel, V., Weil, D., Lepelletier, L., Bizard, E., Wolfrum, U., Hardelin, J. P., and Petit, C. (2008). A core cochlear phenotype in USH1 mouse mutants implicates fibrous links of the hair bundle in its cohesion, orientation and differential growth. *Development* **135,** 1427–1437.

Liberman, M. C. (1980). Morphological differences among radial afferent fibers in the cat cochlea: An electron-microscopic study of serial sections. *Hear. Res.* **3,** 45–63.

Liberman, M. C., Gao, J., He, D. Z., Wu, X., Jia, S., and Zuo, J. (2002). Prestin is required for electromotility of the outer hair cell and for the cochlear amplifier. *Nature* **419,** 300–304.
Liberman, M. C., Zuo, J., and Guinan, J. J., Jr. (2004). Otoacoustic emissions without somatic motility: Can stereocilia mechanics drive the mammalian cochlea? *J. Acoust. Soc. Am.* **116,** 1649–1655.
Liu, X., Bulgakov, O. V., Darrow, K. N., Pawlyk, B., Adamian, M., Liberman, M. C., and Li, T. (2007). Usherin is required for maintenance of retinal photoreceptors and normal development of cochlear hair cells. *Proc. Natl. Acad. Sci. USA* **104,** 4413–4418.
Liu, X. Z., Hope, C., Liang, C. Y., Zou, J. M., Xu, L. R., Cole, T., Mueller, R. F., Bundey, S., Nance, W., Steel, K. P., and Brown, S. D. (1999). A mutation (2314delG) in the Usher syndrome type IIA gene: High prevalence and phenotypic variation. *Am. J. Hum. Genet.* **64,** 1221–1225.
Liu, X. Z., Xia, X. J., Xu, L. R., Pandya, A., Liang, C. Y., Blanton, S. H., Brown, S. D., Steel, K. P., and Nance, W. E. (2000). Mutations in connexin31 underlie recessive as well as dominant non-syndromic hearing loss. *Hum. Mol. Genet.* **9,** 63–67.
Loundon, N., Marcolla, A., Roux, I., Rouillon, I., Denoyelle, F., Feldmann, D., Marlin, S., and Garabedian, E. N. (2005). Auditory neuropathy or endocochlear hearing loss? *Otol. Neurotol.* **26,** 748–754.
Malassiné, A., and Cronier, L. (2005). Involvement of gap junctions in placental functions and development. *Biochim. Biophys. Acta* **1719,** 117–124.
Maw, M. A., Allen-Powell, D. R., Goodey, R. J., Stewart, I. A., Nancarrow, D. J., Hayward, N. K., and Gardner, R. J. (1995). The contribution of the DFNB1 locus to neurosensory deafness in a Caucasian population. *Am. J. Hum. Genet.* **57,** 629–635.
Mburu, P., Mustapha, M., Varela, A., Weil, D., El-Amraoui, A., Holme, R. H., Rump, A., Hardisty, R. E., Blanchard, S., Coimbra, R. S., Perfettini, I., Parkinson, N., *et al.* (2003). Defects in whirlin, a PDZ domain molecule involved in stereocilia elongation, cause deafness in the whirler mouse and families with mutations in DFNB31. *Nature Genet.* **34,** 421–428.
McGee, J., Goodyear, R. J., McMillan, D. R., Stauffer, E. A., Holt, J. R., Locke, K. G., Birch, D. G., Legan, P. K., White, P. C., Walsh, E. J., and Richardson, G. P. (2006). The very large G-protein-coupled receptor VLGR1: A component of the ankle link complex required for the normal development of auditory hair bundles. *J. Neurosci.* **26,** 6543–6553.
McMillan, D. R., and White, P. C. (2004). Loss of the transmembrane and cytoplasmic domains of the very large G-protein-coupled receptor-1 (VLGR1 or Mass1) causes audiogenic seizures in mice. *Mol. Cell. Neurosci.* **26,** 322–329.
Mellado Lagarde, M. M., Drexl, M., Lukashkina, V. A., Lukashkin, A. N., and Russell, I. J. (2008). Outer hair cell somatic, not hair bundle, motility is the basis of the cochlear amplifier. *Nat. Neurosci.* **11,** 746–748.
Merchan-Perez, A., and Liberman, M. C. (1996). Ultrastructural differences among afferent synapses on cochlear hair cells: Correlations with spontaneous discharge rate. *J. Comp. Neurol.* **371,** 208–221.
Michalski, N., Michel, V., Bahloul, A., Lefèvre, G., Barral, J., Yagi, H., Chardenoux, S., Weil, D., Martin, P., Hardelin, J. P., Sato, M., and Petit, C. (2007). Molecular characterization of the ankle link complex in cochlear hair cells and its role in the hair bundle functioning. *J. Neurosci.* **27,** 6478–6488.
Michel, V., Hardelin, J. P., and Petit, C. (2003). Molecular mechanism of a frequent genetic form of deafness. *N. Engl. J. Med.* **349,** 716–717.
Mustapha, M., Weil, D., Chardenoux, S., Elias, S., El-Zir, E., Beckmann, J. S., Loiselet, J., and Petit, C. (1999). An alpha-tectorin gene defect causes a newly identified autosomal

recessive form of sensorineural pre-lingual non-syndromic deafness, DFNB21. *Hum. Mol. Genet.* **8,** 409–412.
Naz, S., Griffith, A. J., Riazuddin, S., Hampton, L. L., Battey, J. F., Jr., Khan, S. N., Riazuddin, S., Wilcox, E. R., and Friedman, T. B. (2004). Mutations of ESPN cause autosomal recessive deafness and vestibular dysfunction. *J. Med. Genet.* **41,** 591–595.
Nelles, E., Butzler, C., Jung, D., Temme, A., Gabriel, H. D., Dahl, U., Traub, O., Stumpel, F., Jungermann, K., Zielasek, J., Toyka, K. V., Dermietzel, R., *et al.* (1996). Defective propagation of signals generated by sympathetic nerve stimulation in the liver of connexin32-deficient mice. *Proc. Natl. Acad. Sci. USA* **93,** 9565–9570.
Nowotny, M., and Gummer, A. W. (2006). Nanomechanics of the subtectorial space caused by electromechanics of cochlear outer hair cells. *Proc. Natl. Acad. Sci. USA* **103,** 2120–2125.
Pakarinen, L., Karjalainen, S., Simola, K. O., Laippala, P., and Kaitalo, H. (1995). Usher's syndrome type 3 in Finland. *Laryngoscope* **105,** 613–617.
Perez-Moreno, M., Jamora, C., and Fuchs, E. (2003). Sticky business: Orchestrating cellular signals at adherens junctions. *Cell* **112,** 535–548.
Petit, C. (1996). Genes responsible for human hereditary deafness: *Symphony of a thousand. Nature Genet.* **14,** 385–391.
Petit, C. (2001). Usher syndrome: From genetics to pathogenesis. *Annu. Rev. Genomics Hum. Genet.* **2,** 271–297.
Petit, C. (2006). From deafness genes to hearing mechanisms: Harmony and counterpoint. *Trends Mol. Med.* **12,** 57–64.
Pickles, J. O., Comis, S. D., and Osborne, M. P. (1984). Cross-links between stereocilia in the guinea pig organ of Corti, and their possible relation to sensory transduction. *Hear. Res.* **15,** 103–112.
Plum, A., Winterhager, E., Pesch, J., Lautermann, J., Hallas, G., Rosentreter, B., Traub, O., Herberhold, C., and Willecke, K. (2001). Connexin31-deficiency in mice causes transient placental dysmorphogenesis but does not impair hearing and skin differentiation. *Dev. Biol.* **231,** 334–347.
Qu, Y., Tang, W., Dahlke, I., Ding, D., Salvi, R., Sohl, G., Willecke, K., Chen, P., and Lin, X. (2007). Analysis of connexin subunits required for the survival of vestibular hair cells. *J. Comp. Neurol.* **504,** 499–507.
Raphael, Y., and Altschuler, R. A. (2003). Structure and innervation of the cochlea. *Brain Res. Bull.* **60,** 397–422.
Ren, T., and Gillespie, P. G. (2007). A mechanism for active hearing. *Curr. Opin. Neurobiol.* **17,** 498–503.
Richard, G., White, T. W., Smith, L. E., Bailey, R. A., Compton, J. G., Paul, D. L., and Bale, S. J. (1998). Functional defects of Cx26 resulting from a heterozygous missense mutation in a family with dominant deaf-mutism and palmoplantar keratoderma. *Hum. Genet.* **103,** 393–399.
Rivolta, C., Sweklo, E. A., Berson, E. L., and Dryja, T. P. (2000). Missense mutation in the *USH2A* gene: Association with recessive retinitis pigmentosa without hearing loss. *Am. J. Hum. Genet.* **66,** 1975–1978.
Robles, L., and Ruggero, M. A. (2001). Mechanics of the mammalian cochlea. *Physiol. Rev.* **81,** 1305–1352.
Rodriguez-Ballesteros, M., del Castillo, F. J., Martin, Y., Moreno-Pelayo, M. A., Morera, C., Prieto, F., Marco, J., Morant, A., Gallo-Teran, J., Morales-Angulo, C., Navas, C., Trinidad, G., *et al.* (2003). Auditory neuropathy in patients carrying mutations in the otoferlin gene (OTOF). *Hum. Mutat.* **22,** 451–456.
Rouiller, E. M. (1997). Functional organization of the auditory pathways. *In* "In The Central Auditory System" (G. Ehret and R. Romand, eds.), pp. 3–96. Oxford University Press, Oxford.

Roux, I., Safieddine, S., Nouvian, R., Grati, M., Simmler, M. C., Bahloul, A., Perfettini, I., Le Gall, M., Rostaing, P., Hamard, G., Triller, A., Avan, P., *et al.* (2006). Otoferlin, defective in a human deafness form, is essential for exocytosis at the auditory ribbon synapse. *Cell* **127,** 277–289.

Ruel, J., Emery, S., Nouvian, R., Bersot, T., Amilhon, B., Van Rybroek, J. M., Rebillard, G., Lenoir, M., Eybalin, M., Delprat, B., Sivakumaran, T. A., Giros, B., *et al.* (2008). Impairment of SLC17A8 encoding vesicular glutamate transporter-3, VGLUT3, underlies nonsyndromic deafness DFNA25 and inner hair cell dysfunction in null mice. *Am. J. Hum. Genet.* **83,** 278–292.

Safieddine, S., and Wenthold, R. J. (1999). SNARE complex at the ribbon synapses of cochlear hair cells: Analysis of synaptic vesicle- and synaptic membrane-associated proteins. *Eur. J. Neurosci.* **11,** 803–812.

Seal, R. P., Akil, O., Yi, E., Weber, C. M., Grant, L., Yoo, J., Clause, A., Kandler, K., Noebels, J. L., Glowatzki, E., Lustig, L. R., and Edwards, R. H. (2008). Sensorineural deafness and seizures in mice lacking vesicular glutamate transporter 3. *Neuron* **57,** 263–275.

Snoeckx, R. L., Hassan, D. M., Kamal, N. M., Van Den Bogaert, K., and Van Camp, G. (2005). Mutation analysis of the GJB2 (connexin 26) gene in Egypt. *Hum. Mutat.* **26,** 60–601.

Sobkowicz, H. M., Rose, J. E., Scott, G. L., and Levenick, C. V. (1986). Distribution of synaptic ribbons in the developing organ of Corti. *J. Neurocytol.* **15,** 693–714.

Sokolov, M., Lyubarsky, A. L., Strissel, K. J., Savchenko, A. B., Govardovskii, V. I., Pugh, E. N., Jr., and Arshavsky, V. Y. (2002). Massive light-driven translocation of transducin between the two major compartments of rod cells: A novel mechanism of light adaptation. *Neuron* **34,** 95–106.

Spicer, S. S., and Schulte, B. A. (1996). The fine structure of spiral ligament cells relates to ion return to the stria and varies with place-frequency. *Hear. Res.* **100,** 80–100.

Starr, A., Picton, T. W., Sininger, Y., Hood, L. J., and Berlin, C. I. (1996). Auditory neuropathy. *Brain* **119**(Pt 3), 741–753.

Sudhof, T. C. (2002). Synaptotagmins: Why so many? *J. Biol. Chem.* **277,** 7629–7632.

Sun, J., Ahmad, S., Chen, S., Tang, W., Zhang, Y., Chen, P., and Lin, X. (2005). Cochlear gap junctions coassembled from Cx26 and 30 show faster intercellular Ca2+ signaling than homomeric counterparts. *Am. J. Physiol. Cell. Physiol.* **288,** C613–623.

Tang, W., Zhang, Y., Chang, Q., Ahmad, S., Dahlke, I., Yi, H., Chen, P., Paul, D. L., and Lin, X. (2006). Connexin29 is highly expressed in cochlear Schwann cells, and it is required for the normal development and function of the auditory nerve of mice. *J. Neurosci.* **26,** 1991–1999.

Teubner, B., Michel, V., Pesch, J., Lautermann, J., Cohen-Salmon, M., Sohl, G., Jahnke, K., Winterhager, E., Herberhold, C., Hardelin, J. P., Petit, C., and Willecke, K. (2003). Connexin30 (Gjb6)-deficiency causes severe hearing impairment and lack of endocochlear potential. *Hum. Mol. Genet.* **12,** 13–21.

van Wijk, E., Pennings, R. J., te Brinke, H., Claassen, A., Yntema, H. G., Hoefsloot, L. H., Cremers, F. P., Cremers, C. W., and Kremer, H. (2004). Identification of 51 novel exons of the Usher syndrome type 2A (USH2A) gene that encode multiple conserved functional domains and that are mutated in patients with Usher syndrome type II. *Am. J. Hum. Genet.* **74,** 738–744.

Varga, R., Avenarius, M. R., Kelley, P. M., Keats, B. J., Berlin, C. I., Hood, L. J., Morlet, T. G., Brashears, S. M., Starr, A., Cohn, E. S., Smith, R. J., and Kimberling, W. J. (2006). OTOF mutations revealed by genetic analysis of hearing loss families including a potential temperature sensitive auditory neuropathy allele. *J. Med. Genet.* **43,** 576–581.

Varga, R., Kelley, P. M., Keats, B. J., Starr, A., Leal, S. M., Cohn, E., and Kimberling, W. J. (2003). Non-syndromic recessive auditory neuropathy is the result of mutations in the otoferlin (OTOF) gene. *J. Med. Genet.* **40,** 45–50.
Verhoeven, K., Van Laer, L., Kirschhofer, K., Legan, P. K., Hughes, D. C., Schatteman, I., Verstreken, M., Van Hauwe, P., Coucke, P., Chen, A., Smith, R. J., Somers, T., *et al.* (1998). Mutations in the human alpha-tectorin gene cause autosomal dominant non-syndromic hearing impairment. *Nat. Genet.* **19,** 60–62.
Verpy, E., Leibovici, M., Zwaenepoel, I., Liu, X. Z., Gal, A., Salem, N., Mansour, A., Blanchard, S., Kobayashi, I., Keats, B. J. B., Slim, R., and Petit, C. (2000). A defect in harmonin, a PDZ domain-containing protein expressed in the inner ear sensory hair cells, underlies Usher syndrome type 1C. *Nat. Genet.* **26,** 51–55.
von Békésy, G. (1960). "Experiments in Hearing." Mc Graw-Hill, New York.
Wang, A., Liang, Y., Fridell, R. A., Probst, F. J., Wilcox, E. R., Touchman, J. W., Morton, C. C., Morell, R. J., Noben-Trauth, K., Camper, S. A., and Friedman, T. B. (1998). Association of unconventional myosin MYO15 mutations with human nonsyndromic deafness DFNB3. *Science* **280,** 1447–1451.
Wang, Y., and Nathans, J. (2007). Tissue/planar cell polarity in vertebrates: New insights and new questions. *Development* **134,** 647–658.
Wangemann, P. (2006). Supporting sensory transduction: Cochlear fluid homeostasis and the endocochlear potential. *J. Physiol.* **576,** 11–21.
Weil, D., Blanchard, S., Kaplan, J., Guilford, P., Gibson, F., Walsh, J., Mburu, P., Varela, A., Levilliers, J., Weston, M. D., Kelley, P. M., Kimberling, W. J., *et al.* (1995). Defective myosin VIIA gene responsible for Usher syndrome type 1B. *Nature* **374,** 60–61.
Weil, D., El-Amraoui, A., Masmoudi, S., Mustapha, M., Kikkawa, Y., Laine, S., Delmaghani, S., Adato, A., Nadifi, S., Zina, Z. B., Hamel, C., Gal, A., *et al.* (2003). Usher syndrome type I G (USH1G) is caused by mutations in the gene encoding SANS, a protein that associates with the USH1C protein, harmonin. *Hum. Mol. Genet.* **12,** 463–471.
Wenthold, R. J., Safieddine, S., Ly, C. D., Wang, Y. X., Lee, H. K., Wang, C. Y., Kachar, B., and Petralia, R. S. (2002). Vesicle targeting in hair cells. *Audiol. Neurootol.* **7,** 45–48.
Weston, M. D., Luijendijk, M. W., Humphrey, K. D., Moller, C., and Kimberling, W. J. (2004). Mutations in the VLGR1 gene implicate G-protein signaling in the pathogenesis of Usher syndrome type II. *Am. J. Hum. Genet.* **74,** 357–366.
Wilcox, S. A., Osborn, A. H., and Dahl, H. H. (2000). A simple PCR test to detect the common 35delG mutation in the connexin 26 gene. *Mol. Diagn.* **5,** 75–78.
Williams, D. S. (2008). Usher syndrome: Animal models, retinal function of Usher proteins, and prospects for gene therapy. *Vision Res.* **48,** 433–441.
Xia, J. H., Liu, C. Y., Tang, B. S., Pan, Q., Huang, L., Dai, H. P., Zhang, B. R., Xie, W., Hu, D. X., Zheng, D., Shi, X. L., Wang, D. A., *et al.* (1998). Mutations in the gene encoding gap junction protein beta-3 associated with autosomal dominant hearing impairment. *Nat. Genet.* **20,** 370–373.
Yagi, H., Tokano, H., Maeda, M., Takabayashi, T., Nagano, T., Kiyama, H., Fujieda, S., Kitamura, K., and Sato, M. (2007). Vlgr1 is required for proper stereocilia maturation of cochlear hair cells. *Genes. Cells.* **12,** 235–250.
Yamada, S., Pokutta, S., Drees, F., Weis, W. I., and Nelson, W. J. (2005). Deconstructing the cadherin-catenin-actin complex. *Cell* **123,** 889–901.
Yang, J. J., Huang, S. H., Chou, K. H., Liao, P. J., Su, C. C., and Li, S. Y. (2007). Identification of mutations in members of the connexin gene family as a cause of nonsyndromic deafness in Taiwan. *Audiol. Neurootol.* **12,** 198–208.

Yasunaga, S., Grati, M., Chardenoux, S., Smith, T. N., Friedman, T. B., Lalwani, A. K., Wilcox, E. R., and Petit, C. (2000). OTOF encodes multiple long and short isoforms: Genetic evidence that the long ones underlie recessive deafness DFNB9. *Am. J. Hum. Genet.* **67,** 591–600.

Yasunaga, S., Grati, M., Cohen-Salmon, M., El-Amraoui, A., Mustapha, M., Salem, N., El-Zir, E., Loiselet, J., and Petit, C. (1999). A mutation in OTOF, encoding otoferlin, a FER-1-like protein, causes DFNB9, a nonsyndromic form of deafness. *Nat. Genet.* **21,** 363–369.

Yum, S. W., Zhang, J., Valiunas, V., Kanaporis, G., Brink, P. R., White, T. W., and Scherer, S. S. (2007). Human connexin26 and connexin30 form functional heteromeric and heterotypic channels. *Am. J. Physiol. Cell Physiol.* **293,** C1032–1048.

Zelante, L., Gasparini, P., Estivill, X., Melchionda, S., D'Agruma, L., Govea, N., Mila, M., Monica, M. D., Lutfi, J., Shohat, M., Mansfield, E., Delgrosso, K., *et al.* (1997). Connexin26 mutations associated with the most common form of non-syndromic neurosensory autosomal recessive deafness (DFNB1) in Mediterraneans. *Hum. Mol. Genet.* **6,** 1605–1609.

Zhang, Y., Tang, W., Ahmad, S., Sipp, J. A., Chen, P., and Lin, X. (2005). Gap junction-mediated intercellular biochemical coupling in cochlear supporting cells is required for normal cochlear functions. *Proc. Natl. Acad. Sci. USA* **102,** 15201–15206.

Zhao, H. B., Kikuchi, T., Ngezahayo, A., and White, T. W. (2006). Gap junctions and cochlear homeostasis. *J. Membr. Biol.* **209,** 177–186.

CHAPTER NINE

The Value of Mammalian Models for Duchenne Muscular Dystrophy in Developing Therapeutic Strategies

Glen B. Banks *and* Jeffrey S. Chamberlain

Contents

Abstract

Duchenne muscular dystrophy (DMD) is the most common form of muscular dystrophy. There is no effective treatment and patients typically die in approximately the third decade. DMD is an X-linked recessive disease caused by mutations in the dystrophin gene. There are three mammalian models of DMD that have been used to understand better the pathogenesis of disease and develop therapeutic strategies. The *mdx* mouse is the most widely used model of DMD that displays some features of muscle degeneration, but the pathogenesis of disease is comparatively mild. The severity of disease in mice lacking both dystrophin and utrophin is similar to DMD, but one has to account for the discrete functions of utrophin. Canine X-linked muscular dystrophy (*cxmd*) is the best representation of DMD, but the phenotype of the most widely used golden retriever (GRMD) model is variable, making functional endpoints

Department of Neurology, University of Washington, Seattle, Washington

Current Topics in Developmental Biology, Volume 84
ISSN 0070-2153, DOI: 10.1016/S0070-2153(08)00609-1

difficult to ascertain. Although each mammalian model has its limitations, together they have been essential for the development of several treatment strategies for DMD that target dystrophin replacement, disease progression, and muscle regeneration.

1. Introduction

Duchenne muscular dystrophy (DMD) is a progressive muscle wasting disease caused by X-linked recessive mutations in the dystrophin gene (Chamberlain, 1991; Chamberlain and Rando, 2006; Emery, 1990; Engel and Franzini-Armstrong, 2004). Approximately 1 in 3500 boys has DMD. Early clinical features include a delay in learning to walk, inability to run properly, and pseudohypertrophy of the gastrocnemius muscles. During this stage, the skeletal muscle fibers begin to degenerate and a prominent inflammatory response exists within the skeletal musculature. This is followed by severe muscle weakness and joint contractures that render children unable to walk by ~9–13 years of age. As the disease progresses, many of the skeletal muscle fibers are replaced by fibrotic tissue and adipose cells. DMD is also occasionally associated with cognitive deficits and smooth muscle abnormalities that affect digestion. Ultimately, severe cardiac abnormalities and respiratory failure in the second to third decade lead to death.

Dystrophin is a large 2.2 Mb gene composed of 79 primary exons plus 6 alternate first exons (Hoffman *et al.*, 1987; Muntoni *et al.*, 2003). The gene encodes a 427 kDa protein that has an N-terminal actin binding domain, a large central rod domain, a cysteine-rich region, and a C-terminal domain (Abmayr and Chamberlain, 2004) (Fig. 9.1). The central rod domain contains 24 spectrin-like repeats, 4 hinge regions, and a central actin-binding domain (Abmayr and Chamberlain, 2004) (Fig. 9.1). Internal promoters encode shorter isoforms of dystrophin in nonmuscle tissues, including Dp260, Dp140, Dp116, and Dp71, where the names correspond to the size of the protein (Chelly *et al.*, 1990b; D'Souza *et al.*, 1995; Gorecki *et al.*, 1992; Lidov *et al.*, 1995) (Fig. 9.1). Only full-length dystrophin is expressed in neonatal and adult skeletal muscle, where it provides a flexible connection between the cytoskeleton and the dystrophin–glycoprotein complex (DGC) (Ervasti, 2007) (Fig. 9.2). Dystrophin is located at the sarcolemma in connection with the costameric lattice at Z- and M-lines of peripheral sarcomeres (Ervasti, 2007). Dystrophin is also concentrated at the neuromuscular and myotendinous junctions (Ervasti, 2007).

A comprehensive list of dystrophin mutations that cause DMD can be found on the Leiden muscular dystrophy pages© at www.dmd.nl. There are two hot spots for mutations that encode regions within the N-terminal actin-binding domain and regions surrounding hinge 3 in the central rod

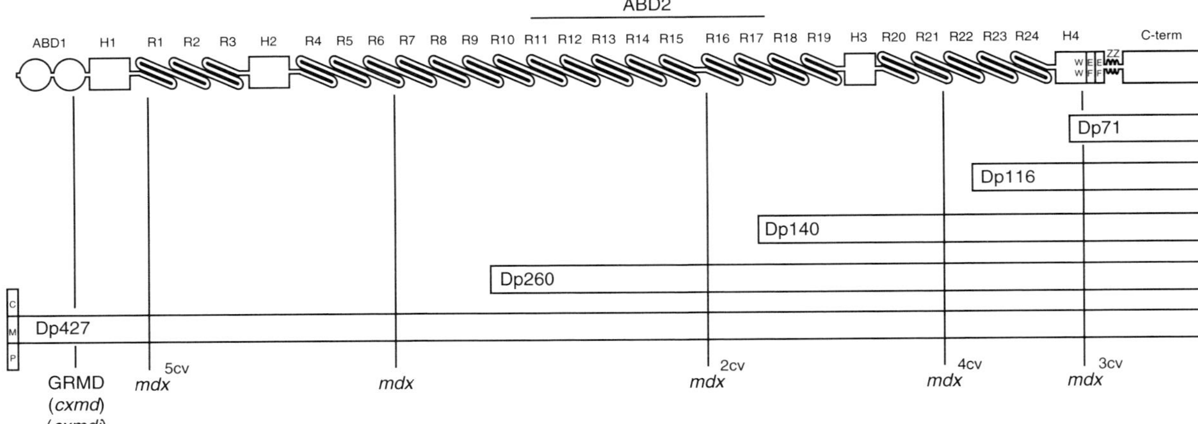

Figure 9.1 The molecular structure of dystrophin. Dystrophin contains an N-terminal actin binding domain (ABD1), a large central rod domain, a cysteine-rich region and a C-terminal domain. ABD1 at the N-terminus is composed of two calponin homology domains denoted by the two circles. The central rod domain contains 24 spectrin-like repeats (R1–24), 4 hinge domains, a 20 amino acid insertion between spectrin repeats 15 and 16, and a central actin-binding domain (ABD2). A cluster of basic repeats forms ABD2 that bind to actin through an electrostatic interaction (Amann *et al.*, 1998). The spectrin-like repeats vary in terms of their helicity (Legardinier *et al.*, 2008). The hinge domains also vary in that hinge 2 contains a polyproline site and hinge 4 contains a WW motif that is required for binding to dystroglycan (Ervasti *et al.*, 1990; Koenig and Kunkel, 1990; Rentschler *et al.*, 1999). The cysteine-rich region contains two EF hands and a ZZ domain that is also required for binding to dystroglycan. The dystrophin gene contains seven promoters that express either the full length dystrophin (C; central nervous system: M; muscle: P; purkinje) or truncated dystrophins (Dp260, Dp140, Dp116, or Dp71). The sites of mutations that lead to the five different *mdx* mouse models are shown below. Also shown is the site of a mutation that causes canine linked muscular dystrophy (*cxmd*) in the golden retriever muscular dystrophy (GRMD) and beagle (*cxmdj*). Note that certain mutations would not disrupt the expression of truncated dystrophins in nonmuscle tissues (Im *et al.*, 1996).

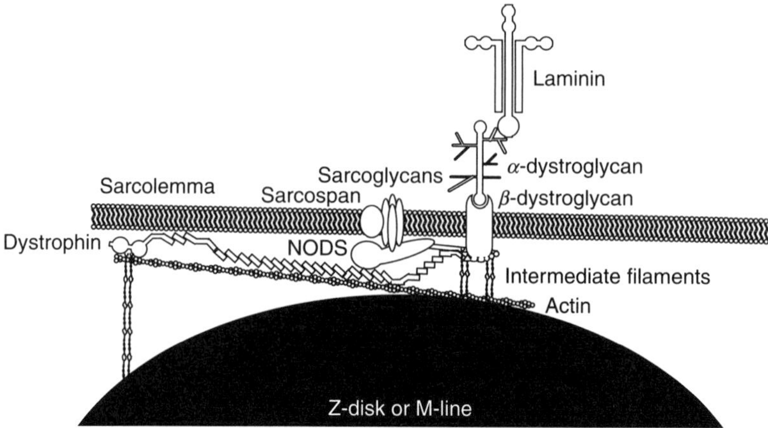

Figure 9.2 Dystrophin forms a flexible connection between the cytoskeleton and the dystrophin–glycoprotein complex, and the NODS (nNOS, dystrobrevin, and syntrophin) complex. This model is based on evidence that suggests a single actin filament binds to dystrophin through its two actin binding domains (Rybakova *et al.*, 2006). The hinges likely allow spectrin-like repeats 1–3 to interact with the sarcolemma and the central actin-binding domain to interact with actin (Legardinier *et al.*, 2008). The more flexible nature of spectrin repeats 6 and 7 probably help expose the central actin-binding domain to actin (Saadat *et al.*, 2006). For a detailed description of the intermediate filaments that bind dystrophin to the Z-disks or M-lines (see Ervasti 2003, 2007).

domain (Baumbach *et al.*, 1989; Chamberlain *et al.*, 1992; Gillard *et al.*, 1989; Hoffman *et al.*, 1988; Koenig *et al.*, 1989; Monaco *et al.*, 1988). Deletions are the most common type of mutation in the dystrophin gene and can be rapidly detected during DNA diagnostics using multiplex PCR (Chamberlain *et al.*, 1988; Koenig *et al.*, 1989; Miller and Hoffman, 1994). As a general rule, out-of frame deletions lead to DMD (Koenig *et al.*, 1989). In-frame deletions can also lead to DMD if the region being deleted is essential for dystrophin expression or function, such as the cysteine-rich region (Koenig *et al.*, 1989). In-frame deletions in the central rod domain usually lead to a more mild form of dystrophy, called Becker muscular dystrophy (BMD) (Koenig *et al.*, 1989; Monaco *et al.*, 1988), unless they are within the N-terminal actin binding domain where protein levels are usually reduced to 10–20% of wild-type, which causes a more severe BMD (Beggs *et al.*, 1991; Chelly *et al.*, 1990a; Le *et al.*, 1993; Matsumura *et al.*, 1994; Muntoni *et al.*, 1994; Novakovic *et al.*, 2005; Prior *et al.*, 1993; Takeshima *et al.*, 1994; Winnard *et al.*, 1995). However, there are exceptions: Some out-of-frame deletions in the N-terminal actin-binding domain can lead to BMD, due to translational reinitiation at an internal Kozak sequence (Winnard *et al.*, 1995). Furthermore, in-frame deletions around hinge 3 can lead to highly variable phenotypes, including DMD and mild BMD (Baumbach *et al.*, 1989).

The identification of mutations in the dystrophin gene that can cause DMD and BMD gave hope to many families for finding a quick cure. However, the mammalian models of DMD quickly served to highlight the complexity of this disease and required several breakthroughs in technology to understand some of the pathogenic mechanisms and develop prospective treatment strategies (Chamberlain and Rando, 2006). Here, we provide an overview of the most commonly used mammalian models for DMD: the *mdx* mice, *mdx*:utrophin double mutant mice (*mdx*:*utrn*$^{-/-}$), and canine x-linked muscular dystrophy (*cxmd*). We describe the mutations in dystrophin that lead to muscle degeneration in each mammalian model. We also describe how these mammalian models have been used to develop several prospective treatment strategies and the limitations of these models for predicting the effectiveness of these therapeutic agents.

2. The *mdx* Mouse Models of DMD

Our understanding of the molecular mechanisms that underlie the pathogenesis of DMD has benefited enormously from five *mdx* mouse models (Fig. 9.1). The original *mdx* mutant contains a premature stop codon in exon 23 (Bulfield *et al.*, 1984; Sicinski *et al.*, 1989). The *mdx*^{2cv} model results from a mutation in intron 42, which affects RNA splicing (Im *et al.*, 1996). The *mdx*^{3cv} allele arises from a mutant splice acceptor site in intron 65 (Cox *et al.*, 1993). The *mdx*^{4cv} mouse has a C- to T-transition in exon 53, creating a nonsense ochre codon (Im *et al.*, 1996). And finally, the *mdx*^{5cv} allele has an A to T transition in exon 10, creating a new splice donor site that generates a premature stop codon in RNA transcripts (Im *et al.*, 1996). Each mutation leads to a loss of dystrophin protein expression in skeletal muscles. The mice vary in the number of revertant muscle fibers in which some muscle fibers express partially functional truncated dystrophins (Danko *et al.*, 1992). Both *mdx*^{4cv} and *mdx*^{5cv} have the fewest revertant fibers (Im *et al.*, 1996). In contrast, the *mdx*^{3cv} strain expresses very low levels of an internally truncated dystrophin arising from aberrant splicing that disrupts the dystroglycan-binding domain (Cox *et al.*, 1993). The strains of *mdx* mice also vary in which isoforms of dystrophin are expressed in nonmuscle tissues (Haenggi and Fritschy, 2006; Im *et al.*, 1996) (Fig. 9.1).

2.1. Contraction-induced injury

Dystrophin provides a flexible connection between the costameric cytoskeleton and the DGC at the sarcolemma (Ervasti, 2007) (Fig. 9.2). In turn, the DGC binds directly to the basal lamina and the extracellular matrix (Ervasti, 2007). This molecular scaffold is thought to transfer lateral forces from the sarcomeres, to the extracellular matrix, and ultimately, to the tendon (Bloch and Gonzalez-Serratos, 2003; Ervasti, 2003). Dystrophin also directly binds to the membrane and could stabilize it during contraction (Le Rumeur *et al.*, 2003; Legardinier *et al.*, 2008). The absence of dystrophin leads to profound reductions in the DGC at the sarcolemma (Grady *et al.*, 1997). The lack of structural support at the sarcolemma leaves *mdx* muscles more susceptible to contraction-induced injury, especially during lengthening contractions (Brooks, 1998; Dellorusso *et al.*, 2001; Faulkner *et al.*, 2008; Moens *et al.*, 1993; Petrof *et al.*, 1993). Contraction-induced injury is thought to initiate muscle degeneration (Ervasti, 2007; Lynch, 2004) (Fig. 9.3).

The most efficient mechanism to protect skeletal muscles from contraction-induced injury in DMD is to replace or restore dystrophin expression. Gene replacement/correction includes viral delivery of therapeutic cassettes to skeletal and cardiac muscles (Chamberlain and Rando, 2006). The most proficient of these is recombinant adeno-associated virus (rAAV), which delivers highly functional truncated dystrophins to cardiac and skeletal muscles body wide with a single intravenous injection when pseudotyped with serotype 1, 6, 8, or 9 capsids (Bostick *et al.*, 2008; Gregorevic *et al.*, 2004; Judge and Chamberlain, 2005; Rodino-Klapac *et al.*, 2007; Wang *et al.*, 2005). RNA manipulation is also an effective therapeutic strategy to restore dystrophin expression in *mdx* mice (Muntoni

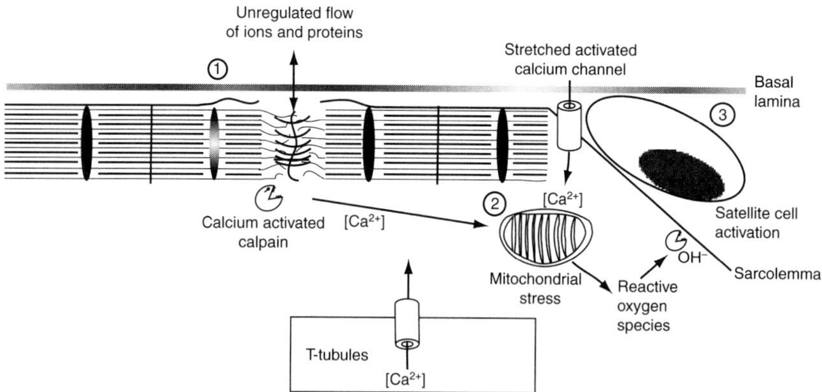

Figure 9.3 Model of pathogenesis of muscular dystrophy. Contraction-induced injury leads to the unregulated flow of ions and proteins through the sarcolemma. The influx of calcium activates calpains and causes mitochondrial stress that can lead to muscle degeneration and activation of satellite cells. Prospective targets for therapy include: (1) Dystrophin replacement or increasing the expression of utrophin at Z-disks to prevent contraction-induced injury; (2) Mitigate mitochondrial related stressors; and (3) Utilize the regenerative potential of skeletal muscle. See also Ozawa (2006) in Chamberlain and Rando (2006).

and Wells, 2007). Viral-directed exon skipping (Denti *et al.*, 2006; Goyenvalle *et al.*, 2004), oligonucleotide exon skipping (Alter *et al.*, 2006; Fletcher *et al.*, 2007; Lu *et al.*, 2003), and pharmacological read through of premature stop codons (Welch *et al.*, 2007) each manipulate the RNA to restore dystrophin expression. *mdx* mice have proven extremely useful for maximizing the efficacy of each treatment in terms of generating functional truncated dystrophins and improving the dissemination of therapeutic agents to skeletal and cardiac muscles. Each of these therapeutic agents has been shown to significantly ameliorate muscle degeneration, improve muscle force production, and partially protect muscles from contraction-induced injury in *mdx* mice.

Utrophin-mediated therapies can also prevent muscle degeneration in *mdx* mice (Odom *et al.*, 2008; Rybakova *et al.*, 2002; Tinsley *et al.*, 1996; Tinsley *et al.*, 1998). Utrophin is a large (376 kDa) dystrophin paralogue (Love *et al.*, 1989). Utrophin has two full-length isoforms that are transcribed by different promoters and differ in the initial exons at the N-termini (Burton *et al.*, 1999). Utrophin A is expressed in striated muscles, choroid plexus, pia matter, and renal glomerulus (Weir *et al.*, 2002). Utrophin B is expressed in blood vessels (Weir *et al.*, 2002). Utrophin is also expressed in the lung and presynaptic nerve terminal, but the isoform specificity has not been described. Utrophin A is located throughout the muscle membrane during embryonic development (Takemitsu *et al.*, 1991), and is restricted to the neuromuscular and myotendinous junctions during early postnatal development (Zhao *et al.*, 1992). In *mdx* mice, utrophin A protein expression is increased and is found in low levels at the sarcolemma (Law *et al.*, 1994; Matsumura *et al.*, 1992). Utrophin functionally compensates for the lack of dystrophin in *mdx* mice by maintaining an interaction between the actin cytoskeleton and the DGC at the sarcolemma (Matsumura *et al.*, 1992; Rybakova *et al.*, 2002). An increase in expression levels of utrophin correlates with a delay to wheelchair use in DMD patients, validating this type of therapy in humans (Kleopa *et al.*, 2006). Possibly, the greatest advantage of a utrophin-mediated therapy is that it has the potential to circumvent an immune response to dystrophin, because utrophin would not be viewed as a foreign antigen as it has previously been produced in the recipient (Wakefield *et al.*, 2000). Several laboratories are looking for pharmacological agents that could increase utrophin expression at the sarcolemma to protect the muscles from contraction-induced injury (reviewed in Hirst *et al.*, 2005; Khurana and Davies, 2003).

2.2. Degeneration

The skeletal muscles of *mdx* mice begin to degenerate at approximately 3 weeks of age (McGeachie *et al.*, 1993). Large regions of muscle necrosis are evident by approximately 7 weeks of age beyond which there is a low level of ongoing muscle degeneration and regeneration (McGeachie *et al.*, 1993). The molecular mechanisms of how the loss of dystrophin leads to muscle degeneration remain unclear, although some reasonable hypotheses exist (Deconinck and Dan, 2007) (Fig. 9.3). Contraction-induced injury leads to tears in the sarcolemma, allowing unregulated passage of proteins and ions through the membrane (Lynch, 2004). Creatine kinase leaves the muscles through these tears and enters the blood stream (Ozawa *et al.*, 1999). Creatine kinase is used as an indicator of muscle degeneration in the diagnosis of various muscular dystrophies. Evans blue dye enters dystrophic muscles through tears in the skeletal muscle and is a common marker of sarcolemma integrity (Matsuda *et al.*, 1995). It is thought that the unregulated passage of proteins and ions through the sarcolemma could lead to muscle degeneration (Fig. 9.3).

Calcium is normally tightly controlled in skeletal muscle, where influxes into the cytosol are only allowed during muscle contraction (Gissel, 2005). Tears in the muscle membrane can lead to an unregulated influx of calcium down its electrochemical gradient (Alderton and Steinhardt, 2000; Blake *et al.*, 2002; Bradley and Fulthorpe, 1978). Calcium also enters the muscles through stretch activated ion channels (Suchyna and Sachs, 2007; Yeung *et al.*, 2005) and internal calcium stores (Boittin *et al.*, 2006; Vandebrouck *et al.*, 2002) (Fig. 9.3). The persistence of calcium in the cytosol could lead to muscle degeneration in at least two ways. First, calcium activates calpains in *mdx* muscle fibers (Alderton and Steinhardt, 2000). Calpains are calcium-dependent proteases that have an active role in muscle necrosis in *mdx* mice by digesting myofibrils and other proteins, such as titin, at Z-disks and I bands (Spencer and Mellgren, 2002). The second way calcium can lead to muscle degeneration is *via* disturbances in metabolism through mitochondria. Calcium is sequestered by mitochondria when the cytosolic levels become very high (Robert *et al.*, 2001). This concentration can lead to mitochondrial swelling, loss of the mitochondrial inner membrane, and oxidative stress (Rando *et al.*, 1998). The production of reactive oxygen species such as hydroxyl radicals (OH^-) can self propagate damage to the cell membrane and cause necrotic and/or apoptotic cell death (Schriner *et al.*, 2005). Various pharmacological treatments are being explored that mitigate the progression of dystrophy in *mdx* mice. Treatment of *mdx* mice with the cyclophilin inhibitor Debio-025 prevents calcium-induced mitochondrial swelling and minimizes the progression of dystrophy (Millay *et al.*, 2008). The antioxidant *N*-acetylcysteine can reduce the amount of reactive oxygen species and muscle degeneration (Whitehead *et al.*, 2008). Thus, unregulated calcium entry into skeletal muscle fibers could lead to muscle degeneration through several different mechanisms and some of these can be targeted with pharmacological agents to ameliorate muscle degeneration in *mdx* mice.

Neuronal nitric oxide synthase is absent from the sarcolemma in *mdx* mice and DMD patients (Brenman *et al.*, 1995). Transgenic expression of nNOS in *mdx* mice significantly attenuates muscle degeneration (Wehling *et al.*, 2001), but the mechanism is not clear (Tidball and Wehling-Henricks, 2007). It is possible that the production of nitric oxide could minimize the damage evoked by reactive oxygen species (Tidball and Wehling-Henricks, 2007). Nitric oxide could also prevent macrophages from lysing the myotubes in a free radical-mediated mechanism (Nguyen and Tidball, 2003). Nitric oxide can mediate its affects through cGMP (Hare and Stamler, 2005), and up regulation of cGMP with sildenafil (a PDE5 inhibitor) mitigates the cardiomyopathy in *mdx* mice (Khairallah *et al.*, 2008).

2.3. Regeneration

Skeletal muscles have the ability to grow and regenerate through satellite cells. Satellite cells are a heterogeneous mixture of stem cells and committed myogenic precursors that lie between the basal lamina and the sarcolemma of skeletal muscle fibers (Kuang *et al.*, 2007). Satellite cells contribute to muscle growth and regeneration (Collins *et al.*, 2005) by dividing, fusing, and terminally differentiating into myofibers (Kuang *et al.*, 2007). The mechanisms responsible for causing satellite cells to exit quiescence and enter the cell cycle remain unclear (Le Grand and Rudnicki, 2007).

Satellite cells regenerate skeletal muscles in DMD patients and in *mdx* mice (Partridge, 2006). Ultimately, environmental factors reduce the myogenic capacity of satellite cells with age and contribute to sarcopenia (Brack and Rando, 2007; Carlson *et al.*, 2008; Conboy *et al.*, 2003, 2005; Schultz and Lipton, 1982). This regression is accelerated in DMD patients (Webster and Blau, 1990). Most cell-based therapies for DMD aim to restore skeletal muscles with allogeneic stem cells or autologous stem cells that are genetically corrected for the dystrophin mutation (Parker *et al.*, 2008; Partridge, 2006).

While age-dependent decreases in myogenic potential of satellite cells in *mdx* mice are minor (Schuierer *et al.*, 2005), *mdx* mice have proven extremely useful in gauging the efficacy of cell-based therapies. Unfortunately, the regenerative capacity of satellite cells *in vivo* is not maintained, when they are expanded in culture (Montarras *et al.*, 2005). Furthermore, satellite cells are unable to exit the vasculature after intravenous administration, and they do not migrate far after intramuscular injection (Morgan *et al.*, 1993). And finally, many of the satellite cells are lost in the first day after injection (Beauchamp *et al.*, 1999). Thus, the therapeutic efficacy of satellite cells in regenerating large regions of muscle is limited. For these reasons, development of cell-based therapies for DMD has largely benefited from stem cells other than satellite cells. Embryonic stem cells (Darabi *et al.*, 2008), bone marrow derived myogenic progenitors (Ferrari *et al.*, 1998), and bone marrow stromal cells (Dezawa *et al.*, 2005) can be expanded, differentiated into muscle precursors, and delivered systemically to regenerate and repair muscles in *mdx* mice. Pericytes and mesoangioblasts are myogenic precursors that are also capable of being expanded to large numbers in cell culture, delivered systemically to the muscles and differentiate into muscles with high efficacy (Dellavalle *et al.*, 2007; Sampaolesi *et al.*, 2003).

2.4. Deficiencies of the *mdx* mouse models

It is unclear why muscle degeneration in the *mdx* mouse model is mild in comparison to DMD patients (although some would argue that progression of dystrophy in the diaphragm is similar in *mdx* mice and DMD patients (Faulkner *et al.*, 2008)). For instance, *mdx* mice are mobile, they do not have significant fibrosis or joint contractures, and the skeletal myofibers are only partially replaced by adipose cells later in life. The myotendinous junctions are severely impaired in DMD patients (Bell and Conen, 1968; Hasegawa *et al.*, 1992; Nagao *et al.*, 1991), but only have minor alterations in maturation and maintenance in *mdx* mice (Law and Tidball, 1993). In addition, the loss of synaptic folds in the neuromuscular synapse has little effect on synaptic transmission in *mdx* mice (Banks *et al.*, 2003; Carlson and Roshek, 2001; Lyons and Slater, 1991), but could have a greater effect in DMD patients where the safety factor is more reliant on the postsynaptic folds (Slater, 2003). Furthermore, it is not clear by using *mdx* mice that different therapeutic strategies can prolong the lifespan of *mdx* mice, because it is only moderately shortened (~20%) (Chamberlain *et al.*, 2007). Thus, therapeutic strategies may benefit from examining more severe mammalian models of DMD.

3. *MDX*:Utrophin Double Knockout Mouse Model

One possibility for the mild phenotype of *mdx* mice is that the functional requirement of dystrophin to transmit muscle forces is less in mice because they are smaller and weaker than in humans. Satellite cells also retain their regenerative potential better in *mdx* mice than in DMD patients. Another possibility is that homologous proteins compensate more effectively for the absence of dystrophin in mice. Consistent with this hypothesis, two independent laboratories generated mice lacking both dystrophin and utrophin to generate a more severe model of DMD (Deconinck *et al.*, 1997a; Grady *et al.*, 1997). *mdx*:*utrn*$^{-/-}$ mice are smaller than wild-type mice, develop severe kyphosis, and become less mobile with age (Deconinck *et al.*, 1997a; Grady *et al.*, 1997). They develop an inflammatory response in the skeletal musculature (Deconinck *et al.*, 1997a; Grady *et al.*, 1997). Many of the muscle fibers are replaced by fibrotic tissue that contributes to joint contractures (Deconinck *et al.*, 1997a; Grady *et al.*, 1997).

Regions of the muscles also turn into adipose tissue (Deconinck *et al.*, 1997a; Grady *et al.*, 1997). The skeletal muscles are smaller and weaker than both wild-type and *mdx* mice and are more susceptible to contraction-induced injury (Gregorevic *et al.*, 2006). Neuromuscular junctions are fragmented and the postsynaptic folds are significantly reduced when compared with wild-type and *mdx* mice (Deconinck *et al.*, 1997a; Grady *et al.*, 1997; Rafael *et al.*, 2000). The terminal sarcomeres rarely make direct contact with the tendons at myotendinous junctions (Deconinck *et al.*, 1997a). *mdx:utrn*$^{-/-}$ mice also present with a severe cardiomyopathy (Grady *et al.*, 1997; Janssen *et al.*, 2005). Ultimately, *mdx:utrn*$^{-/-}$ mice die within 20 weeks of age (Deconinck *et al.*, 1997a; Grady *et al.*, 1997). The pathology and shortened lifespan of *mdx:utrn*$^{-/-}$ mice is more representative of DMD patients and is therefore a more rigorous model for testing novel treatment strategies (Deconinck *et al.*, 1997a; Grady *et al.*, 1997).

Gene therapy using either truncated dystrophins or utrophins ameliorate the dystrophic pathology in *mdx:utrn*$^{-/-}$ mice (Gregorevic *et al.*, 2006; Odom *et al.*, 2008; Wakefield *et al.*, 2000; Yue *et al.*, 2006). Systemic injection of rAAV6-microutrophin and rAAV6-microdystrophin mitigates muscle degeneration, inflammation, fibrosis, joint contractures, and kyphosis (Gregorevic *et al.*, 2006; Odom *et al.*, 2008). Treated muscles are stronger and more resistant to contraction-induced injury (Gregorevic *et al.*, 2006; Odom *et al.*, 2008). rAAV6-microdystrophin has also been shown to attenuate the cardiomyopathy in *mdx:utrn*$^{-/-}$ mice (Gregorevic *et al.*, 2006). Both rAAV6-microutrophin and -microdystrophin significantly prolong the lifespan of *mdx:utrn*$^{-/-}$ mice, but they eventually die from mega-esophagus (Gregorevic *et al.*, 2006; Odom *et al.*, 2008). Mesoangioblasts have also been shown to mitigate disease progression in *mdx:utrn*$^{-/-}$ mice (Berry *et al.*, 2007). Thus, *mdx:utrn*$^{-/-}$ mice have the advantage over *mdx* mice of being able to reveal whether a treatment can maintain muscle mass, prevent severe fibrosis and prolong survival.

While dystrophin and utrophin have many overlapping functions in *mdx* mice, it is possible that some of their functions are distinct. For instance, utrophin is expressed in satellite cells and could localize the DGC to the sarcolemma to achieve normal muscle regeneration (Cohn *et al.*, 2002). In addition, the utrophin B isoform is expressed in blood vessels that could affect muscle metabolism in *mdx:utrn*$^{-/-}$ mice (Rafael *et al.*, 2000). Furthermore, bidirectional communication between the muscle and nerve could affect the progression of dystrophy through increased expression of utrophin (Handschin *et al.*, 2007). The loss of both dystrophin and utrophin also elicits unusual changes in expression of shorter dystrophin isoforms in nonmuscle tissues (Deconinck *et al.*, 1997b). Thus, the *mdx:utrn*$^{-/-}$ mice phenotype may not faithfully reproduce the DMD phenotype and interpretations of therapeutic interventions in these mice should consider broad aspects of utrophin function.

4. The *cxmd* Canine Model

There are several *cxmd* models of DMD, including the golden retriever (GRMD) (Cooper *et al.*, 1988), rottweiler (Partridge, 1997), and German short-haired pointer (Schatzberg *et al.*, 1999). The GRMD dogs have been bred to beagles in Japan, which they call *cxmdj* (Shimatsu *et al.*, 2003). The GRMD dogs and the *cxmdj* have the same point mutation and similar phenotype (Shimatsu *et al.*, 2005). A feline model of DMD also exists (Kohn *et al.*, 1993), but it does not closely follow the pathogenesis of DMD and is rarely studied. The most studied canine model is the GRMD (Cooper *et al.*, 1988).

The pathogenesis of *cxmd* is similar to DMD (Valentine *et al.*, 1992). There is a high mortality rate of early neonatal GRMD dogs from selective muscle degeneration (Valentine and Cooper, 1991; Valentine *et al.*, 1988). For dogs that live through the neonatal period, muscle degeneration is followed by muscle regeneration and a large inflammatory response (Nguyen *et al.*, 2002). Some of the muscles have high concentrations of crystalline calcium and hyaline (Cooper *et al.*, 1988; Nguyen *et al.*, 2002). The muscle fibers begin to be replaced by fibrotic tissue and adipose cells at approximately 2 months of age (Nguyen *et al.*, 2002). Joint contractures are prominent by 6 months and mobility is severely impaired. The muscles are atrophic, weaker, and more susceptible to contraction-induced injury (Childers *et al.*, 2002; Nguyen *et al.*, 2002). GRMD dogs develop cardiomyopathy (Chetboul *et al.*, 2004a,b) and respiratory distress that can lead to death (Valentine and Cooper, 1991).

There is significant variability between *cxmd(j)*(j) dogs with the same mutation (Cooper *et al.*, 1988; Shimatsu *et al.*, 2003). The dystrophic pathology in these dogs results from a point mutation in the 3′ consensus splice site of exon 6 in dystrophin (Sharp *et al.*, 1992) (Fig. 9.1). Dystrophin expression can be restored, when there is a point mutation in the N-terminal actin binding domain in humans (Winnard *et al.*, 1995), and this is also evident in the GRMD model (Schatzberg *et al.*, 1998). Expression of truncated dystrophins with deletions from exons 2–10 and 4–13 can be found in the GRMD dogs (Schatzberg *et al.*, 1998). The GRMD model has mosaic expression of dystrophin that becomes more uniform with age (Cooper *et al.*, 1990). These truncated dystrophins would lack part of the N-terminal actin binding domain, hinge 1, spectrin repeat 1, and part of spectrin repeat 2. The N-terminal actin-binding domain of dystrophin is important for dystrophin expression and function (Banks *et al.*, 2007; Beggs *et al.*, 1991; Chelly *et al.*, 1990a; Le *et al.*, 1993; Matsumura *et al.*, 1994; Muntoni *et al.*, 1994; Novakovic *et al.*, 2005; Prior *et al.*, 1993; Takeshima *et al.*, 1994; Winnard *et al.*, 1995). Deletions within hinge 1 usually lead to a more severe BMD than do deletions of adjacent sequences (Novakovic *et al.*, 2005). In addition, deletions within the second spectrin-like repeat could affect the interaction of dystrophin with lipid (Legardinier *et al.*, 2008). Although these regions are important for dystrophin expression and function, the central actin-binding domain can partially compensate for deletions in the N-terminal actin-binding domain (Warner *et al.*, 2002; Rybakova *et al.*, 2006). Dp260 (Fig. 9.1) can mitigate muscle degeneration similar to a mild BMD phenotype when expression levels are near normal (Warner *et al.*, 2002). Thus, variations in expression of the truncated dystrophins in the *cxmd* model could explain the variability in phenotype between dogs with the same mutation.

The *cxmd* dogs have several advantages over other mammalian models. They have an immune system that is similar to the human immune system. The prevalent immune response in skeletal muscles in *cxmd* dogs also primes the immune system toward foreign substances (Wang *et al.*, 2007). Immune responses against the capsid of rAAV6 has been examined in the *cxmd* model and transient immune suppression was observed to allow sustained expression of the dystrophin transgene following delivery using AAV (Wang *et al.*, 2007). The *cxmd* model could also be used to examine whether novel dystrophin polypeptide boundaries (caused by exon skipping or rational design of truncated dystrophins) are recognized by the immune system as foreign antigens. However, it is unlikely that the *cxmd* model would elicit an immune response to most parts of dystrophin because the *cxmd* dogs have had prior exposure to this protein in revertant fibers. *cxmd* dogs can also be used to examine whether a therapy can be scaled up to treat the significant muscle mass of humans (Sampaolesi *et al.*, 2006). However, establishing appropriate functional endpoints of a therapy has proven to be challenging because the clinical course of disease can vary considerably between dogs (Bretag, 2007; Sampaolesi *et al.*, 2007). Another limitation is that *cxmd* dogs are expensive and difficult to breed and maintain.

5. Conclusions

Mammalian models of DMD have proven invaluable in defining the complexity of muscle disease. Each mammalian model has been used extensively to generate several promising therapeutic strategies for DMD. Mammalian models also limit the number of unexpected challenges everyone must face in clinical trials. Translational science in DMD is paving the way for potential treatment strategies for the more than 30 other forms of muscular dystrophy.

ACKNOWLEDGMENTS

We are grateful to Chamberlain lab members for critical review of this manuscript. This work was supported by grants from the National Institutes of Health. GBB was supported by a CJ Martin Postdoctoral Fellowship from the National Health and Medical Research Council of Australia (372212).

REFERENCES

Abmayr, S., and Chamberlain, J. S. (2004). The structure and function of dystrophin. *In* "The Molecular Mechanisms of Muscular Dystrophy" (S. Winder, ed.), Landes Bioscience, Georgetown.

Alderton, J. M., and Steinhardt, R. A. (2000). How calcium influx through calcium leak channels is responsible for the elevated levels of calcium-dependent proteolysis in dystrophic myotubes. *Trends Cardiovasc. Med.* **10,** 268–272.

Alter, J., Lou, F., Rabinowitz, A., Yin, H., Rosenfeld, J., Wilton, S. D., Partridge, T. A., and Lu, Q. L. (2006). Systemic delivery of morpholino oligonucleotide restores dystrophin expression bodywide and improves dystrophic pathology. *Nat. Med.* **12,** 175–177.

Amann, K. J., Renley, B. A., and Ervasti, J. M. (1998). A cluster of basic repeats in the dystrophin rod domain binds F-actin through an electrostatic interaction. *J. Biol. Chem.* **273,** 28419–28423.

Banks, G. B., Fuhrer, C., Adams, M. E., and Froehner, S. C. (2003). The postsynaptic submembrane machinery at the neuromuscular junction: Requirement for rapsyn and the utrophin/dystrophin-associated complex. *J. Neurocytol.* **32,** 709–726.

Banks, G. B., Gregorevic, P., Allen, J. M., Finn, E. E., and Chamberlain, J. S. (2007). Functional capacity of dystrophins carrying deletions in the N-terminal actin-binding domain. *Hum. Mol. Genet.* **16,** 2105–2113.

Baumbach, L. L., Chamberlain, J. S., Ward, P. A., Farwell, N. J., and Caskey, C. T. (1989). Molecular and clinical correlations of deletions leading to Duchenne and Becker muscular dystrophies. *Neurology* **39,** 465–474.

Beauchamp, J. R., Morgan, J. E., Pagel, C. N., and Partridge, T. A. (1999). Dynamics of myoblast transplantation reveal a discrete minority of precursors with stem cell-like properties as the myogenic source. *J. Cell Biol.* **144,** 1113–1122.

Beggs, A. H., Hoffman, E. P., Snyder, J. R., Arahata, K., Specht, L., Shapiro, F., Angelini, C., Sugita, H., and Kunkel, L. M. (1991). Exploring the molecular basis for variability among patients with Becker muscular dystrophy: Dystrophin gene and protein studies. *Am. J. Hum. Genet.* **49,** 54–67.

Bell, C. D., and Conen, P. E. (1968). Histopathological changes in Duchenne muscular dystrophy. *J. Neurol. Sci.* **7,** 529–544.

Berry, S. E., Liu, J., Chaney, E. J., and Kaufman, S. J. (2007). Multipotential mesoangioblast stem cell therapy in the *mdx/utrn*$^{-/-}$ mouse model for Duchenne muscular dystrophy. *Regen. Med.* **2,** 275–288.

Blake, D. J., Weir, A., Newey, S. E., and Davies, K. E. (2002). Function and genetics of dystrophin and dystrophin-related proteins in muscle. *Physiol. Rev.* **82,** 291–329.

Bloch, R. J., and Gonzalez-Serratos, H. (2003). Lateral force transmission across costameres in skeletal muscle. *Exerc. Sport Sci. Rev.* **31,** 73–78.

Boittin, F. X., Petermann, O., Hirn, C., Mittaud, P., Dorchies, O. M., Roulet, E., and Ruegg, U. T. (2006). Ca^{2+}-independent phospholipase A2 enhances store-operated Ca2+ entry in dystrophic skeletal muscle fibers. *J. Cell Sci.* **119,** 3733–3742.

Bostick, B., Yue, Y., Lai, Y., Long, C., Li, D., and Dongsheng, D. (2008). AAV-9 micro-dystrophin gene therapy ameliorates electrocardiographic abnormalities in mdx mice. *Hum. Gene Ther.*

Brack, A. S., and Rando, T. A. (2007). Intrinsic changes and extrinsic influences of myogenic stem cell function during aging. *Stem. Cell Rev.* **3,** 226–237.

Bradley, W. G., and Fulthorpe, J. J. (1978). Studies of sarcolemmal integrity in myopathic muscle. *Neurology* **28,** 670–677.

Brenman, J. E., Chao, D. S., Xia, H., Aldape, K., and Bredt, D. S. (1995). Nitric oxide synthase complexed with dystrophin and absent from skeletal muscle sarcolemma in Duchenne muscular dystrophy. *Cell* **82,** 743–752.

Bretag, A. H. (2007). Stem cell treatment of dystrophic dogs. *Nature* **450,** E23; discussion E23–E25.

Brooks, S. V. (1998). Rapid recovery following contraction-induced injury to *in situ* skeletal muscles in *mdx* mice. *J. Muscle Res. Cell Motil.* **19,** 179–187.

Bulfield, G., Siller, W. G., Wight, P. A., and Moore, K. J. (1984). X chromosome-linked muscular dystrophy (*mdx*) in the mouse. *Proc. Natl. Acad. Sci. USA* **81,** 1189–1192.

Burton, E. A., Tinsley, J. M., Holzfeind, P. J., Rodrigues, N. R., and Davies, K. E. (1999). A second promoter provides an alternative target for therapeutic up-regulation of utrophin in Duchenne muscular dystrophy. *Proc. Natl. Acad. Sci. USA* **96,** 14025–14030.

Carlson, C. G., and Roshek, D. M. (2001). Adult dystrophic (*mdx*) endplates exhibit reduced quantal size and enhanced quantal variation. *Pflugers Arch.* **442,** 369–375.

Carlson, M. E., Hsu, M., and Conboy, I. M. (2008). Imbalance between pSmad3 and Notch induces CDK inhibitors in old muscle stem cells. *Nature.*

Chamberlain, J. S. (1991). Duchenne muscular dystrophy. *Curr. Opin. Genet. Dev.* **1,** 11–14.

Chamberlain, J. C., and Rando, T. A. (2006). "Duchenne muscular dystrophy: Advances in therapeutics." New York: Taylor and Francis, New York.

Chamberlain, J. S., Gibbs, R. A., Ranier, J. E., Nguyen, P. N., and Caskey, C. T. (1988). Deletion screening of the Duchenne muscular dystrophy locus via multiplex DNA amplification. *Nucleic Acids Res.* **16,** 11141–11156.

Chamberlain, J. S., Chamberlain, J. R., Fenwick, R. G., Ward, P. A., Caskey, C. T., Dimnick, L. S., Bech-Hansen, N. T., Hoar, D. I., *et al.* (1992). Diagnosis of Duchenne and Becker muscular dystrophies by polymerase chain reaction: A multicenter study. *JAMA* **267,** 2609–2615.

Chamberlain, J. S., Metzger, J., Reyes, M., Townsend, D., and Faulkner, J. A. (2007). Dystrophin-deficient *mdx* mice display a reduced life span and are susceptible to spontaneous rhabdomyosarcoma. *FASEB J.* **21,** 2195–2204.

Chelly, J., Gilgenkrantz, H., Lambert, M., Hamard, G., Chafey, P., Recan, D., Katz, P., de la Chapelle, A., Koenig, M., Ginjaar, I. B., *et al.* (1990a). Effect of dystrophin gene deletions on mRNA levels and processing in Duchenne and Becker muscular dystrophies. *Cell* **63,** 1239–1248.

Chelly, J., Hamard, G., Koulakoff, A., Kaplan, J. C., Kahn, A., and Berwald-Netter, Y. (1990b). Dystrophin gene transcribed from different promoters in neuronal and glial cells. *Nature* **344,** 64–65.

Chetboul, V., Carlos, C., Blot, S., Thibaud, J. L., Escriou, C., Tissier, R., Retortillo, J. L., and Pouchelon, J. L. (2004a). Tissue Doppler assessment of diastolic and systolic alterations of radial and longitudinal left ventricular motions in Golden Retrievers during the preclinical phase of cardiomyopathy associated with muscular dystrophy. *Am. J. Vet. Res.* **65,** 1335–1341.

Chetboul, V., Escriou, C., Tessier, D., Richard, V., Pouchelon, J. L., Thibault, H., Lallemand, F., Thuillez, C., Blot, S., and Derumeaux, G. (2004b). Tissue Doppler imaging detects early asymptomatic myocardial abnormalities in a dog model of Duchenne's cardiomyopathy. *Eur. Heart J.* **25,** 1934–1939.

Hasegawa, T., Matsumura, K., Hashimoto, T., Ikehira, H., Fukuda, H., and Tateno, Y. (1992). [Intramuscular degeneration process in Duchenne muscular dystrophy–investigation by longitudinal MR imaging of the skeletal muscles]. *Rinsho Shinkeigaku* **32,** 333–335.

Hirst, R. C., McCullagh, K. J., and Davies, K. E. (2005). Utrophin upregulation in Duchenne muscular dystrophy. *Acta. Myol.* **24,** 209–216.

Hoffman, E. P., Brown, R. H., Jr., and Kunkel, L. M. (1987). Dystrophin: The protein product of the Duchenne muscular dystrophy locus. *Cell* **51,** 919–928.

Hoffman, E. P., Fischbeck, K. H., Brown, R. H., Johnson, M., Medori, R., Loike, J. D., Harris, J. B., Waterston, R., Brooke, M., Specht, L., *et al.* (1988). Characterization of dystrophin in muscle-biopsy specimens from patients with Duchenne's or Becker's muscular dystrophy. *N. Engl. J. Med.* **318,** 1363–1368.

Im, W. B., Phelps, S. F., Copen, E. H., Adams, E. G., Slightom, J. L., and Chamberlain, J. S. (1996). Differential expression of dystrophin isoforms in strains of *mdx* mice with different mutations. *Hum. Mol. Genet.* **5,** 1149–1153.

Janssen, P. M., Hiranandani, N., Mays, T. A., and Rafael-Fortney, J. A. (2005). Utrophin deficiency worsens cardiac contractile dysfunction present in dystrophin-deficient *mdx* mice. *Am. J. Physiol. Heart Circ. Physiol.*

Judge, L. M., and Chamberlain, J. S. (2005). Gene therapy for Duchenne muscular dystrophy: AAV leads the way. *Acta. Myol.* **24,** 184–193.

Khairallah, M., Khairallah, R. J., Young, M. E., Allen, B. G., Gillis, M. A., Danialou, G., Deschepper, C. F., Petrof, B. J., and Des Rosiers, C. (2008). Sildenafil and cardiomyocyte-specific cGMP signaling prevent cardiomyopathic changes associated with dystrophin deficiency. *Proc. Natl. Acad. Sci. USA* **105,** 7028–7033.

Khurana, T. S., and Davies, K. E. (2003). Pharmacological strategies for muscular dystrophy. *Nat. Rev. Drug Discov.* **2,** 379–390.

Kleopa, K. A., Drousiotou, A., Mavrikiou, E., Ormiston, A., and Kyriakides, T. (2006). Naturally occurring utrophin correlates with disease severity in Duchenne muscular dystrophy. *Hum. Mol. Genet.* **15,** 1623–1628.

Koenig, M., and Kunkel, L. M. (1990). Detailed analysis of the repeat domain of dystrophin reveals four potential hinge segments that may confer flexibility. *J. Biol. Chem.* **265,** 4560–4566.

Koenig, M., Beggs, A. H., Moyer, M., Scherpf, S., Heindrich, K., Bettecken, T., Meng, G., Muller, C. R., Lindlof, M., Kaariainen, H., *et al.* (1989). The molecular basis for Duchenne versus Becker muscular dystrophy: Correlation of severity with type of deletion. *Am. J. Hum. Genet.* **45,** 498–506.

Kohn, B., Guscetti, F., Waxenberger, M., and Augsburger, H. (1993). [Muscular dystrophy in a cat]. *Tierarztl Prax* **21,** 451–457.

Kuang, S., Kuroda, K., Le Grand, F., and Rudnicki, M. A. (2007). Asymmetric self-renewal and commitment of satellite stem cells in muscle. *Cell* **129,** 999–1010.

Law, D. J., and Tidball, J. G. (1993). Dystrophin deficiency is associated with myotendinous junction defects in prenecrotic and fully regenerated skeletal muscle. *Am. J. Pathol.* **142,** 1513–1523.

Law, D. J., Allen, D. L., and Tidball, J. G. (1994). Talin, vinculin and DRP (utrophin) concentrations are increased at *mdx* myotendinous junctions following onset of necrosis. *J. Cell Sci.* **107**(Pt 6), 1477–1483.

Le, T. T., Nguyen, T. M., Love, D. R., Helliwell, T. R., Davies, K. E., and Morris, G. E. (1993). Monoclonal antibodies against the muscle-specific N-terminus of dystrophin: Characterization of dystrophin in a muscular dystrophy patient with a frameshift deletion of exons 3–7. *Am. J. Hum. Genet.* **53,** 131–139.

Legardinier, S., Hubert, J. F., Le Bihan, O., Tascon, C., Rocher, C., Raguenes-Nicol, C., Bondon, A., Hardy, S., and Le Rumeur, E. (2008). Sub-domains of the dystrophin rod domain display contrasting lipid-binding and stability properties. *Biochim. Biophys. Acta.* **1784,** 672–682.

Le Grand, F., and Rudnicki, M. A. (2007). Skeletal muscle satellite cells and adult myogenesis. *Curr. Opin. Cell Biol.* **19,** 628–633.

Le Rumeur, E., Fichou, Y., Pottier, S., Gaboriau, F., Rondeau-Mouro, C., Vincent, M., Gallay, J., and Bondon, A. (2003). Interaction of dystrophin rod domain with membrane phospholipids. Evidence of a close proximity between tryptophan residues and lipids. *J. Biol. Chem.* **278,** 5993–6001.

Lidov, H. G., Selig, S., and Kunkel, L. M. (1995). Dp140: A novel 140 kDa CNS transcript from the dystrophin locus. *Hum. Mol. Genet.* **4,** 329–335.

Love, D. R., Hill, D. F., Dickson, G., Spurr, N. K., Byth, B. C., Marsden, R. F., Walsh, F. S., Edwards, Y. H., and Davies, K. E. (1989). An autosomal transcript in skeletal muscle with homology to dystrophin. *Nature* **339,** 55–58.

Lu, Q. L., Mann, C. J., Lou, F., Bou-Gharios, G., Morris, G. E., Xue, S. A., Fletcher, S., Partridge, T. A., and Wilton, S. D. (2003). Functional amounts of dystrophin produced by skipping the mutated exon in the *mdx* dystrophic mouse. *Nat. Med.* **9,** 1009–1014.

Lynch, G. S. (2004). Role of contraction-induced injury in the mechanisms of muscle damage in muscular dystrophy. *Clin. Exp. Pharmacol. Physiol.* **31,** 557–561.

Lyons, P. R., and Slater, C. R. (1991). Structure and function of the neuromuscular junction in young adult *mdx* mice. *J. Neurocytol.* **20,** 969–981.

Matsuda, R., Nishikawa, A., and Tanaka, H. (1995). Visualization of dystrophic muscle fibers in *mdx* mouse by vital staining with Evans blue: Evidence of apoptosis in dystrophin-deficient muscle. *J. Biochem. (Tokyo)* **118,** 959–964.

Matsumura, K., Ervasti, J. M., Ohlendieck, K., Kahl, S. D., and Campbell, K. P. (1992). Association of dystrophin-related protein with dystrophin-associated proteins in *mdx* mouse muscle. *Nature* **360,** 588–591.

Matsumura, K., Burghes, A. H., Mora, M., Tome, F. M., Morandi, L., Cornello, F., Leturcq, F., Jeanpierre, M., Kaplan, J. C., Reinert, P., *et al.* (1994). Immunohistochemical analysis of dystrophin-associated proteins in Becker/Duchenne muscular dystrophy with huge in-frame deletions in the NH2-terminal and rod domains of dystrophin. *J. Clin. Invest.* **93,** 99–105.

McGeachie, J. K., Grounds, M. D., Partridge, T. A., and Morgan, J. E. (1993). Age-related changes in replication of myogenic cells in *mdx* mice: Quantitative autoradiographic studies. *J. Neurol. Sci.* **119,** 169–179.

Millay, D. P., Sargent, M. A., Osinska, H., Baines, C. P., Barton, E. R., Vuagniaux, G., Sweeney, H. L., Robbins, J., and Molkentin, J. D. (2008). Genetic and pharmacologic inhibition of mitochondrial-dependent necrosis attenuates muscular dystrophy. *Nat. Med.* **14,** 442–447.

Miller, R. G., and Hoffman, E. P. (1994). Molecular diagnosis and modern management of Duchenne muscular dystrophy. *Neurol. Clin.* **12,** 699–725.

Moens, P., Baatsen, P. H., and Marechal, G. (1993). Increased susceptibility of EDL muscles from *mdx* mice to damage induced by contractions with stretch. *J. Muscle Res. Cell Motil.* **14,** 446–451.

Monaco, A. P., Bertelson, C. J., Liechti-Gallati, S., Moser, H., and Kunkel, L. M. (1988). An explanation for the phenotypic differences between patients bearing partial deletions of the DMD locus. *Genomics* **2,** 90–95.

Montarras, D., Morgan, J., Collins, C., Relaix, F., Zaffran, S., Cumano, A., Partridge, T., and Buckingham, M. (2005). Direct isolation of satellite cells for skeletal muscle regeneration. *Science* **309,** 2064–2067.

Morgan, J. E., Pagel, C. N., Sherratt, T., and Partridge, T. A. (1993). Long-term persistence and migration of myogenic cells injected into pre-irradiated muscles of *mdx* mice. *J. Neurol. Sci.* **115,** 191–200.

Muntoni, F., and Wells, D. (2007). Genetic treatments in muscular dystrophies. *Curr. Opin. Neurol.* **20,** 590–594.

Muntoni, F., Gobbi, P., Sewry, C., Sherratt, T., Taylor, J., Sandhu, S. K., Abbs, S., Roberts, R., Hodgson, S. V., Bobrow, M., *et al.* (1994). Deletions in the 5′ region of dystrophin and resulting phenotypes. *J. Med. Genet.* **31,** 843–847.

Muntoni, F., Torelli, S., and Ferlini, A. (2003). Dystrophin and mutations: One gene, several proteins, multiple phenotypes. *Lancet. Neurol.* **2,** 731–740.

Nagao, H., Morimoto, T., Sano, N., Takahashi, M., Nagai, H., Tawa, R., Yoshimatsu, M., Woo, Y. J., and Matsuda, H. (1991). [Magnetic resonance imaging of skeletal muscle in patients with Duchenne muscular dystrophy–serial axial and sagittal section studies]. *No To Hattatsu* **23,** 39–43.

Nguyen, H. X., and Tidball, J. G. (2003). Interactions between neutrophils and macrophages promote macrophage killing of rat muscle cells *in vitro*. *J. Physiol.* **547,** 125–132.

Nguyen, F., Cherel, Y., Guigand, L., Goubault-Leroux, I., and Wyers, M. (2002). Muscle lesions associated with dystrophin deficiency in neonatal golden retriever puppies. *J. Comp. Pathol.* **126,** 100–108.

Novakovic, I., Bojic, D., Todorovic, S., Apostolski, S., Lukovic, L., Stefanovic, D., and Milasin, J. (2005). Proximal dystrophin gene deletions and protein alterations in becker muscular dystrophy. *Ann. N. Y. Acad. Sci.* **1048,** 406–410.

Odom, G. L., Gregorevic, P., Doremus, C., Allen, J., and Chamberlain, J. S. (2008). Microutrophin delivery via rAAV6 increases lifespan and improves muscle function in dystrophic *mdx:utrn*$^{-/-}$ mice. *Mol. Ther.* **9,** 1539–1545.

Ozawa, E., Hagiwara, Y., and Yoshida, M. (1999). Creatine kinase, cell membrane and Duchenne muscular dystrophy. *Mol. Cell Biochem.* **190,** 143–151.

Parker, M. H., Kuhr, C., Tapscott, S. J., and Storb, R. (2008). Hematopoietic cell transplantation provides an immune-tolerant platform for myoblast transplantation in dystrophic dogs. *Mol. Ther.* **16,** 1340–1346.

Partridge, T. A. (1997). Models of dystrophinopathy, pathological mechanisms and assessment of therapies. *In* "Dystrophin: Gene Protein and Cell Biology" (S. C. Brown and J. A. Lucy, eds.), pp. 310–311. Cambridge University Press, Cambridge.

Partridge, T. (2006). Disciplining the stem cell into myogenesis. *N. Engl. J. Med.* **354,** 1844–1845.

Petrof, B. J., Shrager, J. B., Stedman, H. H., Kelly, A. M., and Sweeney, H. L. (1993). Dystrophin protects the sarcolemma from stresses developed during muscle contraction. *Proc. Natl. Acad. Sci. USA* **90,** 3710–3714.

Prior, T. W., Papp, A. C., Snyder, P. J., Burghes, A. H., Bartolo, C., Sedra, M. S., Western, L. M., and Mendell, J. R. (1993). A missense mutation in the dystrophin gene in a Duchenne muscular dystrophy patient. *Nat. Genet.* **4,** 357–360.

Rafael, J. A., Townsend, E. R., Squire, S. E., Potter, A. C., Chamberlain, J. S., and Davies, K. E. (2000). Dystrophin and utrophin influence fiber type composition and post-synaptic membrane structure. *Hum. Mol. Genet.* **9,** 1357–1367.

Rando, T. A., Disatnik, M. H., Yu, Y., and Franco, A. (1998). Muscle cells from *mdx* mice have an increased susceptibility to oxidative stress. *Neuromuscul. Disord.* **8,** 14–21.

Rentschler, S., Linn, H., Deininger, K., Bedford, M. T., Espanel, X., and Sudol, M. (1999). The WW domain of dystrophin requires EF-hands region to interact with beta-dystroglycan. *J. Biol. Chem.* **380,** 431–442.

Robert, V., Massimino, M. L., Tosello, V., Marsault, R., Cantini, M., Sorrentino, V., and Pozzan, T. (2001). Alteration in calcium handling at the subcellular level in *mdx* myotubes. *J. Biol. Chem.* **276,** 4647–4651.

Rodino-Klapac, L. R., Janssen, P. M., Montgomery, C. L., Coley, B. D., Chicoine, L. G., Clark, K. R., and Mendell, J. R. (2007). A translational approach for limb vascular delivery of the micro-dystrophin gene without high volume or high pressure for treatment of Duchenne muscular dystrophy. *J. Transl. Med.* **5,** 45.

Rybakova, I. N., Patel, J. R., Davies, K. E., Yurchenco, P. D., and Ervasti, J. M. (2002). Utrophin binds laterally along actin filaments and can couple costameric actin with sarcolemma when overexpressed in dystrophin-deficient muscle. *Mol. Biol. Cell* **13,** 1512–1521.

Rybakova, I. N., Humston, J. L., Sonnemann, K. J., and Ervasti, J. M. (2006). Dystrophin and Utrophin Bind Actin through Distinct Modes of Contact. *J. Biol. Chem.* **281,** 9996–10001.

Saadat, L., Pittman, L., and Menhart, N. (2006). Structural cooperativity in spectrin type repeats motifs of dystrophin. *Biochim. Biophys. Acta.* **1764,** 943–954.

Sampaolesi, M., Torrente, Y., Innocenzi, A., Tonlorenzi, R., D'Antona, G., Pellegrino, M. A., Barresi, R., Bresolin, N., De Angelis, M. G., Campbell, K. P., Bottinelli, R., and Cossu, G. (2003). Cell therapy of alpha-sarcoglycan null dystrophic mice through intra-arterial delivery of mesoangioblasts. *Science* **301,** 487–492.

Sampaolesi, M., Blot, S., D'Antona, G., Granger, N., Tonlorenzi, R., Innocenzi, A., Mognol, P., Thibaud, J. L., Galvez, B. G., Barthelemy, I., Perani, L., Mantero, S., *et al.* (2006). Mesoangioblast stem cells ameliorate muscle function in dystrophic dogs. *Nature* **444,** 574–579.

Sampaolesi, M., Blot, S., Bottinelli, R., and Cossu, G. (2007). Sampaolesi *et al.* reply. *Nature* **450,** E23–E25.

Schatzberg, S. J., Anderson, L. V., Wilton, S. D., Kornegay, J. N., Mann, C. J., Solomon, G. G., and Sharp, N. J. (1998). Alternative dystrophin gene transcripts in golden retriever muscular dystrophy. *Muscle Nerve* **21,** 991–998.

Schatzberg, S. J., Olby, N. J., Breen, M., Anderson, L. V., Langford, C. F., Dickens, H. F., Wilton, S. D., Zeiss, C. J., Binns, M. M., Kornegay, J. N., Morris, G. E., and Sharp, N. J. (1999). Molecular analysis of a spontaneous dystrophin 'knockout' dog. *Neuromuscul. Disord.* **9,** 289–295.

Schriner, S. E., Linford, N. J., Martin, G. M., Treuting, P., Ogburn, C. E., Emond, M., Coskun, P. E., Ladiges, W., Wolf, N., Van Remmen, H., Wallace, D. C., and Rabinovitch, P. S. (2005). Extension of murine life span by overexpression of catalase targeted to mitochondria. *Science* **308,** 1909–1911.

Schuierer, M. M., Mann, C. J., Bildsoe, H., Huxley, C., and Hughes, S. M. (2005). Analyses of the differentiation potential of satellite cells from myoD$^{-/-}$, mdx, and PMP22 C22 mice. *BMC Musculoskelet. Disord.* **6,** 15.

Schultz, E., and Lipton, B. H. (1982). Skeletal muscle satellite cells: Changes in proliferation potential as a function of age. *Mech. Ageing Dev.* **20,** 377–383.

Sharp, N. J., Kornegay, J. N., Van Camp, S. D., Herbstreith, M. H., Secore, S. L., Kettle, S., Hung, W. Y., Constantinou, C. D., Dykstra, M. J., Roses, A. D., *et al.* (1992). An error in dystrophin mRNA processing in golden retriever muscular dystrophy, an animal homologue of Duchenne muscular dystrophy. *Genomics* **13,** 115–121.

Shimatsu, Y., Katagiri, K., Furuta, T., Nakura, M., Tanioka, Y., Yuasa, K., Tomohiro, M., Kornegay, J. N., Nonaka, I., and Takeda, S. (2003). Canine X-linked muscular dystrophy in Japan (*CXMDJ*). *Exp. Anim.* **52,** 93–97.

Shimatsu, Y., Yoshimura, M., Yuasa, K., Urasawa, N., Tomohiro, M., Nakura, M., Tanigawa, M., Nakamura, A., and Takeda, S. (2005). Major clinical and histopathological characteristics of canine X-linked muscular dystrophy in Japan, *CXMDJ*. *Acta. Myol.* **24,** 145–154.

Sicinski, P., Geng, Y., Ryder-Cook, A. S., Barnard, E. A., Darlison, M. G., and Barnard, P. J. (1989). The molecular basis of muscular dystrophy in the *mdx* mouse: A point mutation. *Science* **244,** 1578–1580.

Slater, C. R. (2003). Structural determinants of the reliability of synaptic transmission at the vertebrate neuromuscular junction. *J. Neurocytol.* **32,** 505–522.

Spencer, M. J., and Mellgren, R. L. (2002). Overexpression of a calpastatin transgene in *mdx* muscle reduces dystrophic pathology. *Hum. Mol. Genet.* **11,** 2645–2655.

Suchyna, T. M., and Sachs, F. (2007). Mechanosensitive channel properties and membrane mechanics in mouse dystrophic myotubes. *J. Physiol.* **581,** 369–387.

Takemitsu, M., Ishiura, S., Koga, R., Kamakura, K., Arahata, K., Nonaka, I., and Sugita, H. (1991). Dystrophin-related protein in the fetal and denervated skeletal muscles of normal and *mdx* mice. *Biochem. Biophys. Res. Commun.* **180,** 1179–1186.

Takeshima, Y., Nishio, H., Narita, N., Wada, H., Ishikawa, Y., Ishikawa, Y., Minami, R., Nakamura, H., and Matsuo, M. (1994). Amino-terminal deletion of 53% of dystrophin results in an intermediate Duchenne-Becker muscular dystrophy phenotype. *Neurology* **44,** 1648–1651.

Tidball, J. G., and Wehling-Henricks, M. (2007). The role of free radicals in the pathophysiology of muscular dystrophy. *J. Appl. Physiol.* **102,** 1677–1686.

Tinsley, J. M., Potter, A. C., Phelps, S. R., Fisher, R., Trickett, J. I., and Davies, K. E. (1996). Amelioration of the dystrophic phenotype of *mdx* mice using a truncated utrophin transgene. *Nature* **384,** 349–353.

Tinsley, J., Deconinck, N., Fisher, R., Kahn, D., Phelps, S., Gillis, J. M., and Davies, K. (1998). Expression of full-length utrophin prevents muscular dystrophy in *mdx* mice. *Nat. Med.* **4,** 1441–1444.

Valentine, B. A., and Cooper, B. J. (1991). Canine X-linked muscular dystrophy: Selective involvement of muscles in neonatal dogs. *Neuromuscul. Disord.* **1,** 31–38.

Valentine, B. A., Cooper, B. J., de Lahunta, A., O'Quinn, R., and Blue, J. T. (1988). Canine X-linked muscular dystrophy. An animal model of Duchenne muscular dystrophy: Clinical studies. *J. Neurol. Sci.* **88,** 69–81.

Valentine, B. A., Winand, N. J., Pradhan, D., Moise, N. S., de Lahunta, A., Kornegay, J. N., and Cooper, B. J. (1992). Canine X-linked muscular dystrophy as an animal model of Duchenne muscular dystrophy: A review. *Am. J. Med. Genet.* **42,** 352–356.

Vandebrouck, C., Martin, D., Colson-Van Schoor, M., Debaix, H., and Gailly, P. (2002). Involvement of TRPC in the abnormal calcium influx observed in dystrophic (*mdx*) mouse skeletal muscle fibers. *J. Cell Biol.* **158,** 1089–1096.

Wakefield, P. M., Tinsley, J. M., Wood, M. J., Gilbert, R., Karpati, G., and Davies, K. E. (2000). Prevention of the dystrophic phenotype in dystrophin/utrophin-deficient muscle following adenovirus-mediated transfer of a utrophin minigene. *Gene. Ther.* **7,** 201–204.

Wang, Z., Zhu, T., Qiao, C., Zhou, L., Wang, B., Zhang, J., Chen, C., Li, J., and Xiao, X. (2005). Adeno-associated virus serotype 8 efficiently delivers genes to muscle and heart. *Nat. Biotechnol.* **23,** 321–328.

Wang, Z., Kuhr, C. S., Allen, J. M., Blankinship, M., Gregorevic, P., Chamberlain, J. S., Tapscott, S. J., and Storb, R. (2007). Sustained AAV-mediated dystrophin expression in a canine model of Duchenne muscular dystrophy with a brief course of immunosuppression. *Mol. Ther.* **15,** 1160–1166.

Warner, L. E., DelloRusso, C., Crawford, R. W., Rybakova, I. N., Patel, J. R., Ervasti, J. M., and Chamberlain, J. S. (2002). Expression of Dp260 in muscle tethers the actin cytoskeleton to the dystrophin-glycoprotein complex and partially prevents dystrophy. *Hum. Mol. Genet.* **11,** 1095–1105.

Webster, C., and Blau, H. M. (1990). Accelerated age-related decline in replicative life-span of Duchenne muscular dystrophy myoblasts: Implications for cell and gene therapy. *Somat. Cell Mol. Genet.* **16,** 557–565.

Wehling, M., Spencer, M. J., and Tidball, J. G. (2001). A nitric oxide synthase transgene ameliorates muscular dystrophy in *mdx* mice. *J. Cell Biol.* **155,** 123–131.

Weir, A. P., Burton, E. A., Harrod, G., and Davies, K. E. (2002). A- and B-utrophin have different expression patterns and are differentially up-regulated in *mdx* muscle. *J. Biol. Chem.* **277,** 45285–45290.

Welch, E. M., Barton, E. R., Zhuo, J., Tomizawa, Y., Friesen, W. J., Trifillis, P., Paushkin, S., Patel, M., Trotta, C. R., Hwang, S., Wilde, R. G., Karp, G., *et al.* (2007). PTC124 targets genetic disorders caused by nonsense mutations. *Nature* **447,** 87–91.

Whitehead, N. P., Pham, C., Gervasio, O. L., and Allen, D. G. (2008). N-Acetylcysteine ameliorates skeletal muscle pathophysiology in *mdx* mice. *J. Physiol.* **586,** 2003–2014.

Winnard, A. V., Mendell, J. R., Prior, T. W., Florence, J., and Burghes, A. H. (1995). Frameshift deletions of exons 3–7 and revertant fibers in Duchenne muscular dystrophy: Mechanisms of dystrophin production. *Am. J. Hum. Genet.* **56,** 158–166.

Yeung, E. W., Whitehead, N. P., Suchyna, T. M., Gottlieb, P. A., Sachs, F., and Allen, D. G. (2005). Effects of stretch-activated channel blockers on [Ca2+]i and muscle damage in the *mdx* mouse. *J. Physiol.* **562,** 367–380.

Yue, Y., Liu, M., and Duan, D. (2006). C-terminal-truncated microdystrophin recruits dystrobrevin and syntrophin to the dystrophin-associated glycoprotein complex and reduces muscular dystrophy in symptomatic utrophin/dystrophin double-knockout mice. *Mol. Ther.* **14,** 79–87.

Zhao, J., Yoshioka, K., Miyatake, M., and Miike, T. (1992). Dystrophin and a dystrophin-related protein in intrafusal muscle fibers, and neuromuscular and myotendinous junctions. *Acta. Neuropathol. (Berl)* **84,** 141–146.

Index

Q

R

S

Contents of Previous Volumes

Volume 47

Volume 48

Volume 49

Volume 50

Volume 51

Volume 52

Volume 53

Volume 54

Volume 55

Volume 56

Volume 57

Volume 60

Volume 61

Volume 62

Volume 63

Volume 64

Volume 65

Volume 66

Volume 67

Volume 68

Volume 69

Volume 70

Volume 71

Volume 72

Volume 73

Volume 74

Volume 75

Volume 76

Volume 77

Volume 78

Volume 79

Volume 80

Volume 81

Volume 82

Volume 83

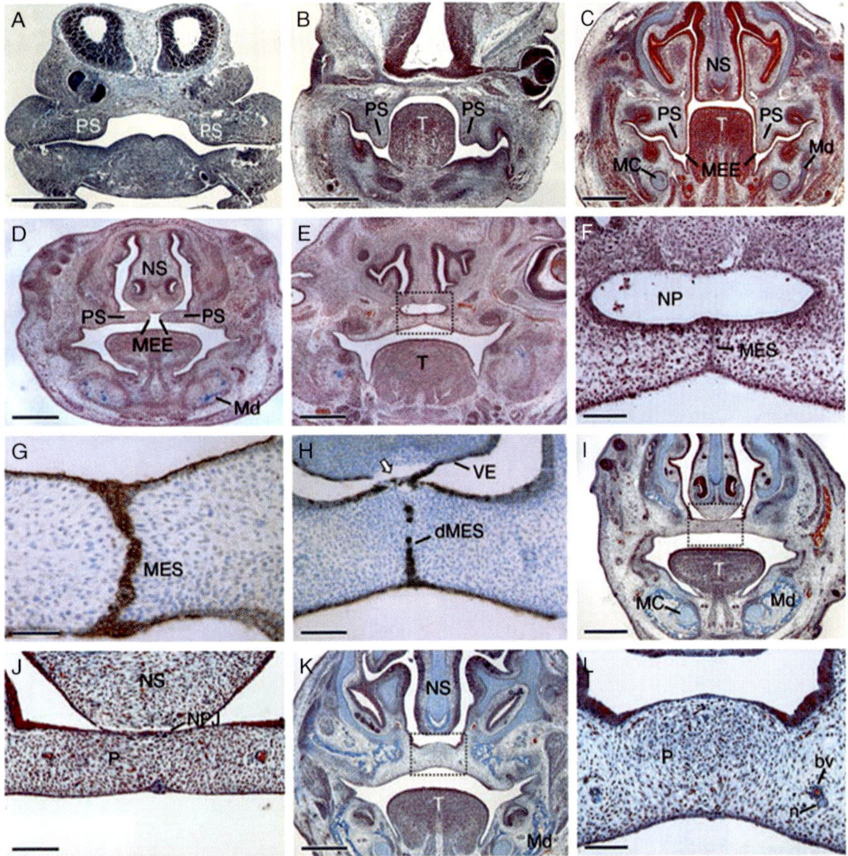

Amel Gritli-Linde, Figure 2.2 Histological sections showing the different steps of development of the murine secondary palate. At E11.5, the palatal shelves (PS) appear as outgrowths from the internal side of the maxillary processes (MxP) (A). During their growth phase at E13.5 (B) and E14.5 (C), the PS are vertical. The tongue (T) is elevated (B and C). At E14.5–E15, the PS have elevated above the tongue and are oriented horizontally (D). At E15–E15.5 (E and F), adhesion of the opposing medial edge epithelia (MEE) following further extension of the PS forms the transient medial epithelial seam (MES). Panel (F) is a high magnification view of the area indicated in (E). Sections immunostained with an anti-E-cadherin antibody which highlights the MES before (G) and during (H) its progressive regression. Note the epithelial islands, transient remnants of the degenerating MES (dMES) and the site of adhesion between the vomerine epithelium (VE) and palate (arrow in H). At E16–E16.5 (I–L), disappearance of the MES allows mesenchymal confluence and successful palate (P) fusion (I–L). The epithelial seam along the nasoplatine junction (NPS) is a result of adhesion between the vomerine epithelium (VE) and PS epithelium. This seam will eventually degenerate allowing successful fusion of the palate with the nasal septum (NS). Additional abbreviations: bv, blood vessel; MC, Meckel's cartilage; Md, mandibular bone; n, nerve. Scale bars: 500 μm (A–E, I, K), 100 μm (F, H, J, L), 50 μm (G).

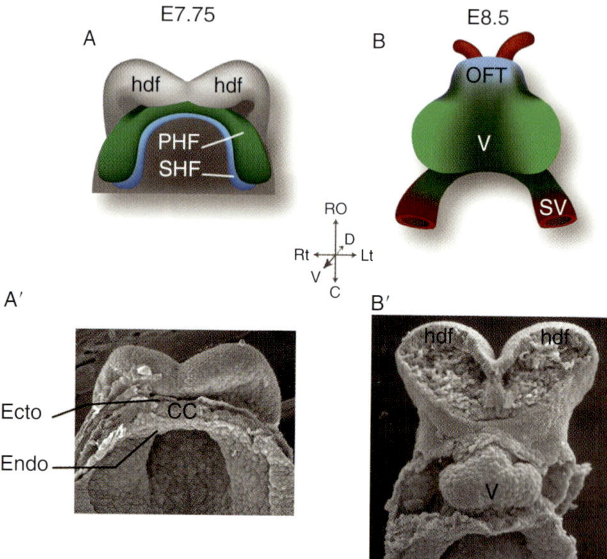

Anne Moon, Figure 4.1 (A and A′) Schematic (A) and scanning electron micrograph (A′) showing the rostral portion of a mouse embryo in a ventral view at approximately 7.75 days gestation (E7.75), prior to formation of the heart tube. The head is at top (hdf, headfold), the primary and second heart fields (PHF, SHF) are shown in green and blue, respectively and constitute the mesoderm of the cardiac crescent (CC) between the ectodermal (Ecto) and endodermal (Endo) layers. Note the relatively more dorsal location of the SHF. (B and B′) By E8.5, the primary heart tube consists of a primitive ventricle (V) and atrium (not visible as it is dorsal to the ventricle) in series with the sinus venosus (sv, inflow, posterior pole). Cells that will contribute to the right ventricle (RV) and outflow tract (OFT) are beginning to accrue to the rostral (anterior pole) end of the primary heart tube. Axis diagram: RO, rostral; C, caudal; Rt, right; Lt, left; V, ventral; D, dorsal.

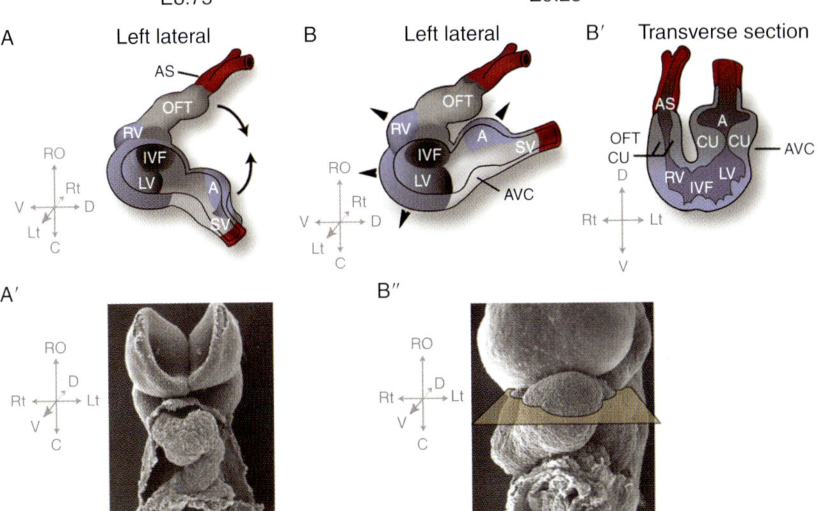

Anne Moon, Figure 4.2 (A) Schematic left lateral view of E8.75 heart tube with left outer curvature removed. (A′) EM of similar stage embryo. The heart is beginning to loop as indicated by the black arrows indicating movement of the venous pole (SV) rostrally toward the outflow concurrent with addition of myocardium to both poles. Blue represents working myocardium that is beginning to expand. Gray regions are myocardium that maintains a nonworking myocardial phenotype. Red regions are vascular connections at the poles and include the sinus venosus (SV) and aortic sac (AS). OFT, outflow tract; RV, right ventricle; LV, left ventricle; A, atrium; IVF, interventricular foramen. (A′) EM of early looping stage embryo in ventral view; note rightward "S" curve of the heart tube. (B) E9.25, looping has brought the inflow to approximately the same rostrocaudal level as the outflow. Ventricular ballooning and atrial growth (symbolized by black arrowheads) have increased the size of these chambers. More OFT myocardium has been added. AVC, atrioventricular canal. (A) and (B) are adapted from Christoffels *et al.* (2004a,b). (B′) Schematic of transverse section (plane denoted in B″); the endocardial cushions (cu) are beginning to form in the OFT and AVC. (B″) EM of approximately E9.25 embryo in ventral view. The schematic overlay indicates the approximate plane of section in (B′).

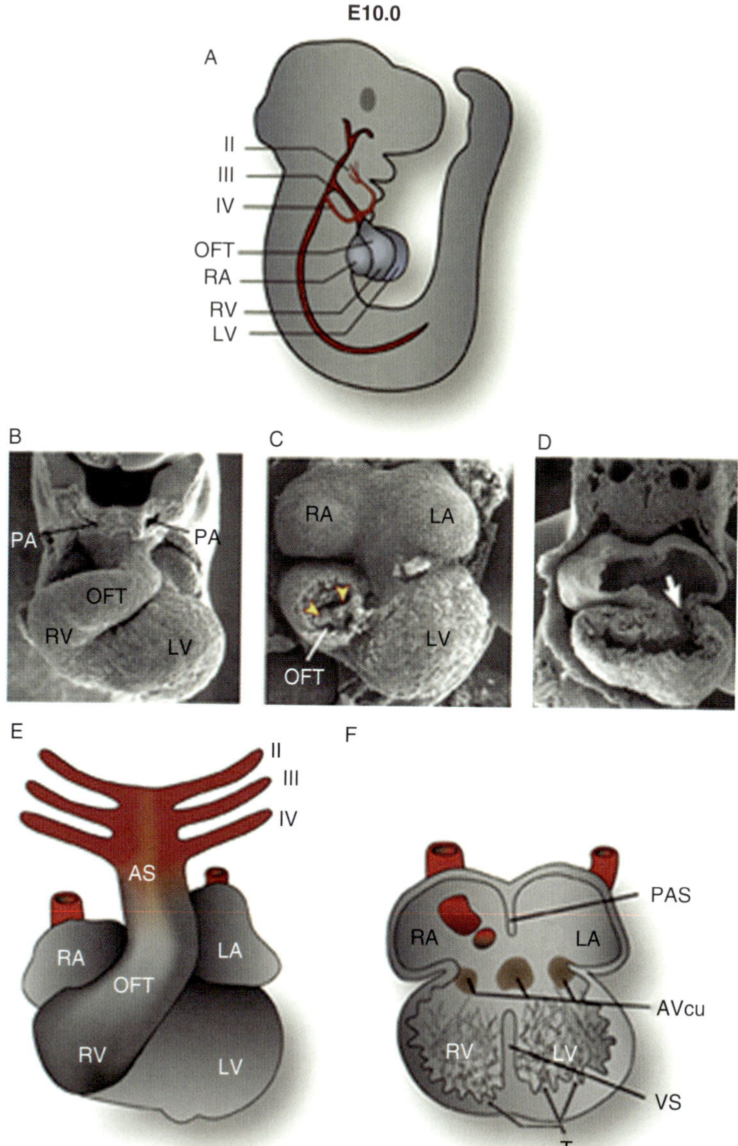

Anne Moon, Figure 4.3 (A) Right lateral view of E10.0 embryo. The pharyngeal arches and the pharyngeal arch arteries that run through them to connect with the dorsal aorta are numbered (II–IV). The heart is in blue and the segments are labeled as in previous figures. (B) EM of similar stage embryo as in A. The view is ventral but slightly oblique; the pharyngeal arch and artery arising from the aortic sac are labeled PA. (C) Same view and stage embryo, distal OFT removed. The OFT cushions are marked with yellow arrowheads. The external walls of the right and left atria (RA, LA)

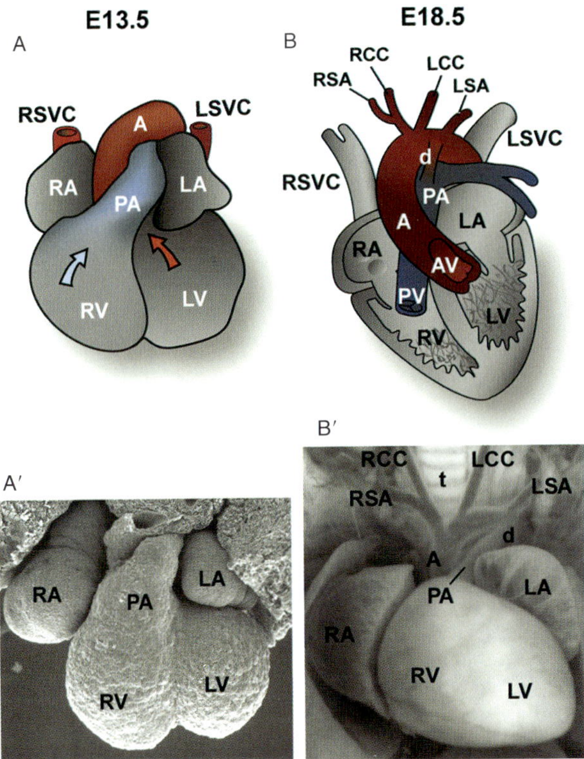

Anne Moon, Figure 4.4 (A and A′) Schematic and EM of E13.5 heart in ventral view. Note that the OFT has septated and rotated such that the pulmonary artery (PA) arises from the RV and the aorta from the LV. (B nad B′) Schematic (with ventral half of heart removed) and whole mount photograph of E18.5 heart. The pharyngeal arch arterial system has asymmetrically remodeled to form the left arch of the aorta, right subclavian (RSA), right common carotid (RCC), left common carotid (LCC), left subclavian (LSA) arteries, and the ductus arterious (d). The ductus is a communication vital for the fetal circulation that allows relatively well-oxygenated blood returning to the RV to bypass the nonfunctioning, high resistance pulmonary circuit, and enter the aorta for distribution to the periphery. This conduit normally closes in the first few postnatal days. Egress of blood from the RV is via the pulmonic valve (PV) and from the LV, the aortic valve (AV). The AV valves are not labeled but are shown as flaps in gray between the atria and the ventricles. In mice the left superior vena cava (LSVC) persists and empties into the right atrium (not visible). t, trachea.

---

and AVC are visible. (D) Same view and stage but more of the rostral heart has been removed so the AV canal (white arrow) is visible, as are the atrial and ventricular chambers. (E) Schematic showing the symmetrically paired pharyngeal arch arteries (II–IV) arising from the aortic sac at this stage. Both sides are extensively remodeled and there is regression of the right-sided structures. The left fourth pharyngeal arch artery contributes to the arch of the aorta. (F) Schematic showing details similar to plane in (D). The venous inflow into the right atrium is visible in the dorsal wall. The pulmonary pit as the site of the future entrance of the pulmonary veins is shown in the left atrial wall in blue. There are 4 AV cushion (brown, Avcu), but one is out of the plane of the section. The primary atrial septum is forming (PAS) as is the interventricular septum (VS). Trabeculae (T) are forming in both ventricular chambers.

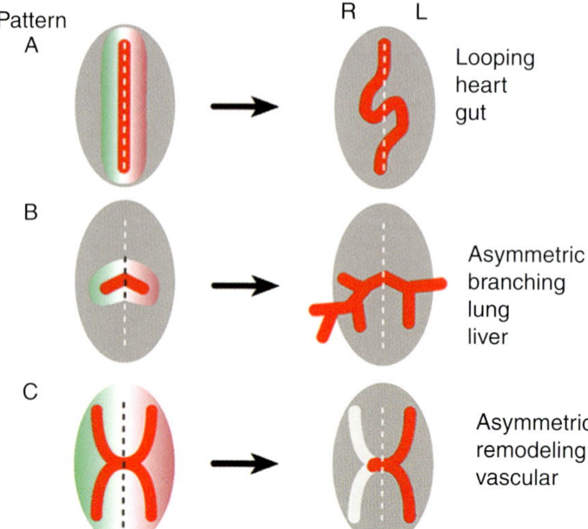

Anne Moon, Figure 4.5 Patterns of organ left–right asymmetry. (A) A linear or tubular structure may bend asymmetrically, as in heart and gut looping to the right. (B) An initially symmetric budding tube can be converted by asymmetric branching morphogenesis as in the lung and liver. (C) An initially symmetric array can be remodeled by differential regression as occurs in the pharyngeal arch arterial system.

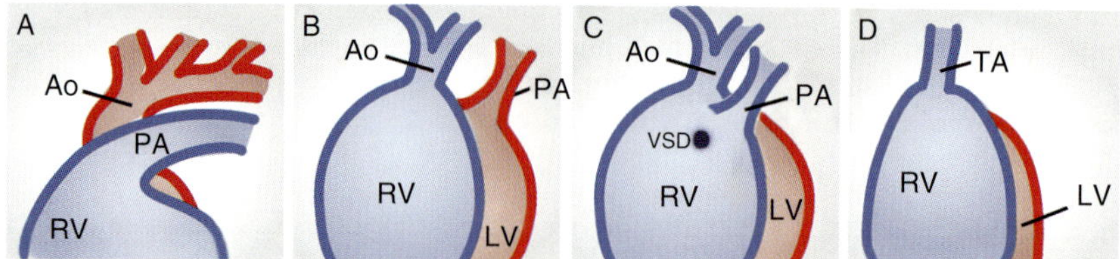

Anne Moon, Figure 4.6 (A) Schematic showing the positions of the great arteries (aorta, Ao and pulmonary artery, PA) relative to the ventricles and to one another in the normal situation. When alignment/rotation of the OFT does not occur properly, defects such as complete transposition of the great arteries (B) or Double outlet RV with VSD (C) occur. When the OFT does not septate, persistent truncus arteriosus occurs (D)o. The case shown is the most severe type, in which the OFT is unseptated along its entire extent and there is only a singe great vessel (TA).

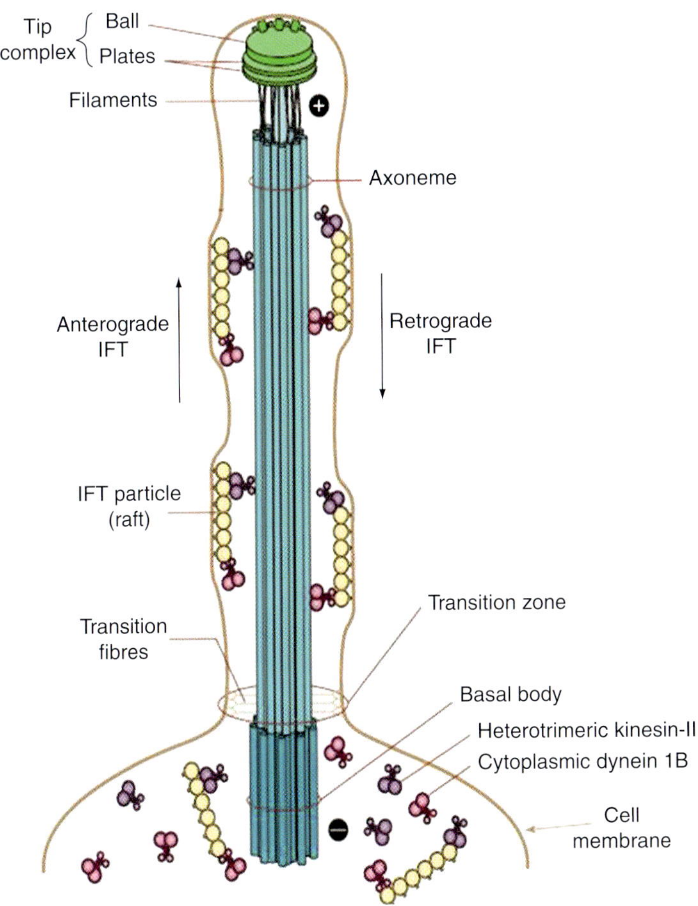

Robyn J. Quinlan *et al.*, Figure 5.2 Structure of the cilium illustrating intraflagellar transport. Adapted from Eley *et al.* (2005).

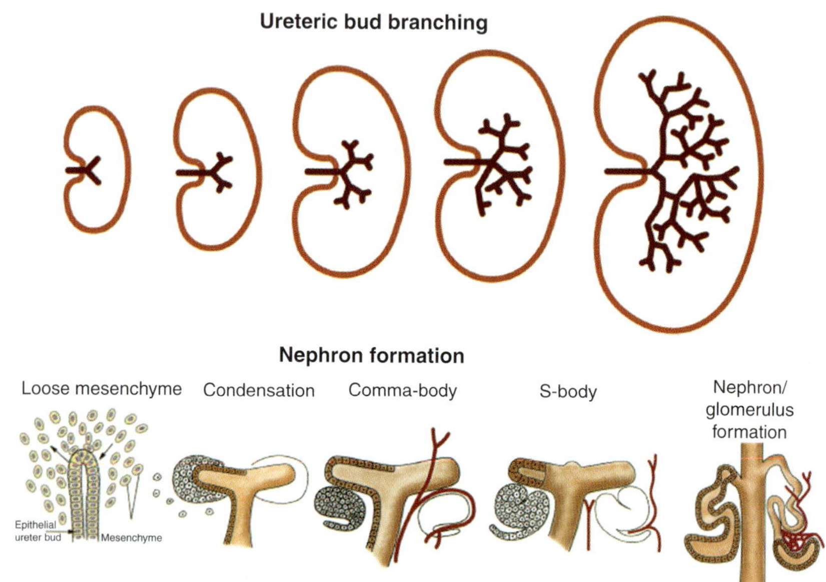

Patricia D. Wilson, Figure 6.2 Diagram of renal development. Upper panel: ureteric bud branching. Lower panel: induction of metanephric mesenchyme to differentiate renal tubules.

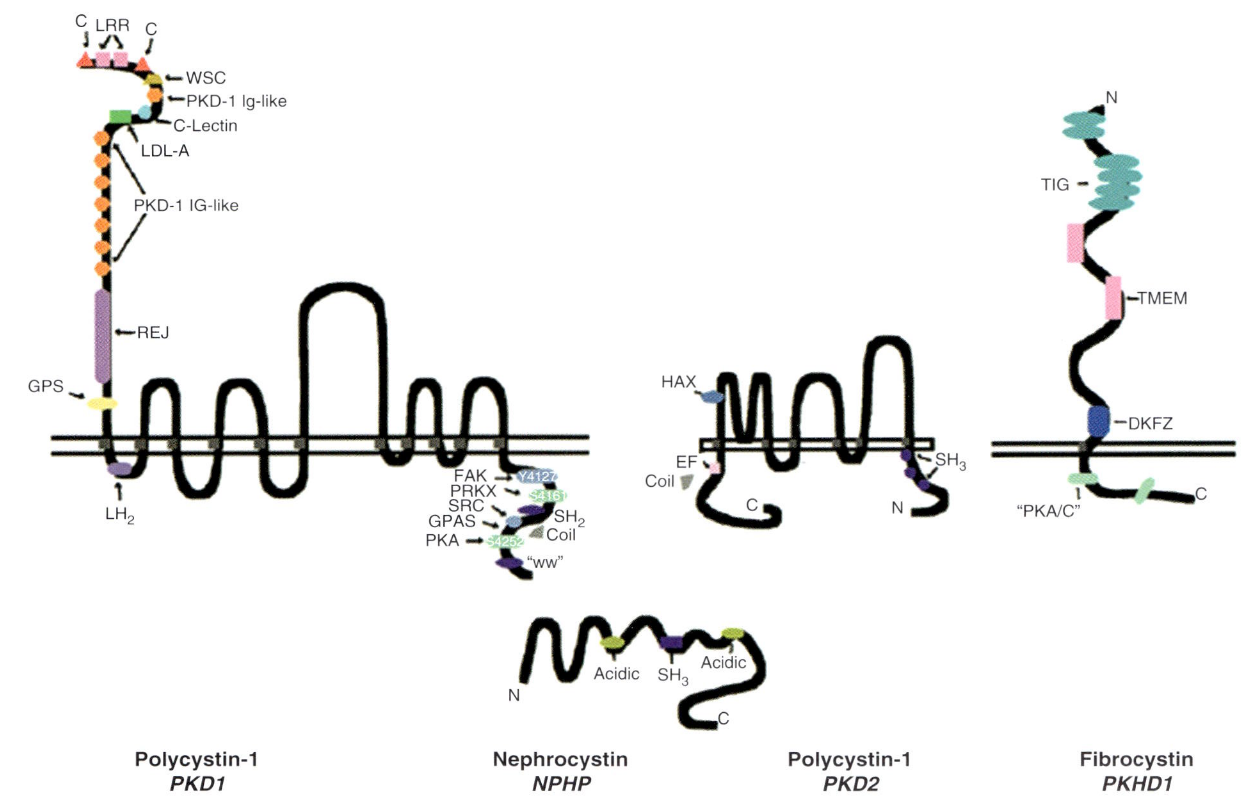

Patricia D. Wilson, Figure 6.3 Domain structure of cystic proteins: Polycystin-1, Polycystin-2, Fibrocystin-1, and Nephrocystin-1.

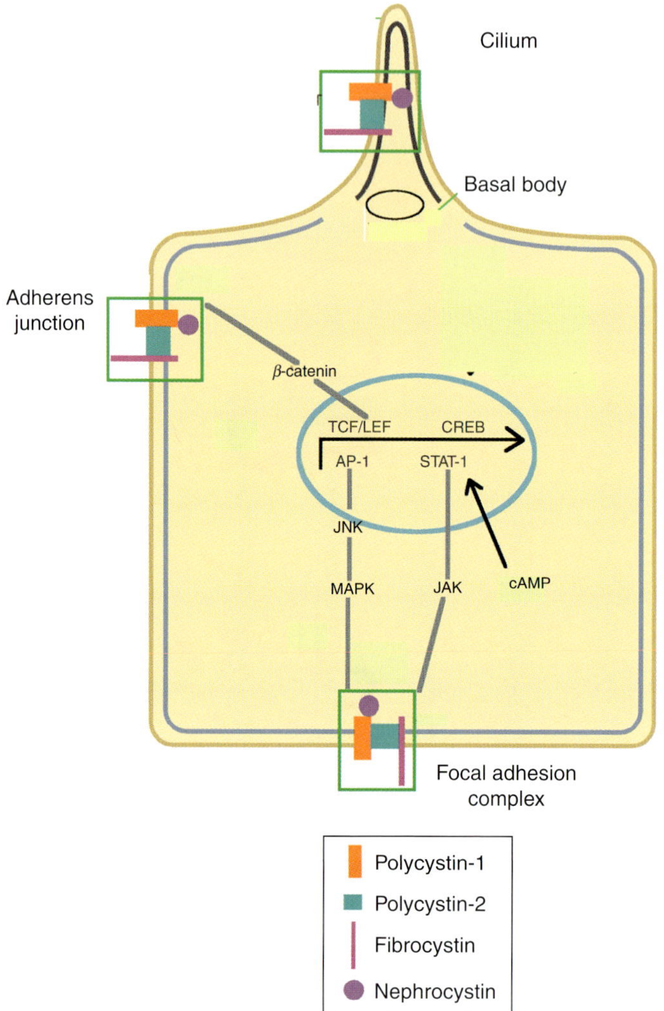

Patricia D. Wilson, Figure 6.4 Diagram of a renal epithelial cell showing cystic protein localization sites and interacting proteins.

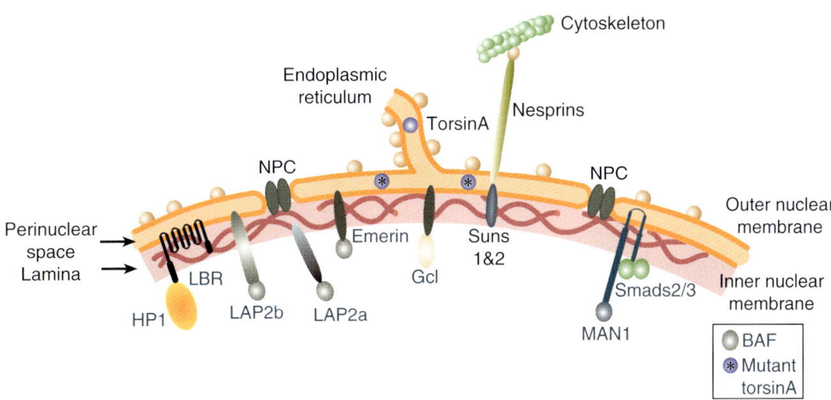

Tatiana V. Cohen and Colin L. Stewart, Figure 7.1 A diagram outlining the principal components of the NE and lamina to which 24 diseases and anomalies have been linked. (NPC - Nuclear Pore comples, LAP- Lamin associated protein, HP- heterochromatin protein, Gcl- germ cell lethal, LBR- Lamin B receptor.)

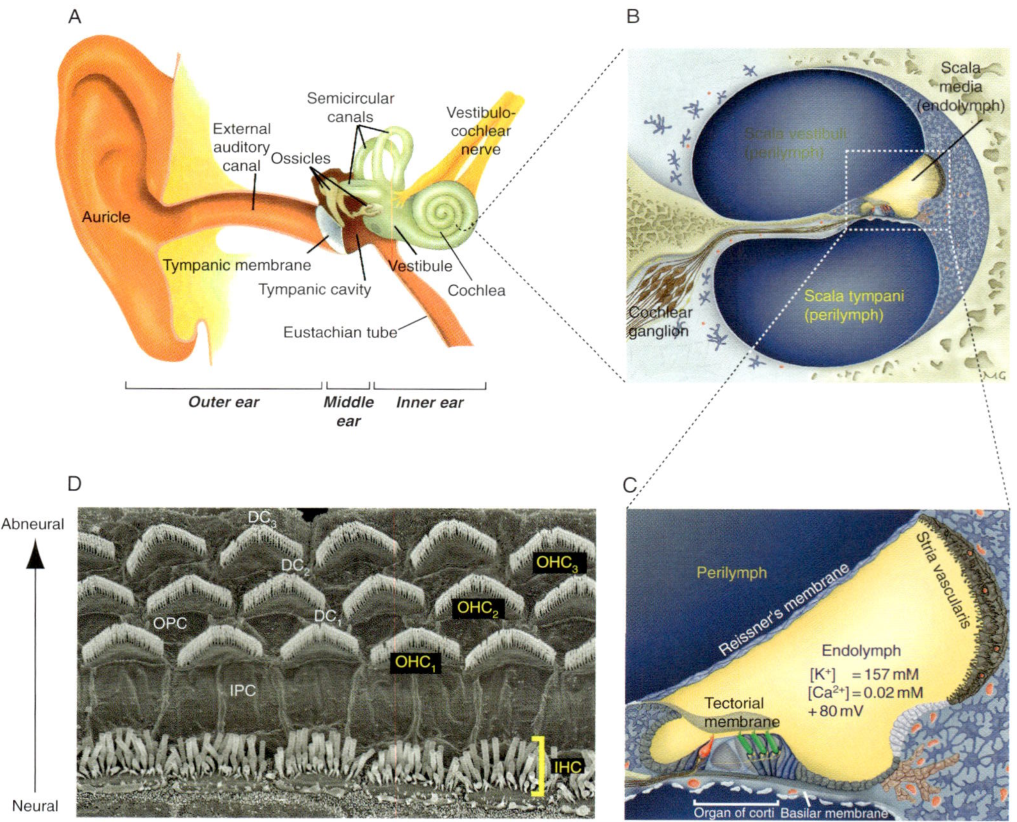
A
Auricle
External auditory canal
Ossicles
Semicircular canals
Vestibulo-cochlear nerve
Tympanic membrane
Tympanic cavity
Vestibule
Cochlea
Eustachian tube
Outer ear
Middle ear
Inner ear
B
Scala vestibuli (perilymph)
Scala media (endolymph)
Scala tympani (perilymph)
Cochlear ganglion
C
Perilymph
Reissner's membrane
Stria vascularis
Endolymph
[K+] = 157 mM
[Ca2+] = 0.02 mM
+ 80 mV
Tectorial membrane
Organ of corti
Basilar membrane
D
Abneural
Neural
DC3
DC2
DC1
OPC
IPC
OHC3
OHC2
OHC1
IHC

Michel Leibovici *et al.*, Figure 8.1 The mammalian ear is composed of three compartments: outer, middle, and inner ear. (A) The outer ear collects incoming sound vibration, while the ossicles of the middle ear transmit the vibration of the tympanic membrane to the inner ear, a fluid-filled organ. The auditory organ of the inner ear, a coiled-shape duct named cochlea, transduces sound waves into nerve impulses. (B) A cross section through the cochlear duct shows the three fluid chambers filled with perilymph (scala vestibuli and scala tympani), or endolymph, an extracellular fluid of unusual ion composition, with high K^+ (~157 mM) and low Ca^{2+} (~0.02 mM) concentrations (scala media). Tight junctions between cells of the scala media prevent ion leaking between the perilymphatic and endolymphatic compartments. (C) An enlargement of the scala media displays the main structures responsible for sound perception. The stria vascularis is a secretory epithelium that accounts for the high K^+ ion concentration of the endolymph and the positive potential of the scala media (+80 mV to +100 mV) compared to scalae tympani and vestibuli. The organ of Corti is the sensory epithelium of the cochlea, which contains outer (green) and inner (red) sensory cells as well as supporting cells. The organ of Corti is sandwiched between the tectorial membrane at the top and the basilar membrane (blue) at the bottom. (D) A scanning electron micrograph of the organ of Corti (without the tectorial membrane) shows the three rows of outer hair cells (OHC) and the single row of inner hair cells (IHC) with their apical hair bundles. OHCs and their supporting cells (Deiters cells, DC) are denoted from 1 to 3 from the neural to the abneural edge of the organ of Corti. Inner pillar cells (IPC) and outer pillar cells (OPC) separate the IHC row from the three OHC rows. Note that all hair bundles are orientated in the same direction (vertices of the "V"-shaped hair bundles pointing to the abneural edge), so that the sound pressure wave produces a coherent mechanical stimulation of the hair bundles.

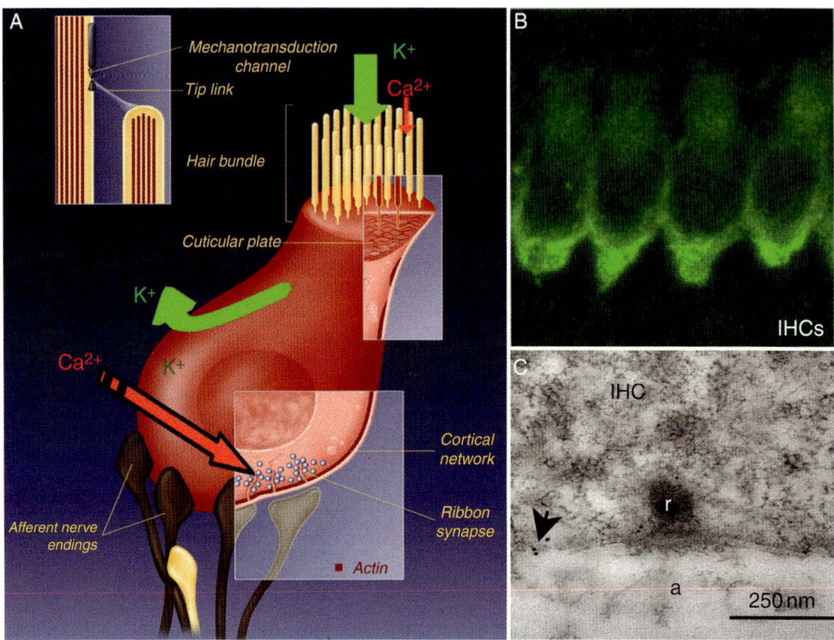

Michel Leibovici *et al.*, Figure 8.4 The IHC. (A) Schematic representation of the IHC and its innervation. Deflection of the hair bundle opens the mechanotransduction channels. Calcium channels open upon cell depolarization resulting in Ca^{2+} influx at the active zone where otoferlin is enriched, as shown in IHC stained for otoferlin (B and C). The Ca^{2+} ions bind to the calcium sensor, probably otoferlin, thereby triggering the fusion of synaptic vesicles with the presynaptic plasma membrane and neurotransmitter release onto the glutamate receptors located on the auditory nerve fibers. (B) Immunogold electron microscopy localized otoferlin around the ribbon (r) facing and afferent nerve fiber (a) and to the presynaptic plasma membrane (arrowhead).

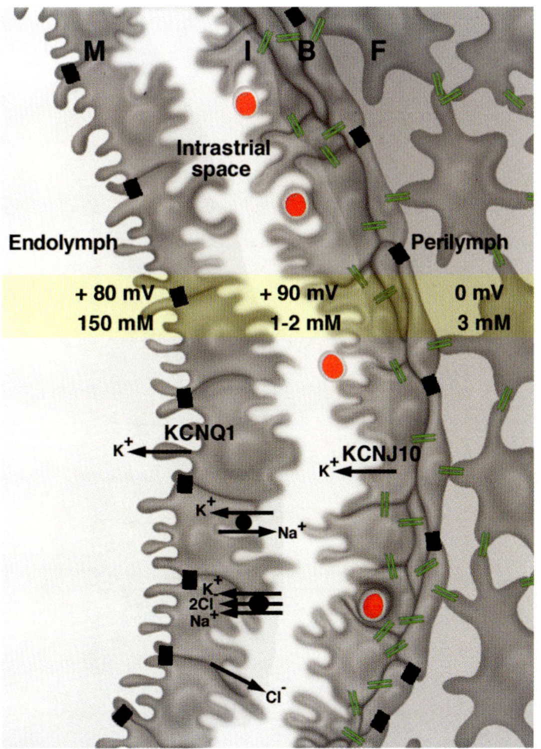

Michel Leibovici *et al.*, Figure 8.5 The stria vascularis and its electrogenic machinery: the stria vascularis consists of two epithelial cell layers; the basal (B) and intermediate (I) cell layer faces the spiral ligament connective tissue, and the marginal cell layer (M) faces the cochlear duct filled with endolymph. These layers limit the intrastrial space, an extracellular fluid compartment that contains a dense capillary network (red circles). Tight junctions between marginal cells and between basal cells are denoted by black boxes. Gap junctions (green double bars) connect intermediate and basal cells of the stria vascularis, and fibrocytes (F) of the spiral ligament. The endocochlear potential is generated across the basal and intermediate cell layer. The K^+ concentration and the electric potential are indicated in the three extracellular fluid spaces, namely the perilymph, intrastrial fluid, and endolymph. Each electric potential is indicated relative to that of the perilymph, which is taken as reference (0 mV). Adapted from (Cohen-Salmon *et al.*, 2007), Copyright 2007 National Academy of Sciences, U.S.A.

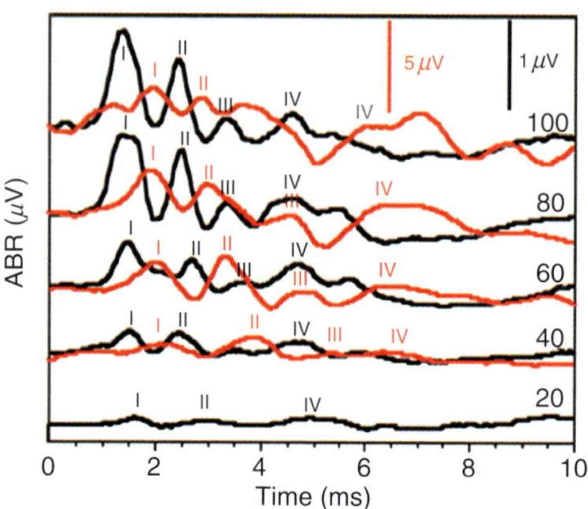

Michel Leibovici *et al.*, Figure 8.6 ABR waveforms rom wild type (black) and *Dfnb59* knockin (red) mice. Waves I–IV are marked. Labels at right indicate stimulus levels (from 20 to 100 dB SPL). Note that the vertical scales are different for the *Dfnb59* knockin mouse and the wild type mouse. In the *Dfnb59* knockin mouse, ABR waves I–II (auditory nerve) and III–IV (brainstem) have reduced amplitudes compared to the wild type mouse and the latencies of all four waves are significantly augmented, indicating a dysfunction in synaptic transmission or neuroual conduction. Reproduced from (Delmaghani *et al.*, 2006).